AF249006

FOOD PROTEINS

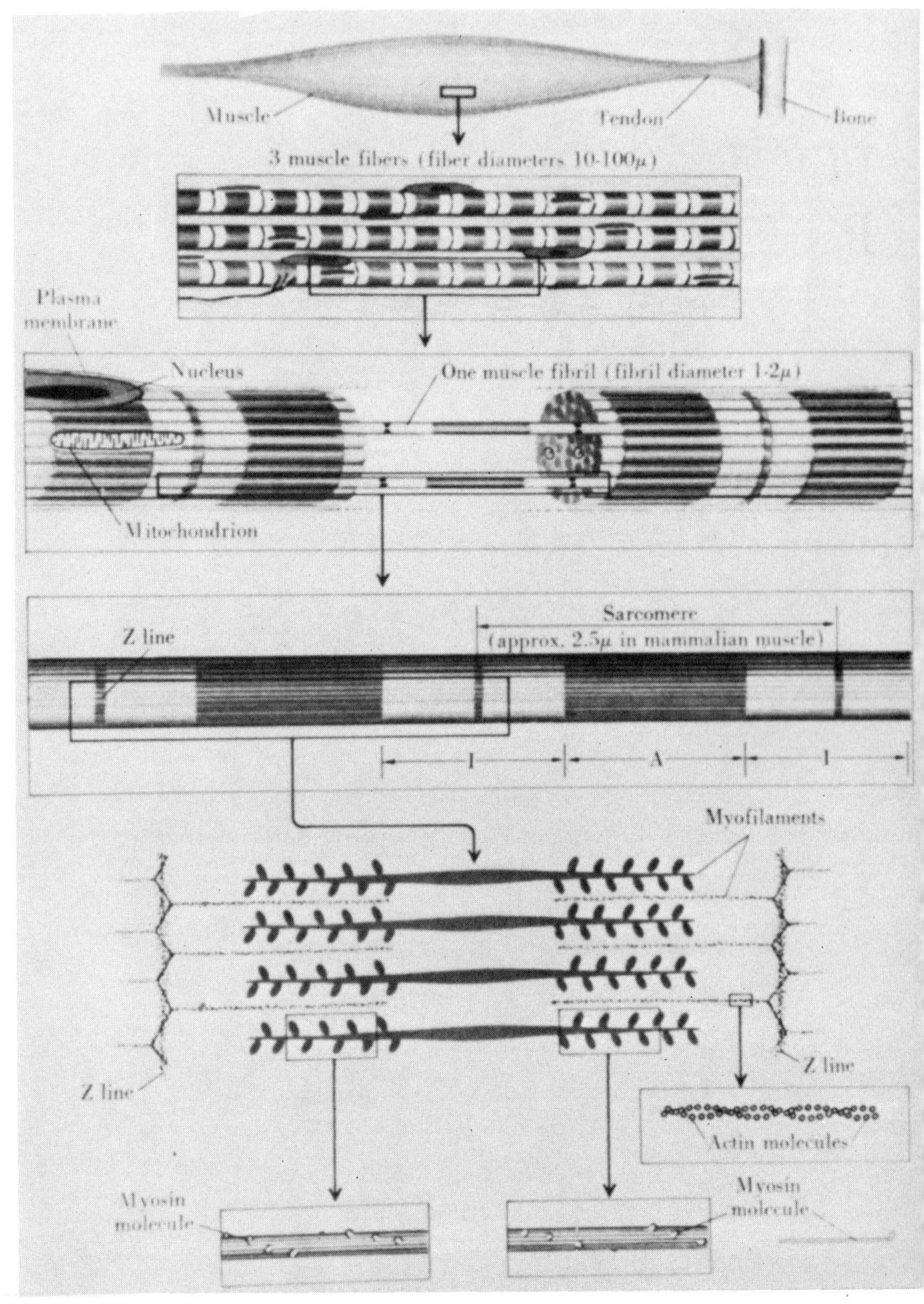

Muscle
Tendon
Bone
3 muscle fibers (fiber diameters 10-100μ)
Plasma membrane
Nucleus
One muscle fibril (fibril diameter 1-2μ)
Mitochondrion
Z line
Sarcomere
(approx. 2.5μ in mammalian muscle)
I
A
I
Myofilaments
Z line
Z line
Actin molecules
Myosin molecule
Myosin molecule

FOOD PROTEINS

John R. Whitaker
University of California

and

Steven R. Tannenbaum
*Massachusetts Institute
of Technology*

AVI PUBLISHING COMPANY, INC.
Westport, Connecticut

Library of Congress Catalog Card Number: 77-70646
ISBN-0-87055-230-9

Printed in the United States of America

Preface

This is the first in what we hope will be a series of annual basic symposia on selected topics of major importance to food scientists and food technologists. It is appropriate that this first symposium has as its central theme proteins in foods as the word protein is derived from the Greek word *proteios* meaning "holding the first place." Proteins are of primary importance in the germination, growth and maturation of all foodstuffs as well as being of primary importance in the growth, nutrition and well-being of the consumer.

It is appropriate that we embark on these basic symposia during the bicentennial year of our country. It is also appropriate and right that the Institute of Food Technologists and the International Union of Food Science and Technology should sponsor these basic symposia. The foundations of food science and technology are rooted in the fundamental principles of biochemistry, chemistry, engineering, microbiology, nutrition and physics. Contributions to these areas are made by scientists in universities, USDA laboratories, public and private research organizations and industry laboratories both in and outside the field of food science. In the early developmental periods of these disciplinary areas before they were clearly defined and later became fragmented it was the food scientist and the agricultural chemist that made many of the basic findings. It is encouraging to note that in the last ten to twenty years the food and agricultural scientist has again assumed a professional status in the above areas and has contributed substantially to our fundamental knowledge in these areas. With the present focus on world food problems which are likely to become more intensified rather than

v

diminished, the food and agricultural scientist will assume an ever increasing leadership role in the scientific community.

The idea for an annual basic symposium sponsored through the collaboration of the Institute of Food Technologists and the International Union of Food Science and Technology is the the brain child of Dr. Ernest Briskey, Vice-President of Campbell Soup and president of the Institute of Food Technologists, 1975–76. It is through his untiring efforts in obtaining financial and scientist support for the concept and his excellence in managerial and organizational skills that the first annual symposium became reality.

A Planning Committee to develop a basic symposium around the title "Fundamental Aspects of Proteins Basic to Foods" was set up and immediately began its work shortly after October 1974. The membership of this Planning Committee consisted of Steven Tannenbaum and John Whitaker as co-chairmen, Charles Beck, Robert Brunner, Masao Fujimaki, Darrel Goll, Kent Stewart and L. D. Williams. Through frequent correspondence, planning for the symposium proceeded.

In order to assure that the concept of an annual basic symposium would be carried forward, a Basic Symposia Advisory Committee was established under the chairmanship of Bernard Schweigert. The other members of this committee were Howard Baumann, E. M. Foster, John Hawthorne, Paul Hopper, Junius McIntire, Gilbert Leveille, Max Milner and Eric von Sydow. The Planning Committee is indebted to the Basic Symposia Planning Committee for advice and assistance and it wishes the Committee well in its coordination of future symposia.

The success of the first Basic Symposium was the result also of the expert assistance of Calvert L. Willey, Executive Director of IFT and John B. Klis, Director of Publications and their staff who gave unselfishly of themselves in providing publicity for the symposium, coordinating the registration and taking care of the many details of setting up the meeting rooms, hotel reservations and the numerous other details that go into a successful meeting. John Klis has served as coordinator and Anna May Schenck as copy editor for publication of the manuscripts presented at the symposium and it is through their patience and diligence that this monograph has being.

Success of the Symposium is also due to the excellent cooperation of the many food industries in providing financial assistance and moral support. Planning and execution of this Basic Symposium has truly been a magnificent joint effort on the part of industry people, scientists and professional people.

May the seed of the idea of an annual basic symposium survive the winters of uncertainty and doubt and spring forth with renewed vigor each year to bear fruit which will have a major role in the progress of food science and technology over the next 100 years.

JOHN R. WHITAKER
Professor,
Department of Food Science,
University of California,
Davis, California

STEVEN R. TANNENBAUM
Professor,
Massachusetts Institute of
Technology,
Cambridge, Massachusetts

December 1976

Contributors

BRISKEY, E. J., President, Technological Resources, Inc., Camden, New Jersey

BRUNNER, J. ROBERT, Professor, Department of Food Science and Human Nutrition. Michigan State University, East Lansing, Michigan

CATSIMPOOLAS, NICHOLAS, Professor, Biophysics Laboratory, Department of Nutrition and Food Science, Massachusetts Institute of Technology, Cambridge, Massachusetts

CHEFTEL, J. CLAUDE, Professor, Laboratoire de Biochimie et Technologie Alimentaires, Université des Sciences et Techniques, 34060 Montpellier, France

FENNEY, ROBERT E., Professor, Department of Food Science and Technology, University of California, Davis, California

FENNEMA, OWEN, Professor, Department of Food Science, University of Wisconsin-Madison, Madison, Wisconsin

FORSYTHE, R. H., Vice President, Basic Research, Campbell Institute for Food Research, Camden, New Jersey

FRIEDMAN, MENDEL, Western Regional Research Laboratory, Agricultural Research Service, U. S. Department of Agriculture, Berkeley, California

GOLL, DARREL E., Professor, Department of Nutrition and Food Science, University of Arizona, Tucson, Arizona

HARPER, A. E., E. V. McCollum Professor of Nutritional Sciences, University of Wisconsin-Madison, Madison, Wisconsin

HEGSTED, D. M., Professor of Nutrition, Harvard School of Public Health, Boston, Massachusetts

HORAN, F. E., Archer Daniels Midland Company, Decatur, Illinois

INGLETT, G. E., Chief, Cereal Science and Foods Laboratory, Northern Regional Research Center, Agricultural Research Service, U. S. Department of Agriculture, Peoria, Illinois

KOHLER, G. O., Research Leader, U. S. Department of Agriculture, Albany, California

LYON, C. K., U. S. Department of Agriculture, Albany, California

MILNER, MAX, Formerly: Office of Technology Assessment, U. S. Congress, Washington, D. C.

McINTIRE, J. M., General Manager of Research, Carnation Research Laboratories, Carnation Company, Van Nuys, California

OSUGA, D. T., Staff Research Associate, Department of Food Science and Technology, University of California, Davis, California

PIMENTEL, DAVID, Professor, New York State College of Agriculture and Life Sciences, Cornell University, Ithaca, New York

PIMENTEL, MARCIA, Lecturer, New York State College of Human Ecology, Cornell University, Ithaca, New York

QVIST, INGMAR H., Professor, SIK- The Swedish Food Institute, S-400 21 Göteborg 16, Sweden

ROBSON, RICHARD M., Muscle Biology Group, Iowa State University, Ames, Iowa

SCHOEN, HERBERT M., Director, Basic Sciences, General Foods Corporation, Technical Center, White Plains, New York

SCRIMSHAW, N. S., Department of Nutrition and Food Sciences, Massachusetts Institute of Technology, Cambridge, Massachusetts

STROMER, MARVIN H., Muscle Biology Group, Iowa State University, Ames, Iowa

TANNENBAUM, STEVEN R., Professor, Department of Nutrition and Food Science, Massachusetts Institute of Technology, Cambridge, Massachusetts

TAPPEL, A., Professor, Department of Food Science and Technology, University of California, Davis, California

VAN HOLDE, K. E., Professor, Department of Biochemistry and Biophysics, Oregon State University, Corvallis, Oregon

VON SYDOW, C. F. ERIC, Professor, SIK- The Swedish Food Institute, S-400 21 Göteborg 16, Sweden

WANG, D. I. C., Department of Nutrition and Food Sciences, Massachusetts Institue of Technology, Cambridge, Massachusetts

WHITAKER, JOHN R., Professor, Department of Food Science and Technology, University of California, Davis, California

WOLF, W. J., Cereal Properties Laboratory, Northern Regional Research Center, Agricultural Research Service, U. S. Department of Agriculture, Peoria, Illinois

Contents

Effects of Amino Acid Composition and Microenvironment on Protein Structure

K. E. Van Holde

The native structures of globular proteins are both precise and enormously complex. They involve, in most cases, linear chains of more than 100 amino acid residues, folded in a complicated way, and often associated with other similar or dissimilar chains to form higher-order structures. In general, such proteins serve specific biological functions, and their structures have been adapted, through evolution of the polypeptide chain sequence, to such functions. So far as we know, the only information transmitted from the gene is the sequence of amino acid residues. Even in cases where post-translational modification of the protein occurs (such as phosphorylation, proteolytic cleavage, etc.) the sites of such modification appear to be dictated by the folded structures of the protein, and hence once again by the primary sequence. The only further changes in conformation that will occur will be consequences of the interaction of the protein with its microenvironment; *i.e.*, the macromolecules with which it interacts, the surrounding solvent, etc. Yet even these conformational changes are only those allowed to the structure by virtue of its amino acid sequence. The overall situation is depicted schematically in Fig. 1.1.

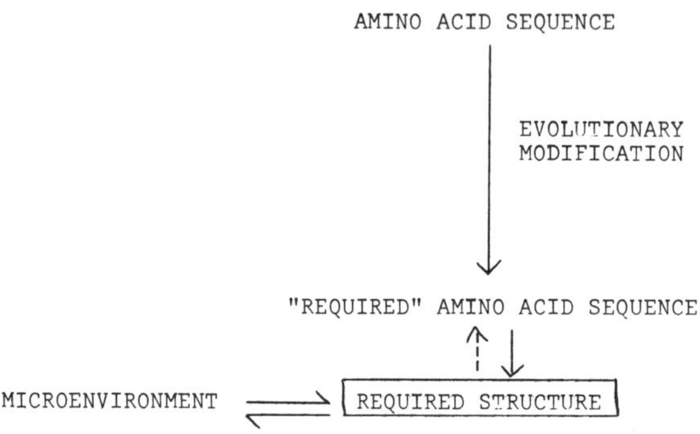

FIG. 1.1. FACTORS INFLUENCING PROTEIN CON-
FORMATION

A "GLOBAL" MINIMUM?

This view that protein's three-dimensional structure is entirely
predetermined by the amino acid sequence and microenvironment
has dominated modern protein chemistry, and does not appear to be
seriously challenged. However, there is a corollary that is often
drawn that is by no means self-evident, and perhaps not correct. It is
often assumed that the conformation adopted by a given polypeptide
sequence is that of minimum free energy for the system: protein +
microenvironment. That is, of all of the minimal energy states, the
native protein inhabits the *lowest* one or "global" minimum. (See
Fig. 1.2A.) This concept has been strongly supported by studies of
reversible denaturation: that is, there are many cases in which it can
be shown that a globular protein can be transformed by changes in
temperature, pressure, or solvent environment, into a more-or-less
"random coil" form, a form devoid of at least most of the precise
three dimensional structure of the native protein, and capable of
existing in many microstates. A common (but by no means universal)
observation is that such denatured proteins, upon restoration of the
"native" environment, can regain quite precisely and *spontaneously*
the native conformation. It is argued that even if the original
conformation were not that of a global free energy minimum, the
denatured protein would be likely to seek and find that minimum
when "native" conditions were reimposed. (See Fig. 1.2B, 1.2C.)
Thus, it is argued, the fact that the conformations found before and
after reversible denaturation are indistinguishable must imply that
both (and hence by inference the native conformation) represent the
global minimum. This point of view has been termed the "thermodynamic hypothesis" (Anfinsen 1973).

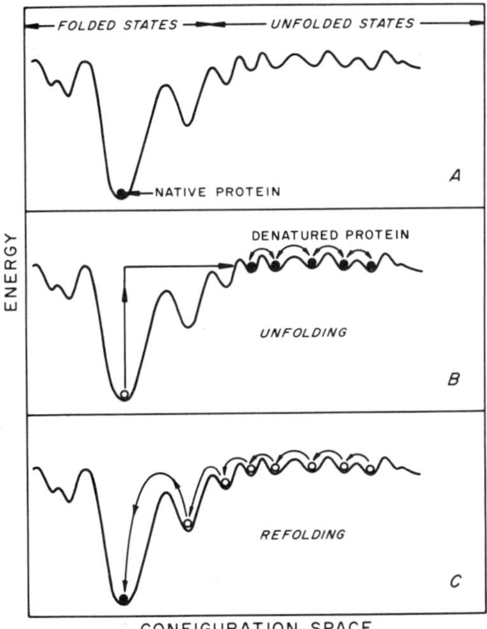

FIG. 1.2. HIGHLY SCHEMATIC DIAGRAMS OF THE ENERGY OF A PROTEIN MOLECULE AS A FUNCTION OF CHAIN CONFORMATION

The "configuration space" depicted here as one dimensional must really be a many hundred dimensional space in which a point depicts one of the many possible conformations of the molecule. As usual in such diagrams, we graph the energy (internal energy or enthalpy), rather than the free energy. The free energy will also contain an entropic contribution from the multiplicity of the states. Thus, when sufficient thermal energy is available to denature a protein (as in B) the multiple denatured states are preferred statistically to the small number of folded states. (A) The traditional view: of all of the possible energy minima, the native protein inhabits the *lowest* one, the "global minimum." (B) Protein thermal denaturation: the protein is displaced from its global minimum and adopts the higher entropy state (unfolded state), which has many substates. (C) The traditional view of renaturation: the protein "searches" until it finds once more the conformation corresponding to the global minimum. It somehow avoids getting trapped in other deep minima.

THE KINETICS OF FOLDING, AND "LINCS"

While many scientists accept the thermodynamic hypothesis in its strongest sense, there are substantial reasons for doubt. These stem from a consideration of kinetic pathways from the random coil state (either as produced by translation off ribosomes or by denaturing condition) toward the native state. The problem is simply this: If the randomly coiled polypeptide chain were to go through a *random* search for the global minimum, the number of conformations to be explored is so enormous that a conservative estimate indicates that times of the order of 10^{26} yr would be required (Anfinsen and Scheraga 1975). This calculation, even if incorrect by many orders of magnitude, *compels* us to believe that the folding of a protein chain occurs along certain kinetically accessible pathways that cover only a very small fraction of the potentially possible conformations. It follows that we may at least doubt the thermodynamic hypothesis; for there is no guarantee that the global minimum will lie in a region of conformation space that is ever kinetically accessible to the protein. The crucial point is diagrammed in Fig. 1.3: once into the preferred pathway, the protein may be "trapped" into a free-energy well that leads to *a* minimum, but not the global minimum. The argument from the reversibility of denaturation thus loses its power, for the random coil may always be led down the same path, simply because kinetically favored events early in folding commit it to this path. Such events may be the "nucleation" steps proposed by Levinthal (1968).

A somewhat similar argument can be based on the likely events during protein synthesis on ribosomes. It is unreasonable to expect the protein chain to "wait" to begin folding until it is all synthesized.

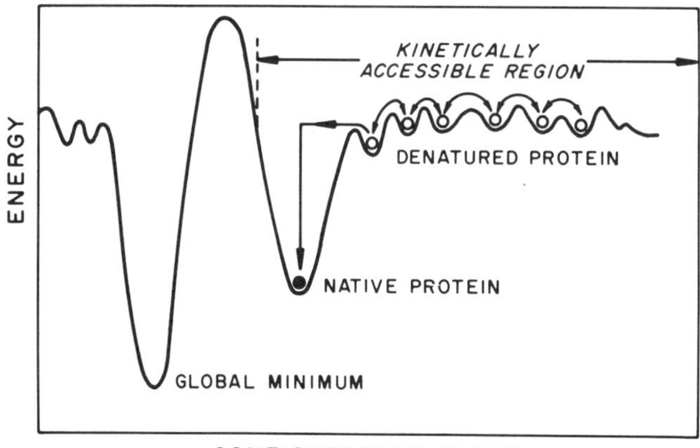

FIG. 1.3. "KINETIC" VIEW OF RENATURATION OR FOLDING
The protein is confined by local folding to a small region of conformation space. It then finds the lowest overall minimum *in the region*. The deepest well may lie elsewhere but be inaccessible.

Rather, portions of the N-terminal region will be expected to begin to fold as soon as they are free of the ribosome. The pathway of folding dictated by these first segments to fold may then largely decide which final state is accessible to the protein.

Folding is determined by short sequences of amino acids, and these units retain their structure in the final native conformation. This view has been developed in some detail by Rose (1975) and Rose *et al.* (1976). Rose calls short segments of defined local conformation LINCS (Local Independently Nucleated Continuous Segments), and presumes them to consist of such known structures as α-helices, β-folds and β-turns. The picture of folding is then shown in Fig. 1.4. The amino acid sequence in the immediate vicinity of each LINC region predisposes that region to adopt a particular kind of secondary structure, in a rapid process that is independent of the rest of the polypeptide chain. These folded LINCS then encounter one another, via what is probably a random diffusion process, to form an approximation to the final structure. This structure is stabilized by side chain interactions (side chain hydrogen bonds, salt bridges, hydrophobic interactions, etc.) between the LINCS. The fact that the units which now must search for the final conformation are larger, and few in number, makes this process reasonably fast. Karplus and Weaver (1976) in considering a very similar model, estimated a folding time between 0.01 sec and 10 sec. This is

I. POLYPEPTIDE CHAIN
   AS SYNTHESIZED

II. LOCAL FOLDING AS
    DICTATED BY LOCAL
    SEQUENCE - FORMATION
    OF LINCS

III. TERTIARY FOLDING
     OF CHAIN AT INTER-LINC
     JOINTS TO MINIMIZE
     FREE ENERGY OF LINC
     STRUCTURE

FIG. 1.4. PROTEIN FOLDING IN TERMS OF THE
LINCS HYPOTHESIS

certainly of the order of magnitude of the time observed for protein folding reactions (Anfinsen 1973).

THE SUCCESS OF THE PREDICTORS

Indirect support for the LINCS concept comes from the successes of those who have attempted to predict protein secondary structures from sequence data. A large number of empirical schemes have been devised to predict α-helical, β-sheet, and β-turn regions in native protein structures, *on the basis of local amino acid sequences.* In their basic assumption, such schemes assume that the formation of local regions is a local, not a global, event. Such methods are doomed to failure if the conformation of the protein is dominated by a global minimization process. Let us consider a hypothetical example: suppose that the stability of an α-helical region between residues 10 and 20 in some protein was strongly determined by the nature of the amino acid residues in some other part of the sequence (say residues 40–50) which make close contact with the 10–20 sequence in the folded structure. Since the predictors of secondary structure do not pretend to know the tertiary folding of the molecule, they could not possibly know how residue 10–20 would behave. If the details of local folding are dictated by the *overall* structure of the molecule, attempting to predict even local secondary structure from sequence data is a hopeless task.

In fact, some wholly empirical prediction schemes work remarkably well. Recently, a "blind" test was conducted (Schulz *et al.* 1974). A number of "predictors" were asked to predict the secondary structure of the enzyme, adenyl kinase, *before* its structure was revealed by x-ray diffraction studies. Fig. 1.5 depicts the results. On the basis of the amino acid sequence alone, a number of workers were able to predict much of the secondary structure of this protein. It is not important that the predictors do not all agree, nor that none is wholly accurate; the relatively high success shown in Fig. 1.5 is very impressive.

Thus, we are led to the rather strong conclusion that the conformation of globular proteins is determined in a two-state process: rapid local formation of secondary structural elements, followed by the folding together of these LINCS. This is a kinetically determined process, not a random search. It means that the native state need not be the thermodynamically most stable. It may well be, for selective pressure would be expected to favor those mutations of the polypeptide sequence that stabilize the native structure. We do not know at the present time whether or not the native state represents the global free energy minimum. Even if it does, we do not know whether this is a cause, or a consequence, a result of evolutionary adaptation of a kinetically determined state.

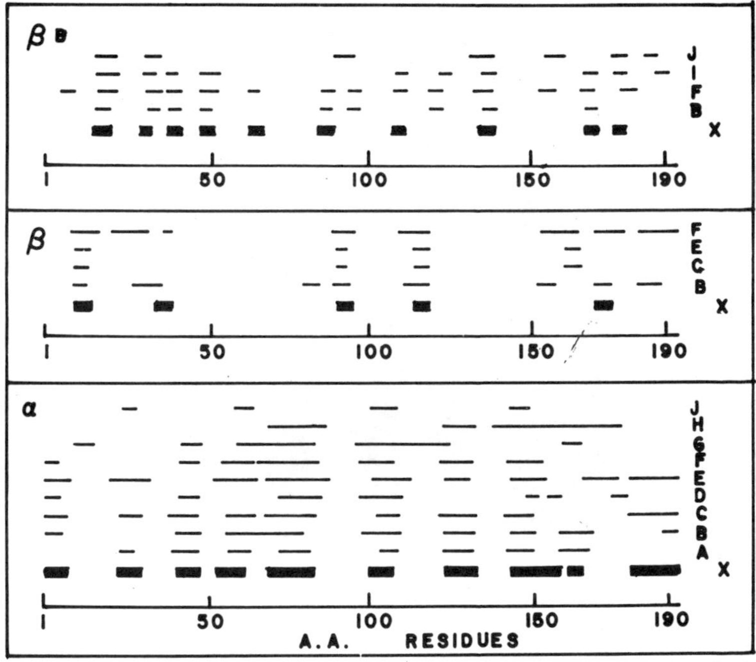

Adapted from Schulz et al. (1974)

FIG. 1.5. RESULTS OF A "BLIND" TEST OF PREDICTION OF PROTEIN
SECONDARY STRUCTURE
The predictions by a number of investigators (indicated by letters) of α-helical,
β-sheet, and β-turn regions in the protein adenyl kinase are illustrated by lines.
The (subsequently revealed) results of x-ray diffraction studies are shown by the
shaded blocks (indicated by X).

QUATERNARY STRUCTURE

It is now well recognized that very many globular proteins
function not as individual polypeptide chains, but as multi-subunit
structures. Such structures range in complexity from simple dimers
to large, hierarchically organized molecules like the hemocyanin
depicted in Fig. 1.6. On the other hand, a large number of proteins
are known that steadfastly refuse to self-associate in aqueous
solution. We may ask: "Is there any simple rule to predict how any
particular protein will behave?"

A number of years ago I proposed a simple correlation between
the tendency to self-associate and the content of certain nonpolar
amino acids: valine, proline, leucine, isoleucine, and phenylalanine
(Van Holde 1966). At the time that paper was written (1964), few
amino acid sequences were known, and even much of the amino acid
composition data were uncertain. I have now reexamined this
correlation, using only proteins for which both amino acid data are
exact (from known sequences) and the state of association in
solution is definitely established. The results are shown in Fig. 1.7.

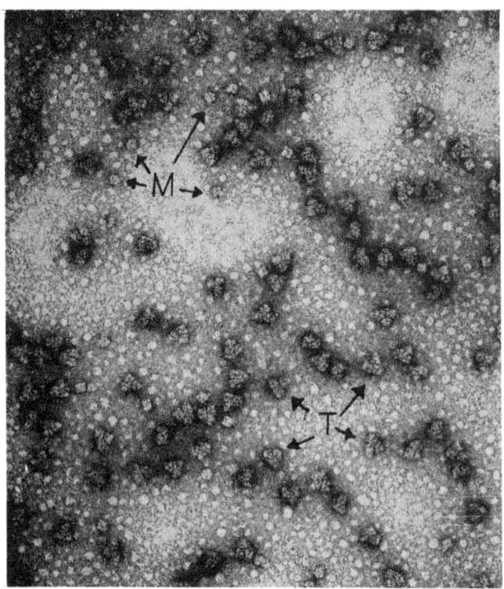

FIG. 1.6. ELECTION MICROGRAPH OF THE HEMOCYANIN FROM THE GHOST SHRIMP, *callianassa*
The largest, triangular-appearing structures are actually tetrahedral particles of molecular weight 1.7×10^6. These are each made up of four particles of weight 4.3×10^5. *These*, in turn are each made up of six polypeptide chains of molecular weight 7.2×10^4. There are probably six variants of each chain.

While it is probable that diligent search would yield exceptions, the rule can be stated as follows: A protein will self-associate if and only if it contains more than about 28 mole percent of the hydrophobic amino acids val, pro, leu, ile, phe. The rule allows some neat distinctions. Consider, for example, the heme proteins. The α and β chains of mammalian hemoglobin, which *do* associate, fall above the cutoff, whereas the closely related, but monomeric, myoglobin and insect hemoglobins fall below, as they should. A *general* explanation for the rule is not hard to find: below a critical fraction of hydrophobic residues it is possible to bury these completely within the molecule; if this fraction is exceeded, some must appear on its surface. If enough such residues appear on the surface, "hydrophobic" patches will be created; these can be buried only by self-association. The surprising thing is the regularity of the behavior, for one would expect that subunit size, which will influence the surface area/volume ratio, would play some role. Furthermore, there is abundant evidence that forces other than hydrophobic interactions

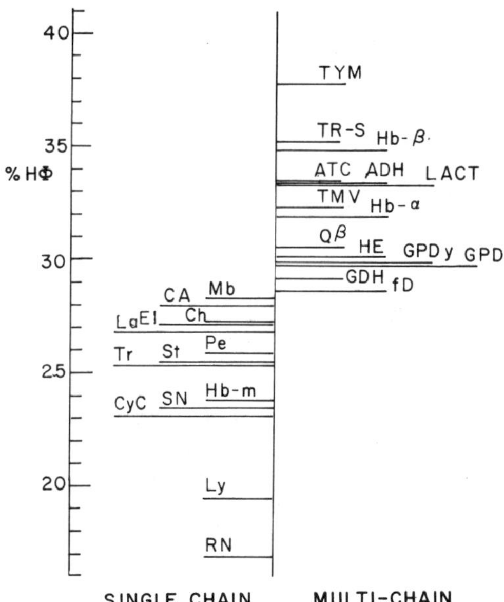

Data from Dayhoff (1972, 1973)

FIG. 1.7. MOLE PERCENT OF HYDROPHOBIC AMINO ACIDS (VAL, PRO, LEU, ILE, PHE) FOR A NUMBER OF GLOBULAR PROTEINS OF EXACTLY KNOWN COMPOSITION AND KNOWN QUATERNARY STRUCTURE

Single-chain proteins are listed to the left, multi chain proteins to the right. [Code: TYM-turnip yellow mosaic virus coat protein; TR-S-tryptophan synthetase (A chain, *E. coli*); Hb-β-hemoglobin β chain (human); ATC-asparatyl transcarbamylase (R chain, *E. coli*); ADH alcohol dehydrogenase (horse); LACT-β-lactoglobulin (bovine); TMV-tobacco mosaic virus coat protein; Hb-α-hemoglobin α chain (human); Qβ-phage Qβ coat protein; He-hemerythrin (sipunculid); GPDγ-glycerol-3-phosphate dehydrogenase (yeast); GPD-same (pig); GDH-glutamate dehydrogenase (bovine); fD-phage fD coat protein; Mb-myoglobin (human); CA-carbonic anhydrase (human); Ch-chymotrypsinogen (bovine); El-elastase (pig); La-lactalbumin (bovine); Pe-pepsinogen (bovine); St-subtilisin (bacteria); Tr-trypsin (bovine); Hb-m-hemoglobin (monomeric, blowfly); SN-staphylococcal nuclease; CyC-cytochrome C (human); Ly-lysozyme (chicken); RN-ribonuclease (bovine).

are involved in maintaining multi-subunit structure [such a role for salt bridges in hemoglobin, for example, has been well documented (Perutz 1965)]. Again one might suggest that the explanation lies in evolution. The simplest way for a protein to evolve from a nonassociating to an associating form is by a partial replacement of polar amino acids by nonpolar ones. Such evolution does not require the concerted development of a number of specific inter-subunit bonds. Once such an evolved protein does exhibit self-association, further mutation can provide specific interactions (salt bridges, side chain hydrogen bonds, etc.) that will further stabilize the structure. This possible path of development is curiously similar to that proposed above for the evolution of stability in tertiary structure. It emphasizes once again that we must think of protein structure in a quasi-teleological sense: the structures we perceive today have been *adaptively* formed over at least a billion years of natural selection.

HISTONES — "SCHIZOPHRENIC" STRUCTURES

In conclusion, I would like to draw attention to the interesting structures and behavior of histones. It has long been recognized that these proteins are tightly bound to the nuclear DNA in the chromatin of higher organisms. More recently it has been discovered that histones are involved in *two* kinds of interactions: electrostatic binding to DNA, and specific interaction with one another to form the *nucleosomes* that are now recognized to be the fundamental units of chromatin structure. Each nucleosome involves a tight globular complex of eight histone molecules, about which a coil of DNA is wrapped. (For a recent review see Van Holde and Isenberg 1975.) Four kinds of histones (designated f2a2, f2b, f3, and f2al) take part in this complex. How can one molecule interact with two so very different kinds of microenvironment—the negatively charged DNA and the companion histone molecules? Some time ago, it was pointed out (Bradbury and Rattle 1972, Bonner and Li 1971) that these histones all have very unusual amino acid sequences: the

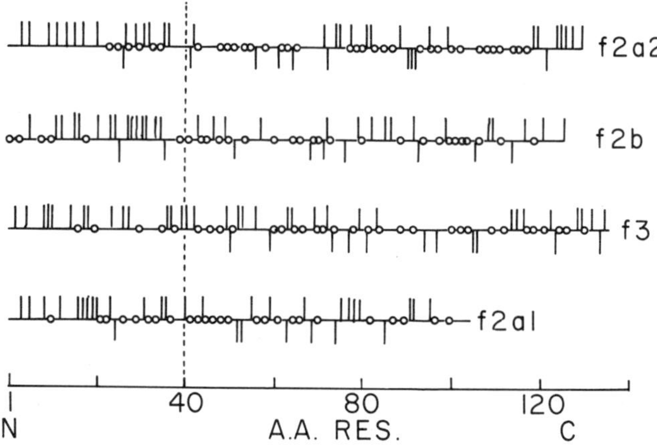

FIG. 1.8. DIAGRAM SHOWING FEATURES OF THE AMINO ACID COMPOSITION OF THE NUCLEOSOMAL HISTONES
In each case the N-terminus is to the left. Upward vertical lines correspond to basic amino acids (lys, arg, his); downward lines to acidic amino acids (gly, asp); and dots to the "hydrophobic" residues listed in Fig. 1.7. The somewhat arbitrary 40 residue cutoff used in Fig. 1.9 is shown as a vertical broken line.

N-terminal region is highly basic and binds tightly to DNA, whereas the C-terminal portion of the molecule contains few basic amino acids and more hydrophobic ones (see Fig. 1.8). If we compare, as in Fig. 1.9, the hydrophobic content in the N-terminal and C-terminal regions of these four histones, we see that in most cases the C-terminal regions exhibit the properties of associating proteins, whereas the N-terminal regions most clearly do not. It is known (Bradbury and Rattle 1972, Bonner and Li 1971) that the C-terminal halves of these histones exhibit a strong tendency to self-association, yet in such associations, the N-terminal regions remain in an extended random coil conformation (Lilley *et al.* 1975). These basic N-terminal segments presumably adopt a defined conformation only upon binding to the negatively charged DNA. Thus, in this case, to respond to a complex and specific microenvironment, a "schizo-phrenic" amino acid composition has evolved. The demands of this microenvironment are very specific; in corollary, the histones are among the most evolutionarily stable proteins known. For example,

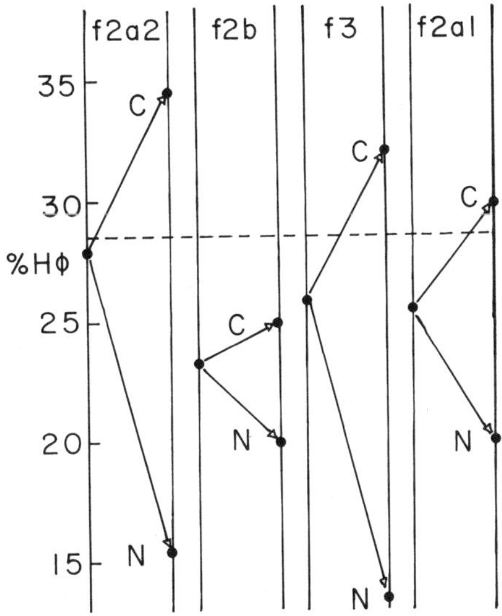

FIG. 1.9. A COMPOSITION DIAGRAM LIKE THAT IN FIG. 1.7, BUT HERE RESTRICTED TO THE HISTONES INVOLVED IN NUCLEO-SOME FORMATION IN CHROMATIN

To the left in each case is given the overall histone composition. To the right in each diagram are shown the compositions of the 40 amino acid N-terminal segment (N) and the remainder of the molecule (about 60–100 residues), designated (C). The N-terminal portions are like single chain proteins, the C-terminal portions more like those that form multi-subunit structures. The broken line corresponds to the "transition composition" of 28.5% deduced from Fig. 1.7.

the amino acid sequences of histone H4 from organisms as different as cow and pea differ at only two residues (Dayhoff 1972, 1973).

SUMMARY

We are beginning to see that the interaction between a protein molecule and its microenvironment involves a continued adaptation of the protein, through evolution of the amino acid sequence, to the demands of that environment. Although most scientists still feel that the *only* information built into the molecule is via the sequence, it seems likely that the folding of the molecule into its native conformation is a much simpler process than hereto believed. The assembly of polypeptide chains into larger multi-subunit proteins may parallel, on a larger scale, the mechanism of folding of local regions in the individual chains.

BIBLIOGRAPHY

ANFINSEN, C. B. 1973. Principles that govern the folding of protein chains. Science *181*, 223–230.

ANFINSEN, C. B., and SCHERAGA, H. A. 1975. Experimental and theoretical aspects of protein folding. Adv. Protein Chem. *29*, 205–300.

BONNER, J., and LI, H. J. 1971. Interaction of histone half-molecules with deoxyribonucleic acid. Biochemistry *10*, 1461–1470.

BRADBURY, E. M., and RATTLE, H. W. E. 1972. Simple computer-aided approach for the analyses of the NMR spectra of histones. Fractions F1, F2a1, F2B, cleaved halves of F2B and F2B°DNA. Eur. J. Biochem. *27*, 270–281.

DAYHOFF, M. O. 1972. Atlas of Protein Sequence and Structure, Vol. 5, Nat. Biomed. Res. Found., Washington; also Suppl. 1, 1973.

KARPLUS, M., and WEAVER, D. L. 1976. Protein-folding dynamics. Nature (London) *260*, 404–406.

LEVINTHAL, C. 1968. Are there pathways for protein folding? J. Chim. Phys. *65*, 44–45.

LILLEY, D. M. J., HOWARTH, O. W., CLARK, V. M., PARDON, J. F., and RICHARDS, B. M. 1975. Investigation of the conformational and self-aggregational processes of histones using hydrogen and carbon-13 nuclear magnetic resonance. Biochemistry *14*, 4590–4600.

PERUTZ, M. F. 1965. Structure and function of hemoglobin. I. A tentative atomic model of horse oxyhemoglobin. J. Mol. Biol. *13*, 646–668.

ROSE, G. 1976. Calculations relating to the structure of biopolymers. Ph.D. Thesis, Oregon State University.

ROSE, G., WINTERS, R. H., and WETLAUFER, D. B. 1976. A testable model for protein folding. FEBS Letters *63*, 10–16.

SCHULZ, G. E., BARRY, C. D., FRIEDMAN, J., CHOU, P. Y., FASMAN, G. D., FINKELSTEIN, A. V., LIM, V. I., PTITSYN, O. B., KABAT, E. A., WU, T. T., LEVITT, M., ROBSON, B., and NAGANO, K. 1974. Comparison of predicted and experimentally determined secondary structure of adenyl kinase. Nature (London) *250*, 140–142.

VAN HOLDE, K. E. 1966. The molecular architecture of multichain proteins. *In* Molecular Architecture in Cell Physiology, T. Hayashi and A. Szent-Györgyi (Editors). Prentice-Hall, Inc., Englewood Cliffs, N. J.

VAN HOLDE, K. E., and ISENBERG, I. 1975. Histone interactions and chromatin structure. Acc. Chem. Res. *8*, 327-335.

2

Denaturation and Renaturation of Proteins

John R. Whitaker

Structural studies on proteins were given a major boost when Sanger and coworkers in 1955 completed the primary sequence of insulin, a polypeptide of 58 amino acids and a molecular weight of 6000 (Ryle *et al.* 1955). Combining the techniques developed by Sanger and coworkers with those of Edman degradation, development of the protein sequentator, etc., led to rapid progress in primary sequence determination of proteins. Sequence determination is now routine in many laboratories and the complete sequence of an average size protein can be determined in one or two months.

In 1967, the secondary and tertiary structures of ribonuclease were elucidated by X-ray crystallography (Kartha *et al.* 1967). While the technique of X-ray crystallography was developed in the 1930's and used to determine the three-dimensional structures of small molecules, including small peptides, it required development of high speed computers for data processing and isomorphorous heavy metal labeling before X-ray crystallography could be applied successfully to proteins. The three-dimensional structures of several proteins are known to 2 Å resolution (Dayhoff 1969; Dickerson 1972; Haschemeyer and de Harven 1974).

The quaternary structure of macromolecular organization of several proteins into biologically more efficient units have been elucidated by a combination of techniques including X-ray crystallography and electron microscopy.

FOLDING OF PROTEINS

With the advances listed above it was natural to begin to ask detailed questions about the factors determining the precise and reproducible folding of proteins into native structures. Answers to these questions are important in (a) folding of proteins following biosynthesis on the ribosomes; (b) denaturation of proteins by heat, pH or chemical reagents; (c) regain of activity of enzymes such as peroxidase and alkaline phosphatase following heat denaturation; (d) aggregation of proteins as in blood clotting and cheese manufacture; (e) functional properties of proteins in foods, particularly with respect to emulsifying properties, whippability and foamability; (f) thermostability of certain microbial enzymes such as the pectic enzyme(s) produced by the brown rot fungus of apricots and enzyme inhibitors found in food materials; (g) changes induced by conversion of zymogens to active enzymes or combination of antigens and substrates with antibodies and enzymes, respectively; (h) predicting protein structure from the amino acid sequence; (i) in refining the secondary and tertiary structures of proteins determined by X-ray crystallography; and (j) in activation of inactive enzymes in certain diseases such as Lesch-Nyhan syndrome.

Importance of Primary Structure in Protein Folding

The brilliant work of Anfinsen and coworkers (Sela *et al.* 1957; Anfinsen *et al.* 1961) on ribonuclease, as well as that of more recent investigators on other proteins, has shown unequivocally that the sequence of amino acids in the primary structure determines the exact way in which the protein is to be folded. Ribonuclease, a protein of 124 amino acids and a molecular weight of 13,683, was unfolded by the combined action of mercapto-ethanol and urea. Mercaptoethanol reduced the -SS- bonds while urea caused the chain to open up by rupture of hydrophobic and hydrogen bonds. On removal of the urea and mercaptoethanol by dialysis or gel filtration and letting the protein stand in air, the protein regained its original structure as measured by X-ray crystallography, by UV and fluorescence spectrophotometry and by enzymatic activity. This included the correct formation of the four disulfide bonds (Anfinsen and Haber 1961) even though there are 105 possible ways of recombining the eight sulfhydryl groups to form the four disulfide bonds (Kauzmann 1959; Sela and Lifson 1959). An immunoglobulin containing 23 -SS- bonds would have on the order of 2×10^{28} possible pairings of the 46 sulfhydryl groups. Yet nearly quantitative recovery of activity can be achieved by refolding of the reduced and unfolded protein (Freedman and Sela 1966)!

Nature of Bonds Stabilizing Protein Structure

Before detailed consideration of the denaturation and renaturation of proteins it will be helpful to look at the nature of the bonds which stabilize the structures of proteins. The primary sequence is formed by condensation of one amino acid residue via its α-carboxyl group with the α-amino group of another amino acid as shown in Eq (2.1).

$$
\overset{\overset{+}{NH_3}}{\underset{\underset{H}{|}}{R-C-COO^-}} + \overset{\overset{+}{NH_3}}{\underset{H}{R''-C-COO^-}} \rightarrow H_3{}^+N-\overset{R}{\underset{H}{C}}-\overset{O}{C}-\underset{H}{N}-\overset{H}{\underset{R''}{C}}-COO^- \tag{2.1}
$$

Thus the important bond in primary structure is the covalent peptide bond with a heat of formation of about 100 kcal/mol (Table 2.1). This bond is not broken in protein denaturation which is defined as a change in the secondary and tertiary structures of the protein. Breakage of the peptide bond is a hydrolytic reaction involving catalysis by strong acid or base plus heat or by proteolytic enzymes.

The secondary structure of a protein consists of helical regions (usually α-helices), and β-pleated sheet regions. The α-helix is stabilized by 3.6 hydrogen bonds (on the average) per turn. While

TABLE 2.1
ENERGIES AND BOND DISTANCES OF SEVERAL TYPES OF
BONDS FOUND IN PROTEINS[1]

Type	Energy (kcal/mol)	Distance of Interaction (Å)
Covalent bond	30–100	1–2
Electrostatic bond	10–20	2–3
Hydrogen bond	1–5[2]	2–3
Van der Waals attractive forces	1–3	3–5

[1] Jones 1964.
[2] Cited reference gives 2–10 kcal/mol which is generally considered too large.

each hydrogen bond is weak, having a heat of formation of 1–5 kcal/mol, the multiple hydrogen bonds found in an helical region gives stability to the helix. A hydrogen bond is readily disrupted by water competing for binding; however, in the α-helix and in the interior of the native molecule, the hydrogen bonds are not in contact with water.

The β-pleated sheet regions are stabilized by hydrogen bonds and to a lesser extent by hydrophobic and electrostatic bonds between the side chains of the amino acid residues (Table 2.1, Fig. 2.1). The β-pleated sheet regions are formed through interchain interaction or intrachain interaction where folding is brought about by β-turn regions to permit the segments of the chain to come together. Individual hydrophobic bonds are quite weak but collectively the numerous hydrophobic bonds are an important stabilizing factor in maintaining the secondary, tertiary and quaternary structures of a protein. The electrostatic bond is quite strong having a heat of formation of some 10–20 kcal/mol.

The types of bonds involved in maintaining the tertiary structures of proteins are shown in Fig. 2.1. The covalent disulfide bond is the strongest of these with a heat of formation of $\sim$50 kcal/mol. Not all proteins contain disulfide bonds. Proteins excreted out of the cells frequently have disulfide bonds, but not always (as for example myoglobin). Proteins designed to function intracellularly do not have disulfide bonds. Some of the most heat resistant proteins have several disulfide bonds (as well as being small in size). Examples are pancreatic ribonuclease (mol wt 13,683, 4 disulfide bonds), hen egg white lysozyme (mol wt 14,000, 4 disulfide bonds), and Bowman-Birk soybean proteinase inhibitor (mol wt 8,000, 7 disulfide bonds, shown in Fig. 2.2). In some proteins such as alkaline phosphatase, lactate dehydrogenase and α-amylase, divalent metal ions contribute to the stability of the tertiary and/or quaternary structure. Other nonprotein parts such as the heme group contribute to stability of myoglobin, hemoglobin, peroxidase, etc.

The quaternary structures of proteins are stabilized by the same types of bonds listed above for the tertiary structure. Interchain disulfide bonds are less common in proteins although they occur in such proteins as α-chymotrypsin, insulin, and the immunoglobulins. In α-chymotrypsin and insulin the disulfide bonds are formed via an intrachain mechanism in the precursor molecules. Subsequent limited proteolysis leads to formation of three polypeptide chains in α-chymotrypsin and two in insulin. In the quaternary assemblage of procollagen, interchain disulfide bonds, located near the amino terminal end, play an important role; this portion of the procol-

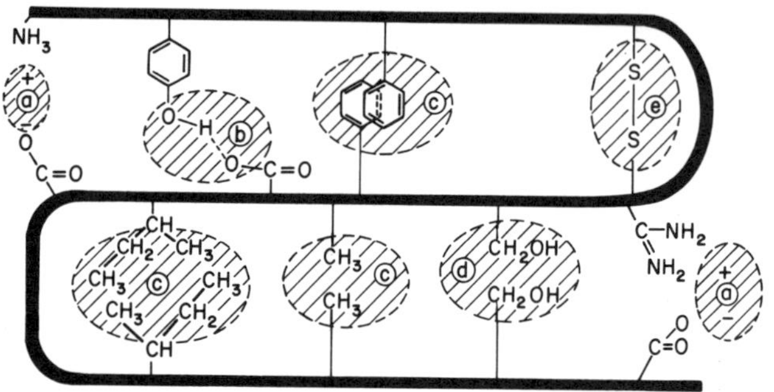

From Anfinsen (1959)

FIG. 2.1. BONDS WHICH STABILIZE PROTEIN STRUCTURE
(a) electrostatic interaction; (b) hydrogen bonding; (c) hydrophobic interaction; (d) dipole-dipole interaction; (e) disulfide bond.

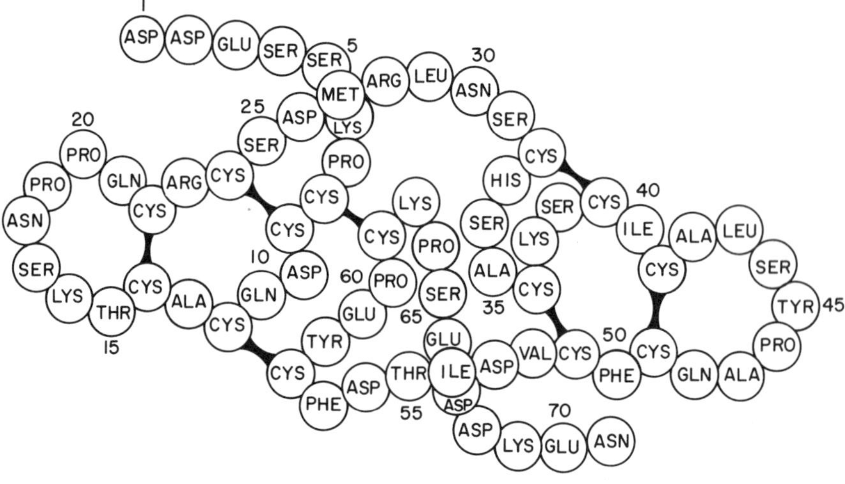

From Ikenaka et al. (1974)

FIG. 2.2. PRIMARY STRUCTURE OF BOWMAN-BIRK PROTEASE INHIBITOR FROM SOYBEANS

lagen molecule is subsequently hydrolyzed away by a specific procollagen peptidase to give collagen. We shall return later to the importance of disulfide bonds in the correct assemblage of the structure of proteins. In some highly specialized proteins, such as collagen and elastin, other interchain covalent crosslinkages are important in maintaining the superstructure of the protein.

Role of the Solvent in Protein Folding

A discussion of the forces maintaining the native structure of a protein molecule as well as the structure itself is not complete without considering the role of the solvent. Both entropic as well as energetic factors contribute to the free energy of the native molecule, generally considered to be a minimum. The two major sources of entropy are: (1) from the various possible configurations of the solvent molecules for a given conformation of the protein, and (2) from the translational and rotational motions of the protein as well as the stretching and bending of bonds within the protein. In water, the protein folds so as to put the majority of the hydrophobic groups on the surface of the molecule. One of the easiest ways to denaturate a protein is to expose it to an hydrophobic interface, such as in formation of an emulsion or interaction with a detergent. The protein structure can, in some cases, be "turned inside out" by forcing it into a nonaqueous environment.

The number of water molecules associated with a protein molecule, *i.e.*, freedom of movement is restricted by interaction with the protein, is large since many proteins have 20–30% hydration. This would represent 200–300 water molecules per protein molecule of 18,000 daltons.

The role of the solvent in maintaining the native structure of the protein is depicted schematically in Fig. 2.3. Water attempts to maintain its own structured arrangement and to resist changes in shape of the protein molecule. This forces the hydrophobic groups into the interior of the protein where they react with each other.

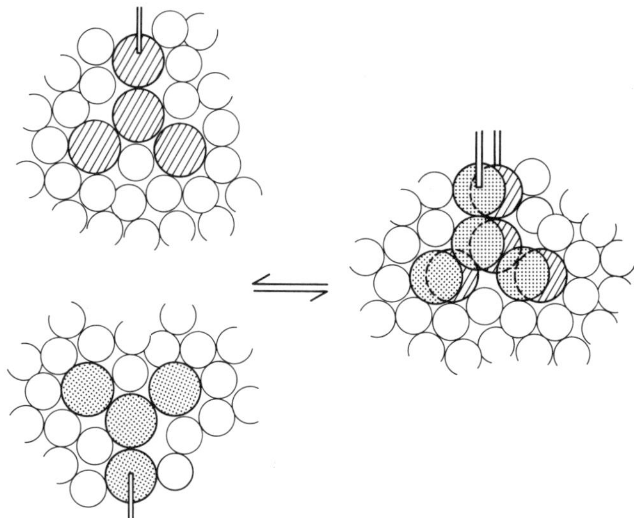

FIG. 2.3. SCHEMATIC REPRESENTATION OF INTERACTION OF HYDROPHOBIC GROUPS IN PROTEINS WITH WATER
Shaded circles represent side chains of the protein; white circles represent water molecules.

This permits the entropy value for the solvent molecules to assume a minimum value. Changes in shape require a disruption of water structure, with a required subsequent reorganization. When a protein has undergone a major unfolding, aggregation involving the hydrophobic regions of the molecule would be favored as the water reorganizes to maintain its structure. The importance of this entropy factor involving the solvent can be seen from the activation parameters for denaturation of some proteins (Table 2.2). Note the very large $\Delta S\ddagger$ values, the entropic term. Protein denaturation is a highly cooperative process, meaning that once a few bonds are broken by having achieved sufficient activation energy, the secondary and tertiary structures are disrupted at a very fast rate. Small changes in temperature have a tremendous effect on the rate of denaturation as can be seen from the data in Table 2.2. For example, a change in temperature from $60°$–$70°$C would increase the rate of denaturation of peroxidase by 3400-fold.

TABLE 2.2
TRANSITION STATE DENATURATION CONSTANTS FOR VARIOUS PROTEINS[1]

Protein	$\Delta H\ddagger$ (cal/mol)	$\Delta S\ddagger$ (e.u.)[2]	$\Delta F\ddagger(25°C)$ (cal/mol)
Trypsin	40,200	44.7	26,900
Pepsin	55,600	113.3	21,800
Hemoglobin	75,600	152.7	30,100
Egg albumin	132,000	315.7	37,900
Peroxidase, milk	185,300	466.0	46,400

[1] Modified from Stearn 1949.
[2] In cal/mol/degree at $25°$C.

UNFOLDING AND REFOLDING OF PROTEINS

In this discussion renaturation will be defined as the refolding of an unfolded protein back to the original native structure insofar as experimental methods permit determination including recovery of full biological activity. Methods used for determination of regain of original structure include measurement of biological activity including antigenic properties, complementation, hydrogen-tritium exchange rates, nuclear magnetic resonance, ultraviolet spectrophotometry, optical rotatory dispersion, circular dichroism, hydrodynamic properties such as intrinsic viscosity and sedimentation rates, and rate of proteolytic digestion. Rate of proteolytic digestion is an extremely valuable tool in that the native protein is resistant to proteolysis while slight changes of the peptide chain from the native structure lead to proteolysis. We shall illustrate use of this technique later for ribonuclease. The limited proteolysis of

κ-casein, proenzymes and other proteins is thought to occur because a small region of the precursor is in a random coil (denaturated) state.

It has been known since 1902 (Woods 1902) when regeneration of activity of tobacco peroxidase was reported that under appropriate conditions proteins can be reversibly denatured. In a classic set of experiments Northrop (1932) showed that trypsin could be reversibly denatured by heat. The denatured trypsin was separated from the remaining native trypsin in 0.25 saturated $(NH_4)_2SO_4$ where it was insoluble. On raising the temperature to $20°C$, the denatured trypsin regained essentially full activity. Similar results on other enzymes were obtained by other workers (Schwimmer 1944).

In his classic article on proteins in 1945, Anson wrote

> "Hemoglobin which has been denatured in a variety of ways can be converted back into native protein which has the same solubility as the original protein, is crystallizable, has the characteristic spectrum of native hemoglobin, can combine loosely with oxygen, has the same relative affinities for carbon monoxide and oxygen, and is not readily digested by trypsin."

This statement is all the more amazing when taken in the context that there was almost total lack of knowledge of the chemical structure of proteins at that time.

Major advances in understanding protein structure were made by elucidation of the primary sequence of insulin (Ryle *et al.* 1955) and the demonstration that reduced and denatured ribonuclease could be refolded to a state possessing full biological activity (Sela *et al.* 1957; Anfinsen and Haber 1961). The latter results became the basis for the concept that the primary amino acid sequence determines the higher order conformations of proteins. Many other examples of reversible denaturation of proteins are available (Sund and Weber 1966). More recently, the effect of environmental factors and structural changes on renaturation of the enzymes fumarase, enolase, aldolase, glyceraldehyde phosphate dehydrogenase, lactic dehydrogenase, and malic dehydrogenase following denaturation in 6 *M* guanidine hydrochloride have been reported (Teipel and Koshland 1971A, B). Similar experiments have been performed recently with cytochrome c and lysozyme (Ikai *et al.* 1973; Tanford *et al.* 1973).

Current intensive interest in the folding and unfolding of protein chains is evidenced by several recent excellent reviews on protein unfolding and refolding (Tanford 1968, 1970; Brandts 1969; Lumry and Biltonen 1969; Hermans *et al.* 1972; Ptitsyn *et al.* 1972; Anfinsen 1973, Wetlaufer and Ristow 1973; Nagano 1974;

Anfinsen and Scheraga 1975; Baldwin 1975; Tanaka and Scheraga 1975).

In the remainder of this paper we shall deal with the reversible denaturation of alkaline phosphatase and peroxidase, enzymes of major importance in food science, with a few other selected examples and end with a postulated mechanism of protein folding.

Alkaline Phosphatase

Absence of alkaline phosphatase (orthophosphoric monoester phosphohydrolase, EC 3.1.3.1) activity in pasteurized bovine milk and cream is used to determine adequacy of heat treatment. The first indication that alkaline phosphatase could regain activity following heat denaturation came soon after introduction of the high temperature-short time (HTST) pasteurization method (Wright and Tramer 1953A). This had not been noted with the long-hold method (57°C for 30 min; Fram 1957A). The amount of regain of activity was dependent on the holding temperature of the milk subsequent to pasteurization (Wright and Tramer 1953A, B; Fram 1957B; McFarren $et\ al.$ 1960) (Fig. 2.4) with maximum regain occurring when the milk was held at 30–37°C. Little or no regain of activity occurred at 4°C or below in one week. The temperature of heating was also important in determining the extent of regain of activity (Fram 1957B). It is interesting that pasteurization at the higher temperature for 16 sec (Table 2.3) led to recovery of more activity. This effect of higher temperature, up to a point, is seen also in the data of Fig. 2.5. The same effect has been reported for peroxidase (Schwimmer 1944; see section on peroxidase). One possible explanation would be a subsequent aggregation reaction occurring after denaturation which is relatively slow and less temperature dependent so that a longer time at lower temperatures would favor irreversible loss of activity such as has been seen for the acid denaturation of metmyoglobin (Shen and Hermans 1972A, B, C).

Regain of phosphatase activity was observed to be quite variable from one batch of milk to another (Wright and Tramer 1953B). The variability did not correlate with levels of fat, ascorbic acid, amino acids or original level of phosphatase present. However, absence of O_2 and presence of reducing conditions generally increased substantially the regain of activity. Added Zn^{2+} at 1 ppm increased the extent of reactivation in 2 hr by 50% while 10 ppm increased the extent of reactivation by 100% (Wright and Tramer 1956). Mg^{2+} also markedly increased the regain of activity but was required at higher levels (McFarren $et\ al.$ 1960; Fig. 2.4).

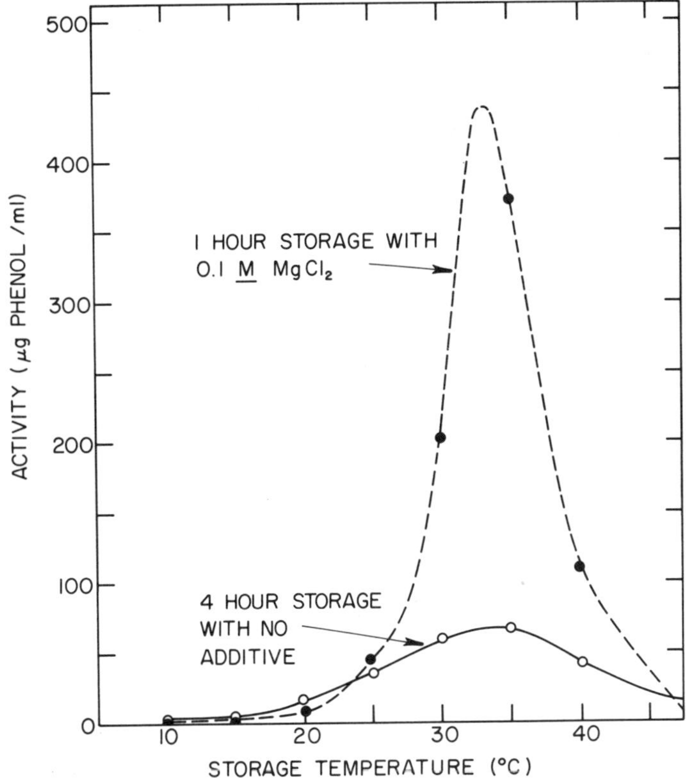

From McFarren et al. (1960)

FIG. 2.4. EFFECT OF STORAGE TEMPERATURE AND MgCl$_2$ ON
REGENERATION OF MILK ALKALINE PHOSPHATASE ACTIVITY
Samples were heat treated at 106°C for 2.8 sec, cooled and stored at
indicated temperatures in presence or absence of 0.1 M MgCl$_2$.

TABLE 2.3

PHOSPHATASE ACTIVITY IN 20% HTST PASTEURIZED CREAM
AFTER STORAGE AT 31°C[1]

Treatment	Activity[2]		
	Initially	2 hr at 31°C	4 hr at 31°C
None, raw cream	3.8		
74°C/16 sec	0	1.2	6.3
82°C/16 sec	0	6.6	16.8
98°C/16 sec	0	9.6	21.6

[1] Fram 1957B.
[2] μg phenol produced/ml cream under standard conditions.

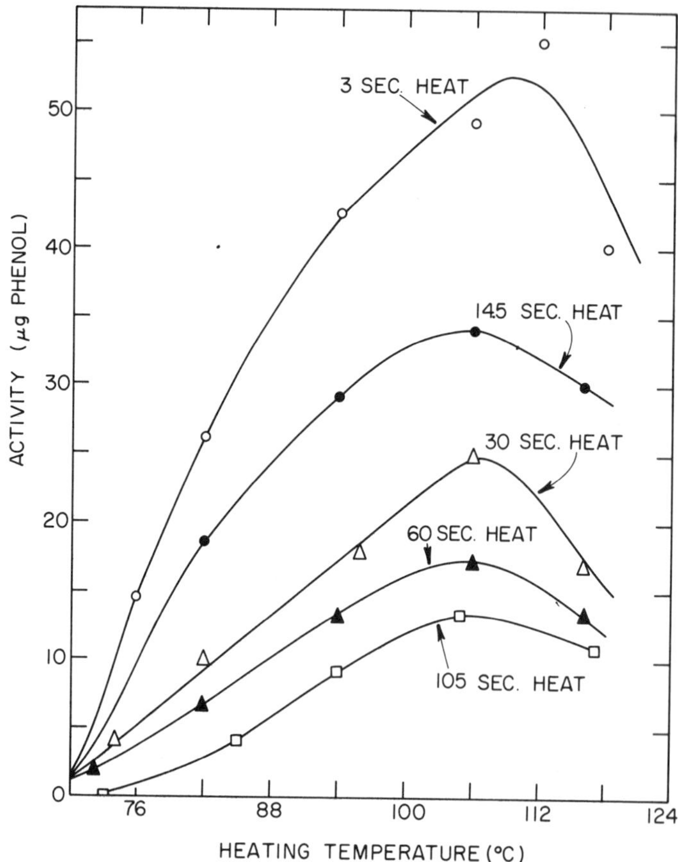

From McFarren et al. (1960)
FIG. 2.5. EFFECT OF HEATING TEMPERATURE AND TIME ON
REACTIVATION OF MILK ALKALINE PHOSPHATASE
Storage was at 30°C for 4 hr in the absence of added $MgCl_2$.

Experiments with purified milk alkaline phosphatase clarified some of the conditions leading to reactivation (Lyster and Aschaffenburg 1962). In the presence of 20 mM β-glycerophosphate (substrate), 1 mM $MgSO_4$ and 0.1% β-lactoglobulin, at pH 6.8, the purified enzyme regained 26–30% of its original activity when held at 37°C. Omission of any one of the components resulted in very low regain of activity. Maximum regain of activity was obtained on incubation at pH 6.8; at pH 5.5 and 8.3 there was no regain of activity. Other substrates, α-glycerophosphate and phenyl phosphate, were as effective as β-glycerophosphate; however, the competitive inhibitor, inorganic orthophosphate gave only 3.5% regain of activity. $MgSO_4$ was by far the most effective inorganic compound in restoring activity. Mg^{2+} had a major effect (8–14X) on reactivation of milk alkaline phosphatase but no effect on activity of residual phosphatase left from inadequate pasteurization (McFarren *et al.* 1960). This has been proposed as a criterion for distinguishing between reactivable phosphatase and residual phosphatase. $HgCl_2$, $CdSO_4$ and $ZnSO_4$ as well as *p*-chloromercuribenzoate, were very effective inhibitors of reactivation, possibly as a result of reaction with the two sulfhydryl groups of β-lactoglobulin (see later discussion on the role of sulfhydryl compounds in regain of activity). The authors (Lyster and Aschaffenburg 1962) reported preliminary evidence for a dialyzable, heat labile compound in milk which markedly decreased the regain of activity. Milk was also reported (Lyster and Aschaffenburg 1962) to contain a heat stable, nondialyzable compound (a protein?) which could replace β-lactoglobulin in the model experiments.

It is appropriate to stress that the regain of activity following pasteurization is indirectly proportional to the time of heating. Milk that has been heated at $71.7°C$ for 15 sec (accepted HTST process) does not show regain of activity although this can be a problem in cream heated under the same conditions.

Essentially nothing can be said about the mechanism of reactivation of milk alkaline phosphatase because its detailed molecular properties aren't known (a serious deficiency for so important an enzyme to the dairy industry). The enzyme has been purified to essentially homogeneity (Morton 1953) and its molecular weight determined to be 190,000 (Barman and Gutfreund 1966).

A great deal more data at the molecular level are available for *Escherichia coli* alkaline phosphatase; therefore, we shall use it as a model for bovine milk alkaline phosphatase. Although most assuredly different proteins there are likely to be marked similarities between them.

E. coli alkaline phosphatase is a zinc-containing protein composed of two subunits of about 45,000 each (mol wt of 89,000; Bosron *et al.* 1975). When isolated in the absence of $(NH_4)_2SO_4$ and at pH 7.2–7.5, the enzyme has four g-atoms of Zn^{2+} and 1.3 g-atoms of Mg^{2+} per molecular weight 89,000 (Bosron *et al.* 1975). Mg^{2+} is not able to activate the apoenzyme (metal-free protein) alone but the presence of stoichiometric amounts of Mg^{2+} regulates the mode of Zn^{2+} binding and thereby increases enzymatic activities over those attained with Zn^{2+} only. Two of the Zn^{2+} are required for activity of the enzyme (Simpson *et al.* 1968; Csopak and Szajn 1973) while all four appear to be important in maintaining the tertiary (Reynolds and Schlesinger

1969A) or quaternary structures (Simpson and Vallee 1969; Applebury and Coleman 1969: Brown *et al.* 1974). Each Zn^{2+} appears to be liganded to a histidyl and a tyrosyl residue (Reynolds and Schlesinger 1969A). The molecule contains two disulfide bonds per monomer (Reynolds and Schlesinger 1967). The two monomers (subunits) are identical. In the presence of $>1 \times 10^{-5}$ M Zn^{2+} the dimer forms a tetramer between pH 7 and 8 which is fully active (Reynolds and Schlesinger 1969B). Reynolds and Schlesinger (1969B) suggest, without data, that the tetrameric form of mol wt 174,000 might be the active form *in vivo* in *E. coli*. If so, the molecular weight would be similar to that of bovine milk alkaline phosphatase (mol wt 190,000).

Extensive studies on the mechanism of refolding of the monomers and their reassociation (Reynolds and Schlesinger 1967; 1969A) and the dissociation of *E. coli* alkaline phosphatase (Applebury and Coleman 1969) have been performed. Crystallographic studies of the enzyme at 7.7 Å resolution have been reported (Knox and Wyckoff 1973); this is not sufficient resolution to completely delineate the tertiary and quaternary structures. In the absence of such data the schematic depiction of the molecule shown in Fig. 2.6 should suffice. The diagram shows the two subunits, each with two disulfide bonds and each with a Zn^{2+} in the active site and the two bridging Zn^{2+} helping to maintain the quaternary structure. The molecule is essentially spherical as shown by an intrinsic viscosity of 3.4 (Reynolds and Schlesinger, 1967). The native molecule contains approximately 60% β-pleated sheet, about 30% random coil and about 10% α-helix (Applebury and Coleman 1969).

We shall consider the dissociation and unfolding route first. At Zn^{2+} concentrations $>1 \times 10^{-5}$ M, the protein is a tetramer between pH 7 and 8 and is fully active (Reynolds and Schlesinger 1969B) (not shown in Fig. 2.6). As usually isolated between pH 7 and 8, the molecule is a dimer with four Zn^{2+}. Treatment with 8-hydroxyquinoline-5-sulfonic acid (HQSA) at neutral pH removes the two Zn^{2+} associated with stabilization of the tertiary and quaternary structures of the molecule (Simpson and Vallee 1968). The dimer is still fully active enzymatically. Treatment with ethylenediaminetetraacetic acid (EDTA), a stronger chelating agent that HQSA, removes the remaining two Zn^{2+} to give an inactive protein, still in dimeric form. Alternatively, the four Zn^{2+} can be removed by adjusting the pH to 4. The quaternary and tertiary structures remain in the native form. Therefore, the hydrogen ion at pH 4 effectively competes with the Zn^{2+} binding to the histidyl and tyrosyl residues.

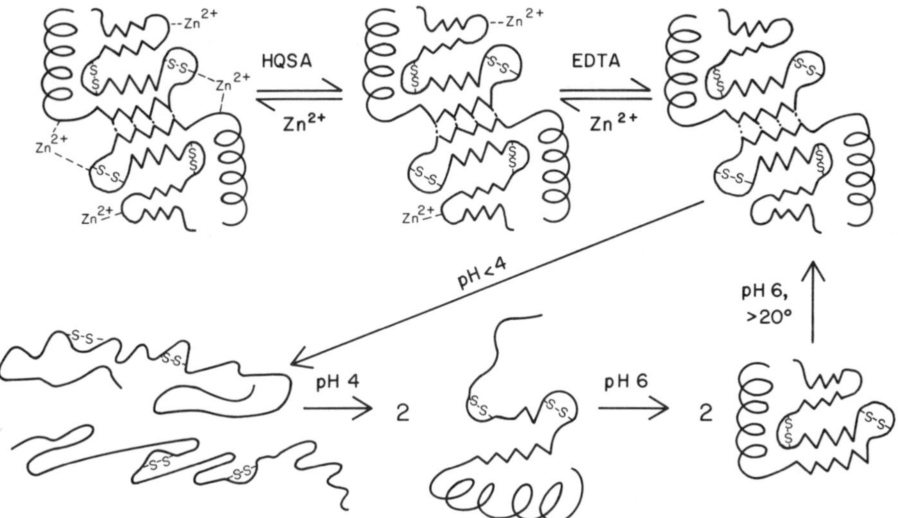

FIG. 2.6. SCHEMATIC REPRESENTATION OF THE DENATURATION AND RE-
NATURATION OF *Escherichia coli* ALKALINE PHOSPHATASE
HQSA, 8-hydroxyquinoline-5-sulfonic acid; EDTA, ethylenediaminetetraacetic acid.

Adjustment of pH below 4 leads to dissociation of the dimer and unfolding of the monomers to random coils with intrinsic viscosity of 32 vs 3.4 for the native dimer (Reynolds and Schlesinger 1967). Since no reducing compounds are added, the disulfide bonds remain intact. They appear to involve only small segments of the polypeptide chain since the hydrodynamic properties of the subunits at pH 2 indicate essentially a completely random coil. The investigators did not do experiments in which the disulfide bonds were broken.

Refolding and reassociation of the molecule to a native structure appears to follow a different pathway back to the dimer (Fig. 2.6). When the pH is adjusted from 2 to 4, the intrinsic viscosity drops from 32 to 3.2 (Reynolds and Schlesinger 1967) but there is no change in the spectral properties indicating that aromatic amino acid residues have not yet been removed from water. Thus, the data indicate a globular structure in which much of the structure is open to the solvent, and occupying a volume and shape close to that of the native dimer. When the pH is adjusted to 6–8 and the solution held at $\geqslant 20°$C for approximately 1 min, the globular, open monomer at pH 4 folds to the native state. By returning the temperature to 15°C, the protein can be held as the refolded subunit which has spectral properties nearly identical, but not quite, to that of the native molecule.

When a solution containing the refolded subunit is held at 20°C or above the subunits associate to dimers identical to those obtained from the native protein by removal of the four Zn^{2+} with EDTA. Examination of the optical rotatory dispersion of the solution at pH 4.5 showed approximately 75% dimer species (Applebury and Coleman 1969). Addition of Zn^{2+} converts the remaining monomer to dimer. Therefore, it does not appear that Zn^{2+} is necessary for dimer formation; rather, by combining with dimer it shifts the equilibrium between monomer and dimer to the dimer species. Addition of two equivalents of Zn^{2+} leads to a fully active dimer with two bound Zn^{2+} and addition of more Zn^{2+} leads to the dimer with four Zn^{2+}. Hydrogen-tritium exchange rates show there are 60–70 more hydrogens which exchange more rapidly in the apoenzyme (metal-free protein) than in the enzyme (Brown *et al.* 1974). Therefore, the structure of the apoenzyme must be a little more "open" than in the enzyme although other measurements are unable to detect the difference. Enzyme containing two Zn^{2+} per molecule shows an intermediate number of rapidly exchanging hydrogens between that of the apoenzyme and enzyme with four Zn^{2+}.

It is most interesting that the pathway for dissociation of the dimer and unfolding of the monomer is different from that for its refolding and reassociation. That such might be the case was indicated by the hysteresis seen in titrating the protein with acid or base (Reynolds and Schlesinger 1968). Apparently, there is a kinetic barrier preventing the random coils from refolding and forming the dimer at pH 4 where it is stable once formed as shown by the dissociation data. Whatever the explanation, it has been fortunate for the investigator that the difference exists.

Peroxidase

Peroxidase (donor: hydrogen peroxide oxidoreductase; EC 1.11.1.7) is the indicator enzyme used to determine adequacy of heat treatment in the blanching of fruits and vegetables. Peroxidase is particularly useful for this purpose since it is found in all fruits and vegetables, its activity is easy to determine and its relative heat stability is such that by the time it is completely inactivated most other enzymes and microorganisms will have been inactivated. It is not clear whether it contributes in a major way to off-flavors and off-colors which develop in underblanched vegetables and fruits.

Peroxidase regains activity rather readily and rapidly when the tissue has not been adequately heat-treated. This can be demonstrated quite readily by addition of substrate to an extract of a plant tissue heat-treated until about 1–5% residual activity remains and following the rate of the reaction in a recording spectro-photometer for 5-10 min. Some of the denatured enzyme re-natures quite rapidly in the presence of substrate and the rate of the reaction increases with time.

Reactivation of peroxidase in heat-treated extracts from tobacco were reported in 1902 (Woods 1902). Short-time treatment at high temperatures led to the greatest regeneration of activity (Schwimmer 1944) as reported above for milk alkaline phosphatase. For example, a 0.2 ml sample of turnip extract held at 105°C for 30 sec reduced the original peroxidase activity to 10%. Holding at 26°C for 20 hr resulted in regain to 65% of the original activity. On the other hand, heating for 3 min at 90°C decreased the original activity to 8% and holding at 25°C for 20 hr resulted in a regain of 22% activity. The effect of rate of heating on regain of turnip peroxidase activity is shown in Fig. 2.7. Within the experi-mental limits used, the faster the heating rate, the better regain of activity on storage at 26°C. This is a very important observation in use of HTST processing.

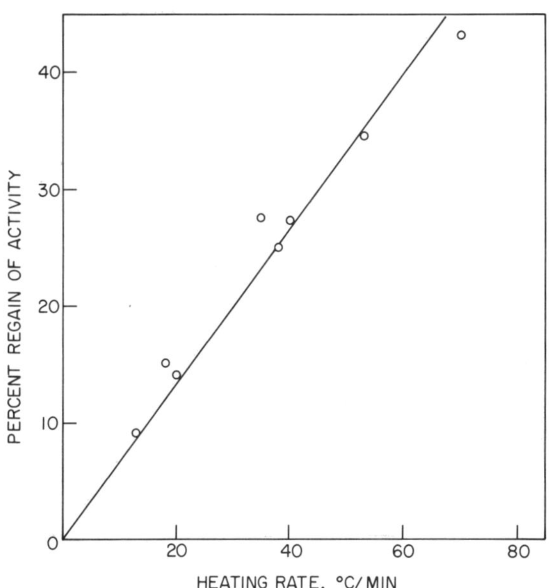

From Schwimmer (1972)
FIG. 2.7. EFFECT OF HEATING RATE ON THE REGEN-ERATION OF TURNIP PEROXIDE ACTIVITY
The same quantity of heat was used (temperature X time).
Storage following heating was at 26°C.

Regain of activity on holding heat-treated extracts at 25°C was appreciably better than at 6°C (Schwimmer 1944). Heat treatment of a turnip extract led to formation of a slight precipitate. When this precipitate was removed by centrifugation there was no reactivation. This is evidence that the apoprotein and hemin dissociated on heating and that the protein was denatured thus becoming insoluble while the hemin remained in solution. Separation of the two prevented recovery of activity. Thus, the following minimum events probably occur during the denaturation of peroxidase as shown in Eq (2.2).

$$\text{Apoenzyme-hemin} \to \text{hemin-denatured apoenzyme} \to$$
$$\text{denatured apoenzyme}$$
$$+ \text{ hemin} \qquad\qquad (2.2)$$

Other workers have reported on reactivation of peroxidase when low acid foods were processed by high temperature-short time methods or in *in vitro* experiments (Guyer and Holmquist 1954; Esselen and Anderson 1956; Vetter *et al.* 1958; Zoueil and Esselen 1959; Joffe and Ball 1962 and others).

Peroxidases have been purified from milk (Theorell and Åkeson 1943), Japanese radish (Morita and Kondo 1954), horseradish (Theorell 1943) and fig latex (Kon and Whitaker 1966). Almost nothing is known about the molecular properties of the peroxidases of English peas, sweet corn, broccoli sprouts, cauliflower, etc. Therefore, we must limit our discussion on the possible molecular events occurring on denaturation to horseradish peroxidase. Even then, data on denaturation and renaturation are far from adequate. Most of the studies have been complicated by use of impure enzyme in that a number of isozymes of horseradish peroxidase are present (Klapper and Hackett 1965).

Horseradish peroxidase is a glycoprotein ($\sim 18\%$ carbohydrate) of mol wt 40,200. Apparently, variation in the carbohydrate moiety accounts for the isozymes of peroxidase. The enzyme has been crystallized (Theorell 1943). It is composed of a single polypeptide chain and contains one ferriprotoporphyrin III prosthetic group per molecule. The prosthetic group can be removed from the protein moiety by acidic acetone below 0°C. [Other types of peroxidases include the verdoperoxidases such as lactoperoxidase in which the prosthetic group is an iron porphyrin nucleus but is not ferriprotoporphyrin III. It cannot be removed by treatment with acidic acetone. The flavoprotein peroxidases contain flavin adenine dinucleotide (FAD) as a prosthetic group.] Four of the six coordination bonds of iron in ferriprotoporphrin III are involved in interaction with the pyrrole ring nitrogens, the fifth is probably coordinated to the imidazole group of a histidyl residue in the protein (Brill and Sandberg 1968) and the sixth is occupied by water in the "resting" enzyme. The prosthetic group is also held to the protein via electrostatic bonds involving one or both of the side chain propionyl groups of the prosthetic group.

In several molecular properties, peroxidase resembles myoglobin and hemoglobin. Below we shall discuss the denaturation-renaturation of myoglobin as a possible model for peroxidase.

Several factors affecting the inactivation and reactivation of horseradish peroxidase have been reported (Lu and Whitaker 1974). As shown in Fig. 2.8A the rate of heat inactivation at 76°C is pH dependent. Maximum stability occurs at pH 6.4. Below pH 5 the rate of inactivation is strongly dependent on hydrogen ion concentration suggesting that splitting of the hemin from the apoenzyme is a major factor in inactivation (Maehly 1953). The slight acceleration in rate of inactivation in alkaline solution may be due to changes in the hemin prosthetic group (Maehly 1953).

Reactivation of heat denatured peroxidase is quite pH dependent (Fig. 2.8B). Below pH 4, there is no regain of activity. Maximum reactivation after 10 min at 35°C was found to be at pH 9.

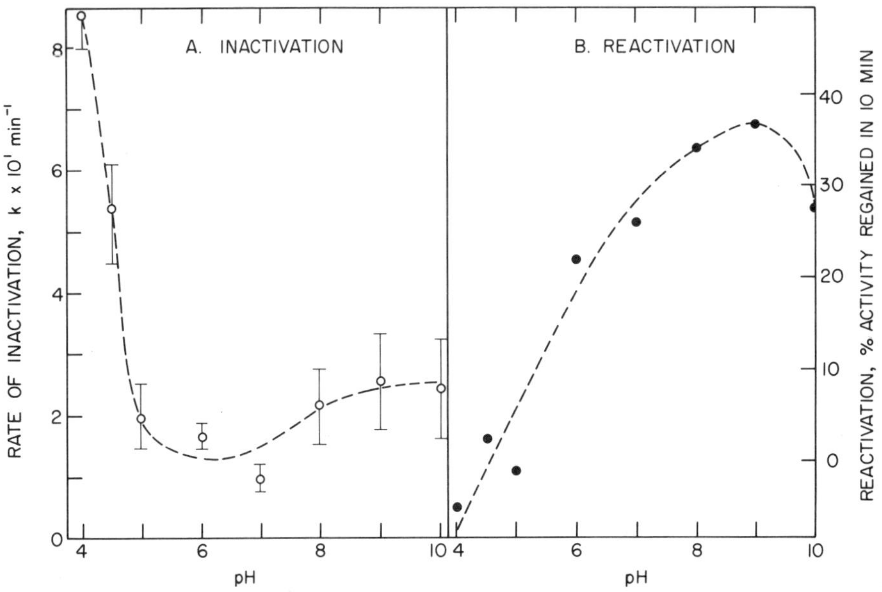

(From Lu and Whitaker (1974)

FIG. 2.8. EFFECT OF pH ON RATE OF INACTIVATION AND EXTENT OF REACTIVATION OF HORSERADISH PEROXIDASE

Inactivation as a function of pH was done at 76°C. Reactivation was at the same pH used for inactivation with a storage temperature of 35°C.

Following partial inactivation of peroxidase at 76°C and pH 7.0 the regain of activity on storage at 20–35°C was found to be biphasic (Fig. 2.9A). There was an initial rapid regain of activity to about 80% of reconvertible activity within the first 8–10 min followed by a slower increase in activity over a period of several hours. It is possible the slow regain in activity is due to a rate controlling dissociation of aggregated protein as reported for myoglobin (Shen and Hermans 1972A; see section on myoglobin). The regain of activity did not follow either first or second order kinetics indicating that several reactions (either consecutive or simultaneous) occurred. Reactivation did not occur at 0°C in agreement with that reported for milk alkaline phosphatase (Wright and Tramer 1953A, B; McFarren *et al.* 1960) while the rates and extent of reactivation at 20°, 24°, 30°, and 35°C were essentially the same (in contrast to that for milk alkaline phosphatase; Fig. 2.4). Within the experimental conditions used, the extent of inactivation had only a small, but reproducible, effect on the rate of regain of activity (Fig. 2.9B). The half-life for maximum reactivation at 35°C was 0.5, 1.0, 1.0 and 1.8 min following heat treatment at 76°C for 4, 6, 8, and 10 min, respectively. Time of holding at 76°C affected the final amount of activity at maximum reactivation although there was a constant percent recovery of activity as shown by data of Table 2.4.

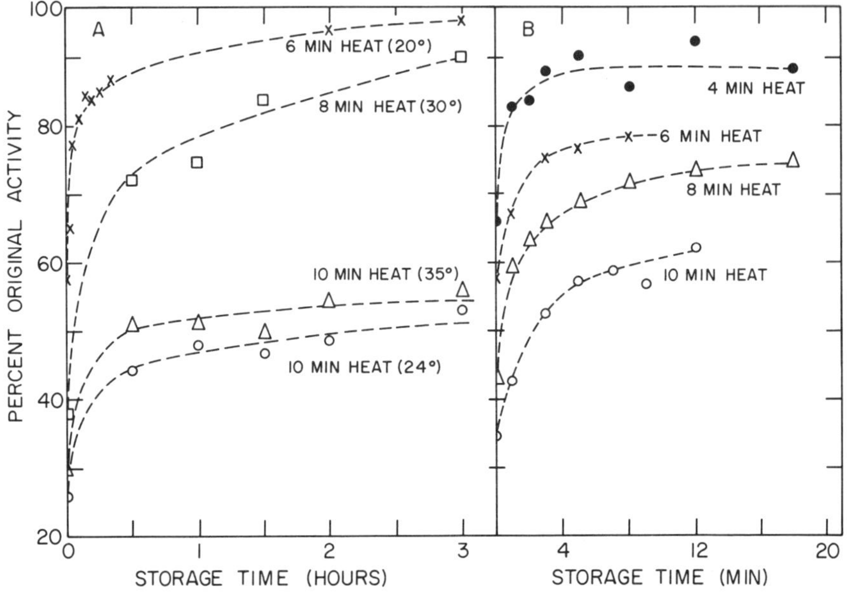

From Lu and Whitaker (1974)

FIG. 2.9. RATE AND EXTENT OF REGAIN OF PEROXIDASE ACTIVITY

Heat treatment was at 76°C for the times indicated on the curves. Storage for regeneration of activity was at 35°C or the temperatures in parenthesis on Fig. 2.9A.

TABLE 2.4
EFFECT OF TIME OF HEATING AT 76°C ON RECOVERY OF
PEROXIDASE ACTIVITY[1]

Time of Heating (min)	Initial Activity Left (%)[2]	Activity after 8 min at 35°C (%)[2]	Difference[3] (%)
4	66.1	88.5	22.4
6	57.6	78.5	20.9
8	43.1	71.5	28.4
10	35.0	59.5	24.5

[1] Adapted from Lu and Whitaker 1974.
[2] All percentages are based on original activity prior to heat treatment of 100%.
[3] Difference = column 3 − column 2.

Hematin added to a solution of peroxidase slowed down the rate of loss of activity (Lu and Whitaker 1974). A tenfold molar excess of added hematin decreased the rate of inactivation by fourfold. There was no effect of added hematin on rate of reactivation; however, the amount of activity regained was affected. Following 10 min heat treatment at 76°C and incubation of solution at 35°C for 8 min, the percentage recovery of activity was 29% and 60%, in the absence and presence of tenfold molar excess of added hematin, respectively. Presence of substrates or cofactors is known to decrease the rate of thermal or pH inactivation of enzymes. Addition of substrate or cofactor following inactivation of several enzymes by urea or guanidine hydrochloride gave higher rates and extent of regain of activity but the effect varied from one enzyme to another (Teipel and Koshland 1971A). For example, the presence of L-malate increased the extent of reactivation of fumarase by threefold over that in absence of L-malate while the presence of NAD^+ and arsenate increased the extent of reactivation of malic dehydrogenase by only 0.2. Under certain conditions 2.5 times the original activity of peroxidase could be achieved following heating and reactivation in the presence of added hematin (Lu and Whitaker 1974) in agreement with other observations on this enzyme (Maehly 1955; Mitsui et $al.$ 1956).

Inactivation and reactivation of peroxidase could proceed by one or both of the pathways shown in Eq (2.3) and (2.4)

$$\overset{1}{\rightleftharpoons} \quad \overset{2}{\rightleftharpoons} \quad \overset{3}{\rightarrow} \tag{2.3}$$

$$\text{Hem} \cdot \text{Prot} \underset{\substack{+\\\text{Hem}}}{\overset{1}{\rightleftharpoons}} \text{Prot} \overset{2}{\rightleftharpoons} \text{Prot}_R \overset{3}{\rightarrow} \text{Prot}_I$$

$$\text{Hem} + \text{Prot}_R \overset{6}{\rightarrow} \text{Prot}_I$$

$$5 \updownarrow$$

$$\text{Hem} \cdot \text{Prot} \overset{4}{\rightleftharpoons} \text{Hem} \cdot \text{Prot}_R \overset{7}{\rightarrow} \text{Hem} \cdot \text{Prot}_I \overset{8}{\rightarrow} \text{Hem} + \text{Prot}_I \tag{2.4}$$

where Hem·Prot is the active hemin containing enzyme, Hem·Prot$_R$ and Hem·Prot$_I$ are reversibly and irreversibly denatured forms of the intact enzyme and Prot, Prot$_R$, and Prot$_I$ are native, reversibly denatured and irreversibly denatured forms of the apoenzyme, respectively. Regain of activity might be through reversal of Steps 1 and 2 and/or 4 and 5. Loss of activity appears to be in part due to dissociation of hemin from the protein as shown by the effect of added hematin on slowing down the rate of inactivation and on enhancement of extent of reactivation (Lu and Whitaker 1974). Bakh and Oparin (1925) and Schwimmer (1944) observed that when a precipitate, formed on heating of peroxidase-containing plant extracts, was separated from the solution, no activity was regained until the precipitate and supernatant liquid were recombined. Presumably, the precipitate contained denatured apoenzyme while the hemin remained in solution. This interpretation is probably correct but is complicated by the fact that hemin can associate to aggregates which may be insoluble also (Maehly 1952).

Data are not available yet to determine whether hemin dissociates from the protein before or after denaturation of protein [Eq (2.3) vs Eq (2.4)]. If denaturation occurs after dissociation of hemin from apoenzyme, all levels of added hematin would be expected to enhance the stability of enzyme to heating. This was found not to be the case (Lu and Whitaker 1974). The data of Anson (1944) indicate that the prosthetic group of hemoglobin is not released until the protein is denatured (see discussion below on myoglobin). At pHs below 5 hemin is dissociated from the enzyme more readily (Maehly 1952, 1953) which might explain the more rapid denaturation of horseradish peroxidase (Fig. 2.8A). The apoenzyme of Japanese radish peroxidase was reported to be less stable to denaturing conditions than the enzyme (Yoshida and Morita 1970).

The splitting of hemin from peroxidase at low pH occurs in several distinct steps as shown spectrophotometrically (Maehly 1952, 1953). Denaturation of the enzyme and dissociation of the hemin is caused by hydrogen ions. Based on the data available, however, it is not possible to delineate in detail the sequence of events involved in dissociation.

Hemin and apoenzyme, split by holding at pH 1.5–4 in the presence of NaCl, can be recombined to form active horseradish peroxidase (Maehly 1955). The steps involved can be followed spectrophotometrically in the absence and presence of substrates. The sequence of events appear to be as shown in Eq (2.5).

$$\text{Prot} + \text{Hem} \rightleftharpoons \text{Prot·Hem}_a \rightleftharpoons \text{Prot·Hem}_b \rightleftharpoons \text{Prot·Hem}_c \quad (2.5)$$

The first step is a rapid bimolecular reaction involving the combination of apoenzyme (Prot) with hemin to give Prot·Hem$_a$. Formation of Prot·Hem$_a$ results in a change in the hemin absorbance spectrum. Prot·Hem$_b$ formation is a much slower monomolecular reaction having an activation energy of 11.9 kcal/mol and can be distinguished from Prot·Hem$_a$ by its ability to combine with H_2O_2 (determined by spectral changes) although it has no catalytic activity. The rate of formation of Prot·Hem$_c$ is the slowest step in reactivation and leads to formation of catalytic activity.

Myoglobin

While the events occurring in the renaturation of peroxidase can be discussed in a general way, molecular events associated with denaturation and renaturation must await detailed knowledge of the primary, secondary and tertiary structures of peroxidase. Myoglobin has many properties in common with peroxidase (Rakshit and Spiro 1974) and the general details of denaturation and renaturation are likely to be similar. The molecular events occurring in the acid denaturation and reactivation (Shen and Hermans 1972A, B, C) and heat denaturation (Awad and Deranleau 1968) of myoglobin have been carefully investigated. Fortunately, the primary, secondary, and tertiary structures of this protein are known.

Acid-catalyzed denaturation and subsequent renaturation of myoglobin occur in several distinct steps that can be followed by ultraviolet spectrophotometry in the range 350–420 nm and by optical rotation at 233 nm. As shown in Fig. 2.10, the rate of unfolding is pH dependent and occurs via two steps which are quite different in rates (note the log plot). The refolding process also occurs via a minimum of two steps. The difference in rates between the two refolding steps are best seen in Fig. 2.11. In these data, unfolding was performed by holding at pH 3.4 and 25°C for times ranging from 2 min to 1400 min. Renaturation was carried out at various pHs and 25°C for either 80 min or 28 hr. Short periods of holding at pH 3.4 (2 or 16 min) resulted in regain of native structure within 80 min while longer holding times did not. (This is analogous to the results seen with HTST inactivation of alkaline phosphatase and peroxidase described above.) Much longer holding times were required to regain most of the activity when the denaturation at pH 3.4 was 160 min or longer.

The results can be schematically described as shown in Fig. 2.12. Myoglobin is composed of a single polypeptide chain of mol wt 17,000 and contains a single heme group. It has eight α-helical regions which comprise about 75% of the total protein. There are no disulfide bonds. In the faster step in acid denaturation (Fig. 2.12; controlled by k_1) there appears to be a decrease in the compactness of the molecule which leads to dissociation of the heme group in the slower step (k_2). This is the rate-determining step in the denaturation of the molecule. The apoprotein, as a random coil, undergoes a slower rate of aggregation (k_3) to a species that is eluted in the void volume of Sephadex G-200 column (exclusion limits for globular proteins of 800,000). When the pH is adjusted to 6.5, the monomeric apoprotein, in random coil, refolds very rapidly (within time resolution of experiments; k_4) which then combines with the heme group in the rate-determining step (k_5) to give what appears to be native myoglobin. This is the faster of the two steps observed in renaturation (Fig. 2.10). In a much slower step (k_{-3}), the aggregated apoprotein dissociates (rate-determining step) to apoprotein (random coil) which rapidly refolds to native apoprotein to react with the heme group to give native myoglobin. It is interesting that renaturation is not a reversal of the denaturation process just as was seen for alkaline phosphatase.

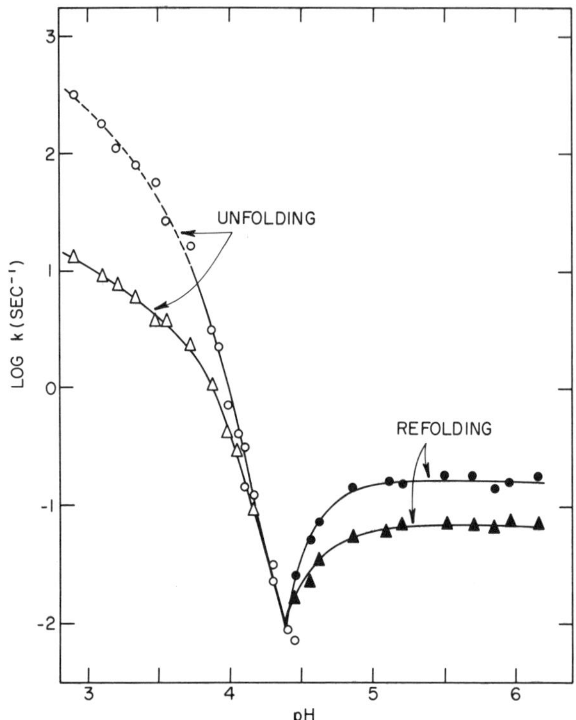

From Shen and Hermans (1972A)
FIG. 2.10. pH DEPENDENCE OF THE RATE CONSTANTS FOR THE UNFOLDING AND REFOLDING OF MYO-GLOBIN AT 25°C
For refolding, myoglobin was unfolded at pH 4.16.

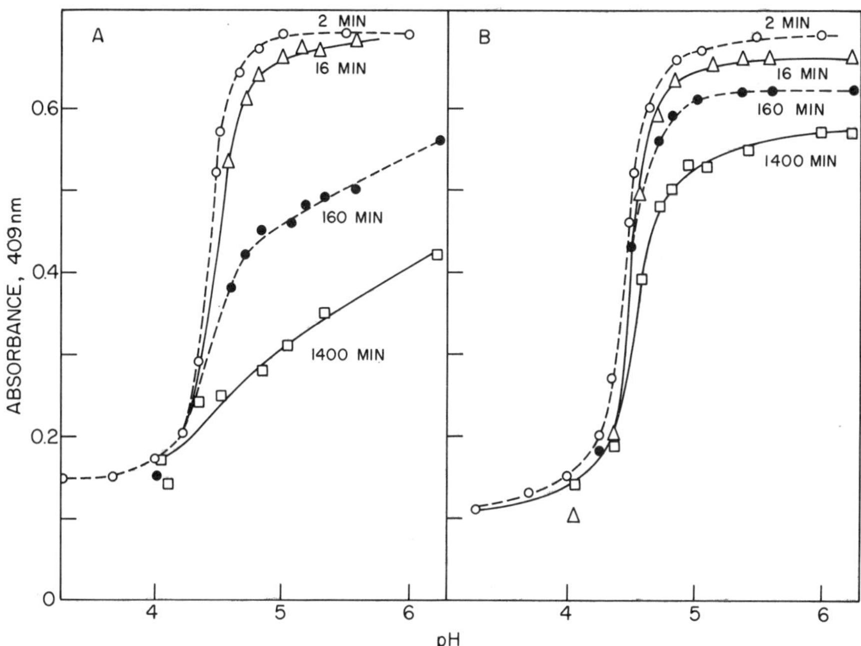

From Shen and Hermans (1972A)

FIG. 2.11. REFOLDING OF DENATURED MYOGLOBIN AS A FUNCTION OF pH AND TIME OF HOLDING AT pH 3.4 (TIMES OF 2–1400 MIN GIVEN ON THE CURVES)
A and B show the results after storage at various pH's (X-axis) for 80 min and 28 hr, respectively.

The thermal denaturation of sperm whale myoglobin at pH 6.85 over the temperature range 25–90°C has been reported (Awad and Deranleau 1968). The first reaction [Eq (2.6)] to occur is characterized by a relatively large change in absorbance at 409 nm while changes in absorbance at 290 and 540 nm are quite small. This indicates some loosening of the protein structure around the heme group but the heme group does not dissociate from the protein as indicated by lack of a large change in absorbance at 540 nm (a measure of heme-ligand interaction). In the second, much slower step there is a large change in absorbance at 290, 409 and 540 nm indicating there are extensive changes in the secondary and tertiary structure of the protein with probable dissociation of the heme group. This step has a $\triangle H\ddagger$ of 109 kcal/mol. Assuming $\triangle F\ddagger$ to be 30 kcal/mol (a reasonable value, see Table 2.2), $\triangle S\ddagger$ is 224 e.u. The third phase is the formation of a precipitate indicating that the random coiled apoprotein undergoes aggregation. These results are qualitatively in agreement with those for acid denaturation of myoglobin (Shen and Hermans 1972A, B, C; Fig. 2.12).

$$N \underset{}{\overset{1}{\rightleftharpoons}} I_a \underset{}{\overset{2}{\rightleftharpoons}} I_b \underset{}{\overset{3}{\rightleftharpoons}} \text{precipitate} \tag{2.6}$$

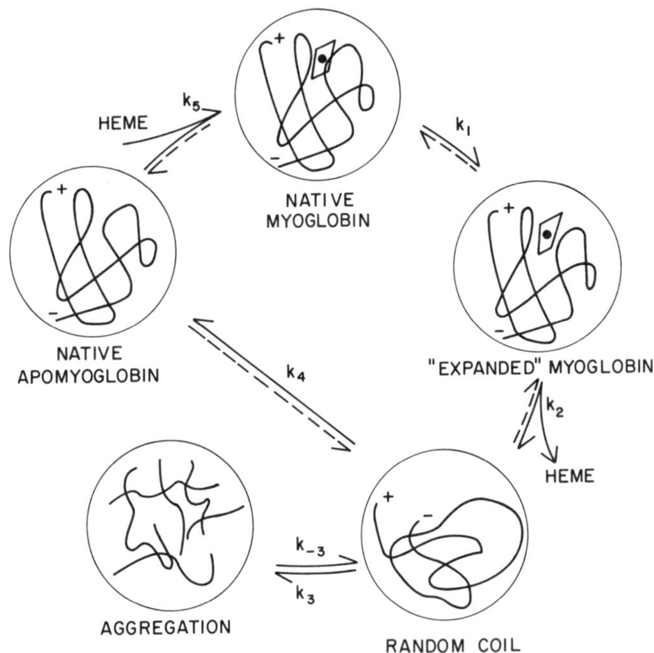

FIG. 2.12. SCHEMATIC REPRESENTATION OF THE STEPS IN THE pH-INDUCED DENATURATION AND RENATURATION OF MYOGLOBIN

Other Proteins

The most detailed understanding of protein unfolding is available for pancreatic ribonuclease. Much of the data has come from selective reduction of the four disulfide bonds and of the selective proteolysis of the molecule at different temperatures.

Native ribonuclease is resistant to proteolysis by α-chymotrypsin, trypsin and carboxypeptidase A. At 45°C and pH 6 or 8, α-chymotrypsin splits the peptide bond between Tyr-25 and Cys-26 (Rupley and Scheraga 1963; Klee 1967) while the protein is still resistant to attack by the other two enzymes. This indicates there is a loosening of the tertiary structure in the region of Tyr-25 and Cys-26 (see Fig. 2.13; structure II). At a somewhat higher temperature, trypsin hydrolyzes first the peptide bonds between Lys-31 and Ser-32 and between Arg-33 and Asn-34 (Ooi et al. 1963; Klee 1967). Thus there is a further loosening of the tertiary structure in the region of residues 31–34 (structure III in Fig. 2.13). At a somewhat higher temperature the C-terminal end becomes susceptible to attack by carboxypeptidase A (structure IV).

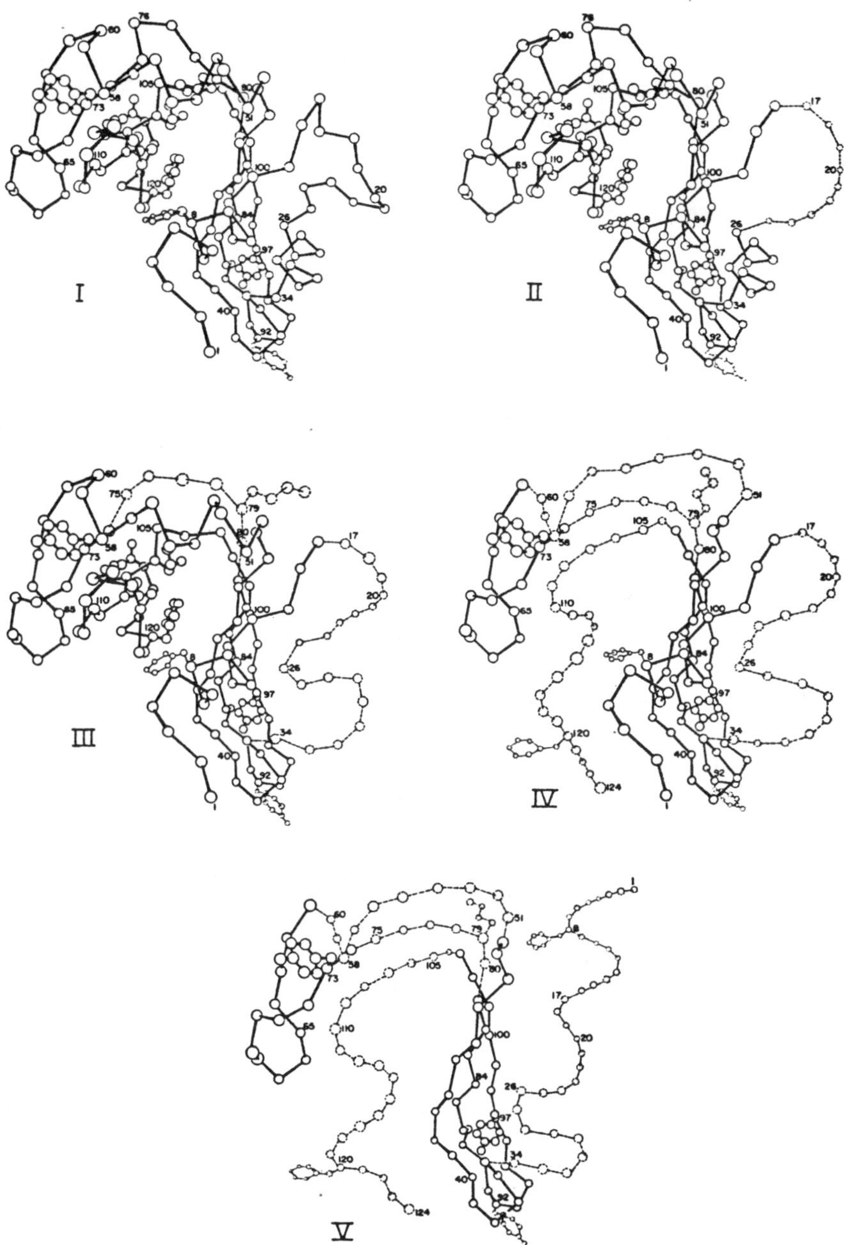

From Burgess and Scheraga (1975)

FIG. 2.13. SCHEMATIC REPRESENTATION OF STEPS IN THE THERMALLY IN-
DUCED UNFOLDING OF RIBONUCLEASE A

The first structure (I) is that of native ribonuclease. The completely unfolded structure (VI)
is not shown.

These data, along with data from NMR and spectral analyses, provides evidence that thermal unfolding of ribonuclease can occur in at least four steps (Burgess and Scheraga 1975). The schematic depiction in Fig. 2.13 should not be taken to imply that a single stable intermediate exists at the end of each stage of unfolding. It is likely that there will be a series of intermediates, slightly different from each other, in equilibrium at each temperature. These results have now been confirmed by laser Raman spectroscopy (Chen and Lord 1976).

The pathway of unfolding of ribonuclease has also been used to depict the pathway of refolding assuming reversibility in detail (Burgess and Scheraga 1975). The results with alkaline phosphatase and myoglobin described above must caution against too ready acceptance of this hypothesis.

The unfolding described above for ribonuclease is with the disulfide bonds intact. The pathway of unfolding could be different in reduced ribonuclease.

Refolding of reduced and denatured ribonuclease has been studied extensively. Although there are 105 possible ways of reforming the four disulfide bonds from the eight sulfhydryl groups close to quantitative correct pairing is obtained (Anfinsen and Haber 1961). When the sulfhydryl groups of reduced ribonuclease are oxidized in the presence of 8 M urea, a large number of the 105 possible pairings are obtained (Haber and Anfinsen 1962) with little or no activity recovery. Removal of the urea and exposure of the "scrambled" material to a catalytic amount of a sulfhydryl group containing reagent, such as mercaptoethanol, at neutral or slightly alkaline pH leads to essentially quantitative recovery of enzymatic activity. Living systems contain a disulfide-interchange enzyme which may play this role *in vivo*.

Even in the absence of urea, correct pairing of disulfide bonds does not occur initially (Hantgan *et al.* 1974) as shown in Fig. 2.14. More than six of the eight free sulfhydryl groups disappeared within 3 min of adding catalytic amounts of a mixture of oxidized and reduced glutathione. There was a very much slower regain of activity and spectral properties which differed in rates also. Incorrect disulfide bond pairings have been shown by peptide analysis (Hantgen *et al.* 1974). The same has been shown to be the case in the early stages of pairing of disulfide bonds in bovine pancreatic trypsin inhibitor (Creighton 1974). In bovine pancreatic trypsin inhibitor about half the correct pairing is made in the very early stages of refolding.

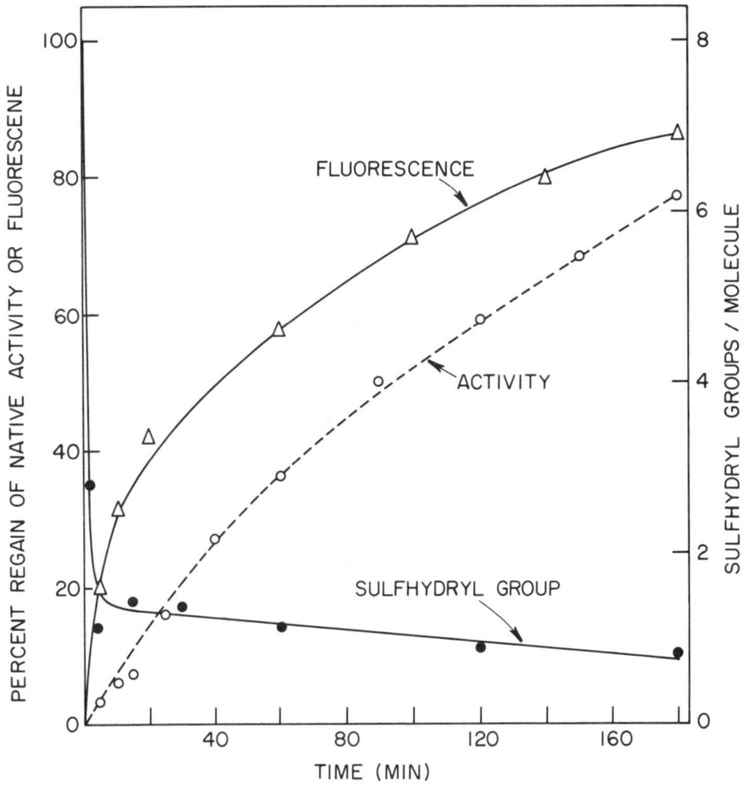

From Hantgan et al. (1974)

FIG. 2.14. TIME DEPENDENCE OF THE REOXIDATION OF REDUCED RIBONUCLEASE A IN THE PRESENCE OF OXIDIZED AND REDUCED GLUTATHIONE AS MEASURED BY LOSS OF SULFHYDRYL GROUPS, CHANGE IN FLUORESCENCE AND REGAIN OF ACTIVITY
Not all data points are included on the fluorescence curve.

Probably all proteins can be reversibly denatured under the correct set of conditions. Under a given set of conditions, the rate and extent of regain of activity varies tremendously among proteins (Teipel and Koshland 1971A). For example, aldolase denatured with 6 M guanidine hydrochloride at 25°C for 2 hr regained at least 70% of its original activity within 20 min at 25°C on dilution to 0.06 M denaturant. Glyceraldehyde phosphate dehydrogenase rapidly regained only 6% of its original activity while 200 min were required to regain 25% (maximum) of the original activity with malic dehydrogenase. In some cases the presence of substrate and cofactor increased the rate and extent of regain of activity; in other cases, there was no effect of added substrate or cofactor.

Optical rotation and spectral changes of six enzymes followed during renaturation showed that reorganization of the secondary and tertiary structures were completed within 1 min (Teipel and Koshland 1971B). Yet, recovery of activity was not 100% in any case and required from 20–200 min to achieve depending on the protein. Thus, there are some quite subtle changes which must take place to achieve activity. Determination of activity is probably the best and most sensitive probe for native structure of the protein.

GENERALIZED SCHEME OF PROTEIN FOLDING

The following generalized scheme for protein refolding is based on results discussed above and material contained in several reviews (Wetlaufer and Ristow 1973; Nagano 1974; Anfinsen and Scheraga 1975; Baldwin 1975; Tanaka and Scheraga 1975) and elsewhere. The following treatment is not intended to be detailed nor quantitative. The reader should consult the review articles for such treatment.

Proteins are synthesized on the ribosome in 2–6 min and correct folding appears to be complete in less than 2–4 min following completion of synthesis. Folding probably begins before complete systhesis of the protein but the final details of folding are incorporated after release from the ribosome. Experiments reported in this paper show even *in vitro* regain of activity from the denatured state can occur within 10 min. There must be some unique mechanism that can permit this remarkable accomplishment. We shall not dwell on whether the control is a thermodynamic step or a kinetic step as it appears to us that certainly both are of importance and are adequately documented in the literature. Thus, we consider the present arguments on this point to be rather fruitless (Baldwin 1975).

It is clear that in folding not all possible conformations of the polypeptide chain are searched before arriving at the correct one. Consider a protein with 150 amino acid residues for example and suppose each amino acid residue along the unfolded chain can exist in two states (five is more likely). The number of possible randomly generated conformations is 10^{45} (Anfinsen and Scheraga 1975). If each conformation could be explored with the frequency of a molecular rotation (10^{12} sec^{-1}) which is unlikely with a large random coil, it would take about 10^{26} yr to examine all possibilities. Contrast this to a folding time of <6 min and one must postulate some unique mechanism for folding.

The key appears to lie in a series of independent nucleation processes occurring simultaneously along the chain (Fig. 2.15).

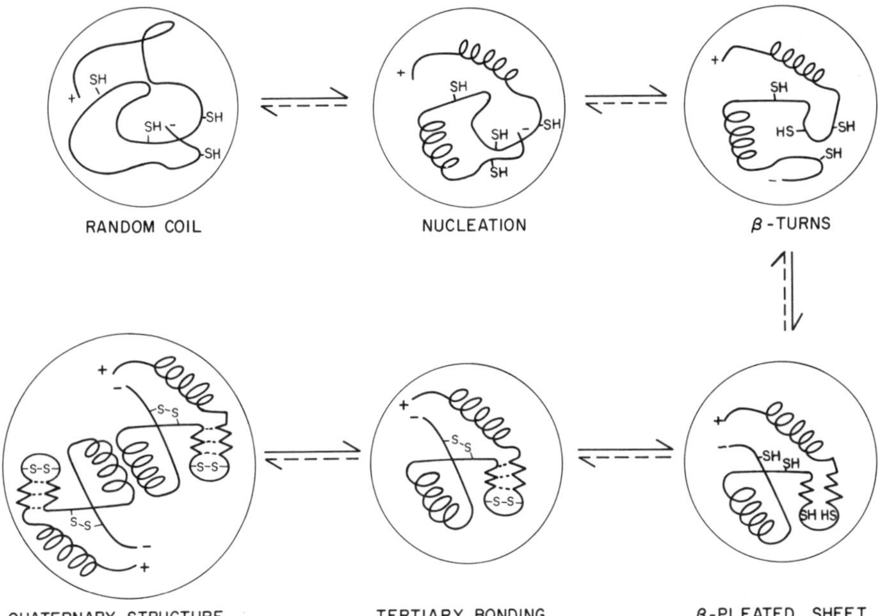

FIG. 2.15. SCHEMATIC REPRESENTATION OF STEPS OCCURRING IN THE FOLDING OF A POLYPEPTIDE CHAIN INTO NATIVE STRUCTURE AND ASSEMBLY INTO QUATERNARY STRUCTURE

Certain amino acids are known to be good helix formers. Should there be several of these in sequence, nucleation may occur. It has been suggested (Chou and Fasman 1974) that when four helix formers out of six residues or three β-formers out of five residues are found clustered together, nucleation starts to give rise to helices and/or β-pleated sheets. Initiation is provided by short-range forces acting between the amino acid side chain and its peptide bond and/or medium-range forces between adjacent residues or those within four residues of each other. Two possible points of nucleation are shown in Fig. 2.13 for ribonuclease (structure V). Other amino acids (such as proline) are important in formation of β-turns in the sequence. These bends bring the peptide chain in proximity to other parts of the chain (Fig. 2.15) where electrostatic bonds, hydrogen bonds, and hydrophobic bonds lead to formation of β-pleated sheet structure.

Formation of disulfide bonds then can "lock" the protein into the final structure of lowest energy minimum. Formation of disulfide bonds certainly doesn't await completion of the correct folding since incorrect disulfide bonds are formed initially (Hantgan et al. 1974) although there is some constraint to form correct pairings as shown by the results on bovine pancreatic trypsin inhibitor (Creighton 1974). A major number of proteins do not contain disulfide bonds so this cannot be the only driving force toward attainment of correct conformation. The presence of substrates and cofactors, such as metal ions, is undoubtedly important in enhancing the rate and extent of folding probably by stabilizing the correct conformation although their possible role in some cases in nucleation has not been excluded. The role of the solvent in promoting folding must not be overlooked (see section on role of solvent).

It is likely that the correct conformation of monomer is achieved prior to assembly of the subunits into the quaternary structure (when this is required). Association of the subunits is a second order process which is very rapid but is not diffusion controlled in all cases. This might indicate some reordering of the conformation on achieving contact which is also indicated by the activation energy for complex formation between proteins and subunits (Koren and Hammes 1976). The assemblage is highly specific and the presence of "foreign" polypeptide chains does not interfere with the assemblage (Cook and Koshland 1969). Effect of temperature on the quaternary structure can account for the cold lability of some enzymes (Bock *et al.* 1975) with resulting changes in kinetic parameters. Evidence that contact between subunits during assembly into the quaternary structure can bring about conformational changes is shown by recent work in which it was reported that mixing active hypoxanthine-guanine phosphoribosyl transferase with the deficit enzyme of patients with hyperuricemia or with Lesch-Nyhan syndrome leads to activiation of the deficient enzyme (Bakay and Nyhan 1972). Successful activation required conditions leading to dissociation and reassociation of the subunits of the enzyme.

Unfolding and refolding of proteins is a fascinating area of research that has importance for a number of areas including food science as pointed out at the beginning of this paper. A great deal is known about the processes involved but much is yet to be learned. Tools for accomplishing complete understanding of the process are now available. Of no less importance is the availability of high-speed computers which can search among the many possible conformations and arrive at one or a few that are unique and of lowest minimum energy for any given protein. It is the ultimate dream of the protein chemist to be able to predict the higher orders of protein structure from knowledge of the primary amino acid sequence.

ACKNOWLEDGMENTS

The author is most appreciative of assistance from Mrs. Linda Geiger in the literature search and checking of references and from Mrs. Clara Robison in arranging and typing the manuscript.

BIBLIOGRAPHY

ANFINSEN, C. B. 1959. The Molecular Basis of Evolution. John Wiley and Sons, New York.

ANFINSEN, C. B. 1973. Principles that govern the folding of protein chains. Science *181*, 223–230.

ANFINSEN, C. B., and HABER, E. 1961. Studies on the reduction and re-formation of protein disulfide bonds. J. Biol. Chem. *236*, 1361–1363.

ANFINSEN, C. B., HABER, E., SELA, M., and WHITE, F. H., JR. 1961. The kinetics of formation of native ribonuclease during oxidation of the reduced polypeptide chain. Proc. Nat. Acad. Sci. USA *47*, 1309–1314.

ANFINSEN, C. B., and SCHERAGA, H. A. 1975. Experimental and theoretical aspects of protein folding. Advan. Protein Chem. *29*, 205–300.

ANSON, M. L. 1944. The coagulation of proteins. *In* The Chemistry of the Amino Acids and Proteins, C. L. A. Schmidt (Editor). Charles C. Thomas, Springfield, Ill.

ANSON, M. L. 1945. Protein denaturation and the properties of protein groups. Advan. Protein Chem. *2*, 361–386.

APPLEBURY, M. L., and COLEMAN, J. E. 1969. *Escherichia coli* alkaline phosphatase. Metal binding, protein conformation, and quaternary structure. J. Biol. Chem. *244*, 308–318.

AWAD, E. S., and DERANLEAU, D. A. 1968. Thermal denaturation of myoglobin. 1. Kinetic resolution of reaction mechanism. Biochemistry 7, 1791–1795.

BAKAY, B., and NYHAN, W. L. 1972. Activation of variants of hypoxanthine-guanine phosphoribosyl transferase by the normal enzyme. Proc. Nat. Acad. Sci. USA *69*, 2523–2527.

BAKH, A. N., and OPARIN, A. I. 1925. Regeneration of enzymes made inactive by heating. Trans. Karpov Inst. Chem. *4*, 217–231.

BALDWIN, R. L. 1975. Intermediates in protein folding reactions and the mechanism of protein folding. Annu. Rev. Biochem. *44*, 453–475.

BARMAN, T. E., and GUTFREUND, H. 1966. The catalytic-centre activity and kinetic properties of bovine milk alkaline phosphatase. Biochem. J. *101*, 460–466.

BOCK, P. E., GILBERT, H. R., and FRIEDEN, C. 1975. Analysis of the cold lability of rabbit muscle phosphofructokinase. Biochem. Biophys. Res. Commun. *66*, 564–569.

BOSRON, W. F., KENNEDY, F. S., and VALLEE, B. L. 1975. Zinc and magnesium content of alkaline phosphatase from *Escherichia coli*. Biochemistry *14*, 2275–2282.

BRANDTS, J. F. 1969. Conformational transitions of proteins in water and in aqueous solutions. *In* Structure and Stability of Biological Macromolecules, S. N. Timasheff, and G. E. Fasman (Editors). Marcel Dekker, New York.

BRILL, A. S., and SANDBERG, H. E. 1968. Spectral studies of iron coordination in hemeprotein complexes: Difference spectroscopy below 250 mμ. Biophys. J. *8*, 669–690.

BROWN, E. M., ULMER, D. D., and VALLEE, B. L. 1974. Hydrogen-tritium exchange of partially and fully reconstituted zinc and cobalt alkaline phosphatase of *Escherichia coli*. Biochemistry *13*, 5328–5334.

BURGESS, A. W., and SCHERAGA, H. A. 1975. A hypothesis for the pathway of thermally-induced unfolding of bovine pancreatic ribonuclease. J. Theor. Biol. *53*, 403–420.

CHEN, M. C., and LORD, R. C. 1976. Laser Raman spectroscopic studies of the thermal unfolding of ribonuclease A. Biochemistry *15*, 1889–1897.

CHOU, P. Y., and FASMAN, G. D. 1974. Prediction of protein conformation. Biochemistry *13*, 222–245.

COOK, R. A., and KOSHLAND, D. E., JR. 1969. Specificity in the assembly of multisubunit proteins. Proc. Nat. Acad. Sci. USA *64*, 247–254.

CREIGHTON, T. E. 1974. Refolding of the reduced pancreatic trypsin inhibitor. *In* Peptides, Polypeptides and Proteins, E. R. Blout, F. A. Bovey, M. Goodman, and N. Lotan (Editors). John Wiley and Sons, New York.

CSOPAK, H., and SZAJN, H. 1973. Factors affecting the zinc content of *E. coli* alkaline phosphatase. Arch. Biochem. Biophys. *157*, 374–379.

DAYHOFF, M. O. 1969. Atlas of Protein Sequence and Structure. The National Biomedical Research Foundation, Silver Spring, Md.

DICKERSON, R. E. 1972. X-Ray studies of protein mechanisms. Annu. Rev. Biochem. *41*, 815–842.

ESSELEN, W. B., and ANDERSON, E. E. 1956. Thermal destruction of peroxidase in vegetables at high temperatures. J. Food Sci. *21*, 322–325.

FRAM, H. 1957A. The reactivation of phosphatase in H.T.S.T. pasteurized dairy products. J. Dairy Sci. *40*, 19–27.

FRAM, H. 1957B. Phosphatase reactivation in high-temperature short-time pasteurized cream. J. Dairy Sci. *40*, 1649.

FREEDMAN, M. H., and SELA, M. 1966. Recovery of antigenic activity upon reoxidation of completely reduced polyalanyl rabbit immunoglobulin G. J. Biol. Chem. *241*, 2383–2396.

GUYER, R. B., and HOLMQUIST, J. W. 1954. Enzyme regeneration in high temperature-short time sterilized canned foods. Food Technol. *8*, 547–550.

HABER, E., and ANFINSEN, C. B. 1962. Side-chain interactions governing the pairing of half-cystine residues in ribonuclease. J. Biol. Chem. *237*, 1839–1844.

HANTGAN, R. R., HAMMES, G. G., and SCHERAGA, H. A. 1974. Pathways of folding of reduced bovine pancreatic ribonuclease. Biochemistry *13*, 3421–3431.

HASCHEMEYER, R. H., and de HARVEN, E. 1974. Electron microscopy of enzymes. Annu. Rev. Biochem. *43*, 279–301.

HERMANS, J., JR., LOHR, D., and FERRO, D. 1972. Treatment of the folding and unfolding of protein molecules in solution according to a lattice model. Advan. Polym. Sci. *9*, 229–283.

IKAI, A., FISH, W. W., and TANFORD, C. 1973. Kinetics of unfolding and refolding of proteins. 2. Results for cytochrome c. J. Mol. Biol. *73*, 165–184.

IKENAKA, T., ODANI, S., and KOIDE, T. 1974. Chemical structure and inhibitory activities of soybean proteinase inhibitors. *In* Proteinase Inhibitors, H. Fritz, H. Tschesche, L. J. Greene, and E. Truscheit (Editors). Springer-Verlag, New York.

JOFFE, F. M., and BALL, C. O. 1962. Kinetics and energetics of thermal inactivation and the regeneration rates of a peroxidase system. J. Food Sci. *27*, 587–592.

JONES, R. T. 1964. The structure of proteins. *In* Symposium on Foods: Proteins and Their Interactions, H. W. Schultz, and A. F. Anglemeir (Editors). Avi Publishing Co., Westport, Conn.

KARTHA, G., BELLO, J., and HARKER, D. 1967. Tertiary structure of ribonuclease. Nature *213*, 862–865.

KAUZMANN, W. 1959. Relative probabilities of isomers in cystine-containing randomly coiled polypeptides. *In* Sulfur in Proteins, R. Benesch, R. E. Benesch, P. D. Boyer, I. M. Klotz, W. R. Middlebrook, A. G. Szent-Györgyi, and D. R. Schwarz (Editors). Academic Press, New York.

KLAPPER, M. H., and HACKETT, D. P. 1965. Investigations on the multiple components of commercial horseradish peroxidase. Biochim. Biophys. Acta *96*, 272–282.

KLEE, W. A. 1967. Intermediate stages in the thermally induced trans-conformation reactions of bovine pancreatic ribonuclease A. Biochemistry *6*, 3736–3742.

KNOX, J. R., and WYCKOFF, H. W. 1973. A crystallographic study of alkaline phosphatase at 7.7 Å resolution. J. Mol. Biol. *74*, 533–545.

KON, S., and WHITAKER, J. R. 1966. Separation and partial characterization of the peroxidases of *Ficus glabrata* latex. J. Food Sci. *30*, 977–985.

KOREN, R., and HAMMES, G. G. 1976. A kinetic study of protein-protein interactions. Biochemistry *15*, 1165–1171.

LU, A. T., and WHITAKER, J. R. 1974. Some factors affecting rates of heat inactivation and reactivation of horseradish peroxidase. J. Food Sci. *39*, 1173–1178.

LUMRY, R., and BILTONEN, R. 1969. Thermodynamic and kinetic aspects of protein conformations in relation to physiological function. *In* Structure and Stability of Biological Macromolecules, S. N. Timasheff, and G. E. Fasman (Editors). Marcel Dekker, New York.

LYSTER, R. L. J., and ASCHAFFENBURG, R. 1962. The reactivation of milk alkaline phosphatase after heat treatment. J. Dairy Res. *29*, 21–35.

MAEHLY, A. C. 1952. Splitting of horseradish peroxidase into prosthetic group and protein as a means of studying the linkages between hemin and protein. Biochim. Biophys. Acta *8*, 1–17.

MAEHLY, A. C. 1953. Experiments with peroxidase. 2. Further observations on splitting the enzyme at extreme pH values. Arch. Biochem. Biophys. *44*, 430–442.

MAEHLY, A. C. 1955. Experiments with peroxidase. 3. Recombination of hemin and protein to the active enzyme. Arch. Biochem. Biophys. *56*, 507–524.

McFARREN, E. F., THOMAS, R. C., BLACK, L. A., and CAMPBELL, J. E. 1960. Differentiation of reactivated from residual phosphatase in high temperature-short time pasteurized milk and cream. J. Assoc. Off. Agric. Chem. *43*, 414–426.

MITSUI, T., KATO, K., ENDO, K., and KAWASAKI, K. 1956. Activating the peroxidase reaction by means of hemoglobin. 1. Igaku to Seibutsugaku *40*, 106–109; Chem. Abstr. *52*, 7384g (1958).

MORITA, Y., and KONDO, K. 1954. Phytoperoxidase. 3. Isolation and purification of Japanese-radish peroxidase and protein component. Bull. Research Inst. Food Sci. Kyoto Univ. *13*, 60–66.

MORTON, R. K. 1953. Alkaline phosphatase of milk. 2. Purification of the enzyme. Biochem. J. *55*, 795–800.

NAGANO, K. 1974. Logical analysis of the mechanism of protein folding. 2. The nucleation process. J. Mol. Biol. *84*, 337–372.

NORTHROP, J. H. 1932. Crystalline trypsin. 4. Reversibility of the inactivation and denaturation of trypsin by heat. J. Gen. Physiol. *16*, 323–348.

OOI, T., RUPLEY, J. A., and SCHERAGA, H. A. 1963. Structural studies of ribonuclease. 8. Tryptic hydrolysis of ribonuclease A at elevated temperatures. Biochemistry *2*, 432–437.

PTITSYN, O. B., LIM, V. I., and FINKELSTEIN, A. V. 1972. Secondary structure of globular proteins and the principle of concordance of local and long-range interactions. Fed. Eur. Biochem. Soc. *25*, 421–429.

RAKSHIT, G., and SPIRO, T. G. 1974. Resonance Raman spectra of horseradish peroxidase: Evidence for anomalous heme structure. Biochemistry *13*, 5317–5323.

REYNOLDS, J. A., and SCHLESINGER, M. J. 1967. Conformational states of the subunit of *Escherichia coli* alkaline phosphatase. Biochemistry *6*, 3552–3559.

REYNOLDS, J. A., and SCHLESINGER, M. J. 1968. Hydrogen ion equilibria of conformational states of *Escherichia coli* alkaline phosphatase. Biochemistry *7*, 2080–2085.

REYNOLDS, J. A., and SCHLESINGER, M. J. 1969A. Alterations in the structure and function of *Escherichia coli* alkaline phosphatase due to Zn^{2+} binding. Biochemistry *8*, 588–593.

REYNOLDS, J. A., and SCHLESINGER, M. J. 1969B. Formation and properties of a tetrameric form of *Escherichia coli* alkaline phosphatase. Biochemistry *8*, 4278–4282.

RUPLEY, J. A., and SCHERAGA, H. A. 1963. Structural studies of ribonuclease. 7. Chymotryptic hydrolysis of ribonuclease A at elevated temperatures. Biochemistry *2*, 421–431.

RYLE, A. P., SANGER, F., SMITH, L. F., and KITAI, R. 1955. The disulphide bonds of insulin. Biochem. J. *60*, 541–556.

SCHWIMMER, S. 1944. Regeneration of heat-inactivated peroxidase. J. Biol. Chem. *154*, 487–495.

SCHWIMMER, S. 1972. Cell disruption and its consequences in food processing. J. Food Sci. *37*, 530–535.

SELA, M., and LIFSON, S. 1959. On the reformation of disulfide bridges in proteins. Biochim. Biophys. Acta *36*, 471–478.

SELA, M., WHITE, F. H., JR., and ANFINSEN, C. B. 1957. Reductive cleavage of disulfide bridges in ribonuclease. Science *125*, 691–692.

SHEN, L. L., and HERMANS, J., JR. 1972A. Kinetics of conformation change of sperm-whale myoglobin. 1. Folding and unfolding of metmyoglobin following pH jump. Biochemistry *11*, 1836–1841.

SHEN, L. L., and HERMANS, J., JR. 1972B. Kinetics of conformation change of sperm-whale myoglobin. 2. Characterization of the rapidly and slowly formed denatured species (D and D*). Biochemistry *11*, 1842–1844.

SHEN, L. L., and HERMANS, J., JR. 1972C. Kinetics of conformation change of sperm-whale myoglobin. 3. Folding and unfolding of apomyoglobin and the suggested overall mechanism. Biochemistry *11*, 1845–1849.

SIMPSON, R. T., and VALLEE, B. L. 1968. Two differentiable classes of metal atoms in alkaline phosphatase of *Escherichia coli*. Biochemistry *7*, 4343–4350.

SIMPSON, R. T., and VALLEE, B. L. 1969. Zinc and cobalt alkaline phosphatases. Ann. N.Y. Acad. Sci. *166*, 670–695.

SIMPSON, R. T., VALLEE, B. L., and TAIT, G. H. 1968. Alkaline phosphatase of *Escherichia coli*. Composition. Biochemistry *7*, 4336–4342.

STEARN, A. E. 1949. Kinetics of biological reactions with special reference to enzymic processes. Advan. Enzymol. *9*, 25–74.

SUND, H., and WEBER, K. 1966. Quaternary structure of proteins. Angew. Chem., Intern. Ed. Engl. *5*, 231–245.

TANAKA, S., and SCHERAGA, H. A. 1975. Model of protein folding: Inclusion of short-, medium-, and long range interactions. Proc. Nat. Acad. Sci. USA *72*, 3802–3806.

TANFORD, C. 1968. Protein denaturation. Advan. Protein Chem. *23*, 121–282.

TANFORD, C. 1970. Protein denaturation. Part C. Theoretical models for the mechanism of denaturation. Advan. Protein Chem. *24*, 1–95.

TANFORD, C., AUNE, K. C., and IKAI, A. 1973. Kinetics of unfolding and refolding of proteins. 3. Results from lysozyme. J. Mol. Biol. *73*, 185–197.

TEIPEL, J. W., and KOSHLAND, D. E., JR. 1971A. Kinetic aspects of conformational changes in proteins. 1. Rate of regain of enzyme activity from denatured proteins. Biochemistry *10*, 792–798.

TEIPEL, J. W., and KOSHLAND, D. E., JR. 1971B. Kinetic aspects of conformational changes in proteins. 2. Structural changes in renaturation of denatured proteins. Biochemistry *10*, 798–805.

THEORELL, H. 1943. The preparation and some properties of crystalline horseradish peroxidase. Arkiv. Kemi, Mineral. Geol. *16A*, No. 2, 1–11.

THEORELL, H., and ÅKESON, Å. 1943. Highly purified milk peroxidase. Arkiv. Kemi, Mineral. Geol. *17B*, No. 7, 1–6.

VETTER, J. L., NELSON, A. I., and STEINBERG, M. P. 1958. Heat inactivation of sweet corn peroxidase in the temperature range of 210° to 310°F. Food Technol. *12*, 244–247.

WETLAUFER, D. B., and RISTOW, S. 1973. Acquisition of three-dimensional structure of proteins. Annu. Rev. Biochem. *42*, 135–158.

WOODS, A. F. 1902. Observations on the Mosaic disease of tobacco. U. S. Dept. Agr., Bureau of Plant Industry, Bull. *18*.

WRIGHT, R. C., and TRAMER, J. 1953A. Reactivation of milk phosphatase following heat treatment. 1. J. Dairy Res. *20*, 177–188.

WRIGHT, R. C., and TRAMER, J. 1953B. Reactivation of milk phosphatase following heat treatment. 2. J. Dairy Res. *20*, 258–273.

WRIGHT, R. C., and TRAMER, J. 1956. Reactivation of milk phosphatase following heat treatment. 4. The influence of certain metallic ions. J. Dairy Res. *23*, 248–256.

YOSHIDA, C., AND MORITA, Y. 1970. Studies of phytoperoxidase. 23. Spectrophotometric and fluorospectrophotometric studies on denaturation of Japanese-radish peroxidase by urea and guanidine hydrochloride. Mem. Res. Inst. Food Sci., Kyoto Univ. 1970, No. 31, 1–9; Chem. Abstr. *73*, 127159w (1970).

ZOUEIL, M. E., and ESSELEN, W. B. 1959. Thermal destruction rates and regeneration of peroxidase in green beans and turnips. Food Research *24*, 119–133.

3

Water and Protein Hydration

O. Fennema

Little need to be said to justify consideration of the topic "Water and Hydration of Proteins." Water and proteins constitute major and vital components of both biological matter and of human diets. Furthermore, their interactions are of immense importance. Proteins influence the properties of vicinal water, and water, in turn, dictates to a considerable degree the properties of proteins (Drost-Hansen 1971; Hagler and Scheraga 1973; Horne 1968; Klotz 1960). Proteins without water are in fact not proteins at all since native structure and normal functional properties do not exist in the absence of water. Details of protein-water relationships have been studied extensively; however, results are often the subject of disagreement. This is not surprising considering the vigorous disagreement that exists with respect to the structure(s) of pure water.

In this paper, all kinds of water that exist in association with proteins at physiological water contents will be discussed and emphasis will be given to general concepts rather than to quantitative details. The properties of pure water will be dealt with in a cursory manner since good reviews are available, and all that is needed here is an understanding of a few noncontroversial concepts (Drost-Hansen 1971; Eisenberg and Kauzmann 1969; Fennema 1973; Forslind 1971; Frank 1970; Franks 1972; Horne 1968, 1972; Rahman and Stillinger 1971).

CHARACTERISTICS OF WATER

The Water Molecule

When the properties of water are compared to the properties of molecules of similar molecular weight and atomic composition (CH_4, NH_3, HF, H_2S, H_2Se, H_2Te) it is evident that water behaves in an abnormal fashion. Water has unusually large values for melting point, boiling point, surface tension, dielectric constant, heat capacity, heat of fusion, heat of vaporization, and heat of sublimation; a moderately low value for density; an unusual density maximum at $3.98°C$; and an unusual attribute of expanding upon solidification (Fennema 1973). These properties suggest the existence of strong attractive forces among water molecules and uncommon structures for water and ice. An explanation is easily derived by considering the properties of individual water molecules.

To form a molecule of water, two hydrogen atoms approach the two sp^3 bonding orbitals of oxygen (0_3^1, 0_4^1) and form two covalent sigma (σ) bonds (40% partial ionic character), each of which has a dissociation energy of 110.2 kcal/mole (Fig. 3.1). The localized molecular orbitals remain symmetrically oriented about the original orbital axes of oxygen, thus the water molecule can be visualized as fitting well in an imaginary tetrahedron (Fig. 3.2A). The bond angle of the isolated water molecule (vapor state) is $104.5°$ and this value is near the perfect tetrahedral angle of $109°28'$ (Fig. 3.2B). The O-H internuclear distance is 0.96 Å, and the van der Waals' radii for oxygen and hydrogen are, respectively, 1.40 and 1.2 Å.

Association of Water Molecules

The V-like form of a HOH molecule and the polarized nature of the O-H bond result in an unsymmetrical charge distribution and a vapor-state dipole moment of 1.84D for pure water. Polarity of this magnitude produces intermolecular attractive forces, and water molecules therefore associate with considerable tenacity. Water's unusually large intermolecular attractive force, however, cannot be fully accounted for on the basis of its large dipole moment. This is not surprising, since dipole moments give no indication of the degree to which charges are exposed or of the geometry of the molecule, and these aspects, of course, have an important bearing on the intensity of molecular association.

Water's large intermolecular attractive forces can be explained quite adequately by its ability to engage in multiple hydrogen bonding on a three-dimensional basis. Compared to covalent bonds

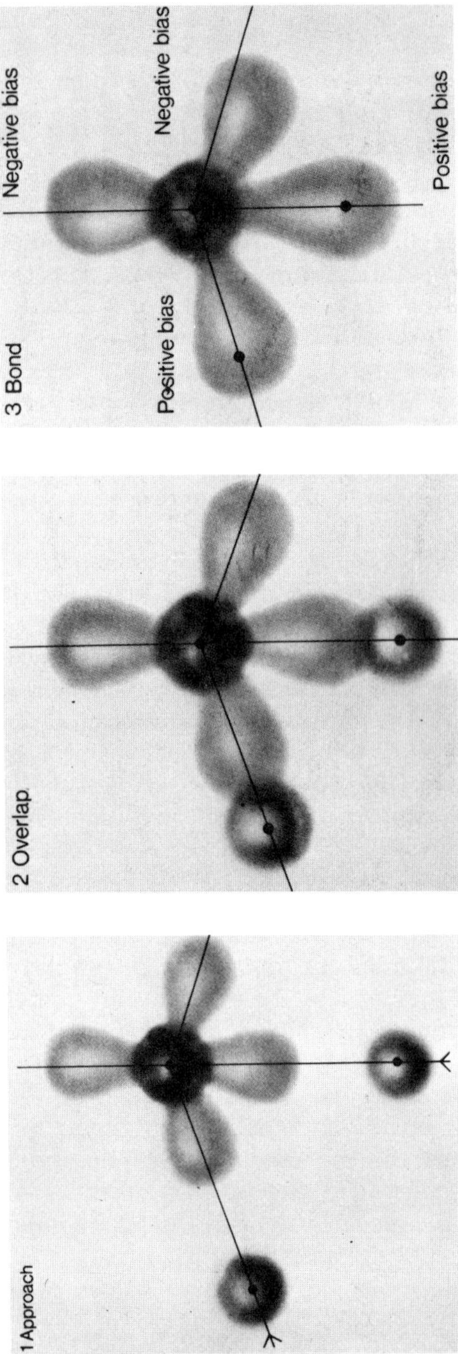

From Taylor (1966); Courtesy of Unilever, Ltd.

FIG. 3.1. SCHEMATIC INTERPRETATION OF THE FORMATION OF A WATER MOLECULE (average bond energy of about 80 kcal/mole), hydrogen bonds are weak (less than 10 kcal/mole) and have greater and more variable lengths. The oxygen-hydrogen hydrogen bond has a dissociation energy of about 3-6 kcal/mole.

Since electrostatic forces make a major contribution to the energy of the hydrogen bond (perhaps the largest contribution), and since an electrostatic model of water is simple and leads to an essentially correct geometric picture of HOH molecules as they are known to exist in ice, further discussion of the geometrical patterns formed by association of water molecules emphasizes electrostatic effects.

The highly electronegative oxygen of the water molecule can be visualized as partially drawing away the single electrons from the two covalently bonded hydrogen atoms, thereby leaving each hydrogen atom with a partial positive charge and a minimal electron shield; i.e., each hydrogen atom assumes some characteristics of a bare proton. Since the hydrogen-oxygen bonding orbitals are located on two of the axes of an imaginary tetrahedron (Fig. 3.2A), these two axes can be thought of as representing areas of positive charge (hydrogen-bond donor sites). Oxygen's two lone-pair orbitals can be pictured as residing along the remaining two axes of the imaginary tetrahedron, and these then represent areas of negative charge (hydrogen-bond acceptor sites). By virtue of these four charged areas, each water molecule is able to hydrogen bond with a maximum of four others. The resulting tetrahedral arrangement is depicted in Fig. 3.3. Because each water molecule has an equal number of hydrogen-bond donor and receptor sites, arranged to permit three-dimensional hydrogen bonding, it is found that the attractive forces among water molecules are unusually large, even when compared to the attractive forces existing among other small molecules which also engage in hydrogen bonding (*e.g.*, NH_3, HF). Hydrogen bonding among ammonia or hydrogen fluoride molecules results only in linear or branched chains involving fewer hydrogen bonds per molecule than water.

Water's ability to engage in three-dimensional hydrogen bonding provides a logical explanation for many of its unusual properties. For example, its large values for heat capacity, melting point, boiling point, surface tension, and heats of fusion, vaporization, and sublimation are all related to the extra energy needed to break intermolecular hydrogen bonds.

Water Structure

Elucidation of the structure of pure water on a scale more extended than that shown in Fig. 3.3 is an extremely complex

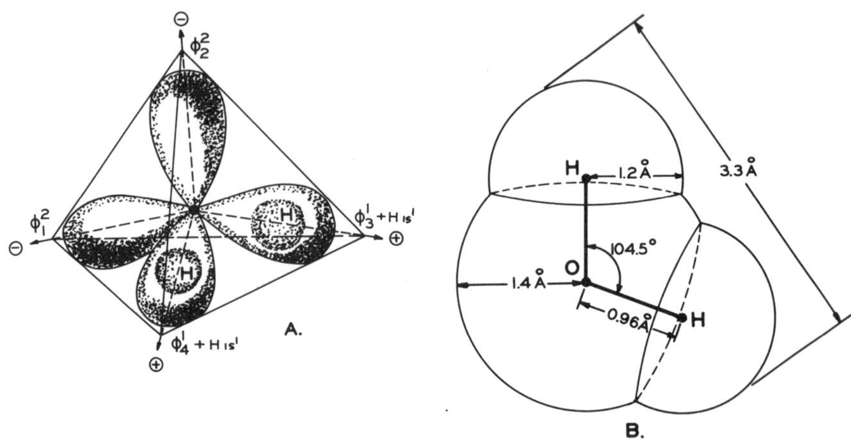

From Fennema (1976); Courtesy of Marcel Dekker, Inc.
FIG. 3.2. SCHEMATIC MODEL OF A SINGLE HOH MOLECULE
(A) sp^3 configuration with molecule enclosed in an imaginary tetrahedron; (B) van der Waals' radii for a HOH molecule in the vapor state.

problem which in recent years has attracted increasing attention from capable investigators. Many theories have been set forth but all are no doubt incomplete, overly simple, and subject to weaknesses which are quickly cited by supporters of rival theories. This, of course, is a healthy situation which will eventually result in an accurate structural picture (or pictures) of water. In the meantime, few statements can be made with any assurance that they will stand essentially unmodified in years to come (Holtzer and Emerson 1969).

To some, it may seem strange to speak of structure in a liquid, when fluidity is the major feature of the liquid state. Yet is is an old and well accepted idea that water has structure, not of the long-range rigid type found in ice but of a short-range order that prevails momentarily over a distance of perhaps several molecular diameters. Direct evidence for this belief is provided by experiments involving X-ray and neutron diffraction, dielectric relaxation times, and infrared, Raman and nuclear magnetic resonance spectroscopy.

Theories of water structure are often classified in two major categories: (1) continuum theories (also called "uniform" or homogeneous theories), and (2) mixture theories. Continuum theories involve the notion that all molecules in cold liquid water are totally hydrogen-bonded (4 bonds/molecule) but that the bonds differ in angle, length and energy (Bernal and Fowler 1933; Kell 1972; Pople 1951). Mixture theories embody the notion that water consists of two or more distinguishable species, these species

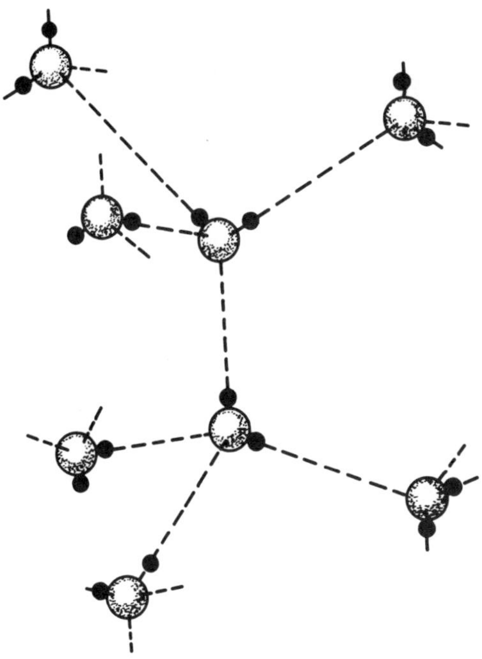

From Fennema (1976);
Courtesy of Marcel Dekker, Inc.
FIG. 3.3. HYDROGEN BONDING OF WATER
MOLECULES IN A TETRAHEDRAL CON-
FIGURATION
Hydrogen bonds are represented by dashed lines.

existing in dynamic equilibrium and differing in the degree of hydrogen bonding (Davis and Jarzynski 1972; Frank 1972). At the present time, scientific opinion favors the mixture theory and more specifically an "interstitial model." According to this model cold liquid water exists partly in the form of bulky framework structures that evolve from a 4-coordinated, approximately tetrahedral arrangement of water molecules. Some of these structures accommodate single water molecules within the cavities formed by the frameworks; the entrapped single molecules being less than 4-coordinated (Narten and Levy 1969).

The low viscosity of water is readily reconcilable with the concept of water structure since the proposed structures are believed to exist for only 10^{-11} to 10^{-12} sec (Cooke and Kuntz 1974; Davis and Jarzynski 1972; Hagler and Scheraga 1973; Kell 1972; Tait and Franks 1971). The rapid formation and collapse of these structural entities would give molecules ample opportunity to move about (flow).

Water structure becomes considerably more complicated when one considers that water contains many isotopic variants as well as hydronium and hydroxyl ions. The hydronium ion (Fig. 3.4), because of its positive charge, would be expected to exhibit a greater hydrogen-bond donating potential than un-ionized water (Lewin 1974). The hydroxyl ion (Fig. 3.5), because of its negative charge, would be expected to exhibit a greater hydrogen-bond acceptor potential than un-ionized water (Lewin 1974).

With respect to protein hydration, the major point that need be remembered is that water associates with other molecules in accord with the charge distribution shown in Fig. 3.2A.

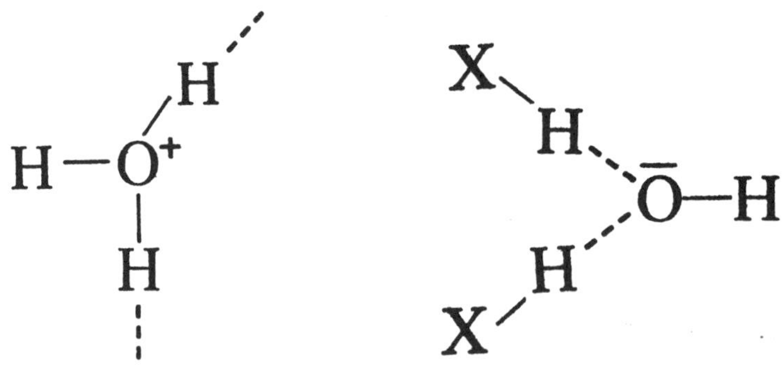

FIG. 3.4. STRUCTURE AND HYDROGEN-BOND POSSIBILITIES FOR A HYDRONIUM ION
Dashed lines are hydrogen bonds.

FIG. 3.5. STRUCTURE AND HYDROGEN-BOND POSSIBILITIES FOR A HYDROXYL ION
Dashed lines are hydrogen bonds and XH represents a solute or a water molecule

GENERAL CONCEPTS OF PROTEIN HYDRATION

Classification

Developing an understanding of protein-water interactions is fraught with difficulty because much of the data is ambiguous, agreement between some experimental methods is not good, and important experimental conditions, most notably temperature and total water content of samples, often differ greatly among studies (Cerbón 1974; Haurowitz 1963; Karmas and DiMarco 1970; Walter and Hope 1971).

Various investigators have attempted to classify and describe the various states of water that are believed to exist in proteinaceous systems. Although there is disagreement on this subject the following classification (see also Table 3.1) appears to be in accord with the most reliable results (Cooke and Kuntz 1974; Drost-Hansen

1971; Finch *et al.* 1971; Hazlewood *et al.* 1969; Kuntz 1975; Kuntz and Kauzmann 1974; Lumry 1973; Ruegg *et al.* 1974; Walter and Hope 1971):

1) Constitutional water
2) Interfacial water (vicinal and multilayer)
3) Bulk phase water (free or entrapped).

Constitutional Water

Constitutional water is located in the interior of the protein molecule at specific sites or simply in tiny interstitial regions. This water is the most immobile type as compared to normal water. A given structure persists on the order of microseconds or longer, as compared to about 10^{-12} sec for normal water-water associations. This fraction of water consists of perhaps several to a few hundred water molecules per protein molecule (about 0.003 g H_2O/g of protein). In a 20% protein solution about 0.1% of the total water is constitutional water (Cooke and Kuntz 1974; Kuntz 1975).

Interfacial Water

Interfacial water, that is, water present at or very near the water-protein interface, can be reasonably divided into two sub-categories: vicinal water, the first one or two layers of water adjacent to the protein molecule, and multilayer water, the next few layers of water molecules beyond vicinal water. Properties of vicinal water are stongly influenced by the nature of the protein surface. Both hydrophilic and hydrophobic surfaces will reduce the mobility and alter the structure of vicinal water as compared to normal water, and water structure near a hydrophilic surface will be different than that near a hydrophobic surface (Drost-Hansen 1971; Franks 1975; Klotz 1965; Kuntz *et al.* 1969). Furthermore, water confined in crevices or pores (*e.g.*, in membranes, between membranes, between protein filaments) of molecular dimensions will exhibit reduced mobility and altered structure as compared to normal water. This water is tightly held in place, largely by physical forces. It is believed, however, to be hydrogen-bonded to a lesser degree than normal water because there is insufficient space to develop "normal" water structure (Clifford 1975).

The vicinal water fraction amounts to perhaps 1,500–3,000 molecules of water per protein molecule with a molecular weight of 100,000 daltons (about 0.3–0.5 g H_2O/g protein; Kuntz and Kauzmann 1974). [Note: This value is greater than the monolayer value of Brunauer, Emmett and Teller (B.E.T.).] In a 20% protein solution this fraction constitutes perhaps 5–10% of the total water content.

TABLE 3.1
CLASSIFICATION OF WATER ASSOCIATED WITH PROTEINS
(PHYSIOLOGICAL WATER CONTENT)

Class	Amount (g H_2O/g Protein)	Description
Constitutional water	About 0.003	Located within the protein molecule. Associated with specific sites or simply located in interstitial regions. A given structure persists for about 10^{-6} sec or longer. Protein-water and water-water binding energies are much greater than those existing in normal water.
Interfacial water: (a) Vicinal (b) Multilayer	Less than about 0.6 ca. 0.3–0.5 (?) ca. 0.1–0.3 (?)	Located at the surface of the protein and in small crevices. Water-solute (dipole-dipole, ion-dipole) and water-water interactions are involved, and the binding energies generally are greater than those found in normal water. This results in decreased mobility of water molecules (intermediate between bulk phase water and constitutional water) and reduced vapor pressure. Part of this fraction is unfreezable (vicinal water) and a given structure persists perhaps 10^{-9} sec (Kuntz and Kauzmann 1974). The multilayer fraction has altered properties for melting point, and a portion of the multilayer fraction has an altered heat of fusion. All interfacial water has structures different from those of normal water and different from that of ice.
Bulk phase water: (a) Free (b) Entrapped	All water except that mentioned above	(a) Properties similar to normal water or a dilute salt solution. Structures persist ca. 10^{-12} sec. (b) Water is physically entrapped in a fashion similar to that found in gels. The structure of entrapped water is the subject of controversy. Many other properties of this water are similar to those of normal water or of a dilute salt solution.

Multilayer water, being situated between vicinal water and "bulk" phase water (bulk phase water is the furthest removed from the protein molecules) is believed by some investigators to have its properties determined by the counteracting influence of these two neighboring water phases. It is not possible with the present state of knowledge to accurately distinguish between the properties of vicinal water and multilayer water, so further consideration will be given only to the larger grouping of interfacial water.

When compared to normal water, interfacial water (both vicinal and multilayer) exhibits reduced vapor pressure, reduced mobility (decreased rotational and translational motion) and a sizeable portion of it is nonfreezable (Cooke and Kuntz 1974; Finch *et al.* 1971; Hazlewood *et al.* 1969). That which does freeze exhibits an altered temperature of fusion and part of this subfraction exhibits an altered heat of fusion (Haly and Snaith 1971; Ruegg *et al.* 1974). Although some authors use the term "ice-like" to describe the structure of interfacial water, this is a misleading term since the most reliable evidence indicates that its structure is not like ice (Bull and Breese 1968; Finch *et al.* 1971; Kuntz *et al.* 1969; Walter and Hope 1971). With respect to structure, the most that can be said is that interfacial water is different from, and probably more structured than, normal water. The water-water binding energies of interfacial water are believed to be greater than those of normal water; however, reliable values are apparently unavailable (Kuntz and Kauzmann 1974).

Bulk Phase Water

Bulk phase water constitutes the major portion of water contained in cellular systems and in dilute protein suspensions. In cellular systems, this water is physically entrapped, apparently in a fashion somewhat analogous to water in some gels (Blanshard and Derbyshire 1975; Hamm 1960). This entrapped water does not flow freely from the cells even when the cellular system has been severely disrupted (Hamm 1960; Ling and Walton 1976). In a dilute protein suspension, bulk phase water sometimes flows readily and is therefore referred to as "free" water (this system is a sol), or it too can be physically restrained in a gel structure and it is then properly regarded as "entrapped" water.

The properties of entrapped water are the subject of considerable controversy. Good investigators using well regarded procedures can be found supporting drastically different views. Some maintain that cell water is "ordered" sufficiently to make it behave differently than a dilute salt solution (Chang *et al.* 1972;

Cope 1969; Hazlewood *et al.* 1969; Ling 1960, 1972; Ling and Walton 1976). Ling, one of the most ardent supporters of this view, regards bulk phase water in cells as existing in the form of polarized multilayers. In muscle, this condition presumably arises because of: (1) a favorable positive-negative charge distribution on fibrous proteins, and (2) the small intervening spaces between protein molecules (Ling 1960, 1972). Ling claims that the structural characteristics of cell water, rather than properties of the cell membrane control the concentrations of potassium and sodium ions in the cell. Acceptance of this view would mean that all cell water with the exception of constitutional water and vicinal water would be categorized as "multilayer" water, *i.e.*, entrapped bulk phase water and multilayer water would be identical.

Other investigators regard the properties of entrapped, bulk phase water in cells as being similar to those of a dilute salt solution (Cooke 1976; Cooke and Kuntz 1974; Cooke and Wien 1973; Drost-Hansen 1971; Finch *et al.* 1971; Szent-Gyorgyi 1971; Woessner and Snowden 1970; Woessner *et al.* 1970). This view is in accord with the traditional concept that the cell membrane controls solute entry and exit from cells. The controversy with respect to the properties of entrapped bulk phase water cannot be resolved here, and fortunately resolution is not absolutely essential for the discussion that follows.

With respect to water in flowable protein suspensions (sols), most would agree that bulk water behaves much like normal water or a dilute salt solution. This reason is sufficient justification for retention of the bulk phase category with its "entrapped" and "free" subclasses.

Protein Hydration in Perspective

Several additional comments, some new and some iterative, are in order before closing this section.

(1) Water associated with proteins is not a homogeneous, easily identifiable entity, and any estimate of the kinds and amounts will vary somewhat with the type of sample being examined, its total water content and the method of measurement.

(2) Although water associated with proteins has been subdivided into several categories, this in no way implies that distinct boundaries exist between these categories. Since water can associate with various kinds of chemical groups, it is intuitively reasonable that a given water molecule would have an opportunity to bind to solutes or other water molecules with a rather wide range of intensities, not just several discrete intensities as might be inferred from the categories in Table 3.1. The two concepts are,

however, not necessarily incompatible since one or two bond types could be more prevalent than others, thereby giving the appearance, with whatever measurement technique that is being employed, of just one or two distinct categories.

(3) Water in association with proteins is not totally immobilized. Water molecules directly bonded to a protein molecule will simply exhibit less mobility than more distant water molecules.

BOUND WATER

Semantics

Nothing yet has been said about "bound water" and some would suggest that abandonment of the term is by far the best course of action (Berendsen 1971). This term is undeniably misused, controversial and, in general, poorly understood. Some follow the inexcusable practice of employing the term without benefit of definition (Webb 1965) and others have created such a broad array of definitions that confusion is rampant. Some examples of definitions will illustrate this point (Berendsen 1971):

(1) Bound water is the equilibrium water content of a sample at some appropriate temperature and relative humidity.

(2) Bound water is that which does not contribute to the dielectric constant at high frequencies and therefore has its rotational mobility restricted by the solute with which it is associated.

(3) Bound water is that which does not freeze at some arbitrary low temperature (usually $-40°C$ or lower).

(4) Bound water is that which is unavailable as a solvent for additional solutes.

(5) Bound water is that which produces a line broadening in experiments involving nuclear magnetic resonance.

(6) Bound water is that which moves with a macromolecule in experiments involving sedimentation rates, viscosity or diffusion.

In spite of the confusion relating to the term bound water, to abandon its use seems questionable, since to do so would not eliminate the phenomenon, it would merely necessitate development of a new terminology. This effort, in my judgment, can be best applied to improving the description of "bound water."

The following general definition, that has value from a conceptual standpoint, was suggested by Kuntz and Kauzmann (1974) as representing the views of most investigators:

"Bound water is that water in the vicinity of a macromolecule whose properties differ detectably from those of the "bulk" water in the same system." To facilitate discussions that follow, it is imperative that a more quantitative definition be developed that relates closely to the scheme in Table 3.1 and to the results obtained from well-regarded analytical methods for measuring water binding. However, before this is done, it is appropriate to consider briefly some of the methods that have been used to determine bound water.

Methods for Measuring Bound Water

Many methods have been employed to measure the characteristics and amounts of bound water in cells and protein suspensions, and a point that should be kept in mind constantly is that they don't all measure the same body of water. The most recent and complete treatments of this topic are those of Cooke and Kuntz (1974) and Kuntz and Kauzmann (1974). They categorized all major methods in the following four groups: (1) thermodynamic, (2) kinetic, (3) spectroscopic, and (4) diffraction techniques.

Thermodynamic methods include moisture sorption isotherms, calorimetric studies, isopiestic experiments, sedimentation studies to determine "buoyant" densities of macromolecules, and measurements of the solvation capabilities of water. Kinetic methods involve measurement of motion of hydrated macromolecules (hydrodynamic measurements) or the motion of water molecules adjacent to macromolecules (proton exchange processes, and translational and rotational motions of water molecules as determined primarily by dielectric and nuclear magnetic resonance (NMR) techniques). Spectroscopic techniques (infrared, Raman and NMR spectroscopy) can be used to determine the sites of water binding and the amounts of bound water. X-ray diffraction techniques are useful for determining the sites and arrangement of water molecules in protein crystals. Techniques such as infrared, NMR and dielectric measurements appear to detect only the most tightly bound water, whereas hydrodynamic measurements provide a somewhat more generous estimate of bound water (Eaglund 1975).

Additional comments with respect to several specific methods used to study bound water are appropriate. Moisture sorption isotherms deserve attention since they are used commonly in food research. Typical sorption isotherms for proteins are shown in Fig. 3.6. These isotherms are, in general, S-shaped, indicating that very dry or very wet specimens adsorb large quantities of water with small increases in water activity (water activity is equivalent to "Relative H_2O vapor pressure" in Fig. 3.6). Useful information that can be gained from moisture sorption isotherms include: heats and entropies of hydration, the water content corresponding to a

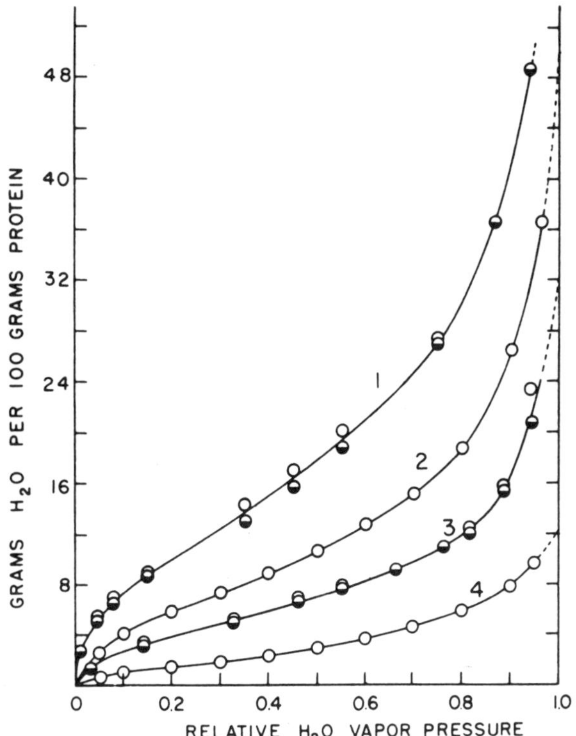

From Bull (1944);
Courtesy of the American Chemical Society
FIG. 3.6. WATER ADSORPTION ISOTHERMS
(1) Wet collagen (open circles) and dry collagen (half circles);
(2) unlyophilized egg albumin; (3) wet silk (open circles) and
dry silk (half circles); and (4) unstretched nylon.

monomolecular layer (and this in turn provides a basis for estimating the moisture content corresponding to optimun stability of
the specimen under study), and the interrelations among water
content, water activity, rates of chemical reactions and rates of
microbial growth (Labuza 1968, 1971, 1975; Schneider and
Schneider 1972). Note should be made, however, of Kuntz and
Kauzmann's (1974) belief that sorption isotherms at low humidi-
ties are not likely to provide useful insight into the problem of
protein hydration in solution. Several reasons were given for this
belief: (1) Unambiguous interpretation of "boundness" of water
cannot be derived from moisture sorption isotherms; (2) at low
moisture levels, proteins probably undergo changes (as indicated by
hysteresis of sorption isotherms) that are not characteristic of
proteins at physiological water contents; and (3) no simple relation

exists between the amount of water adsorbed and the strength ("boundness") of water interactions. Thus the possibility, perhaps probability, of different protein hydration characteristics at low and high moisture contents is a matter that should not be overlooked. Ling (1972), for example, suggested that fibrous proteins at low water activities adsorb water primarily at polar groups, whereas at high water activities, peptide amides may make a substantial contribution to total hydration. Furthermore, small changes in the water content of very dry proteinaceous systems may result in conformational changes in the protein and a change in the number of binding sites (Berlin *et al.* 1970). Again, this behavior would not be characteristic of proteins at physiological water contents.

Measuring the inability of water in a proteinaceous system to dissolve an added solute has been used by some food researchers as a measure of bound water. This technique has at least two serious shortcomings: (1) the value obtained depends on the size of the solute being added (see Fig. 3.7) and (2) unambiguous interpretation of the result is not possible (Berendsen 1971). Because of these problems and the availability of better techniques, this approach to the study of bound water is not highly regarded (Drost-Hansen 1971).

To determine bound water calorimetrically, one can study a series of samples with different water contents. At some low water content none of the water is freezable (bound water) and this is reflected in the heat capacity of the sample. A typical result is shown in Fig. 3.8. This approach, however, appears to be open to many of the same criticisms mentioned with respect to moisture sorption isotherms. Alternatively, the enthalpies of a series of samples differing only in temperature can be determined (*e.g.*, -40 to $0°C$, -39 to $0°C$, -38 to $0°C$, etc.) and by this means the change in ice content per degree can be calculated. From this information one can determine the temperature at which no more freezable water remains. Knowing the total water content of the sample, the amount of unfreezable (bound) water can be determined (Riedel 1969).

NMR spectroscopy can produce valuable information about the characteristics of water in protein suspensions and biological systems. Indications of the motional freedom, and hence "boundness" of water molecules, is provided from NMR spectra. Useful values include relaxation times for water protons (T_1, the spin-spin or traverse relaxation time, and T_2, the spin-lattice or longitudinal relaxation time), self-diffusion coefficients for water molecules, and rotational correlation times for water molecules (James 1975;

Ramirez *et al.* 1974; Walter and Hope 1971). The amount of unfreezable water in various aqueous systems also can be determined accurately by proton NMR spectroscopy. For example, Kuntz *et al.* (1969) froze various protein samples to $-35°C$ and determined the amounts of unfreezable water from line widths of the relatively narrow proton magnetic resonance signals.

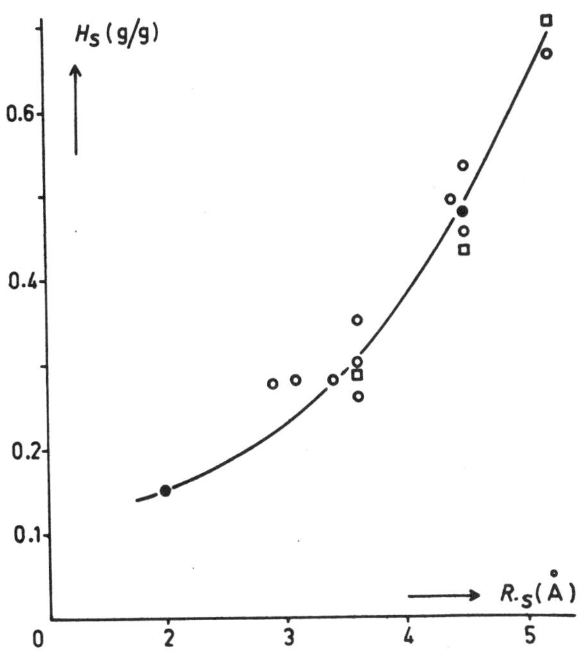

From Walstra (1973);
Courtesy of Dietrich Steinkopff Verlag
FIG. 3.7. AMOUNT OF NONSOLVENT WATER (H_S) IN PARACASEIN AS INFLUENCED BY SOLUTES OF DIF-FERENT MOLECULAR RADII (R_S)

A Quantitative Definition for Bound Water

The matter of a quantitative definition for bound water can now be dealt with. I have chosen to define "bound water" as being equivalent in quantity to water that will not freeze at $-40°C$ (constitutional water plus vicinal water in Table 3.1). Bound water defined in this manner can be accurately measured by either nuclear magnetic resonance spectroscopy or calorimetric procedures (Table 3.2), and the definition is consistent with both the previous conceptual definition of Kuntz and Kauzmann (1974) and with the scheme in Table 3.1. However, equating bound water and un-freezable water is by no means a faultless approach. Riedel (1961) stated that unfreezable water consists of: (1) bound water, which in his view is water that is bound quasi-chemically, and (2) capillary water. This is perhaps a useful theoretical approach, but differences in the two definitions would appear to be of little importance quantitatively, and analytical procedures do not exist for accurately measuring bound water as Riedel defines it. Thus the definition proposed appears to be a workable one, and a superior alternative is not obvious.

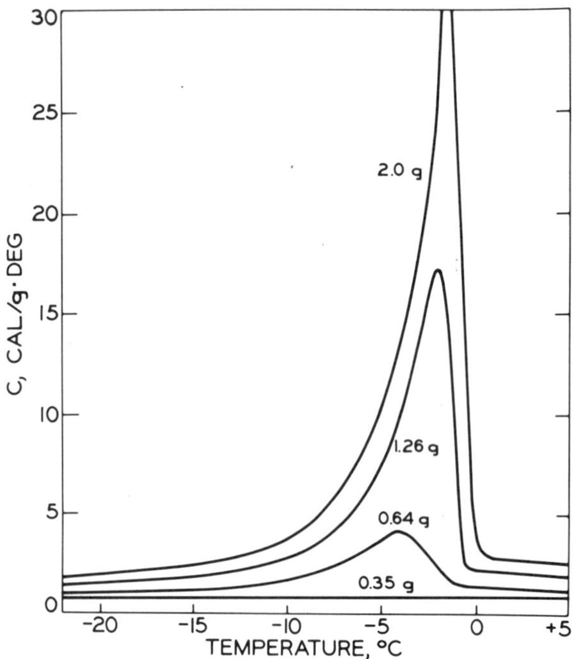

From Mrevlishvili and Privalov (1969);
Courtesy of Consultants Bureau
FIG. 3.8. HEAT CAPACITY OF 1 g OF PROCOLLAGEN AS INFLUENCED BY TEMPERATURE AND WATER CONTENT
Water contents in g H_2O/g protein are indicated on the curves.

TABLE 3.2
AMOUNT OF WATER BOUND TO PROTEINS

Class of Protein	Bound Water[1] (g H_2O/g protein)		Gross Solubility
	Calorimetry	NMR	
Globular	0.3–0.5	0.3–0.4	Soluble
Fibrous	0.3–0.5	ca. 0.5	Insoluble or sl. soluble

[1] Berlin *et al.* (1970), Cooke and Kuntz (1974), Dehl (1970), Fung *et al.* (1974), Haly and Snaith (1971), Kuntz *et al.* (1969), Mrevlishvili and Privalov (1969).

Reliable estimates of bound water attached to proteins range from 0.3-0.5 g H_2O/g protein (Table 3.2). It is interesting and perplexing that similar values are obtained for globular and fibrous proteins since the solubilities of these two classes of proteins differ markedly. This point will receive further consideration later.

PROTEIN GROUPS INVOLVED IN WATER BINDING

General Considerations

It would be highly desirable to approach this subject by summarizing data with respect to the roles that functional groups, amino acids and polypeptides have in water binding. Unfortunately, the reported results cannot be neatly summarized since conflicting results have been reported. A few examples will indicate the problem.

Glasel (1970) studied the hydration characteristics of various biopolymers, including some polypeptides, by means of deuteron magnetic relaxation techniques. He found that uncharged carboxyl and amide groups interacted strongly with water, that imide and carbonyl groups interacted somewhat with water, but not strongly, and that hydrophobic groups did not interact at all with water. In addition, he reported that strong interactions between water and charged polymers occurred only when the charges were neutralized wholly or partly by intra- or interpolymer interactions.

Karmas and DiMarco (1970) used differential scanning calorimetry to study the hydration characteristics of individual amino acids. They reported that cystine, glutamine, tyrosine, the basic amino acids and the acidic amino acids did not bind water, that, surprisingly, nonpolar amino acids (isoleucine, leucine, methionine and valine) exhibited strong water binding properties and that alanine, asparagine, cysteine, glycine, proline, hydroxyproline, phenylalanine, serine, threonine, and tryptophan exhibited small to moderate water binding properties. It should be noted, however, that free amino acids, because of their charged $-NH_3^+$ and $-COO^-$ groups, are not good models for protein hydration.

Bull and Breese (1968) used an isopiestic technique to study the hydration properties of various globular proteins. They found that protein hydration correlated strongly with the sum of the polar residues (hydroxyls, carboxyls and basic groups) minus the amides. Amides were observed to inhibit water binding. Disregarding amides, it was estimated that about six moles of water were bound per mole of polar residue.

Probably the most reliable and complete work on groups involved in protein hydration is that of Kuntz (1971). He used NMR measurements on synthetic polypeptides (usually homogeneous polypeptides) to determine bound water. Measurements were done at $-35°$C and pH was controlled. The results are shown in Table 3.3. It is evident that the charge state (governed by pH) can influence water binding of certain amino acids and that the pattern of hydration follows intuitive expectations, *i.e.*, bound water decreases in the order: ionic groups > polar groups > nonpolar groups. Judging from the results of Hagenmaier (1972; Table 3.4), pH may have less effect on the amount of water bound to some natural proteins than it has on water bound to the smaller polypeptides studied by Kuntz.

Based on the values in Table 3.3, Kuntz (1971) calculated the amount of water that should be bound to nine natural, globular proteins and found that calculated values were slightly greater than, but generally in remarkably good accord with, measured values.

Interaction of Water with Hydrophilic Groups

Interactions between water molecules and hydrophilic solutes are easily visualized since these involve mainly ion-dipole and dipole-dipole interactions, and structural arrangements occur in accord with charges. An example involving hydration of sodium chloride is shown in Fig. 3.9. Hydrogen bonding of water can occur with various potentially eligible groups in proteins (hydroxyl, amino, carbonyl, amide, imino, etc.), and two examples are given in Fig. 3.10. "Water bridges" also have been suggested as possibilities, whereby one or more water molecules form a bridge between two sites on a single macromolecule or between two macromolecules. The example in Fig. 3.11 involves a three-molecule water bridge between backbone peptide units in papain (Berendsen 1975).

It has been noted that hydrophilic groups in many crystalline macromolecules are separated by distances identical to the nearest-neighbor oxygen spacing in pure water (Berendsen 1968; Warner 1965). If this spacing prevails in hydrated macromolecules this could encourage cooperative hydrogen bonding in interfacial water. Based on this concept, Warner (1970) proposed a fascinating mechanism of muscle contraction involving a cooperative relationship between water and myofilaments.

Interactions of Water with Hydrophobic Groups

In most proteins, 20–30% of the amino acids have hydrophobic side chains such as the methyl group of alanine, the isopropyl group of valine, the isobutyl group of leucine, the mercaptomethyl group of cysteine and the benzyl group of phenylalanine (Kauzmann 1959; Ramachandran and Sasisekharan 1968; Schachman, 1963). According to Tanford (1970) and others, any hydrophobic group interferes with the normal structured arrangement of adjacent water molecules, and they do so without compensating effects. The result is an increase in free energy. In order to minimize the increase in free energy, nonpolar groups tend to associate and decrease their exposure to water. These hydrophobic interactions are important since they are largely responsible for stabilizing the compact folded structure that is typical of native proteins (Esipova and Chirgadze 1969; Kauzmann 1959; Klotz 1965; Tanford 1970; Vazina and Lednev 1969). The thermodynamics of hydrophobic interactions have been the subject of some controversy, with Kauzmann (1959), Nemethy *et al.* (1963),

TABLE 3.3

PROPOSED AMINO ACID HYDRATIONS BASED ON NUCLEAR
MAGNETIC RESONANCE STUDIES OF POLYPEPTIDES[1]

Amino Acid Residues[2]	Bound Water (Moles Water/Mole Amino Acid)
Ionic	
Asp^-	6
Glu^-	7
Tyr^-	7
Arg^+	3
His^+	4[3]
Lys^+	4
Polar	
Asn	2
Gln	2
Pro	3
Ser, Thr	2
Trp	2
Asp	2
Glu	2
Tyr	3
Arg	3
Lys	4
Nonpolar	
Ala	1
Gly	1
Phe	0
Val	1
Ile, Leu, Met	1[4]

[1] Kuntz (1971); Kuntz and Kauzmann (1974). Synthetic polypeptides (usually consisting of a single kind of amino acid); temperature $-35°C$; pH controlled.
[2] Standard three-letter code.
[3] Assumed to be equal to the value for lys^+.
[4] Assumed value. Based on one water molecule per amide plus one water molecule per side-chain polar group.

Nemethy and Scheraga (1962), Tanford (1962), Franks (1975), and others arguing that the process is entropy dominated, and Klotz (1960, 1965) arguing that the process is enthalpy dominated. Most investigators favor the first point of view.

Of particular interest here are: (1) that hydrophobic interactions exist because of the aqueous environment, and (2) that something on the order of one-third of the total nonpolar groups in proteins remain exposed to water despite hydrophobic interactions (Wetlaufer 1973). This immediately poses a question concerning the nature of water adjacent to the exposed hydrophobic groups (vicinal water). The arrangement depicted in Fig. 3.12 has been suggested by Lewin (1974) and is compatible with the requirement that water and nonpolar groups are not mutually attracted (other than by weak van der Waals' forces). The fact that this scheme would tend to repel cations and attract anions is in accord with the fact that many proteins above their isoelectric points (net negative charge) do in fact bind some anions. The arrangement proposed is also compatible with the notion advanced by Klotz (1965, 1970) that partial clathrate-type structures form around hydrophobic groups. Although all investigators do not agree as to the details of water structure adjacent to hydrophobic groups, there appears to be no argument that it is more structured than normal water.

How much of the water adjacent to hydrophobic groups is bound water is not accurately known. The results of Kuntz (1971) (Table 3.3) seem to indicate that the amount is small, whereas the results of Karmas and DiMarco (1970) and Klotz (1965, 1970) indicate the amount could be comparatively large.

HYDRATED ION PAIR

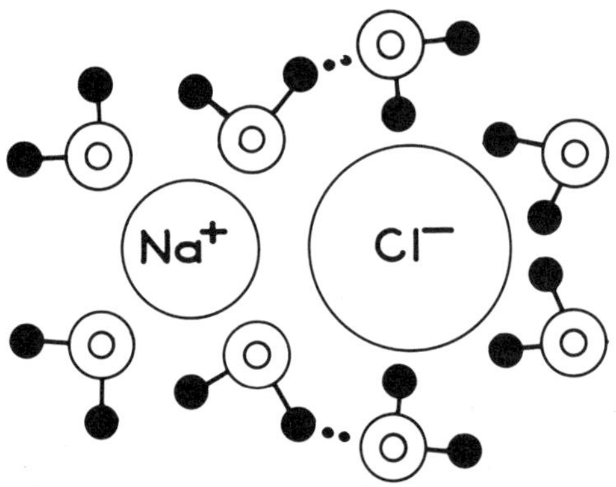

Adapted from Lewin (1974)
FIG. 3.9. LIKELY ARRANGEMENT OF WATER MOLECULES ADJACENT TO SODIUM CHLORIDE

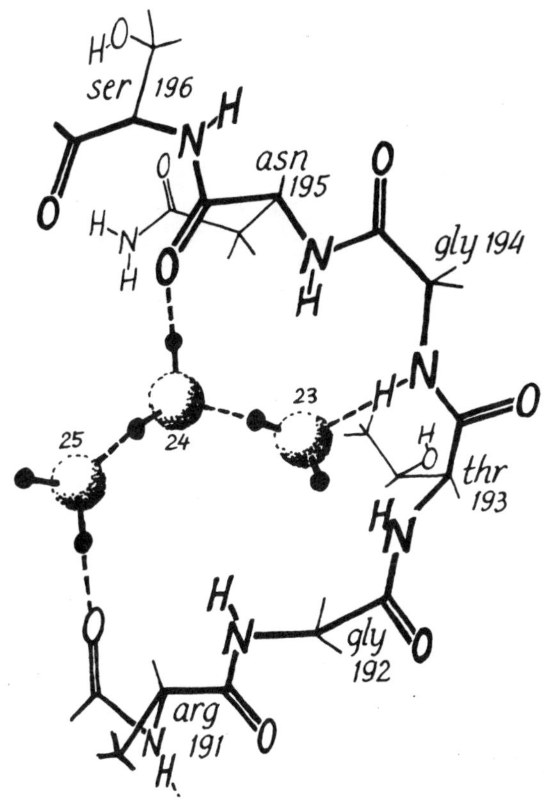

From Lewin (1974); Courtesy of Academic Press

FIG. 3.10. HYDROGEN BONDING (DOTTED LINES) OF WATER TO TWO KINDS OF FUNCTIONAL GROUPS OCCURRING IN PROTEINS

From Berendsen (1975)
Courtesy of Plenum Publishing Corp.

FIG. 3.11. EXAMPLE OF A THREE-MOLECULE WATER BRIDGE IN PAPAIN

EFFECT OF SALTS ON WATER BOUND TO PROTEINS

Water bound to proteins appears to be affected more by salts than by most other factors. Factors which one might assume to have substantial influences on the amount of water bound to proteins, but apparently do not, are: the process of rigor mortis, mastication or homogenization of muscle and denaturation of proteins (Blanshard and Derbyshire 1975; Hamm 1960; Kuntz *et al.* 1969; Mrevlishvili and Privalov 1969). The amount of water bound to some types of homogeneous synthetic polypeptides is influenced by pH (Table 3.3), but this effect is apparently absent in at least some kinds of natural proteins (Table 3.4).

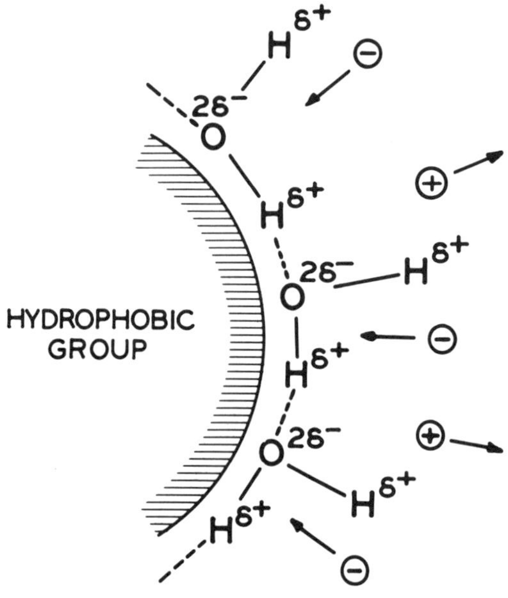

Adapted from Lewin (1974)
FIG. 3.12. PROPOSED WATER ORIENTATION AT
A HYDROPHOBIC SURFACE

TABLE 3.4
EFFECT OF pH ON WATER BOUND TO PROTEINS[1]

Protein	Estimate of Bound Water[2] (g Water/0.16 g N)		
	pH 4.5	pH 6.0	pH 7.5
Cottonseed II	0.16	0.16	0.16
Soybean (Promine D)	0.19	0.19	0.22
Casein	0.21	0.22	0.26

[1] From Hagenmaier (1972). Moisture sorption study.
[2] Samples were dialyzed to remove hydrophilic salts, pH was adjusted to the desired value, samples were freeze-dried, and known amounts were brought to constant weight at 84% relative humidity (presumably at room temperature).

At low concentrations of neutral electrolytes the amount of water bound to proteins increases with increasing salt concentration. This trend continues until an ionic strength of 0.1–0.15 is reached, after which further increases in ionic strength result in marked decreases in bound water. With uni-univalent electrolytes, this maximum in hydration is associated with abrupt changes in conformational structure for a wide variety of proteins (Eaglund 1975; Kuntz and Brassfield 1971). According to Eaglund (1975), "Since this is the region of electrolyte concentration within which suppression of the electric double layer surrounding the macro-molecule will occur, it is perhaps not unexpected that associated changes in conformation and hydration should accompany such a change." It is perhaps significant, then, that biological systems are about 0.1 M in electrolytes (Drost-Hansen 1971).

The effects that high concentrations (2 molal) of some salts have on water bound to proteins is illustrated in Table 3.5. All of the salts studied decreased the amount of water bound to egg albumin as compared to water bound to egg albumin in the absence of salts. Effectiveness of the cations studied are in accord with the Hofmeister or lyotropic series, *i.e.*, the larger the hydrated radius (the smaller the unhydrated radius) of the cation, the greater is its ability to dehydrate the protein. Thus, in order of effectiveness of reducing the quantity of water bound to protein: Li^+, $Na^+ > K^+ > Rb^+ > Cs^+$ (Bull and Breese 1970). Contrary to the behavior of cations, the larger the hydrated radius of an anion the smaller is its ability to dehydrate proteins (Bull and Breese 1970).

At high salt concentrations, electrostatic interactions are apparently of little importance with regard to the amount of water bound to proteins, whereas competition of ions and proteins for the existing water is of considerable importance (Eaglund 1975).

INTERRELATIONSHIPS AMONG BOUND WATER, PROTEIN DENATURATION AND PROTEIN SOLUBILITY

Denaturation of proteins can be defined as any major modification of secondary, tertiary or quaternary structure that does not involve breaking covalent bonds (Anglemier and Montgomery 1976; Wu and Inglett 1974). A schematic illustration depicting the unfolding that occurs during denaturation is given in Fig. 3.13.

TABLE 3.5
EFFECT OF SALTS ON WATER BOUND TO EGG ALBUMIN[1]

Salt (2 molal)	Bound Water (g H_2O/g Protein)[2]
LiCl	0.08
NaCl	0.08
KCl	0.15
RbCl	0.16
CsCl	0.20
No salt	0.30

[1] From Bull and Breese (1970).
[2] Assumes mol wt of 45,000 daltons for egg albumin.

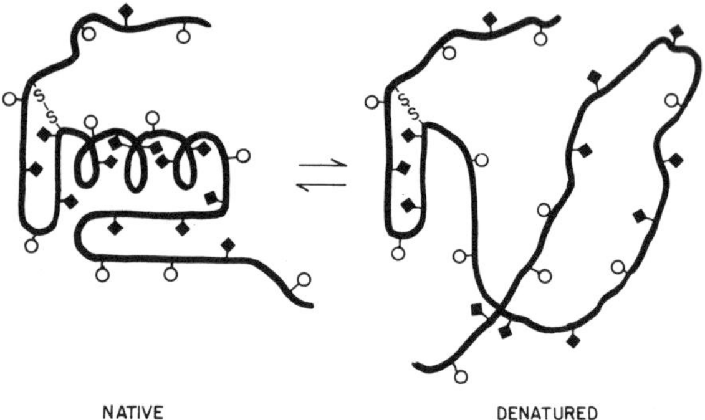

From Brandts (1967) as modified
by Anglemier and Montgomery (1976);
Courtesy of Academic Press and Marcel Dekker
FIG. 3.13. SCHEMATIC INTERPRETATION OF PROTEIN
DENATURATION

It has already been mentioned that denaturation appears to have comparatively little effect on the amount of water bound to proteins. Data with respect to this point are shown in Table 3.6. It is evident that when denaturation does change the amount of bound water the change is almost always an increase. Several possible explanations for this occurrence were suggested by Lewin (1974):

(1) Rupture of solute-solute hydrogen bonds (Fig. 3.14);

(2) Rupture of ionic linkages (Fig. 3.15);

(3) Rupture of water-mediated hydrophobic linkages (Fig. 3.16); and

(4) Dissociation of hydrophobic groups (Fig. 3.17).

Collectively, these changes involve an increase in protein interfacial area, a decrease in protein-protein interactions and an increase in protein-water interactions. In light of these occurrences it may

TABLE 3.6
EFFECT OF DENATURATION ON BOUND WATER

Sample	Method[1]	Bound Water (g H_2O/g Protein)		References
		Native	Denatured	
Tropocollagen	cal	0.46	0.52	2
Serum albumin	cal	0.31	0.33	2
Egg albumin	cal	0.32	0.33	2
Ovalbumin	NMR	0.31	0.3	3
Hemoglobin	cal	0.32	0.34	2
α-Chymotrypsin	NMR	0.37	0.4	3

[1] "cal" is calorimetric; "NMR" is nuclear magnetic resonance.
[2] Mrevlishvili and Privalov (1971).
[3] Kuntz *et al.* (1969).

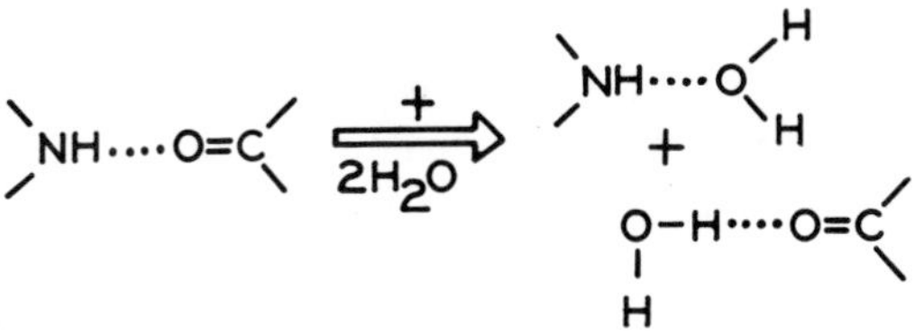

Adapted from Lewin (1974)

FIG. 3.14. INCREASED WATER BINDING RE-
SULTING FROM RUPTURE OF SOLUTE-SOLUTE
HYDROGEN BONDS

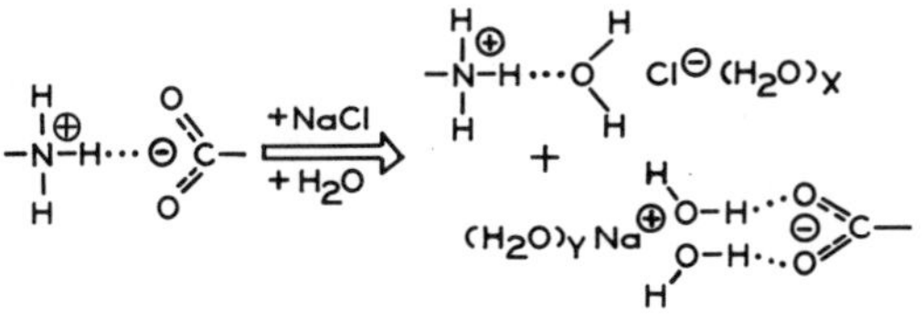

Adapted from Lewin (1974)

FIG. 3.15. INCREASED WATER BINDING RE-
SULTING FROM RUPTURE OF IONIC LINKAGES

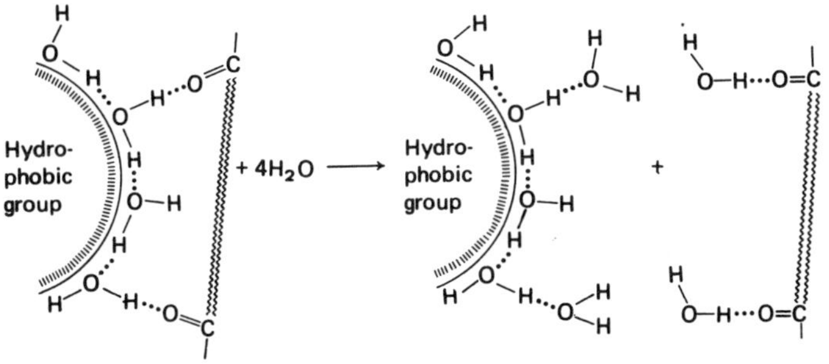

From Lewin (1974); Courtesy of Academic Press

FIG. 3.16. INCREASED WATER BINDING RESULTING FROM
RUPTURE OF WATER-MEDIATED HYDROPHOBIC-
HYDROPHILIC LINKAGES

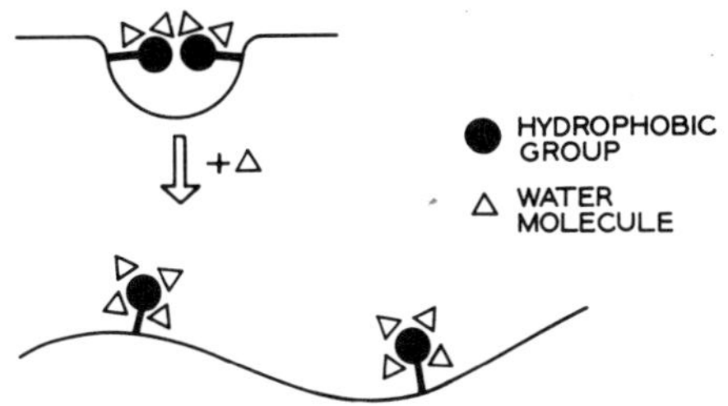

Adapted from Lewin (1974)

FIG. 3.17. INCREASED WATER BINDING
RESULTING FROM DISSOCIATION OF
HYDROPHOBIC GROUPS

seem strange that reports exist in the literature indicating no change in the amount of bound water (w/w) during denaturation (Gordon and Warren 1968). However, logical explanations do exist: (1) The method used may result in values for bound water that are not equivalent to the unfreezable water fraction, (2) some water located in interstitial regions of the protein molecule would be lost during unfolding, and (3) aggregation of unfolded protein molecules could occur resulting in reconversion of protein-water bonds to protein-protein bonds.

That denaturation often results in a slight increase in bound water is not particularly exciting or cause for suspicion until it is recalled that denaturation and aggregation of proteins frequently result in sizeable decreases in water solubility. The example shown in Fig. 3.18 is typical of many studies illustrating this point. It should be emphasized that the indicated changes in solubility resulted from the dual effects of protein unfolding (denaturation) and protein aggregation, with the relative importance of aggregation increasing in an unknown manner with increased heating time.

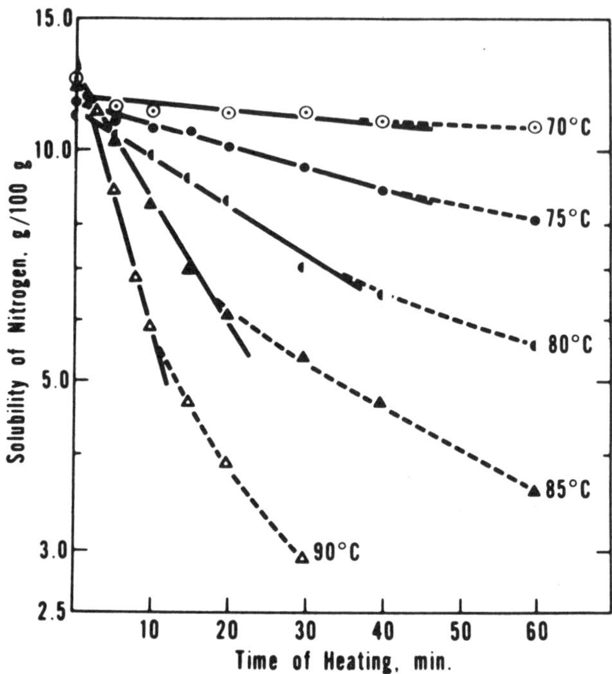

From Pence (1953);
Courtesy of American Association of Cereal Chemists
FIG. 3.18. WATER SOLUBILITY OF WET GUM GLUTEN IN DILUTE ACETIC ACID AS INFLUENCED BY DENATURATION AT VARIOUS TEMPERATURES

Although this interferes with any quantitative determination of the independent effect of protein unfolding on solubility, it does not detract from the validity of the opening statement interrelating denaturation *and aggregation* with reduced solubility and a slight increase in bound water. This is true since some proteins for which bound water has been found to increase during denaturation were, following denaturation, mixtures of unfolded and aggregated molecules (Kuntz *et al.* 1969), whereas others were discrete unfolded molecules (Kuntz and Brassfield 1971).

The situation becomes even more perplexing when one realizes that solubilities of various proteins (different sources; native or denatured) and corresponding values for bound water do not correlate well (Fig. 3.19). Surprisingly, little attention is given to this point in most articles on protein hydration. A possible explanation for this discrepancy is that the means used to express bound water (mass of water per unit mass of dry protein) is inappropriate. Consider the example in Table 3.7. α-Chymotrypsin binds 0.37 g H_2O/g protein in the native state and 0.40 g in the denatured state. When expressed on the more appropriate basis of grams of water per unit of protein interfacial area, these values become 0.90 for native α-chymotrypsin and 0.27 for denatured α-chymotrypsin—clearly more reasonable values since they are in general accord with anticipated changes in protein solubility.

The interfacial area approach to expressing bound water also would be expected to produce far more reasonable relationships than the w/w approach in the following two situations:

(1) Solubilities of globular and fibrous proteins differ greatly, yet values for bound water, when expressed on a w/w basis, are essentially the same (Table 3.2).

TABLE 3.7

MEANS OF EXPRESSING THE AMOUNT OF WATER BOUND TO α-CHYMOTRYPSIN

		Bound Water	
Conformational State	Surface Area per Molecule[1] (Å^2)	g H_2O/g Protein[2]	g H_2O/6.023 $\times$ 10²³ Å^2 of Protein Interfacial Area
Native	10,130	0.37	0.90
Denatured	35,900[3]	0.40[4]	0.27

[1] From Chothia (1975).
[2] From Kuntz and Brassfield (1971) and Kuntz *et al.* (1969).
[3] Molecule in extended conformation with all trans peptide bonds.
[4] Denatured with urea.

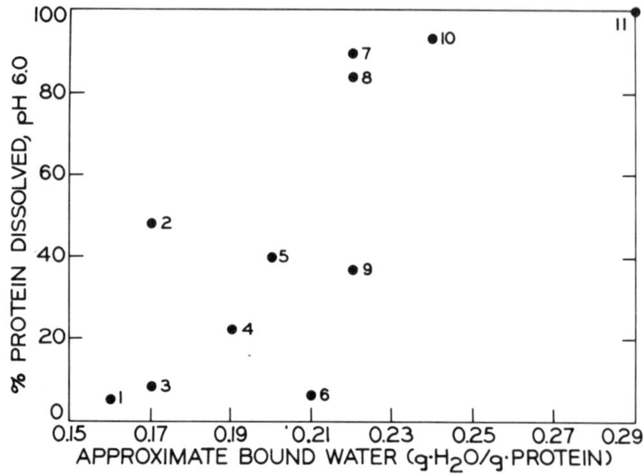

From Hagenmaier (1972)

FIG. 3.19. RELATIONSHIP BETWEEN BOUND WATER AND
PROTEIN SOLUBILITY

(2) Solubilities of various proteins do not correlate well with corresponding values for bound water expressed on a w/w basis (Fig. 3.19). In these two instances, the interfacial area approach to bound water would not be expected to be more than a partial remedy for the poor correlation between bound water and solubility since this approach does not account for important qualitative differences in the nature of the protein surface (*i.e.*, kinds, ratios and distributions of polar and nonpolar groups, and the number and distribution of crevaces of molecular dimensions).

FACTORS INFLUENCING THE AMOUNT OF WATER ENTRAPPED BY CELLULAR PROTEINS

It is appropriate, now, to shift attention from bound water to bulk phase water, especially entrapped bulk phase water since this is the most important type. The obvious example of choice is muscle.

In muscle, the bulk phase water is entrapped in a matrix of filaments in a manner not unlike the situation in some gels (Blanshard and Derbyshire 1975; Hamm 1975). Muscle contains about 75% water, of which perhaps 5% is bound water, and almost all of the remainder is entrapped bulk phase water (Hamm 1975). The ability of muscle to retain its bulk phase water during the application of various stresses such as pressure, heating, grinding, etc., is referred to as water holding capacity (WHC). Measurement of WHC can be accomplished simply and accurately by techniques involving pressing or centrifugation.

The WHC of muscle is closely related to its juiciness, taste, tenderness and color and is therefore of great practical concern to meat processors (Hamm 1960). Contrary to the behavior of bound water (when expressed on a w/w basis) the WHC of muscle is influenced greatly by a variety of processing and handling conditions including rigor mortis, storage, heating, drying, and freezing (Hamm 1970). The remainder of this section will be devoted to consideration of the major factors that influence WHC of muscle. Much of the information presented is derived from the work of Professor Reiner Hamm and associates and from his excellent review papers on this subject (Hamm 1960, 1963, 1970, 1971, 1975).

The spatial arrangement existing among protein filaments is viewed by Hamm (1975) as being of primary importance to WHC. Swelling of the protein matrix is believed to result in improved WHC, whereas shrinkage of the protein matrix is believed to result in decreased WHC. These relationships are depicted schematically in Fig. 3.20. Several factors known to influence swelling or shrinking of the protein matrix and thus WHC are: animal characteristics (species, sex, age, muscle type), pH, rigor mortis, aging and salts. Only the last four factors will be considered here.

The effect of pH on WHC of ground beef is illustrated in Fig. 3.21. Minimum WHC occurs at about pH 5.0, the isoelectric point of actomyosin, the major protein of postrigor muscle. As the pH is elevated from the isoelectric point to pH 10, WHC increases

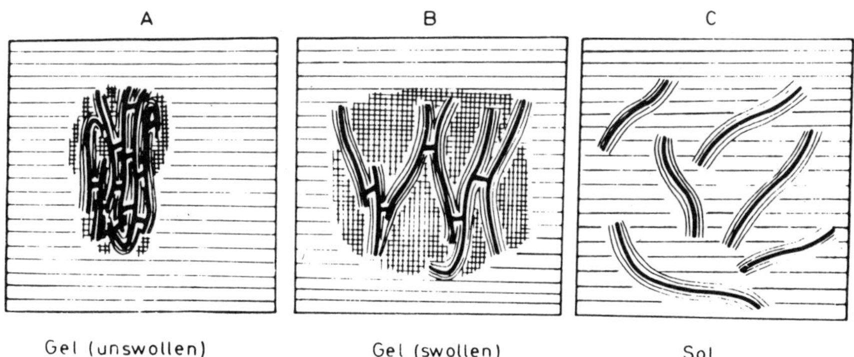

From Hamm (1975);
Courtesy of the Author and Butterworths, London

FIG. 3.20. INFLUENCE OF CROSS-LINKING OF PROTEIN FILAMENTS ON THE SWELLING AND WHC OF MUSCLE

▤ "Free" bulk phase water; ▦ "Entrapped" bulk phase water; ▨ Bound water. A. Strong cross-linking results in a shrunken protein matrix and low WHC. Most of the bulk phase water is "free" rather than "entrapped." B. Few cross-links result in swelling of the protein matrix. Most of the bulk phase water is "entrapped" within the protein matrix rather than "free." C. No cross-linkages and therefore no WHC (sol).

dramatically. The same is true when the pH is lowered from the isoelectric point to about pH 3.5. Adjustment to pH values above 10 or below 3.5 apparently results in protein denaturation and a commensurate decrease in WHC.

At intermediate pH values a logical explanation for the pH-WHC relationship involves the charge state of the protein molecules (Hamm 1960, 1975). At the isoelectric point of a given protein, the net charge is zero, a maximum number of interprotein ionic linkages will exist, the protein matrix will be shrunken and WHC will be at a minimum. On the acid side of the isoelectric point positive charges will dominate, the protein molecules will tend to repel each other, the matrix will swell and WHC will increase. On the basic side of the isoelectric point negative charges will prevail and a similar effect will occur.

The effects of rigor mortis and aging on WHC are clearly evident from Fig. 3.22. Rigor mortis of beef muscle (1 day postmortem) is associated with a substantial decline in pH to about 5.5, a decrease in the solubility of myofibrillar proteins and a considerable decrease in WHC. This decrease in WHC is partly (about 1/3) attributable to the decrease in pH toward the isoelectric point and partly (about 2/3) to degradation of ATP (Bendall 1964; Hamm 1956, 1963).

Aging of meat (2–13 days postmortem in Fig. 3.22) results in moderate increases in pH, solubility of myofibrillar proteins and WHC. Only a part of the improvement on WHC can be explained on the basis of an increase in pH (Fujmaki and Kurabayashi,

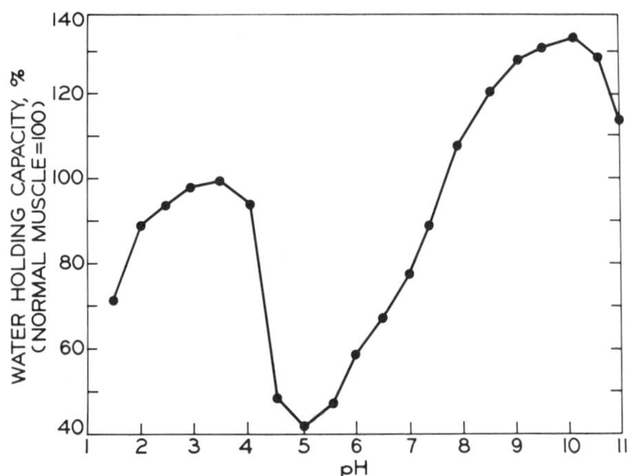

From Hamm (1960);
Courtesy of the Author and Academic Press
FIG. 3.21. EFFECT OF pH ON THE WATER HOLDING
CAPACITY OF BEEF MUSCLE HOMOGENATE

1958; Hamm, 1959; Wierbicki *et al.* 1956). Influences other than pH that may loosen the protein matrix and increase WHC during aging are: changes in ion-protein relationships (adsorption of K^+ and release of Ca^{2+}; Arnold *et al.* 1956) and action of proteolytic enzymes at the Z-line (Busch *et al.* 1972; Suzuki *et al.* 1975), and perhaps some change in the actin-myosin relationship (Goll *et al.* 1970).

The combined effects of pH and sodium chloride (2%) on the WHC of ground beef are shown in Fig. 3.23. At pH values above the isoelectric point, sodium chloride greatly increases the WHC of beef, whereas below the isoelectric point an opposite effect occurs. A satisfactory explanation for this behavior can be devised once it is realized that anions (Cl^- in this instance) generally exhibit stronger interactions with proteins than do cations (Na^+ in this instance) (Hamm 1960; Lewin 1974; Niinivaara and Pohja 1954). Near hydrophobic areas of a protein this behavior is readily understandable from Fig. 3.12. In hydrophilic areas of a protein, the

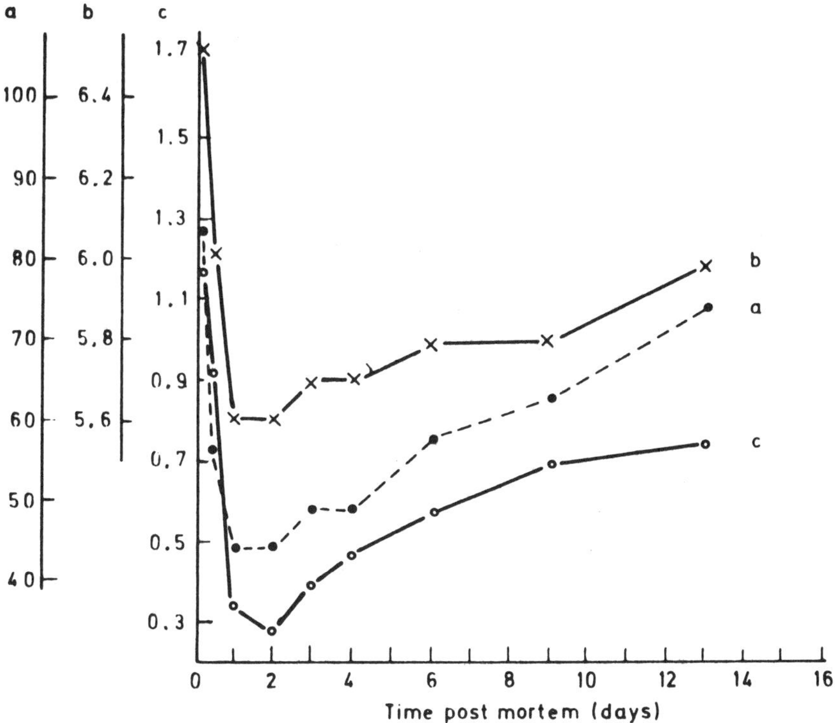

From Brendl and Klein (1970);
Courtesy of the Institute of Chemical Technology, Prague
FIG. 3.22. EFFECT OF TIME POSTMORTEM ON (a) WATER HOLDING CAPACITY, (b) pH, AND (c) SOLUBLE MYOFIBRILLAR PROTEIN-N OF BEEF

reason for preferential interactions with anions is less clear. The preferential binding of the chloride ion over the sodium ion is, however, clearly evident from the decrease in the isoelectric point when sodium chloride is added to ground beef (Fig. 3.23). Knowing this, the effect of sodium chloride on WHC at various pH values can be easily explained (Hamm 1960, 1975). Below the isoelectric point the net positive charge of the protein is neutralized effectively by chloride ions (Fig. 3.24), the protein matrix shrinks and WHC declines. Above the isoelectric point sodium ions adsorb poorly, some chloride ions neutralize positive charges that still exist, net negativity actually increases, swelling of the matrix occurs, and WHC increases.

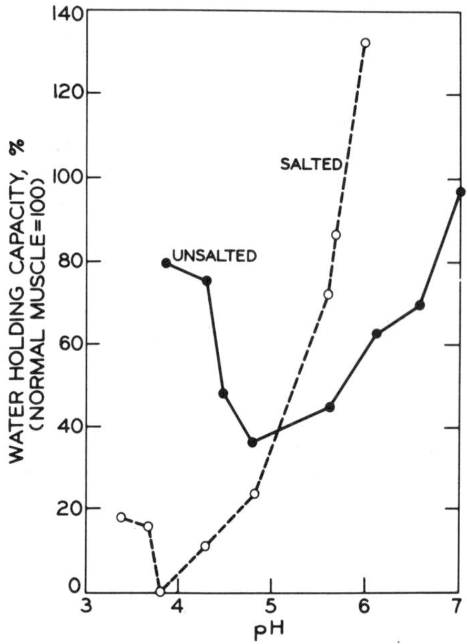

Adapted from Hamm (1960);
Courtesy of the Author
and Academic Press
FIG. 3.23. EFFECT OF SODIUM CHLORIDE (2%) ON THE RELATIONSHIP BETWEEN pH AND WATER HOLDING CAPACITY OF BEEF MUSCLE HOMOGENATE

An additional point of interest is that WHC of muscle is greatly influenced by the ionic strength of the medium (Fig. 3.25). Thus, when sodium chloride is added to ground beef (above the iso-

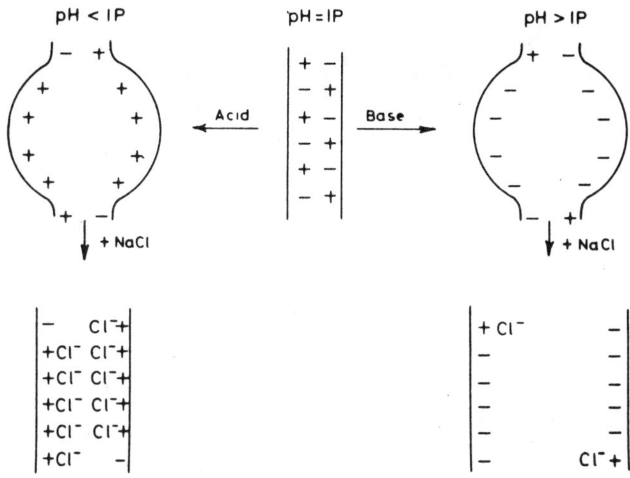

From Hamm (1975);
Courtesy of the Author and Butterworths, London
FIG. 3.24. SCHEMA OF THE INFLUENCE OF NaCl ON SWELL-
ING OF MEAT AT pH VALUES ABOVE AND BELOW THE
ISOELECTRIC POINT

electric point) maximum WHC occurs at an ionic strength of
0.8–1.0 or at about 5% sodium chloride by weight (Grau and
Hamm 1952; Hamm 1957). The decline in WHC at ionic strengths
greater than 1.0 is apparently a "salting-out" effect involving
strong binding of water by the salt and commensurate dehydration
of the protein.

Changes in WHC of muscle resulting from various salts other
than sodium chloride also have been studied, but the results are
too involved for consideration here (Hamm 1960, 1971; Hamm
and Grau 1955).

Before closing this section it is appropriate to comment briefly
on the comparative effects of salts on bound water and WHC. With
neutral salts such as sodium chloride, proteins exhibit maximum
bound water contents at an ionic strength of 0.1–0.15 (Eaglund
1975), whereas muscle exhibits maximum WHC at an ionic
strength of 0.8–1.0 (Fig. 3.25). Anions and cations often affect
water bound to protein and WHC of muscle in accord with the
Hofmeister and lyotropic series except that the order of ion influence is sometimes reversed for bound water as compared to
WHC (Bull and Breese 1970; Hamm 1960, 1971; Hamm and Grau
1958). These apparent inconsistencies (which were also noted between bound water and protein solubility) cannot be adequately

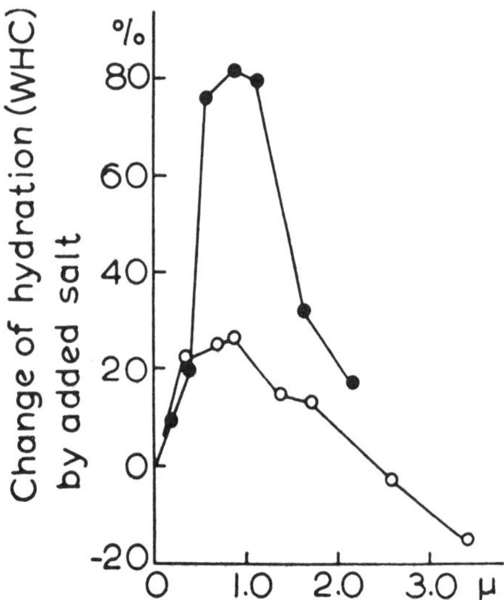

Adapted from Hamm (1957);
Courtesy of Springer-Verlag
FIG. 3.25. INFLUENCE OF SODIUM CHLO-
RIDE ON THE WATER HOLDING CAPAC-
ITY OF GROUND BEEF MUSCLE
Closed circles, no added water; Open circles,
60% added water. Abscissa: ionic strength of
NaCl added to the muscle homogenate.

explained at this time because experimental conditions often differ greatly between studies involving bound water and WHC (pH, ionic strength, muscles vs protein suspensions), because bound water is expressed in terms of protein weight rather than protein interfacial area, and because the state of knowledge with respect to this matter is still inadequate.

CONCLUSIONS

The subject of protein hydration is unquestionably important and it is regrettable that so little is known concerning the nature of bound water and its relation to conformational changes of proteins, to protein solubility, to water holding capacity of muscle and to cell physiology. The complex nature of protein suspensions and cellular systems is largely responsible for this state of affairs, and it is especially disturbing that the lack of accurate values for protein interfacial areas under various environmental conditions has led to the undesirable practice of expressing bound water on a w/w basis rather than on the basis of mass of water per unit of protein interfacial area. It is, however, heartening that the subject of protein hydration has attracted the interest of competent investigators so that valuable information is sure to emerge in the near future.

ACKNOWLEDGMENTS

The helpful suggestions of Professor Dale Ryan are gratefully acknowledged. Contribution from the College of Agricultural and Life Sciences, University of Wisconsin-Madison is appreciated.

BIBLIOGRAPHY

ANGLEMIER, A. F., and MONTGOMERY, M. W. 1976. Amino acids, peptides, and proteins. *In* Principles of Food Science, Part 1. Food Chemistry, O. Fennema (Editor). Marcel Dekker, New York.

ARNOLD, N., WIERBICKI, E., and DEATHERAGE, F. E. 1956. Postmortem changes in the interactions of cations and proteins of beef and their relation to sex and diethylstilbestrol treatments. Food Technol. *10*, 245–250.

BENDALL, J. R. 1964. Meat proteins. *In* Proteins and Their Reactions, H. W. Schultz, and A. F. Angleimer (Editors). AVI Publishing Co., Westport, Conn.

BERENDSEN, H. J. C. 1968. Biological significance of water structure. *In* Biology of the Mouth, P. Person (Editor). Amer. Assoc. Adv. of Sci., Washington, D.C.

BERENDSEN, H. J. C. 1971. The molecular dynamics of water in biological systems. Proc. First European Biophysics Congress, Baden, Austria, 483–488.

BERENDSEN, H. J. C. 1975. Specific interactions of water with biopolymers. *In* Water, a Comprehensive Treatise, Vol. 5, Water in Disperse Systems, F. Franks (Editor). Plenum Publishing Co., London.

BERLIN, E., KLIMAN, P. G., and PALLANSCH, M. J. 1970. Changes in state of water in proteinaceous systems. J. Colloid Interf. Sci. *34*, 488–494.

BERNAL, J. D., and FOWLER, R. H. 1933. A theory of water and ionic solution, with particular reference to hydrogen and hydroxyl ions. J. Chem. Phys. *1*, 515–548.

BLANSHARD, J. M. V., and DERBYSHIRE, W. 1975. The composition of meat: Physicochemical parameters. *In* Meat, D. J. A. Cole and R. A. Lawrie (Editors). Butterworths, London.

BRANDTS, J. F. 1967. Heat effects on proteins and enzymes. *In* Thermobiology, A. H. Rose (Editor). Academic Press, New York.

BRENDL, J., and KLEIN, S. 1970. The effect of autolytic processes on the water holding capacity of chilled meat. Sbornik Vys. Sk. Chem.-Technol. Praze, Potraviny *E28*, 117–130. (German)

BULL, H. B. 1944. Adsorption of water vapor by proteins. J. Amer. Chem. Soc. *66*, 1499–1507.

BULL, H. B., and BREESE, K. 1968. Protein hydration. 1. Binding sites. Arch. Biochem. Biophys. *128*, 488–496.

BULL, H. B., and BREESE, K. 1970. Water and solute binding by proteins. 1. Electrolytes. Arch. Biochem. Biophys. *137*, 299–305.

BUSCH, W. A., STROMER, M. H., GOLL, D. E., and SUZUKI, A. 1972. Ca^{2+}-specific removal of Z-lines from rabbit skeletal muscle. J. Cell Biol. *52*, 367–381.

CERBON, J. 1974. Water in the structure and function of cell membranes. *In* Perspectives in Membrane Biology, S.E-O. C. Gitler (Editor). Academic Press, New York.

CHANG, D. C., HAZLEWOOD, C. F., NICHOLS, B. L., and RORSCHACH, H. E. 1972. Spin echo studies on cellular water. Nature *235*, 170–171.

CHOTHIA, C. 1975. Structural invariants in protein folding. Nature (London) *254*, 304–308.

CLIFFORD, J. 1975. Properties of water in capillaries and thin films. *In* Water, a Comprehensive Treatise, Vol. 5, Water in Disperse Systems, F. Franks (Editor). Plenum Publishing Co., London.

COOKE, R. 1976. Spin label probes of the intracellular aqueous environment. *In* Participation Energetique De L'Eau Solvant Aux Interactions Specifiques Dans Les Systems Biologiques, A. Alfsen (Editor). C.N.R.S., Paris.

COOKE, R., and KUNTZ, I. D. 1974. The properties of water in biological systems. Ann. Rev. Biophys. Bioeng. *3*, 95–126.

COOKE, R., and WIEN, R. 1973. Nuclear magnetic resonance studies of intracellular water protons. Ann. N.Y. Acad. Sci. *204*, 197–209.

COPE, F. W. 1969. Nuclear magnetic resonance evidence using D_2O for structured water in muscle and brain. Biophys. J. *9* 303–319.

DAVIS, C. M., and JARZYNSKI, J. 1972. Mixture models of water. *In* Water and Aqueous Solutions, R. A. Horne (Editor). John Wiley and Sons, New York.

DEHL, R. E. 1970. Collagen: Mobile water content of frozen fibers. Science *170*, 738–739.

DROST-HANSEN, W. 1971. Structure and properties of water at biological interfaces. *In* Chemistry of The Cell Interface, Part B, H. D. Brown (Editor), Academic Press, New York.

EAGLUND, D. 1975. Nucleic acids, peptides, and proteins. *In* Water, A Comprehensive Treatise, Vol. 4, Aqueous Solutions of Amphiphiles and Macromolecules, F. Franks (Editor). Plenum Publishing Co., London.

EISENBERG, D., and KAUZMANN, W. 1969. The Structure and Properties of Water. Oxford Univ. Press, London.

ESIPOVA, N. G. and CHIRGADZE, YU. N. 1969. Structural role of water in fibrillar proteins and polypeptides. *In* Water in Biological Systems, L. P. Kayushin (Editor). Consultants Bureau, New York.

FENNEMA, O. 1973. Water and Ice. *In* Low-Temperature Preservation of Foods and Living Matter, O. Fennema, W. D. Powrie, and E. H. Marth (Editors). Marcel Dekker, New York.

FENNEMA, O. 1976. Water and Ice. *In* Principles of Food Science, Part 1. Food Chemistry, O. Fennema (Editor). Marcel Dekker, New York.

FINCH, E. D., HARMON, J. F., and MULLER, B. H. 1971. Pulsed NMR measurements of the diffusion constant of water in muscle. Arch. Biochem. Biophys. *147*, 299–310.

FORSLIND, E. 1971. Structure of water. Quart. Rev. Biophys. *4*, No. 4, 325–363.

FRANK, H. S. 1970. The structure of ordinary water. Science *169*, 635–641.

FRANK, H. S. 1972. Structural models. *In* Water, A Comprehensive Treatise, Vol. 1, The Physics and Physical Chemistry of Water, F. Franks (Editor). Plenum Publishing Corp., New York.

FRANKS, F. 1972. Introduction—Water, the unique chemical. *In* Water, A Comprehensive Treatise, Vol. 1, The Physics and Physical Chemistry of Water F. Franks (Editor). Plenum Publishing Co., London.

FRANKS, F. 1975. The hydrophobic interaction. *In* Water, A Comprehensive Treatise, Vol. 4, Aqueous Solutions of Amphiphiles and Macromolecules, F. Franks (Editor). Plenum Publishing Co., London.

FUJMAKI, M., and KURABAYASHI, H. 1958. Chemical studies on the autolysis of meats. Part 8. On the influence of aging of meats upon hydration of meat proteins. J. Agr. Chem. Soc. Japan *32*, 775–778.

FUNG, B. M., WITSCHEL, J., JR., and McAMIS, L. L. 1974. The state of water on hydrated collagen as studied by pulsed NMR. Biopolymers *13*, 1767–1776.

GLASEL, J. A. 1970. Participation of water in conformational changes of biopolymers as studied by deuteron magnetic relaxation. J. Amer. Chem. Soc. *92*, 375–381.

GOLL, D. E., ARAKAWA, N., STROMER, M. H., BUSCH, W. A., and ROBSON, R. M. 1970. Chemistry of muscle proteins as a food. *In* The Physiology and Biochemistry of Muscle as a Food, Vol. 2. E. J. Briskey, R. G. Cassens, and B. B. Marsh (Editors). Univ. of Wisconsin Press, Madison.

GORDON, J. A., and WARREN, J. R. 1968. Denaturation of globular proteins. I. The interaction of urea and thiourea with bovine plasma albumin. J. Biol. Chem. *243*, 5663–5669.

GRAU, R., and HAMM, R. 1952. A method for estimating water binding of meat. Fleischwirtschaft *4*, 295–297. (German)

HAGENMAIER, R. 1972. Water binding of some purified oilseed proteins. J. Food Sci. *37*, 965–966.

HAGLER, A. T., and SCHERAGA, H. A. 1973. Current status of the water-structure problem; Application to proteins. Ann. N.Y. Acad. Sci. *204*, 51–78.

HALY, A. R., and SNAITH, J. W. 1971. Calorimetry of rat tail tendon collagen before and after denaturation: The heat of fusion of its absorbed water. Biopolymers *10*, 1681–1699.

HAMM, R. 1956. On the effect of adenosinetriphosphate on the hydration and rigidity of postmortem beef muscle. Biochemische Z. *328*, 309–322. (German)

HAMM, R. 1957. On the water binding capacity of mammalian muscle. 3. Report on the effect of neutral salts. Z. Lebensm. Untersuch. u. Forsch. *106*, 281–297. (German)

HAMM, R. 1959. The biochemistry of meat aging. 1. Report on the hydration and rigidity of beef muscle. Z. Lebensm. Untersuch. u. Forsch. *109*, 113–121. (German)

HAMM, R. 1960. Biochemistry of meat hydration. Advan. Food Res. *10*, 355–363.

HAMM, R. 1963. The water imbibing power of foods. *In* Recent Advances in Food Science, Vol. 3, J. M. Leitch and D. N. Rhodes (Editors). Butterworths, London.

HAMM, R. 1970. Properties of meat proteins. *In* Proteins as Human Food, R. A. Lawrie, (Editor). Avi Publishing Co., Westport, Conn.

HAMM, R. 1971. Interactions between phosphates and meat proteins. *In* Phosphates in Food Processing, J. M. DeMan and P. Melnychyn (Editors). Avi Publishing Co., Westport, Conn.

HAMM, R. 1975. Water-holding capacity of meat. *In* Meat, D. J. A. Cole and R. A. Lawrie (Editors). Butterworths, London.

HAMM, R., and GRAU, R. 1955. The effect of phosphates on the bound water of meat. Deut. Lebensm. Rundschau. *51*, 106–111. (German)

HAMM, R., and GRAU, R. 1958. On the water binding capacity of mammalian muscle. V. Report on the effect of salts of weak acids. Z. Lebensm. Untersuch. u. Forsch. *108*, 280–293. (German)

HAUROWITZ, F. 1963. The Chemistry and Function of Proteins, 2nd Ed. Academic Press, New York.

HAZLEWOOD, C. F., NICHOLS, B. L., and CHAMBERLAIN, N. F. 1969. Evidence for the existence of a minimum of two phases of ordered water in skeletal muscle. Nature (London) *222*, 747–750.

HOLTZER, A., and EMERSON, M. F. 1969. On the utility of the concept of water structure in the rationalization of the properties of aqueous solutions of proteins and small molecules. J. Phys. Chem. *75*, 26–33.

HORNE, R. A. 1968. The structure of water and aqueous solutions. *In* Survey of Progress in Chemistry, Vol. 4, A. F. Scott (Editor). Academic Press, New York.

HORNE, R. A. 1972. Aqueous Solutions. John Wiley and Sons, New York.

JAMES, T. L. 1975. Nuclear Magnetic Resonance in Biochemistry. Academic Press, New York.

KARMAS, E., and DiMARCO, G. R. 1970. Dehydration thermoprofiles of amino acids and proteins. J. Food Sci. *35*, 615–617.

KAUZMANN, W. 1959. Some factors in the interpretation of protein denaturation. Advan. Protein Chem. *14*, 1–63.

KELL, G. S. 1972. Continuum theories of liquid water. *In* Water and Aqueous Solutions, R. A. Horne (Editor). John Wiley and Sons, New York.

KLOTZ, I. M. 1960. Protein molecules in solution. Circulation *21*, 828–844.

KLOTZ, I. M. 1965. Role of water structure in macromolecules. Fed. Proc. *24* (No. 2, Suppl. 15), S-24 through S-33.

KLOTZ, I. M. 1970. Polyhedral clathrate hydrates. *In* The Frozen Cell, G. E. W. Wolstenholme and M. O'Connor (Editors). J & A Churchill, London.

KUNTZ, I. D., JR. 1971. Hydration of macromolecules. 3. Hydration of polypeptides. J. Amer. Chem. Soc. *93*, 514–516.

KUNTZ, I. D., JR. 1975. The physical properties of water associated with biomacromolecules. *In* Water Relations of Foods, R. B. Duckworth (Editor). Academic Press, London.

KUNTZ, I. D., JR., and BRASSFIELD, T. S. 1971. Hydration of macromolecules. 2. Effects of urea on protein hydration. Arch. Biochem. Biophys. *142*, 660–664.

KUNTZ, I. D., JR., and KAUZMANN, W. 1974. Hydration of proteins and polypeptides. Advan. Protein Chem. *28*, 239–345.

KUNTZ, I. D., JR., BRASSFIELD, T. S., LAW, G. D., and PURCELL, G. V. 1969. Hydration of macromolecules. Science *163*, 1329–1331.

LABUZA, T. P. 1968. Sorption phenomena in foods. Food Technol. *22*, No. 3, 263–272.

LABUZA, T. P. 1971. Properties of water as related to the keeping quality of foods. *In* Proceedings Third International Congress of Food Science and Technology, Washington, D.C., 618–635.

LABUZA, T. P. 1975. Interpretation of sorption data in relation to the state of constituent water. *In* Water Relations of Foods, R. B. Duckworth (Editor). Academic Press, London.

LEWIN, S. 1974. Displacement of Water and Its Control of Biochemical Reactions. Academic Press, London.

LING, G. N. 1960. A Physical Theory of the Living State. Blaisdell, New York.

LING, G. N. 1972. Hydration of macromolecules. *In* Water and Aqueous Solutions, R. A. Horne (Editor). John Wiley and Sons, New York.

LING, G. N., and WALTON, C. L. 1976. What retains water in living cells? Science *191*, 293-295.

LUMRY, R. 1973. Some recent ideas about the nature of the interactions between proteins and liquid water. J. Food Sci. *38*, 744-755.

MREVLISHVILI, G. M., and PRIVALOV, P. L. 1969. Calorimetric investigation of macromolecular hydration. *In* Water in Biological Systems, L. P. Kayushin (Editor). Consultants Bureau, New York.

NARTEN, A. H., and LEVY, H. A. 1969. Observed diffraction pattern and proposed models of liquid water. Science *165*, 447-454.

NEMETHY, G., and SCHERAGA, H. A. 1962. The structure of water and hydrophobic bonding in proteins. 3. The thermodynamic properties of hydrophobic bonds in proteins. J. Phys. Chem. *66*, 1773-1789.

NEMETHY, G., STEINBERG, I. Z., and SCHERAGA, H. A. 1963. Influence of water structure and of hydrophobic interactions on the strength of side-chain hydrogen bonds in proteins. Biopolymers *1*, 43-69.

NIINIVAARA, F. P., and POHJA, M. S. 1954. Theory of the water binding of meat. Fleischwirtschaft *6*, 192-193.

PENCE, J. W., MOHAMMAD, A., and MECHAM, D. K. 1953. Heat denaturation of gluten. Cereal Chem. *30*, 115-126.

POPLE, J. A. 1951. Molecular association in liquids. 2. A theory of the structure of water. Proc. Roy. Soc. (London) *A205*, 163-178.

RAHMAN, A., and STILLINGER, F. H. 1971. Molecular dynamics study of liquid water. J. Chem. Phys. *55*, 3336-3359.

RAMACHANDRAN, G. N., and SASISEKHARAN, V. 1968. Conformation of polypeptides and proteins. Advan. Protein Chem. *23*, 283-437.

RAMIREZ, J. E., CAVANAUGH, J. R., and PURCELL, J. M. 1974. Nuclear magnetic resonance studies of frozen aqueous solutions. J. Phys. Chem. *78*, 807-810.

RIEDEL, L. 1961. On the problem of bound water in meat. Kältetechnik *13*, 122-128. (German)

RIEDEL, L. 1969. Physical and physical-chemical changes in freezing of foods. Dechema Monograph *63*, 115-125.

RUEGG, M., LUSCHER, M., and BLANC, B. 1974. Hydration of native and rennin coagulated caseins as determined by differential scanning calorimetry and gravimetric sorption measurements. J. Dairy Sci. *57*, 387-393.

SCHACHMAN, H. K. 1963. Considerations on the tertiary structure of proteins. Cold Spring Harbor Symposium on Quantitative Biology *28*, 409-430.

SCHNEIDER, M. J. T., and SCHNEIDER, A. S. 1972. Water in biological membranes: Adsorption isotherms and circular dichroism as a function of hydration. J. Membrane Biol. *9*, 127-140.

SUZUKI, A., NONAMI, Y., and GOLL, D. E. 1975. Proteins released from myofibrils by CASF (Ca^{2+}-activated sarcoplasmic factor) and trypsin. Agr. Biol. Chem. *39*, 1461–1467.

SZENT-GYÖRGYI, A. 1971. Biology and pathology of water. Perspectives Biol. Med. *15*, 239–249.

TAIT, M. J., and FRANKS, F. 1971. Water in biological systems. Nature *230*, 91–94.

TANFORD, C. 1962. Contribution of hydrophobic interactions to the stability of the globular conformation of protein. J. Amer. Chem. Soc. *84*, 4240–4247.

TANFORD, C. 1970. Protein denaturation. Part C. Theoretical models for the mechanism of denaturation. Advan. Protein Chem. *24*, 1–95.

TAYLOR, R. J. 1966. Water. Unilever Educational Booklet, Advanced Series No. 5, Unilever, Ltd., London.

VAZINA, A. A., and LEDNEV, V. V. 1969. Hydrophobic bonds in pepsin. *In* Water in Biological Systems, L. P. Kayushin (Editor). Consultants Bureau, New York.

WALSTRA, P. 1973. Non-solvent water and steric exclusion of solute. Kollid.-Z. u. Z. Polymere *251*, 603–604.

WALTER, J. A., and HOPE, A. B. 1971. Nuclear magnetic resonance and the state of water in cells. Progr. Biophys. Molec. Biology *23*, 3–19.

WARNER, D. T. 1965. A proposed water-protein interaction and its application to the structure of the tobacco mosaic virus particle. Ann. N.Y. Acad. Sci. *125*, 605–624.

WARNER, D. T. 1970. A proposal for the mechanism of muscle contraction at the molecular level. J. Theor. Biol. *26*, 289–313.

WEBB, S. J. 1965. Bound Water in Biological Integrity. Charles C. Thomas, Springfield, Ill.

WETLAUFER, D. B. 1973. Protein structure and stability: Conventional wisdom and new perspectives. J. Food Sci. *38*, 740–743.

WIERBICKI, E., KUNKLE, L. E., CAHILL, V. R., and DEATHERAGE, F. E. 1956. Postmortem changes in meat and their possible relation to tenderness together with some comparisons of meat from heifers, bulls, steers, and diethylstilbestrol treated bulls and steers. Food Technol. *10*, 80–86.

WOESSNER, D. E. and SNOWDEN, B. S., JR. 1970. Pulsed NMR study of water in agar gels. J. Colloid Interf. Sci. *34*, 290–299.

WOESSNER, D. E., SNOWDEN, B. S., JR., and CHIU, Y.-C. 1970. Pulsed NMR study of the temperature hysteresis in the agar-water system. J. Colloid Interf. Sci. *34*, 283–289.

WU, Y. V., and INGLETT, G. E. 1974. Denaturation of plant proteins related to functionality and food applications. A review. J. Food Sci. *39*, 218–225.

4

Cellular Degradation of Proteins

A. Tappel

The objectives of this chapter are severalfold. Firstly, to indicate the level of present knowledge of the system for cellular degradation of proteins and knowledge of the enzymes that constitute this system. Secondly, it is important to indicate the characteristics of this system of enzymes and the characteristics of some individual enzymes that function in this system. Thirdly, it is appropriate to develop the concepts of concerted action of the proteolytic enzymes in degrading proteins. Finally, this chapter will indicate some applications in food science in this relatively new area of knowledge.

CELLULAR DEGRADATION OF PROTEINS

There is belated recognition among the scientific community of the great importance of cellular degradation of proteins. Some general features of cellular degradation of protein are known based upon information available (Schimke 1975; Schimke and Katunuma 1975) from studies of animal tissues, particularly rat liver. Firstly, the degradation is known to be extensive. For example, rat liver nuclear, supernatant, ribosomal and mitochondrial fractions are degraded and resynthesized with a half-life of five days. Parts of the cell that have even faster turnover times include the endoplasmic reticulum and the plasma membrane, with a half-life of two days. Secondly, degradation of proteins is almost entirely intracellular within animal tissues. Thirdly, the degradation rates for individual proteins vary greatly, with documented half-lives from fractions of hours to over five days.

The fourth characteristic of cellular degradation of protein is that the hydrolysis process for each protein appears to be random. It is widely observed that degradation of labeled enzymes *in vivo* follows first order kinetics. This indicates that for like molecules in a pool, the chance of being degraded is random. A fifth group of characteristics relates to the rates of protein degradation, although the rates are affected by many factors in the cell. The main factor that affects the degradation rate is the molecular weight; it has been observed that large proteins are degraded faster. Sensitivity to proteases is also a function of molecular weight, with the larger proteins being more sensitive.

ENZYMES OF LYSOSOMES AND PROTEIN DIGESTION

Although the focus of this review is specifically on lysosomal proteases and peptidases, it is important to point out a few general characteristics of lysosomes and their enzymes. Lysosomes are a nearly universal cell organelle that contain hydrolytic enzymes and carry out the digestion of the polymeric and complex chemical structures of the cell. Lysosomes are characterized by their content of over 50 enzymes (Barrett 1969; Tappel 1969), some of which have the concerted capability of hydrolyzing proteins and peptides, including glycoproteins and proteoglycans. Also, the lysosomes have a complete enzyme system for hydrolyzing a variety of carbohydrates, lipids and nucleic acids.

To describe the protein digestive enzymes, it is important to introduce the concept of systems of protein digestive enzymes. An example, shown in Table 4.1, is the well-known system involving enzymes for the digestion of proteins taken in as food. In the left-hand side of the table are a number of these enzymes that are very well known. The system action takes place in a number of anatomic locations, including the stomach, production of some enzymes in the pancreas and their function in the small intestine, and further function of other enzymes in the mucosal cells of the intestine. The system is further characterized by a mixture of protease and peptidase activities and is further complicated by the pH differences in the different anatomic regions.

TABLE 4.1
PROTEIN DIGESTIVE ENZYMES

Stomach	Lysosomes
Pepsin	Cathepsin D
Pancreas→Sm. intestine	Cathepsin B1
Trypsin	Dipeptidyl aminopeptidase I
Chymotrypsin	Cathepsin B2
Carboxypeptidase A	Cathepsin A
Carboxypeptidase B	Catheptic carboxypeptidase B
Elastase	Catheptic carboxypeptidase C
Collagenase	Catheptic carboxypeptidase G
Mucosal cells	Other proteases and peptidases
Aminopeptidases	
Dipeptidases	

The enzymes pepsin, trypsin and chymotrypsin are well-known endoproteases (Desnuelle *et al.* 1970); it is a characteristic of protein digestive systems to have one or more endoproteases. These are usually complemented and supplemented by a variety of peptidases, of which there are the pancreatic carboxypeptidases and also, in the mucosal cell, the aminopeptidases. Some of the enzyme complement of the mucosal cell, in fact, might be lysosomal.

Now to look at the lysosomal digestive system, which by analogy has been referred to as the cell's gastrointestinal tract. In common with the protein digestive system of the stomach and intestine, the lysosomes have a characteristic mixture of endoproteases and peptidases (Vaes 1973) shown on the right-hand side of Table 4.1. The main endoprotease is cathepsin D, which, in fact, is an acid protease similar to pepsin. Another endoprotease is cathepsin B1. These two endoproteases would normally initiate the digestion of proteins and their action would be supplemented by the peptidases. Lysosomes have a powerful aminopeptidase and a large variety of carboxypeptidases, among which are the more active enzymes, cathepsin B2 and cathepsin A, and the less active and less well-known activities of the catheptic carboxypeptidases B, C and G. Furthermore, the lysosomal protein digestive system interacts with other proteases and peptidases that are not yet well defined. In the pH function there appears to be some analogy between the food protein digestive system, which starts at low pH in the stomach and goes to a more neutral pH in the intestine, and the pH cycle that might be characteristic of the lysosome.

LYSOSOMAL PROCESSES

Lysosomal processes can best be explained through a figure map that diagrammatically indicates the processes involved (Gordon

1973; Mego 1973). Figure 4.1 portrays lysosomal processes (Vaes 1973). The enclosed area is meant to indicate the cell boundaries. Components inside brackets are those normally recognized as processes of the cellular organelle, lysosomes, and these processes are separated from the cytoplasm by a single membrane. Here we will define these processes mainly with relationship to protein degradation.

Starting at the upper right is the topic of lysosomal enzyme synthesis and formation of primary lysosomes. The synthesis rate is a function of the particular tissue and the metabolic need of that tissue. The need may be driven by starvation or a metabolic action may be controlled by hormones. The lysosomal enzymes are synthesized, as are all proteins, on the ribosomes of the endoplasmic reticulum, and the enzymes flow into a packaging area of the endoplasmic reticulum called the Golgi region. Here, the collection of 50 or more enzymes is packaged as primary lysosomes, which can be recognized through electron microscopy as a cell organelle. When lysosome synthesis is turned on, the whole complement of lysosomal enzymes characteristic of that cell is produced and put into an operational primary lysosome. The complete complement of lysosomal enzymes characteristic of this particular cell is synthesized even though protein may be the only substrate available.

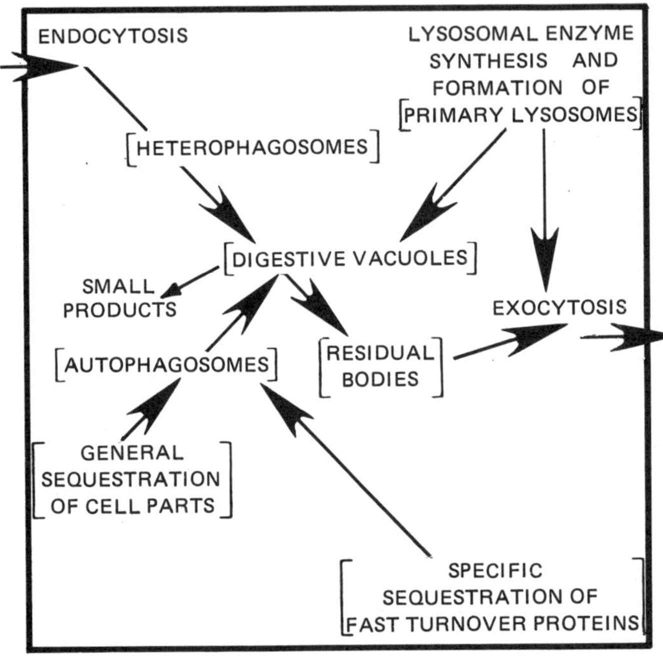

FIG. 4.1. LYSOSOMAL PROCESSES

The main digestive process of lysosomes is initiated when primary lysosomes merge with heterophagosomes or autophagosomes to begin a digestive process in another variant of the lysosomal organelle, the digestive vacuoles.

Next, we can focus on the processes that lead to hydrolysis of extracellular protein. Extracellular protein and other material being brought into the cell for hydrolytic processing comes in through blebing of the plasma membrane and by an intake process called endocytosis. This gives rise to organelles that can be recognized as heterophagosomes, which are united with primary lysosomes to produce digestive vacuoles in which the protein and other material brought into the cell can be digested.

Another important process, the autophagocytic process, is schematically indicated at the bottom left of Fig. 4.1. The main process is general sequestration of cell parts prerequisite to their digestion. The control of this general sequestration is not completely understood but what is recognized is that cell parts, including mitochondria, endoplasmic reticulum and cytosol, are surrounded by single membranes forming organelles recognized as autophagosomes. These in turn unite with primary lysosomes to initiate digestive processes. Of even greater current interest is the more specific sequestration of proteins and enzymes of the cell that are hydrolyzed at faster rates. Again, the initial processes by which these specific enzymes and proteins are recognized in the cell and by which the processing is initiated are not known. Present information is covered in a recent review on intracellular protein turnover edited by Schimke and Katunuma (1975). The second stage in the processing of these proteins, after their biochemical recognition and initial processing, is again the formation of autophagosomes and subsequent digestion.

Now, turning attention to the processes of the main digestive vacuoles as schematically indicated in the center of Fig. 4.1, when primary lysosomes merge with heterophagosomes or autophagosomes or various combinations thereof, the main lysosomal process of digestion is initiated. This process proceeds in a batch operation upon initiation of the digestive vacuole, which is recognized as a cell organelle variant of lysosomes. In the beginning there is apparently production of hydrogen ions to reduce the pH and to activate the main endoprotease cathepsin D, which has a pH optimum of 3.5. As this digestion process proceeds in batch operation, there is evidence that the pH rises to neutrality thus favoring the action of other lysosomal proteases and particularly the peptidases. During this digestive batch cycle small molecular weight products, mainly amino acids and dipeptides and other small molecules, can transfer across the single membrane of the digestive vacuole and enter the cell cytosol for utilization by other cell processes. After one or more batch digestive processes, there remain in the lysosomes enzymes that are not digested by their own protease-peptidase system, and any other indigestible material. These can be recognized as residual bodies and are removed from the cell by the process of exocytosis.

PROTEIN HYDROLYSIS BY LYSOSOMAL PROTEASE-PEPTIDASE SYSTEM

As indicated above, the process of protein hydrolysis by the lysosomal protease-peptidase group of enzymes (Tappel 1969; Vaes 1973) needs to be considered by a systems approach. The concerted action of these lysosomal protease-peptidases is indicated in Fig. 4.2, which constitutes a flow chart or map of the process. From left to right, and in decreasing order of molecular weight, are the commonly recognized biochemical components: proteins, polypeptides, oligopeptides, tripeptides, dipeptides and amino acids. To indicate the present knowledge of how each enzyme is involved, this flow chart or map is useful.

This mapping of enzyme reactions can be illustrated firstly with cathepsin D, the main acid protease of lysosomes. This enzyme mainly processes proteins as reactants to produce various polypeptides and oligopeptides; this is shown in the double line. That the enzyme can also process polypeptides and oligopeptides, but usually does so in smaller amounts, is indicated by the single line of these reactants. Further, that cathepsin D will produce some tripeptides and smaller amounts of dipeptides and amino acids is indicated by the single line and the dashed line.

By use of flow charting or mapping we can illustrate the present knowledge of the lysosomal enzymes in terms of their main and secondary reactants and their production of main, secondary and tertiary products. Thus, cathepsin B1, another major endoprotease of lysosomes, has a similar action in terms of reactants and products as does cathepsin D. Namely, it mainly hydrolyzes proteins and to a smaller extent polypeptides and oligopeptides with the production of polypeptides and oligopeptides as main products, and with some tripeptides, dipeptides and amino acids as secondary products.

Now, having considered the two main endoproteases, we can move on to the next sequential part of protein hydrolysis, which goes on in concert with the above; namely, further processing by peptidases of the products of protease action. Another enzyme of high specific activity in the lysosomes, classically recognized as cathepsin C but more appropriately called dipeptidyl aminopeptidase I, normally processes polypeptides and oligopeptides as major reactants to produce dipeptides. Another major peptidase, carboxypeptidase cathepsin B2, is capable of processing various peptides and also proteins. Its main reactants appear to be polypeptides, oligopeptides and tripeptides and the main products are amino acids, which are cleaved off in sequential manner from the carboxyl end of these reactants. In a similar manner the carboxypeptidase, cathepsin A,

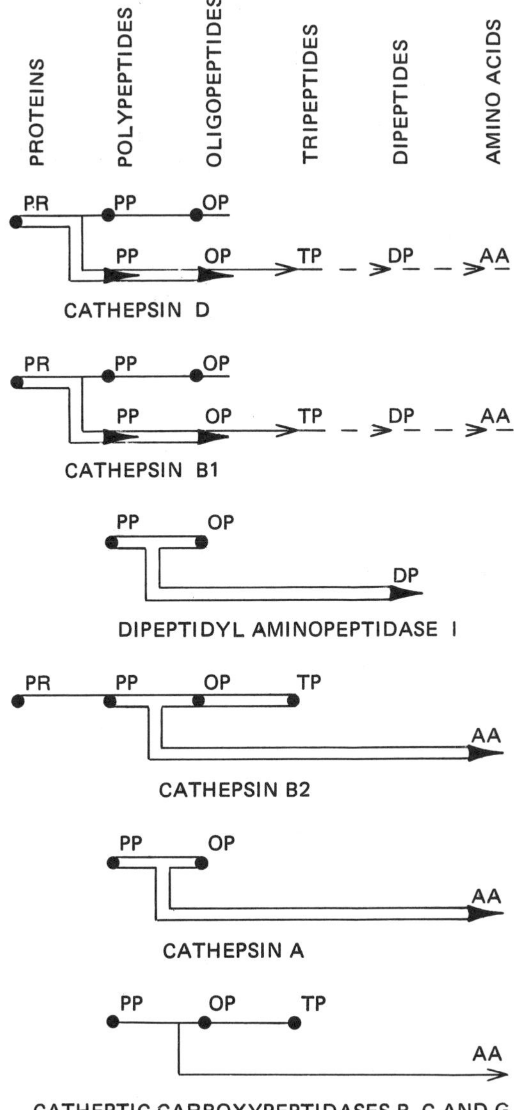

FIG. 4.2. CONCERTED PROTEIN HYDROLYSIS BY LYSOSOMAL PROTEASE-PEPTIDASE SYSTEM

appears to process polypeptides and oligopeptides as the main reactants to produce amino acids. Among other known carboxypeptidases are catheptic carboxypeptidases B, C and G, which may be minor peptidases in activity; they process polypeptides, oligopeptides, and tripeptides to produce amino acids.

This flow charting or mapping of five well-known lysosomal enzymes, which indicates also the probable reaction of other enzymes, seems to provide a reasonable perspective of the lysosomal processes as known to date. But as all of the protease-peptidase system and its interaction is not known, the complete story awaits further research.

CATHEPSIN D AND ITS MEASUREMENT

Cathepsin D (Barrett 1969; Barrett and Dingle 1971; Vaes 1973) is the major acid endoproteinase of lysosomes and many animal tissues. It acts on many proteins. As with many proteinases and peptidases, its specificity has been defined using oxidized insulin B chain, a section of which is shown.

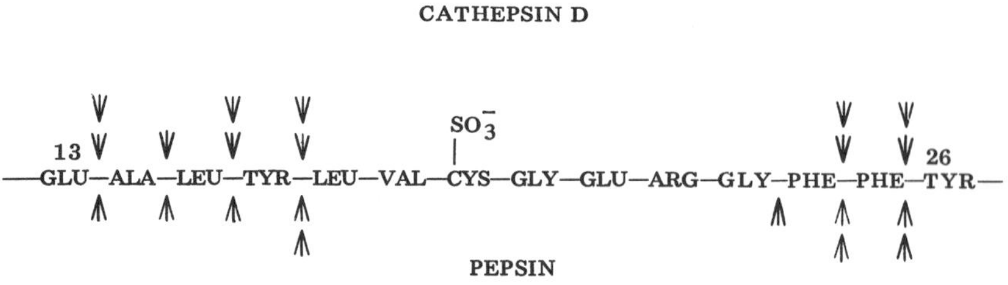

CATHEPSIN D

$$13 \quad \text{—GLU—ALA—LEU—TYR—LEU—VAL—CYS—GLY—GLU—ARG—GLY—PHE—PHE—TYR—} \quad 26$$

with SO_3^- on CYS

PEPSIN

Cathepsin D cleaves the peptide chain of oxidized insulin mainly at the carboxyl side of aromatic amino acids or at some other amino acids with longer side chains, such as leucine. The double arrows on top of the figure show the main points of splitting by cathepsin D, and the double arrows on the bottom show the main points of splitting by pepsin. The single arrows in each case indicate minor points of hydrolytic splitting. Cathepsin D has a specificity very similar to that of pepsin. Cathepsin D is characterized by a very low pH optimum of 3.5. In contrast to many other lysosomal enzymes, it does not require sulfhydryl activation and is not inhibited by sulfhydryl blocking reagents.

A variety of methods have been developed for measurement of activity of cathepsin D or other proteases, and these methods have wide application (Tappel 1972). Hemoglobin is the substrate of choice, as cathepsin D and many other proteases are very active against hemoglobin. For cathepsin D, acid or urea denatured hemoglobin is typically digested at pH 3.8 and 37°C for 10–30 min.

Then, the peptides and amino acids produced are treated with trichloroacetic acid to allow precipitation of remaining hemoglobin and large fragments of hemoglobin. The products of hydrolysis may be measured by the 280 nm absorbance of tryptophan, the fluorescence of tryptophan (280 nm excitation, 350 nm emission), the biuret-phenol reaction using the Lowry procedure, or the ninhydrin reaction.

A sensitive fluorometric assay for cathepsin D and other proteases has been developed (deLumen and Tappel 1970). The method is based upon a hemoglobin substrate modified by being linked with fluorescein isothiocyanate. Trichloroacetic acid soluble hydrolysis products are diluted with 0.1 N NaOH and the fluorescence is read at 515 nm with excitation at 490 nm. The method satisfies the conditions for linearity of reaction velocity with incubation time and for proportionality with enzyme concentration. The main advantages of the assay are high sensitivity and short time requirement.

OTHER DOMINANT CATHEPTIC ENZYMES

Dipeptidyl aminopeptidase I (cathepsin C) (McDonald *et al.* 1971) is the major aminopeptidase of lysosomes and is usually present in high activity in many lysosomes and in many animal tissues. In our research (Liao-Huang and Tappel 1972), rat liver cathepsin C was purified 1790-fold over the homogenate activity by a purification scheme that combined the techniques of centrifugation, acid precipitation, $(NH_4)_2 SO_4$ precipitation, and Sephadex G-200, DEAE-cellulose, and CM-cellulose column fractionations. The specific activity of the most purified CMC-fraction with Gly-Tyr-NH$_2$ as substrate was 92 μmoles of substrate hydrolyzed per mg of protein per min, which is higher than previously reported values. The K_m for Gly-Tyr-NH$_2$ was 6 mM when determined by hydroxamate formation, and that for Val-Leu-NH$_2$ was 2.5 mM.

The requirements for -SH compounds and for halide ions for activity of dipeptidyl aminopeptidase I were absolute. A dithioerythritol concentration of 25 mM and a Cl$^-$ or Br$^-$ concentration of 10 mM were required for full activation. The enzyme catalyzed a hydrolysis reaction at pH 5, but the polymerization reaction catalyzed at pH 7–8 occurred at a greater rate. The substrate specificity of cathepsin C was broader than that reported in older literature and the enzyme was confirmed to be dipeptidyl aminopeptidase that acts upon a number of dipeptide amides and tripeptides. The most favorable substrates were tripeptides or dipeptide amides that have as the NH$_2$-terminal group a small residue and that have the aliphatic residue leucine at the penultimate position.

In recent research (Ninjoor *et al.* 1974), cathepsin B2 was purified 3900-fold over homogenate with respect to the hydrolysis of N-benzoyl-L-arginine amide (Bz-Arg-NH$_2$). Cathepsin B2 was shown to be a relatively nonspecific carboxypeptidase by virtue of its hydrolysis of N-blocked dipeptides. The optimum hydrolysis of Bz-Arg-NH$_2$ and N-blocked dipeptides by cathepsin B2 occurred at pH 5.5–5.6, with K_m values in the range of 10–15 mM. A tetrapeptide (Leu-Trp-Met-Arg), a hexapeptide (Leu-Trp-Met-Arg-Phe-Ala), and glucagon were hydrolyzed by cathepsin B2 with the release of amino acids from the carboxyl terminus. Oxidized insulin A chain and oxidized insulin B chain were not hydrolyzed by cathepsin B2, perhaps owing to the presence of cysteic acid or proline residues near the carboxyl terminus. No endopeptidase activity of cathepsin B2 was found. Cathepsin B2 was activated by sulfhydryl compounds.

Highly purified rat liver cathepsin A1 was shown (Taylor and Tappel 1974A, B) to be a carboxypeptidase with a high degree of specificity for aromatic and hydrophobic amino acids. In lysosomal soluble fractions, cathepsin A1 was partially responsible for the hydrolysis of the model substrates, N-benzyloxycarbonyl (Cbz)-Gly-Phe, Cbz-Glu-Phe, and Ac-Phe-Tyr. N-blocked dipeptides that contain an aromatic or hydrophobic amino acid were hydrolyzed. The optimum hydrolysis of N-blocked dipeptides by cathepsin Al was at pH 5.0–5.8, with K_m values in the range of 1.2–54 mM. The hexapeptide Leu-Trp-Met-Arg-Phe-Ala, glucagon, and oxidized insulin B chain were partially hydrolyzed by cathepsin A1 with the release of amino acids from the C-terminal end. These peptides all contain hydrophobic portions near the C-terminal end. Less hydrophobic peptides, such as oxidized insulin A chain and His-Ser-Gln-Gly-Thr-Phe, were not hydrolyzed. Very little hydrolysis of denatured hemoglobin was found. Cathepsin A1 was activated by halide ions and inhibited by Ag$^+$ and Hg^{2+}.

Four distinct carboxypeptidases were shown (Taylor and Tappel 1974A, B) to exist in rat liver lysosomes: cathepsin A, catheptic carboxypeptidase B, catheptic carboxypeptidase C, and catheptic carboxypeptidase G. These carboxypeptidases were distinguished on the basis of substrate specificity, thiol requirement, and separation by Sephadex G-100 column chromatography. The K_m and pH optimum for each carboxypeptidase were determined. From the Sephadex G-100 elution profile, two peaks of cathepsin A activity were found. The major cathepsin A peak, cathepsin A1, was further purified by ion-exchange chromatography on DEAE-cellulose. The highly purified cathepsin A1 fraction hydrolyzed N-benzyloxycarbonyl (Cbz)-Glu-Phe, Cbz-Gly-Phe, and Ac-Phe-Tyr, which suggests that these hydrolytic activities were due to the same enzyme. Catheptic carboxypeptidase B and catheptic carboxypeptidase G eluted from the Sephadex G-100 column along with cathepsin B2. Cathepsin B2 had carboxypeptidase activity. The purification of the lysosomal carboxypeptidases over homogenate activities was: cathepsin A1, 1280-fold; catheptic carboxypeptidase B, 220-fold; and catheptic carboxypeptidase G, 920-fold. Catheptic carboxypeptidase C was purified 16-fold over light-mitochondrial supernatant activity.

CONCERTED ACTION OF CATHEPSINS

The synergistic and concerted action of cathepsins A, B and D were studied (Goettlich-Riemann *et al.* 1971) after tissue fractionation and partial purification of the enzymes by chromatography on DEAE-Sephadex. Table 4.2 shows the digestion of hemoglobin is greater when cathepsins A, B and D are all present than when cathepsin D and A act alone or in combination. After 24 hr of incubation, hydrolysis of oxidized B chain of insulin was also greatest in the presence of all three of the cathepsins; however, there was less hydrolysis by cathepsin D alone than by cathepsin A or A and B.

TABLE 4.2
CONCERTED ACTION OF CATHEPSINS D, A, AND B[1]

Cathepsins	nmoles Products	
	Hemoglobin, 0.5 Hr	Oxidized Insulin B Chain, 24 Hr
D	26	86
A	5	115
A + B	8	157
D + A	46	212
D + A + B	68	230

[1]Purified cathepsins D, A and B, in ratios similar to those found in lysosomes, digested 0.2 mg protein/ml at pH 3.8 and 37°C. Products were analyzed by the biuret-phenol reaction.

The results of experiments with cathepsins A, B and D show that digestion of hemoglobin and oxidized B chain of insulin is much higher when all three enzymes are present than when cathepsin D acts alone or cathepsin D and A act together. The difference shown between the action of cathepsins A and D on these substrates is interesting. Cathepsin D has a greater affinity for large protein substrates than does cathepsin A; therefore, more hemoglobin was digested by cathepsin D during the 30 min incubation period than by cathepsin A. During the 24 hr incubation used in assay of hydrolysis of oxidized B chain of insulin, more amino acids were removed by the carboxypeptidase cathepsin A than by the endopeptidase cathepsin D.

Partially purified cathepsin D and cathepsin C were used in combination with activator and inhibitor to investigate (Liao-Huang and Tappel 1971) the hydrolysis of hemoglobin. The chromatographic analysis of hydrolysis products is shown in Fig. 4.3. In the presence of iodoacetamide, which inhibits cathepsin C, acid-denatured hemoglobin was only slightly hydrolyzed, owing to the action of cathepsin D. This is indicated by the areas under the curve for cathepsin D. In the presence of Cleland's reagent, which activates cathepsin C, acid-denatured hemoglobin was extensively hydrolyzed. Cathepsin C increased the extent of proteolysis over that obtained with cathepsin D alone, as shown by the area under the curve for cathepsin D + C. Oligopeptides produced by prior action of cathepsin D were degraded by cathepsin C to smaller oligopeptides, dipeptides, and free amino acids.

APPLICATIONS IN FOOD SCIENCE

Many of the studies in the field of cellular degradation of proteins have been done basically to understand the biology and to a lesser extent the enzymology of the process. Investigations of cell biology orientation have been dominant. Biochemical investigation in this field has not been commensurate with its apparent importance. Applications in this field have been dominated by those of medical direction, especially those related to the involvement of lysosomes in pathology and in the lysosomal diseases. Other studies of cellular

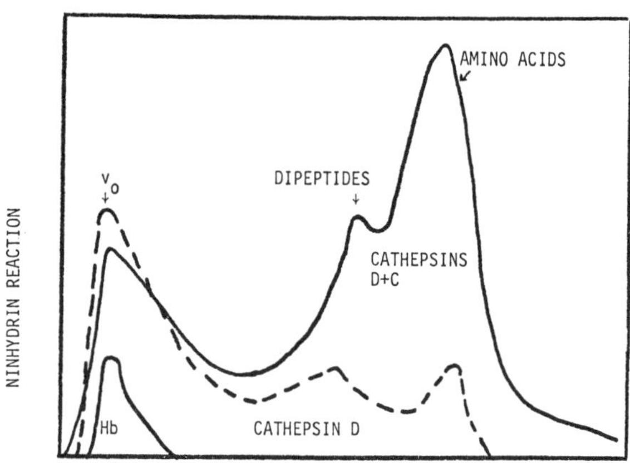

FIG. 4.3. CHROMATOGRAPHIC ANALYSIS OF THE HYDROLYSIS PRODUCTS OF CATHEPSIN D AND D+C

degradation of proteins have been aimed at understanding other phenomena in biology, with studies of cellular degradation of proteins being given secondary importance.

Applications of this newer knowledge of the pathways and enzymes of cellular degradation of proteins in food science (Tappel 1966) have not been large. Thus, only a few examples can be cited here. The main textbooks and reference books (Reed 1975; Whitaker 1972) on enzymes in food science indicate to the food science student and professional that the main concerns in proteolytic enzymes are not about the cathepsins. Of major economic importance are applications of proteolytic enzymes from other sources in solution of food processing problems.

A few papers have been directed toward understanding the role of cathepsins in autolytic deterioration of muscles of aquatic animals. Geist and Crawford (1974) of the Seafoods Laboratory, Astoria, Oregon, studied cathepsins of the muscle of Pacific sole. They established that overall catheptic activity had optimum activity at pH 3.5, indicating that the autolysis was dominated by cathepsin D. Their findings were similar to those of a number of other investigators who have studied various fishes. Whiting *et al.* (1975) of the Food Science Department, Oregon State University, studied the stability of trout muscle lysosomes related to the problem of rigor mortis of that muscle. They found that catheptic activity, mainly cathepsin D, was released when the muscle was held at 4°C. Thus, the lysosomal proteases and other systems would be available for degrading the muscle. Eitenmiller (1974) of the Food Science Department, University of Georgia, studied cathepsin activity of shrimp muscle. It was found that muscle had cathepsin D activity but not cathepsins A, B and C. This is an example of use of specific assays which are available for the various catheptic enzymes.

There are a few studies of catheptic activities in bovine muscle. Laakkonen (1973) has reviewed some of the research of the 1960's on the role of cathepsins in meat tenderness. Lutalo-Bosa and Macrae (1969) of the Animal Science Department, McGill University, studied catheptic enzymes of bovine muscle and compared the activities to those in rat liver. The enzymes found in bovine muscle as percentage of activity in that muscle compared to rat liver were: cathepsin D, 25%; cathepsin C, 2%; and cathepsin B, 52%. In another study related to postmortem action of bovine muscle, West *et al.* (1974) of Texas A&M studied loss of calcium-accumulating ability in the sarcoplasmic reticulum following degradation by cathepsins.

What are the prospects for future research in cellular degradation of proteins in food science? There are many constraints in this topic area. There is a small number of investigators of enzyme action of importance in food science. This is a small area when one considers the many pressing problems in food science. There is not much research funding available for this type of research. From this, we are led to conclude that this newer knowledge of proteolytic enzymes responsible for cellular degradation of proteins will have only modest effects in the applications in food science areas.

BIBLIOGRAPHY

BARRETT, A. J. 1969. Properties of lysosomal enzymes. *In* Lysosomes in Biology and Pathology, J. T. Dingle and H. B. Fell (Editors). North-Holland Publishing Co., Amsterdam.

BARRETT, A. J., and DINGLE, J. T. 1971. Tissue Proteinases. North-Holland Publishing Co., Amsterdam.

DE LUMEN, B. O., and TAPPEL, A. L. 1970. Fluorescein-hemoglobin as a substrate for cathepsin D and other proteases. Anal. Biochem. *36*, 22–29.

DESNUELLE, P., NEURATH, H., and OTTESEN, M. 1970. Structure-Function Relationships of Proteolytic Enzymes. Academic Press, New York.

EITENMILLER, R. R. 1974. Cathepsin activity of Penaeus setiferus muscle. J. Food Sci. *39*, 6–9.

GEIST, G. M., and CRAWFORD, D. L. 1974. Muscle cathepsins in three species of Pacific sole. J. Food Sci. *39*, 548–551.

GOETTLICH-RIEMANN, W., YOUNG, J. O., and TAPPEL, A. L. 1971. Cathepsins D, A and B, and the effect of pH in the pathway of protein hydrolysis. Biochim. Biophys. Acta *243*, 137–146.

GORDON, A. H. 1973. The role of lysosomes in protein digestion. *In* Lysosomes in Biology and Pathology, J. T. Dingle (Editor). North-Holland Publishing Co., Amsterdam.

LAAKKONEN, E. 1973. Factors affecting tenderness during heating of meat. *In* Advances in Food Research, Vol. 20, C. O. Chichester, E. M. Mrak, G. F. Stewart (Editors). Academic Press, New York.

LIAO-HUANG, F., and TAPPEL, A. L. 1971. Action of cathepsins C and D in protein hydrolysis. Biochim. Biophys. Acta *236*, 739–748.

LIAO-HUANG, F., and TAPPEL, A. L. 1972. Properties of cathepsin C from rat liver. Biochim. Biophys. Acta *268*, 527–538.

LUTALO-BOSA, A. J., and MACRAE, H. F. 1969. Hydrolytic enzymes in bovine skeletal muscle. 3. Activity of some catheptic enzymes. J. Food Sci. *34*, 401–404.

McDONALD, J. K., CALLAHAN, P. X., ELLIS, S., and SMITH, R. E. 1971. Polypeptide degradation by dipeptidyl aminopeptidase I (Cathepsin C) and related peptidases. *In* Tissue Proteinases, A. J. Barrett, and J. T. Dingle (Editors). North-Holland Publishing Co., Amsterdam.

MEGO, J. L. 1973. Protein digestion in isolated heterolysosomes. *In* Lysosomes in Biology and Pathology, J. T. Dingle (Editor). North-Holland Publishing Co., Amsterdam.

NINJOOR, V., TAYLOR, S. L., and TAPPEL, A. L. 1974. Purification and characterization of rat liver lysosomal cathepsin B2. Biochim. Biophys. Acta *370*, 308–321.

REED, G. 1975. Enzymes in Food Processing. Academic Press, New York.

SCHIMKE, R. T. 1975. On the properties and mechanisms of protein turnover. *In* Intracellular Protein Turnover, R. T. Schimke and N. Katunuma (Editors). Academic Press, New York.

SCHIMKE, R. T., and KATUNUMA, N. 1975. Intracellular Protein Turnover. Academic Press, New York.

TAPPEL, A. L. 1966. Lysosomes: Enzymes and catabolic reactions. *In* Physiology and Biochemistry of Muscle as a Food, E. J. Briskey, R. G. Cassens, and J. C. Trautman (Editors). University of Wisconsin Press, Madison.

TAPPEL, A. L. 1969. Lysosomal enzymes and other components. *In* Lysosomes in Biology and Pathology, J. T. Dingle and H. B. Fell (Editors). North-Holland Publishing Co., Amsterdam.

TAPPEL, A. L. 1972. Automated methods for measuring lysosomal enzymes. *In* Lysosomes, J. T. Dingle (Editor). North-Holland Publishing Co., Amsterdam.

TAYLOR, S. L., and TAPPEL, A. L. 1974A. Characterization of rat liver lysosomal cathepsin Al. Biochim. Biophys. Acta *341*, 112–119.

TAYLOR, S. L., and TAPPEL, A. L. 1974B. Identification and separation of lysosomal carboxypeptidases. Biochim. Biophys. Acta *341*, 99–111.

VAES, G. 1973. Digestive capacity of lysosomes. *In* Lysosomes and Storage Diseases, H. G. Hers and F. Van Hoof (Editors). Academic Press, New York.

WEST, R. L., MOELLER, P. W., LINK, B. A., and LANDMANN, W. A. 1974. Loss of calcium accumulating ability in the sarcoplasmic reticulum following degradation by cathepsins. J. Food Sci. *39*, 29–31.

WHITAKER, J. R. 1972. Principles of Enzymology for the Food Sciences. Marcel Dekker, New York.

WHITING, R. C., MONTGOMERY, M. W., and ANGLEMIER, A. F. 1975. Stability of rainbow trout (Salmo gairdneri) muscle lysosomes and their relationship to rigor mortis. J. Food Sci. *40*, 854–857.

5

Protein Separation and Analysis by Electrophoretic Methods: An Overview of Principles

Nicholas Catsimpoolas

Study of a food protein at the molecular level can be initiated only after the species of interest has been separated from all other components in the mixture. To accomplish this task, biochemists utilize certain properties of the protein involving primarily electrical charge, size, and more recently biological activity (Catsimpoolas 1975; Morris and Morris 1976). Thus, the techniques of gel filtration (Determann 1968; Fisher 1969; Ackers 1975), ion exchange (Peterson 1970), adsorption (Bernardi 1971), and bio-affinity chromatography (Lowe and Dean 1974; Porath and Kristiansen 1975; Parikh and Cuatrecasas 1975) have been used extensively for the separation of individual proteins from complex mixtures. However, chromatographic methods exhibit relatively low resolution when applied to macromolecules, *i.e.*, a very large difference in size and charge is required to separate two species. Therefore, gel filtration and ion exchange chromatography are usually employed as preliminary steps in a fractionation scheme to provide material for further separation and analysis. Bioaffinity chromatography offers excellent resolution, but the limitation exists that the particular protein of interest should have some biological activity (*e.g.*, enzymatic, hormonal, etc.) or bear prosthetic groups amenable to this type of separation. In the case of most food proteins, antibodies can be prepared which will recognize the corresponding antigens and, thus, biological activity is established indirectly (Catsimpoolas 1976, 1977). However, the preparation of a monospecific antibody requires the availability of a pure protein. Thus, if small amounts of protein can be isolated by other means and antibodies are produced, bioaffinity (immuno-adsorption) chromatography can be used to obtain larger amounts.

Electrophoretic techniques offer very high resolution in the separation of proteins and in addition provide physicochemical parameters pertaining to their characterization in terms of isoelectric point and molecular weight. The differential migration of proteins in an electric field depends on the magnitude of their surface charge density at a particular pH. The ionic groups that are involved are α, β, and γ carboxylic, α and ϵ amino, imidazole, sulfhydryl, phenoxy, and guanidinium.

The behavior of a protein subjected to electrophoresis is governed by its particular *pH vs mobility* curve. At the isoelectric point (pI) the positive charges equal the negative and the protein exhibits zero net charge; therefore, no migration takes place upon the application of an electric field. At any pH value above the pI, the protein has a negative net charge and will migrate toward the positive electrode. The opposite occurs at pH values lower than the pI where the positive net charge of the protein leads to movement toward the negative electrode. By the use of appropriate buffer systems, proteins can be focused at their isoelectric point (isoelectric focusing), migrate with different velocities at a certain pH (electrophoresis), or be stacked next to each other in sequence according to their constituent mobilities (isotachophoresis).

Combination of the above techniques is not only possible, but often desirable in order to achieve higher resolution. Furthermore, since electrophoresis is usually carried out in a supporting medium such as a porous gel which imposes a "sieving" effect on the macromolecules, size differences among various proteins can be utilized for separation either synergistically or independently of charge. In the latter case, the electric charge is utilized only as a means to achieve migration through a porous gel. This is accomplished by interaction of the proteins with ionic detergent molecules such as sodium dodecyl sulfate (SDS) which equalize the charge density on the protein surface. Under these conditions the electrophoretic velocity is totally dependent on the retardation coefficient (K_R) of the protein which is related only to its size and shape.

CONTINUOUS-pH ZONE ELECTROPHORESIS

The term continuous-pH zone electrophoresis (CZE) denotes a
system in which the pH of the buffer is the same in both the
positive and negative electrolyte reservoirs, the sample zone, and
the separation path (Fig. 5.1). Ideally, the pH of the system should

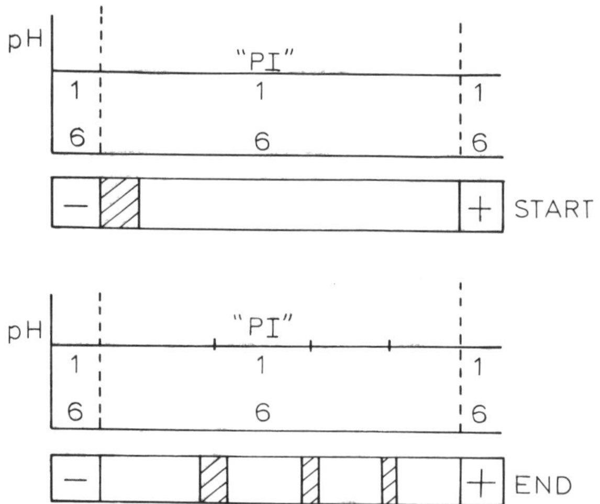

FIG. 5.1. STAGES OF CONTINUOUS-pH ZONE ELECTRO-
PHORESIS (CZE)

remain constant during the electrophoretic run. Zone separation is
achieved among proteins which differ considerably in terms of
charge density and therefore electrophoretic mobility at a specified
pH. Since, for a mixture of proteins the pH-mobility curve for
each component is usually unknown, one cannot predict the pH of
optimum separation for the components of interest. Practically,
various buffer systems covering a wide range of pH values have to
be tried until the desirable results are obtained. At extreme pH
values—either basic or acidic—most proteins are maximally charged
and thus migrate relatively faster than at pHs nearer the pI.
However, possible denaturation, dissociation and association effects
must be kept in mind in interpreting the data. Furthermore, charge
differences among various proteins may not be maximal at extreme
pH.

A large number of buffer systems (>4200) are available for CZE
with exact description of pH, ionic strength, conductance and
mobility relative to the sodium ion (Chrambach *et al.* 1976). These
operate at either 0°C or 25°C. They are prepared according to
formulations described in the *Multiphasic Buffer Systems Output*
from constituents 1 and 6 (PI phase) (see Appendix). The buffer

systems are generated by a computer program based on the multi-phasic zone electrophoresis theory using 45 ionic and buffer constituents for which pK and ionic mobility data are available. The buffers cover the pH range 2.5-11 in 0.5 pH unit intervals from positively and negatively charged species. The computer output and instructions for its use are available from the National Technical Information Service (NTIS). A few selective buffer systems have been described by Chrambach *et al.* (1976) which are sufficient to familiarize the reader with the system.

The most commonly practiced forms of CZE employ cellulose acetate (Chin 1970) and agarose gel (Wieme 1965) as supporting media. Although these media require simple apparatus and procedures, the resolution is very low because of the practical inability to achieve an ultra-thin starting zone and the presence of diffusion spreading during migration. The utilization of starch and poly-acrylamide gels in CZE (Gordon 1969; Maurer 1971) offers considerable improvement in resolution primarily due to the sieving effects of these gels.

An indirect means of improving resolution in agarose gel or cellulose acetate CZE is the employment of immunodiffusion methods (Crowle 1973) in combination with electrophoresis. The technique of immunoelectrophoresis (Grabar and Burtin 1964; Clausen 1969) involves CZE of a mixture of proteins which after separation diffuse against a specific antiserum originally placed in a trough cut parallel to their migration path. The antigenic components interact with their corresponding antibodies forming specific immunoprecipitin arcs. A semiquantitative estimation of each antigenic component can be performed by the technique of electroimmunodiffusion (Laurell 1972; Axelsen *et al.* 1973). The antigenic proteins after separation by electrophoresis in one direction are subsequently forced to migrate electrophoretically in a perpendicular direction and into a gel containing uniformly distributed antiserum. Upon interaction, a series of loops are formed specific for each protein the height of which gives an indication of the relative concentration of the protein under standardized conditions.

ISOTACHOPHORESIS

Isotachophoresis (ITP) involves the separation of proteins into consecutive zones arranged in the order of their constituent mobilities (Fig. 5.2) (Catsimpoolas 1973A, B; Righetti 1976). The zones are in immediate contact with each other and exhibit different pH and temperature in a constant current electrical field. The concentration of material in the zones is adjusted to the

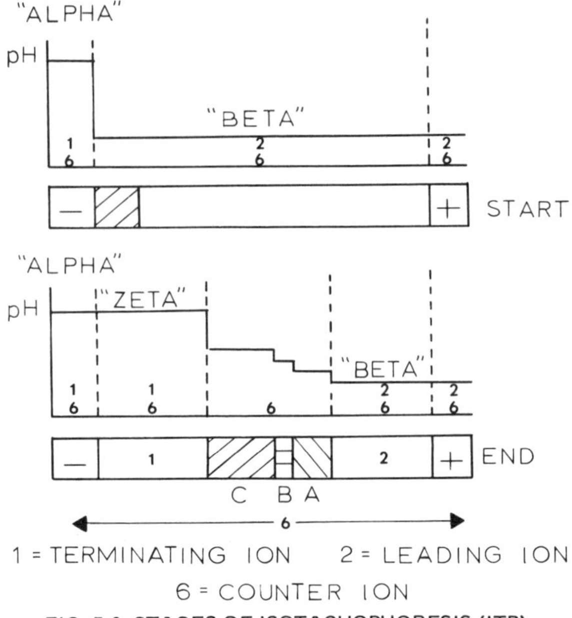

FIG. 5.2. STAGES OF ISOTACHOPHORESIS (ITP)

concentration of the first ion zone, called the leading electrolyte, which must contain the leading ion having the same sign as the sample ions to be separated and a higher mobility than the other ions in the system. The terminating electrolyte which contains the terminating ion—having the lowest mobility than any other ion in the system—plays the role of the last zone. Thus, at equilibrium the species of interest to be separated are "stacked" (steady-state stacking, SSS) between the leading and terminating ions in the presence of a common counterion having opposite sign of charge. The stacked zones migrate with the same velocity. This occurs because the migration velocity of the leading ion regulates that of the terminating ion, *i.e.*, the ion with the lower constituent mobility is accelerated to the constituent mobility of the faster ion by a change in voltage gradient. Since the concentration of ions on either side of a moving boundary between two zones is fixed according to the Kohlrausch principle, proteins tend to be concentrated into thin zones. This principle has been utilized in the initial phase of polyacrylamide gel electrophoresis (discussed below) to achieve stacking (ITP-A) of dilute protein solutions in ultrathin zones prior to separation by CZE (Maurer 1971; Chrambach *et al.* 1976). If stacking of the components of interest is not desirable, separation of the zones can be achieved by the use of "spacers" of intermediate mobility (ITP-C). Ampholine carrier ampholytes (LKB, Inc.), normally used for isoelectric focusing, have been employed successfully as spacers (Catsimpoolas 1973A, B).

In practice, the protein sample is placed between two buffer systems *upper* and *lower*. The *lower* buffer (phase BETA) contains the leading ion (constituent 2) and the counterion (constituent 6) (Fig. 5.2). Constituent 2 can be an ion, or a monovalent or divalent weak electrolyte. Constituent 6 is common to all phases and can be either an ion or a monovalent weak electrolyte. The *upper* buffer (phase ALPHA) contains the terminating ion (constituent 1) and the counterion (constituent 6). Constituent 1 must be a monovalent weak electrolyte. If "spacers" are to be used, these are included in the sample. The sample is dissolved in the *lower* buffer (phase BETA). The supporting medium for the separation is usually a polyacrylamide gel of low gel concentration to allow unrestrictive migration of the proteins. Descriptions of more than 4,000 buffer systems for isotachophoresis are available from the MZE computer output described above (Chrambach *et al.* 1976). Upon application of the electric field, the components in the sample arrange themselves according to their constituent mobilities. At equilibrium all the components are stacked between the BETA phase and the ZETA phase. The width of the stack does not change, but it migrates as a moving pH gradient. The ZETA phase is identical in terms of constituents to the ALPHA phase; however, it is operationally distinguishable because it is formed inside the gel where the mobilities of constituents 1 and 6 may be slightly different than those in free solution.

Isothachophoresis in its most popular form has been used as the stacking phase in disc electrophoresis (Maurer 1971; Chrambach *et al.* 1976), a technique described in the following text under the more general category of multiphasic electrophoresis.

MULTIPHASIC ZONE ELECTROPHORESIS

Multiphasic zone electrophoresis (MZE) involves the separation of proteins by a procedure which consists of three stages: (a) stacking, (b) unstacking and (c) resolution (Chrambach *et al.* 1976). This is achieved by a programmed sequence of developing phases during electrophoresis which are determined by the steady-state boundary conditions imposed on the system by selection of appropriate buffers (Fig. 5.3). The stacking phase (SSS) is identical to isotachophoresis without spacers (ITP-A). The unstacking phase involves the migration of the stack out of the stacking gel and into the resolving gel. The resolving gel has a higher concentration than the stacking gel; therefore the components of the stack are sub-

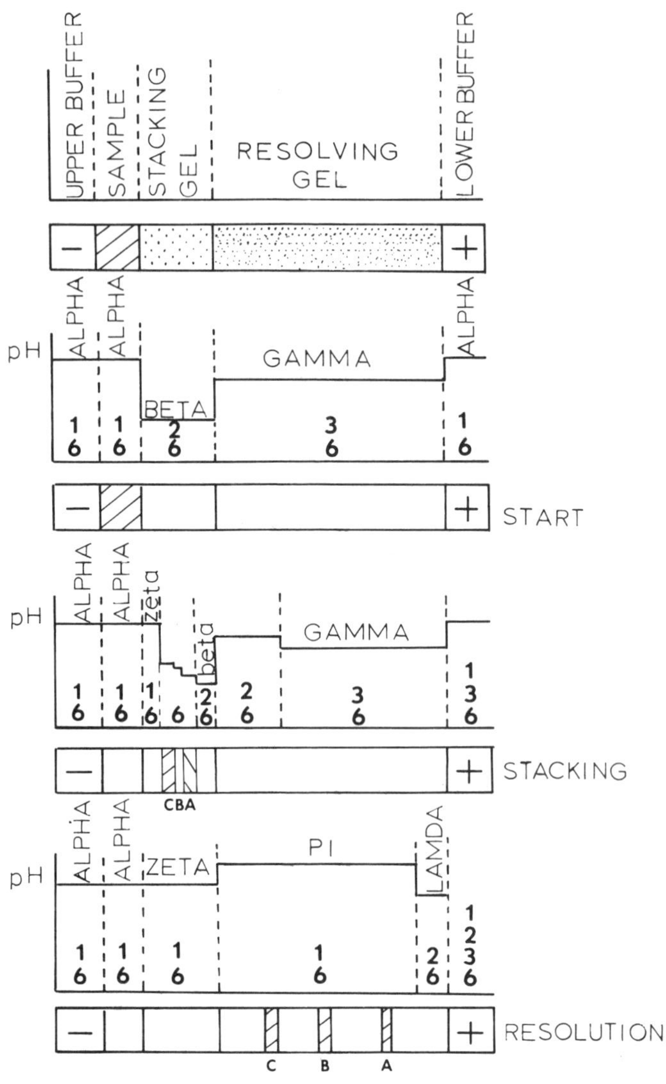

FIG. 5.3. STAGES OF MULTIPHASIC ZONE ELECTRO-
PHORESIS (MZE)

jected to sieving and therefore mobility retardation. This discontinuity in the system introduces an abrupt removal of the steady-state conditions which maintain the isotachophoresis stack.

In addition to the decreased pore size (higher gel concentration), the resolving gel is made in the *resolving* buffer (phase GAMMA) which contains constituent 3 and counterion 6 which has a different pH (higher for anodic migration) than the BETA phase. As the BETA phase migrates into the resolving gel, a new phase

LAMDA is generated. As the new moving boundary (LAMDA/GAMMA) migrates further, constituent 1 of the ZETA phase enters the resolving gel where its mobility changes in such a fashion that allows it to migrate ahead of the protein components. A new phase (PI) and a new moving boundary (PI/LAMDA) is generated through which all the protein zones migrate and are separated according to their characteristic velocity directed by the pH of the PI phase, conductance and applied current. Thus, a high resolution pattern of bands is obtained which is the result of (a) a thin starting zone due to the steady-state-stacking, (b) lower diffusion spreading due to the decreased pore size of the resolving gel, and (c) size separation due to gel sieving effects in addition to charge separation. The latter effect can be utilized effectively for determining the size of globular proteins under controlled conditions of gel polymerization (Rodbard 1976).

ELECTROPHORETIC RETARDATION

In MZE the position of the front (PI/LAMDA boundary) can be visualized by using a tracking dye. The ratio of the distance

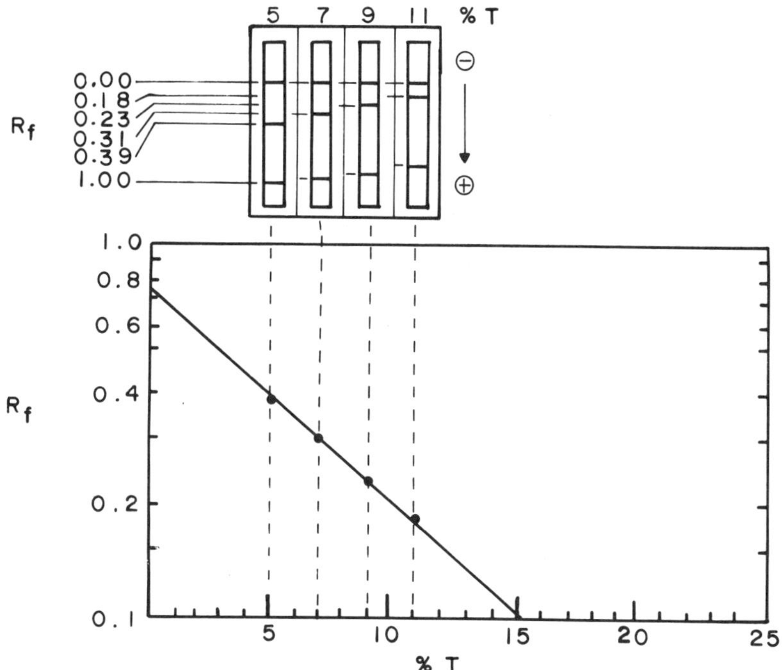

FIG. 5.4. SCHEMATIC DIAGRAM ILLUSTRATING THE MEASUREMENT OF RELATIVE MOBILITY (R_f) VALUES OF A PROTEIN IN REGARD TO THE TRACKING DYE (R_f = 1.0) AT DIFFERENT PERCENT GEL CONCENTRATIONS (%T)

The plot of log R_f vs %T gives the retardation coefficient K_R of a protein which corresponds to the numerical value of the slope.

travelled by a protein zone to that of the dye is designated as the relative mobility (R_f) value (Fig. 5.4). If gels are prepared which vary in the percent total gel concentration, but maintain a constant percent ratio of cross-linking agent to total gel concentration (% C), the retardation coefficient (K_R) of several proteins can be measured simultaneously (Chrambach *et al.* 1976). This is accomplished by plotting log (R_f) vs % T for each species (Fig. 5.4). The numerical value of the slope corresponds to the retardation coefficient K_R and the y-intercept Y_o to the relative free mobility. K_R provides a measure of molecular size and Y_o of molecular net charge. By using globular protein standards of known molecular radius (R), a linear relationship between $(K_R)^{1/2}$ and $\overline{R}$ can be obtained (Fig. 5.5) which provides the molecular radius and there-

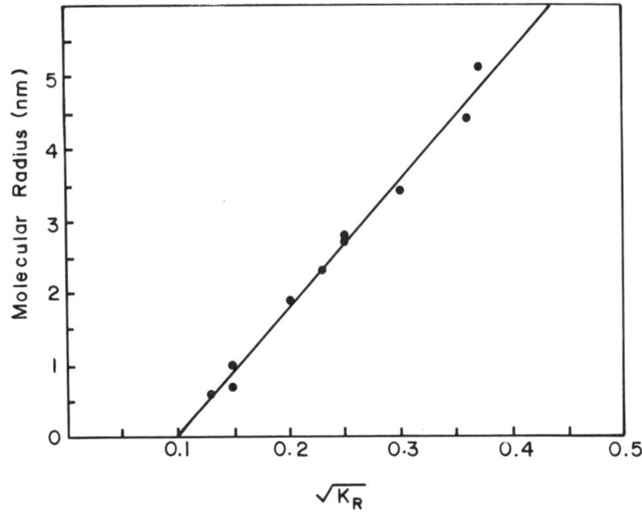

FIG. 5.5. SCHEMATIC DIAGRAM OF THE PLOT OF $K_R^{1/2}$ OF A NUMBER OF "STANDARD" GLOBULAR PROTEINS VS THE MOLECULAR RADIUS ($\overline{R}$)

The molecular weight of an unknown globular protein can be estimated from the standard curve if its K_R value is measured in the same electrophoresis system.

fore molecular weight (MW) of an unknown globular protein (Rodbard 1976). For random-coil proteins (*e.g.*, in SDS), a linear relationship between K_R and MW can be similarly obtained (Weber and Osborn 1975; Rodbard 1976). In the case of polyacrylamide gel electrophoresis in the presence of SDS (SDS-PAGE), the MW of dissociated subunits is determined rather than that of the native undissociated protein. By combination of PAGE and SDS-PAGE

data the number of subunits in a protein molecule can be estimated by electrophoretic retardation techniques.

ISOELECTRIC FOCUSING

Isoelectric focusing (IF) involves the electrophoretic migration of
a protein in a pH gradient until it reaches the pH corresponding to its
isoelectric point (pI) (Catsimpoolas 1973A, B; Arbuthnott and
Beeley 1975; Righetti 1976). At this particular pH, the net charge of
the protein is zero and, therefore, it will be concentrated in that
region. Any movement of the protein by diffusion away from the
isoelectric pH causes restoration of electric charge and consequently
electrophoretic movement toward the pI. Thus, focusing is a
steady-state process between electric mass transport and zonal
diffusion. Since proteins have widely varied isoelectric points, they
are focused at different regions of the pH gradient which results in
their separation (Fig. 5.6). Furthermore, measurement of the pH at

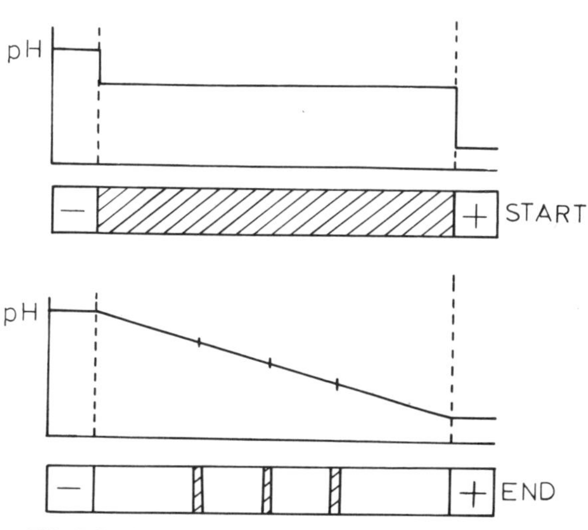

FIG. 5.6. STAGES OF ISOELECTRIC FOCUSING (IF)

the focusing position provides a direct determination of the
isoelectric point of the particular species. Isoelectric focusing is
performed either in density gradients or polyacrylamide and granular
gels.

The pH gradient is formed by the application of a constant voltage
electric field to a mixture of carrier ampholytes which are low
molecular weight amphoteric compounds of closely spaced pI values.

The carrier ampholytes themselves are focused exhibiting a continuous distribution of pI's which direct a pH gradient of sufficient buffering capacity and conductance. This milieu is then used for the focusing of proteins. It is obvious that a protein placed at any point in the pH gradient will acquire either a positive or negative charge and it will migrate electrophoretically away from the electrode of similar charge until it reaches its pI. Thus, in IF it is not required that the sample be applied as a zone of minimal thickness.

The zone width of an isoelectrically focused protein at the steady-state depends on the diffusion coefficient D, the magnitude of the electric field E and the electrofocusing parameter p which in turn depends on the steepness of the pH gradient $(d(pH)/dx)$ and the slope of the pH-mobility curve of the protein at the pI $(dM/d(pH))$ where M is the electrophoretic mobility and x is distance. The standard deviation (σ) of the Gaussian zone is given by:

$$\sigma = (D/pE)^{1/2} \qquad (5.1)$$

where
$$p = -\,(dM/d(pH))\,(d(pH)/dx) \qquad (5.2)$$

It can be seen from Eq 5.1 that low values of D and high values of $dM/d(pH)$ as normally encountered in proteins will produce small values of σ and therefore very sharply focused zones. The sharpness of focusing of a zone can also be increased by the formation of steep pH gradients (*i.e.*, large values of $d(pH)/dx$) and increasing field strength E. However, the latter cannot exceed reasonable limits because of Joule heating effects which result in convective disturbances of the migration path and possible denaturation of the protein.

Isoelectric focusing offers the highest resolution ever achieved in charge separation of proteins. Molecules differing as little as 0.01 pH unit in their isoelectric points can be separated by this method. Often this separation involves not more than one charged amino acid difference between two structures. However, because of its extremely high resolution, isoelectric focusing may detect other charge differences not strictly related to the gross homogeneity of the protein which may be the result of post-synthetic modification of primary structures (*e.g.*, deamidation), ligand binding, variation in nonprotein components (*e.g.*, carbohydrates, lipids, and other prosthetic groups), chemical modification, association and dissociation phenomena, changes of redox states of metalloenzymes, and others. With this caution in mind, interpretation of isoelectric focusing patterns can be rewarding as they reveal the microheterogeneity aspects of protein structures.

Currently, isoelectric focusing is used in combination with other electrophoretic techniques such as SDS-PAGE to produce two-dimensional maps of separated components (Wrigley 1976). The first dimension provides charge differences in terms of pI and the second size differences in terms of MW. Using such techniques thousands of protein components from very complex mixtures can be separated and characterized simultaneously. Such high resolution separations were unthinkable a few years ago. They are, however, now applied increasingly to the investigation of a variety of biochemical problems with overwhelming success.

CONCLUSION

The elucidation of the structure and functional and nutritional properties of proteins basic to foods necessitates their separation into distinct components and further isolation in homogeneous form. The initial utilization of low resolution traditional techniques such as salt and isoelectric precipitation, gel filtration and ion exchange chromatography followed by high resolution electrophoretic methods appears to be the recommended strategy for a successful fractionation route. This paper has provided an overview of underlying principles to stimulate interest to further reading where details of each procedure can be found. The real battle starts in the laboratory.

BIBLIOGRAPHY

ACKERS, K. G. 1975. Molecular sieve methods of analysis. *In* The Proteins, Vol. 1, H. Neurath and R. L. Hill (Editors). Academic Press, New York.
ARBUTHNOTT, J. P. and BEELEY, J. A. 1975. Isoelectric Focusing. Buttersworths, London.
AXELSEN, N. H., KRØLL, J., and WEEKE, B. 1973. A Manual of Quantitative Immunoelectrophoresis. Methods and Applications. Universitetsforlaget, Oslo.
BERNARDI, G. 1971. Chromatography of Proteins on Hydroxyapatite, 325–339. *In* Methods in Enzymology, Vol. 21, part D, L. Grossman and K. Moldave (Editors). Academic Press, New York.
CATSIMPOOLAS, N. 1973A. Isoelectric Focusing and Isotachophoresis. Ann. N.Y. Acad. Sci., Vol. 209, New York.
CATSIMPOOLAS, N. 1973B. Isoelectric focusing and isotachophoresis of proteins: A review. Separ. Sci. *8*, 71–121.
CATSIMPOOLAS, N. 1975. Methods of Protein Separation, Vol. 1. Plenum Publishing Corp., New York.
CATSIMPOOLAS, N. 1976. Isoelectric Focusing. Academic Press, New York.
CATSIMPOOLAS, N. 1977. Immunological Aspects of Food. AVI Publishing Co., Westport, Conn.
CHIN, H. P., 1970. Cellulose Acetate Electrophoresis: Techniques and Applications. Ann Arbor-Humphrey, Ann Arbor, Mich.
CHRAMBACH, A. JOVIN, T. M., SVENDSEN, P. J., and RODBARD, D. 1976. Analytical and preparative polyacrylamide gel electrophoresis. *In* Methods of Protein Separation, Vol. 2, N. Catsimpoolas (Editor). Plenum Publishing Corp., New York.

CLAUSEN, J. 1969. Immunochemical Techniques for the Identification and Estimation of Macromolecules. North-Holland Publishing Co., Amsterdam.

CROWLE, A. J. 1973. Immunodifusion. Academic Press, New York.

DETERMANN, H. 1968. Gel Chromatography. Springer-Verlag, New York.

FISHER, L. 1969. An Introduction to Gel Chromatography. North-Holland Publishing Co., Amsterdam.

GORDON, A. H. 1969. Electrophoresis of Proteins in Polyacrylamide and Starch Gels. North-Holland Publishing Co., Amsterdam.

GRABAR, P., and BURTIN, P. 1964. Immunoelectrophoretic Analysis. Elsevier, Amsterdam.

LAURELL, C.-B. 1972. Electrophoretic and Electro-immunochemical Analysis of Proteins. Universitetsforlaget, Oslo.

LOWE, C. R., and DEAN, P. D. G. 1974. Affinity Chromatography. John Wiley and Sons, New York.

MAURER, H. R. 1971. Disc Electrophoresis. Walter de Gruyter, New York.

MORRIS, C. J. O. R., and MORRIS, P. 1976. Separation Methods in Biochemistry. John Wiley and Sons, New York.

PARIKH, I., and CUATRECASAS, P. 1975. Affinity chromatography, principles and applications. *In* Methods of Protein Separation, Vol. 1, N. Catsimpoolas (Editor). Plenum Publishing Corp., New York.

PETERSON, E. A. 1970. Cellulosic Ion Exchangers. North-Holland Publishing Co., Amsterdam.

PORATH, J., and KRISTIANSEN, T. 1975. Biospecific affinity chromatography and related methods. *In* The Proteins, Vol. 1, H. Neurath and R. L. Hill (Editors). Academic Press, New York.

RIGHETTI, P. G. 1976. Progress in Isoelectric Focusing and Isotachophoresis. North-Holland Publishing Co., Amsterdam.

RODBARD, D. 1976. Determination of molecular weight and radius by gel filtration, polyacrylamide gel electrophoresis (PAGE), and PAGE in sodium dodecyl sulfate. *In* Methods of Protein Separation, Vol. 2, N. Catsimpoolas (Editor). Plenum Publishing Corp., New York.

WEBER, K. and OSBORN, M. 1975. Proteins and sodium dodecyl sulfate: Molecular weight determination on polyacrylamide gels and related procedures. *In* The Proteins, Vol. 1, H. Neurath and R. L. Hill (Editors). Academic Press, New York.

WIEME, R. J. 1965. Agar Electrophoresis. Elsevier, Amsterdam.

WRIGLEY, C. W. 1976. Isoelectric focusing-electrophoresis in gels. *In* Isoelectric Focusing, N. Catsimpoolas (Editor). Academic Press, New York.

APPENDIX

Multiphasic Buffer Systems Output

An example of a computer designed buffer system (No. 4192) for continuous-pH zone electrophoresis (utilizing only the components of the PI phase, *i.e.*, glycine and Tris) and for multiphasic zone electrophoresis is presented below. The recipes for the buffers of the starting phases ZETA, BETA and GAMMA are given in the lower part of the output. The upper and lower buffers in MZE are made using the recipe of the ZETA phase. The pH, ionic strength, conductance, and other physicochemical parameters of each phase are given in detail.

Nomenclature

ALPHA   upper buffer phase prior to electrophoresis
BETA   stacking phase prior to electrophoresis
BV   buffer value
C1   constituent concentration of trailing ion of the stack
C2   constituent concentration of leading ion of the stack
C3   constituent concentration of ion in the separation phase prior to electrophoresis
C6   constituent concentration of common ion of the system
C7   constituent concentration of ion in the restacking phase
CONSTITUENT 1   trailing ion of the stack
CONSTITUENT 2   leading ion of the stack
CONSTITUENT 3   ion of the separation phase prior to electrophoresis
CONSTITUENT 6   counterion common to all phases
GAMMA   separation phase prior to electrophoresis
IS, I, or ION STR.   ionic strength
KAPPA   specific conductance (μmhos/cm)
LAMBDA (8) separation phase after migration of the leading ion (C2) of the stack into the GAMMA phase
NU   boundary displacement (cm^3/Fd)
PHASE 2   stacking phase prior to electrophoresis
PHASE 3   separation phase prior to electrophoresis
PHASE 4   operative stacking phase; also the arbitrary composition of the upper buffer
PHASE 9   operative separation phase
Phase BETA (2)   stacking phase prior to electrophoresis
Phase GAMMA (3)   separation phase prior to electrophoresis
Phase LAMBDA (8)   separation phase after migration of the leading ion of the stack into phase (3)
Phase PI (9)   operative separation phase
Phase ZETA (4)   operative stacking phase
PHI (1)   ratio of ionized to un-ionized CONSTITUTENT 1
PHI (2)   ratio of ionized to un-ionized CONSTITUENT 2
PHI (3)   ratio of ionized to un-ionized CONSTITUTENT 3
PHI (6)   ratio of ionized to un-ionized CONSTITUENT 6
PI (9)   operative separation phase
RM (1)   mobility of constituent 1 relative to Na^+
RM (2)   mobility of constituent 2 relative to Na^+
RM (3)   mobility of constituent 3 relative to Na^+
RM (6)   mobility of constituent 6 relative to Na^+
RM (1, 4)   relative mobility of constituent 1 in phase (4) (Lower Stacking Limit)

RM (2, 2)   relative mobility of constituent 2 in phase (2) (Upper Stacking Limit)

RM (1, 9)   relative mobility of constituent 1 in phase (9) (Unstacking Limit)

Separation gel—gel made in PHASE 3, operative in PHASE 9 (lower gel)

Stacking gel—gel made in PHASE 2, operative in PHASE 4 (upper gel)

SIGMA   relative conductance (C/cm^3)

THETA   ratio of constituents within one phase

ZETA (4)   operative stacking phase

SYSTEM NUMBER 4192

POLARITY = − (MIGRATION TOWARD ANODE)   TEMPERATURE = 25 DEG. C.

CONSTITUENT 1 = NO. 29, GLYCINE
CONSTITUENT 2 = NO. 82, PHOSPHATE-DIBASIC
CONSTITUENT 3 = NO. 99, CHLORIDE
CONSTITUENT 6 = NO. 12, TRIS

PHASES

	ALPHA (1)	ZETA (4)	BETA (2)	PI (9)	LAMBDA (8)	GAMMA (3)
C1	0.0400	0.0400		0.0411		
C2			0.0304		0.0229	
C3						0.0526
C6	0.0380	0.0380	0.0384	0.4187	0.4163	0.4232
THETA	0.950	0.950	1.265	10.196	18.154	8.040
PHI (1)	0.124	0.124		0.365		
PHI (2)			0.236		0.991	
PHI (3)						1.000
PHI (6)	0.131	0.131	0.977	0.036	0.110	0.124
RM (1)	−0.090	−0.090		−0.263		
RM (2)			−0.685		−0.957	
RM (3)						−1.552
RM (6)	0.065	0.065	0.489	0.018	0.055	0.062
PH	8.89	8.89	6.44	9.50	8.99	8.92
ION. STR.	0.0050	0.0050	0.0447	0.0150	0.0684	0.0526
SIGMA	0.586	0.586	4.480	1.766	6.424	10.421
KAPPA	288.	288.	2012.	837.	2810.	4634.
NU	−0.153	−0.153	−0.153	−0.149	−0.149	−0.149
BV	0.020	0.020	0.015	0.055	0.094	0.106

RECIPES FOR BUFFERS OF PHASES ZETA (4), BETA (2), GAMMA (3), PI (9)

CONSTITUENT		1X PHASE 4	4X PHASE 2	4X PHASE 3	4X PHASE 9
GLYCINE	GM	3.00			1.23
1M PHOSPHORIC ACID	ML		12.14		
1N HCL	ML			21.05	
TRIS	GM	4.60	1.86	20.50	20.28
H2O TO		1 LITER	100 ML	100 ML	100 ML
AT FINAL CONCENTRATION =					
PH (25 DEG. C.)		8.89	6.44	8.92	9.50
KAPPA (25 DEG. C.)		288	2012	4634	837

6

Muscle Proteins

Darrel E. Goll
Richard M. Robson
Marvin H. Stromer

Muscle proteins have been a much discussed and sometimes misrepresented class of compounds in food technology. The generous attention bestowed upon muscle proteins by food technologists originates, of course, from the fact that meat is derived directly from living muscle, and meat is perhaps the most conspicuous item in the human diet, at least economically. Although total expenditures for meat and food in the United States have varied with inflation and other costs during the last 20 years, the proportional amount spent for meat has remained almost constant at 25% of total expenditures for food. Consequently, meat is the most expensive single item in the human diet. This cost factor alone has resulted in much consumer emphasis on quality of the meat cut that finally reaches the family table. Because meat is derived directly from living muscle, it is clear that information on the biochemical properties and molecular architecture of living muscle tissue and the changes in these properties and architecture between death of the animal and human consumption is a necessary prerequisite to understanding the causes for variations in meat quality. It is known that living muscle and the edible part of meat generally have the same chemical composition; both contain 55-78% of their weight as water, 15-22% of their weight as protein, 1-15% of their weight as lipid and much smaller amounts (less than 4% of their total weight) of carbohydrates, minerals, and other organic compounds. It is evident from this chemical composition of muscle tissue that protein alone makes up 50 to over 95% of total organic solids in meat, depending on lipid content of the tissue. For this reason, most studies of meat quality have focused on the muscle proteins.

CLASSIFICATION OF THE MUSCLE PROTEINS

Although the uninitiated may think that the term "muscle proteins" is used only in a generic sense to include all proteins in muscle, many biologists have come to accept the term "muscle proteins" as referring exclusively to the contractile or myofibrillar proteins and excluding the cytoplasmic, membrane, and connective tissue proteins also found in muscle. Although this convention may have had some logical basis when it was thought that contractile or myofibrillar proteins were peculiar to muscle tissue and were therefore the only unique "muscle proteins," its use is not justified in light of current knowledge that all animal cells (and perhaps all plant and bacterial cells as well) contain one or more of the myofibrillar or contractile proteins (Pollard and Weihing 1974). Consequently, use of the term "muscle proteins" should be reserved exclusively to refer to all proteins found in muscle cells (*e.g.*, myoglobin is a noncontractile protein found exclusively in muscle cells), and the myofibrillar or contractile proteins should be designated specifically by name.

The Sarcoplasmic Proteins

The muscle proteins may be divided into three major classes (Goll *et al.* 1970, 1974) on the basis of their solubility in aqueous solvents (Table 6.1). The sarcoplasmic protein fraction is the most soluble of the three classes of muscle proteins (Table 6.1) and generally includes those proteins found in the cytoplasm of the muscle cell. Although several different procedures have been described to divide the sarcoplasmic proteins into crude fractions named myogen A, myogen B, globulin X, or myoalbumin (see Dubuisson 1954), these early fractionation schemes used crude separation techniques that were incapable of producing pure proteins from the very complex sarcoplasmic protein fraction. Hence, the protein fractions produced by these schemes were heterogeneous collections containing many different proteins. It should also be clearly understood, as indicated in Table 6.1, that the term myogen refers to the entire sarcoplasmic protein fraction containing at least 100–200 different proteins and not to a single homogeneous protein. Exact protein composition of the sarcoplasmic protein fraction is influenced to a considerable extent by conditions used during extraction and may vary depending

TABLE 6.1
PROTEIN COMPOSITION OF VERTEBRATE MUSCLE

Protein Class	Definition
Sarcoplasmic proteins	Those proteins soluble at ionic strengths of 0.1 or less at neutral pH. Constitute 30–35% of total protein in skeletal muscle and slightly more than this in cardiac muscle. Contains at least 100–200 different proteins. Sometimes called myogen.
Myofibrillar proteins	Those proteins that constitute the myofibril. Make up 52–56% of total protein in skeletal muscle but only 45–50% of total protein in cardiac muscle. Although high ionic strength is required to disrupt the myofibril, many of the myofibrillar proteins are soluble in H_2O once they have been extracted from the myofibril.
Stroma proteins	Those proteins insoluble in neutral aqueous solvents. Constitute 10–15% of total protein in skeletal muscle and slightly more than this in cardiac muscle. Includes lipoproteins and mucoproteins from cell membranes and surfaces as well as connective tissue proteins. Although exact percentage composition can vary widely depending on source of the muscle, collagen frequently makes up 40–60% of total stroma protein and elastin may make up 10–20% of total stroma protein.

on speed and extent of homogenization of the tissue before extraction, pH of the extraction, nature of the extracting solvent, and centrifugal force used to separate the soluble sarcoplasmic protein fraction from unsolubilized proteins and subcellular organelles (see Goll *et al.* 1970). It also is sometimes not appreciated that, if minced muscle is extracted with an equal volume of water, ionic strength of the extraction will be 0.08–0.09 because ionic strength of muscle is 0.16–0.18. Under most extraction conditions used, the sarcoplasmic protein fraction contains all the enzymes associated with glycolysis and most of the enzymes associated with carbohydrate and protein synthesis because these processes occur largely in the cytoplasm of the muscle cell and these enzymes, in addition to being soluble at low ionic strength, are free and readily solubilized when the muscle cell is ruptured. If homogenization is sufficiently severe or the extraction is done at sufficiently low ionic strengths to rupture mitochondria, the sarcoplasmic protein fraction will also contain the enzymes associated with lipid synthesis and the citric acid cycle. It seems likely that most sarcoplasmic protein fractions contain some lysosomal enzymes because muscle lysosomes seem quite labile, and special care is required to preserve them during fractionation of muscle homogenates (Reville *et al.* 1976). Hultin and coworkers (Hultin 1974) have found that some cytoplasmic muscle enzymes may bind to cellular membranes and therefore not be solubilized under some extraction conditions. It is evident from these findings that extraction and homogenization conditions should be carefully standardized and described in detail when discussing studies of the sarcoplasmic proteins.

Although the sarcoplasmic protein fraction from muscle contains many of the same proteins that would be extracted at low ionic strengths from liver or various other cells (for example, most cells contain the glycolytic enzymes in a form extractable at low ionic strengths), this fraction also contains a few proteins unique to muscle cells. Many of these unique sarcoplasmic muscle proteins are present in only very small amounts, such as the initiation factors that Heywood and coworkers (Heywood and Kennedy 1974; Heywood *et al.* 1974; Thompson and Heywood 1974) have found necessary for synthesis of myosin and myoglobin. On the other hand, myoglobin itself is a sarcoplasmic protein that is unique to muscle and is present in fairly large amounts in most muscle cells (Table 6.2). It is well

TABLE 6.2
PROXIMATE MYOGLOBIN CONCENTRATIONS IN DIFFERENT MUSCLES

Source of Muscle	Myoglobin Concentration mg/g Muscle Tissue
Chicken pectoralis	0.05
Porcine skeletal	1–4
Veal skeletal	1–4
Lamb skeletal	6–12
Bovine skeletal (12–24 mo of age)	4–10
Bovine skeletal (4–6 yr of age)	16–20
Whale skeletal	50

known that myoglobin is responsible for most of the observed differences in meat color, and the uniqueness of myoglobin to muscle cells therefore accounts for the uniqueness of meat color.

Studies of protein composition of muscle during animal development and growth have, in general, shown that, when expressed as a percentage of total muscle weight, sacroplasmic protein content increases during prenatal development and also increases postnatally until the animal is approximately half mature. When expressed as a percentage of total muscle protein, however, sarcoplasmic protein content of muscle cells is highest early in the prenatal period, when it may make up as much as 70% of total muscle protein, and decreases during both prenatal and postnatal development.

The Stroma Proteins

In contrast to the sarcoplasmic proteins, the stroma proteins are the least soluble class of proteins in muscle cells (Table 6.1). Experimentally, the stroma proteins are usually measured as the insoluble proteins remaining after exhaustive extraction of all soluble muscle proteins. The insoluble stroma protein fraction is similar to the sarcoplasmic protein fraction in the diversity and number of different proteins constituting the fraction and in the fact that composition of the stroma protein fraction also depends on the extraction conditions used. The stroma protein fraction will usually contain proteins from the sarcolemmal, sarcoplasmic reticular, and mitochondrial membranes in addition to the proteins that constitute the epimysial, perimysial, and endomysial connective tissues surrounding the muscle cell. If prior extraction of myofibrillar proteins has not been done thoroughly or has been done at temperatures above $2°C$ where myofibrillar proteins tend to become denatured and insoluble, the insoluble protein residue measured as stroma proteins may contain large amounts of unextracted myofibrillar proteins that may be erroneously included in the stroma protein fraction. A large number of different conditions, including prolonged storage times, even at temperatures as low as $-20°C$, can make myofibrillar proteins difficult to extract completely. Therefore, careful attention is necessary in studies involving muscle protein composition to ensure that no myofibrillar proteins are included in the stroma protein fraction. Sometimes a 2–4 hr extraction with 0.1 N NaOH at $2°C$ will remove unextracted myofibrillar proteins, although this extraction may also solubilize small amounts of some of the membrane proteins. It is probably prudent when doing muscle protein composition studies to subject the stroma protein fraction to SDS-polyacrylamide gel electrophoresis, using the procedure described by Suzuki *et al.* (1976) to solubilize the myofibrillar proteins. The presence of myosin or other myofibrillar proteins in such SDS-polyacrylamide gels will indicate directly that the stroma protein fraction being examined contains unextracted myofibrillar proteins. Even SDS-polyacrylamide gel electrophoresis is not an infallible safeguard against the presence of unextracted myofibrillar proteins in the stroma protein fraction, and stroma protein contents above 15% of total muscle protein should be viewed with considerable skepticism.

As indicated in Table 6.1, the two connective tissue proteins, collagen and elastin, make up most of the stroma protein fraction, although the exact proportions of collagen and elastin vary widely among different muscles in the same animal species (Bendall 1967).

Many investigators have supposed that the percentage of stroma protein in muscle tissue increases with increasing animal age, but actual analysis shows that stroma protein content of muscle tissue decreases during both prenatal and postnatal development when expressed either as a percentage of total muscle weight or as a percentage of total muscle protein (Goll *et al.* 1963).

We have previously discussed the rather small effects that postmortem aging has on the connective tissue proteins in the stroma protein fraction (Goll *et al.* 1970). On the basis of biochemical studies demonstrating that sarcoplasmic reticular membranes lose their biological activity very quickly postmortem (Eason 1969; Goll *et al.* 1971; Greaser *et al.* 1969) and of biochemical and electron microscope observations showing that mitochondrial and sarcolemmal membranes lose their structural integrity and biochemical activity during postmortem storage (Buege and Marsh 1975; Dawson 1966; Osner 1966; Reed *et al.* 1966), it seems very likely that membrane proteins in the stroma protein fraction undergo large changes during postmortem storage. The exact nature of these changes and their role in meat quality, however, remain completely unclear. The stroma protein fraction clearly is important to food technologists because it has at least four direct effects on meat quality, all of which are deleterious: (1) stroma proteins, especially connective tissue proteins, lower meat tenderness; the effects of connective tissue proteins on meat tenderness depend on both the amount and the degree of cross-linking among the connective tissue proteins (Goll *et al.* 1964A, B, C); (2) stroma proteins, because of their insoluble nature, decrease emulsifying capacity of meat; (3) because of their low content of charged and hydrophilic amino acids, stroma proteins lower water-holding capacity of meat; and (4) the stroma protein fraction contains a low proportion of nutritionally essential amino acids, and the larger the stroma protein content of meat tissue, the lower the nutritive value of that tissue. Consequently, the stroma protein content of muscle and the nature of any postmortem changes in the stroma proteins have egregious significance to food technologists. Postmortem alterations in the sarcoplasmic reticular, mitochondrial, and sarcolemmal membranes may also have far-reaching indirect effects on meat quality because changes in these membranes can produce large changes in intracellular free Ca^{2+}-concentration, and this Ca^{2+} may accelerate glycolysis (Brostrom *et al.* 1971; Heilmeyer *et al.* 1970; Stull and Mayer 1971), initiate contraction (Goll *et al.* 1971), or activate intracellular proteases (Dayton *et al.* 1976A, B; Reville *et al.* 1976). Additional, more detailed studies on the effect of portmortem storage on some of the membrane proteins in the stroma protein fraction may be expected to provide significant new information on postmortem changes in meat quality.

Myofibrillar Proteins

The myofibrillar proteins are the largest fraction of proteins in muscle tissue and are intermediate in solubility between the sarcoplasmic proteins and the stroma proteins (Table 6.1). Although the myofibrillar proteins have frequently been defined as those muscle proteins insoluble in water but soluble in dilute salt solutions (*i.e.*, the classical definition of a globulin), all known myofibrillar proteins except one are soluble in water once they have been extracted. Indeed, at least four of the eight or nine known myofibrillar proteins are extractable with water. Although high ionic strength is required to completely disrupt the myofibril and extract the major myofibrillar protein, myosin, the extracted myosin remains soluble if the ionic strength remains below 0.001. Therefore, it is most accurate to define the myofibrillar proteins as those proteins that constitute the myofibril (Table 6.1). Because the myofibril is an easily observed and well-defined subcellular entity in muscle cells, defining the myofibrillar protein fraction in terms of the myofibril is convenient as well as accurate.

The percentage of myofibrillar protein in muscle tissue increases during both prenatal and postnatal development, whether expressed as a percentage of total muscle weight or as a percentage of total muscle protein. Because they are directly responsible for the contractile properties of muscle cells, myofibrillar proteins have received lavish attention from physiologists and biochemists. Food technologists have also studied the myofibrillar protein fraction extensively because the myofibrillar proteins have large influences on both the culinary qualities and the important commercial properties of meat. For example, it has been estimated that approximately 97% of the water-holding capacity of meat is due to the myofibrillar proteins alone. Although it has been more difficult to estimate accurately the magnitude of the myofibrillar protein contribution to emulsifying capacity of meat, it seems clear the myofibrillar proteins are responsible for over 75% of the emulsifying ability of meat, and, indeed, may account for well over 90% of total emulsifying capacity of meat. It also is difficult to determine the precise extent to which changes in myofibrillar proteins are responsible for variations in meat tenderness (Goll *et al.* 1974), partly because the relative importance of myofibrillar proteins in meat tenderness may be inversely proportional to the connective tissue content of the sample (Goll *et al.* 1974). Hence, in those muscles or animals where connective tissue content is high, connective tissue makes a large contribution to meat toughness, and tenderness is not highly related to state of the myofibrillar proteins. On the other hand, in those muscles of animals where connective tissue content is low or where state of the myofibrillar proteins has been altered to make them large contributors to toughness, tenderness is highly related to state of the myofibrillar proteins and is only poorly related to connective tissue content. Marsh (1972) has described the cold-shortening phenomenon that alters the state of myofibrillar proteins and that can have calamitous effects on meat tenderness. In cold-shortened muscles, connective tissue has little relationship to tenderness, and depending on the extent of the cold shortening, myofibrillar proteins may account for over 90% of the variation in meat tenderness. Purchas (1972) has recently devised an analysis to express different measurements of tenderness on the same scale and has concluded that a 1% change in a measurement that estimates the contribution of myofibrillar proteins to toughness produces the same effect on tenderness as a 10% change in a measurement that estimates the contribution of connective tissue proteins to toughness. Consequently, it seems that myofibrillar proteins may be responsible for anywhere from 50 to nearly 100% of the variation in meat tenderness, depending on connective tissue content and state of the muscle being tested. Finally, because myofibrillar proteins constitute more than 50% of total muscle protein and because the myofibrillar proteins contain relatively high proportions of the nutritionally essential amino acids, especially when compared with the connective tissue proteins, it may be estimated that the myofibrillar proteins contribute more than 70% of the protein nutrition derived from meat.

It is clear, therefore, that myofibrillar proteins have major effects on meat quality and the usefulness of meat in different processing procedures. Moreover, the myofibrillar proteins are a more kaleidoscopic class of proteins than the connective tissue proteins, and are more likely during postmortem storage to undergo large changes that in turn cause large changes in meat quality. The connective tissue proteins, on the other hand, change little during postmortem storage, and effects of the connective tissue proteins on meat quality are, therefore, less amenable to control or alteration during postmortem storage than effects of myofibrillar proteins are. The possibility of controlling or significantly improving meat quality by intervening in the normal effects of postmortem storage on the myofibrillar proteins has stimulated intense interest among food technologists in this class of proteins. Although intelligent intervention in the postmortem changes in myofibrillar proteins requires that the cause and nature of these changes be known, the complexity of the interactions among the myofibrillar proteins has thwarted efforts to understand the postmortem changes in myofibrillar proteins in the molecular detail needed. Because it is important to understand these changes, however, and because several recent findings have provided significant new insights into the nature of the interactions among the myofibrillar proteins and how these interactions are affected by ATP and Ca^{2+}, the remainder of this review will be devoted exclusively to the myofibrillar proteins.

THE MYOFIBRILLAR PROTEINS

Biochemical Properties

Although the sarcoplasmic and stroma protein fractions both contain many different proteins, and the myofibrillar protein fraction is responsible for the complex task of converting chemical energy of ATP into movement, the myofibril is composed of only nine or ten (depending on whether there are one or two M-proteins) different proteins (Table 6.3). The relatively small number of proteins required for biological movement is quite remarkable, especially when compared with the number of proteins required for many other biological processes. For example, muscle cell ribosomes, which contain only part of the proteins required for biosynthesis of muscle proteins, are composed of 67–74 different proteins (Sherton and Wool 1974). The intensive study to which myofibrillar proteins have been subjected and the ability to dissolve myofibrils in sodium dodecyl sulfate (SDS[1]) and separate the myofibrillar proteins on SDS-polyacrylamide gels have provided a great deal of information on protein composition of the myofibril, and it is unlikely that many new myofibrillar proteins will be discovered. Extensive study has shown that, of the nine or ten proteins listed in Table 6.3, actin and myosin are both necessary and completely sufficient for an *in vitro* contractile response. The other seven or eight myofibrillar proteins function either to regulate or control the actin-myosin interaction (tropomyosin, troponin, and possibly α-actinin) so that contraction can be stopped and re-initiated in the continued presence of ATP or to assist in assembly of actin and myosin into the highly organized three-dimensional structure observed in myofibrils (C-protein, α-actinin, β-actinin, M-protein, and paramyosin).

The polypeptide chains constituting the known myofibrillar proteins all have different masses (Table 6.3, last column), and as indicated in the preceding paragraph, it therefore is possible to separate all the known polypeptide chains in myofibrils on a single SDS-polyacrylamide gel. Examples of SDS-polyacrylamide gels of whole myofibrils from porcine skeletal or cardiac muscle are shown in Fig. 6.1. Because SDS-polyacrylamide gel electrophoresis requires

TABLE 6.3
PROPERTIES OF THE MYOFIBRILLAR PROTEINS

Protein	Year of Discovery	% of Myofibril by Wt	Mol Wt (g/mole)	Subunit Polypeptide Composition[1]
Myosin	1941[4]	50–58	475,000	200,000 daltons - two 20,700 daltons - one[5] 19,050 daltons - two[5] 16,500 daltons - one[5]
Actin	1941	15–20	41,785	41,785 daltons - one[5]
Tropomyosin	1946	4–6	70,000	35,000 daltons - one[5] 32,758 daltons - one[5]
Troponin	1965	4–6	72,000	30,503 daltons - one (TN-T)[5] 20,864 daltons - one (TN-I)[5] 17,846 daltons - one (TN-C)[5]
C-Protein	1971	2.5–3	140,000	140,000 daltons - one
α-Actinin	1965	2–3	206,000	103,000 daltons - two
β-Actinin	1965	<1	70,000	?
M-Protein[2]	1974	3–5	160,000	160,000 daltons - one
Paramyosin[3]	1947	2–30	220,000	110,000 daltons - two

[1] Subunit polypeptide mass and composition of myosin, tropomyosin, and troponin differ in "red" (slow-twitch, oxidative) and "white" (fast-twitch, glycolytic) muscles. Figures are given for "white" muscle proteins.

[2] Several reports have indicated that a second M-protein with a 42,000 dalton subunit exists in addition to the M-protein described.

[3] Paramyosin is found only in invertebrate muscles.

[4] The term, "myosin," was first used by Kuhne in 1859 but his "myosin" was a mixture of actin and myosin, and all "myosins" up to 1941 were actually mixtures of actin and myosin.

[5] Amino acid sequence has been determined.

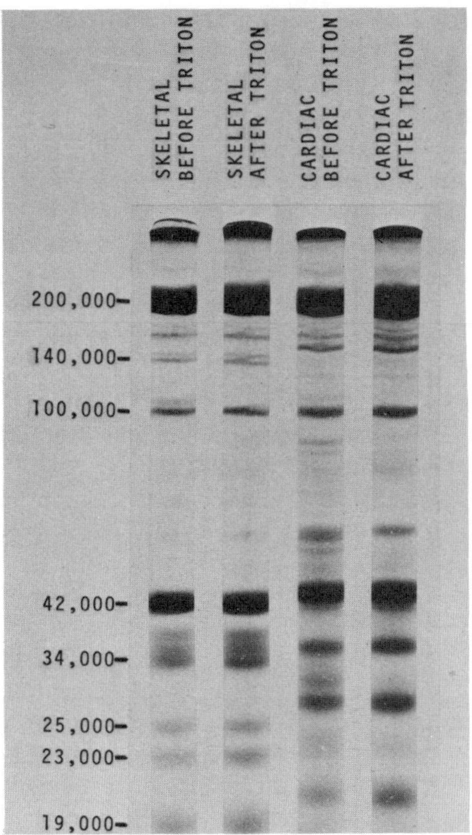

FIG. 6.1 SDS-POLYACRYLAMIDE GELS (7½%)
OF MYOFIBRILS FROM PORCINE SEMI-
TENDINOSUS ("WHITE PORTION") OR
PORCINE CARDIAC MUSCLE BEFORE AND
AFTER TREATMENT WITH 1% TRITON X-100
Gels were run as described by Suzuki *et al.* (1976).
Numbers at the left of the gels designate approxi-
mate migration distances of polypeptides having
the indicated molecular weight. Protein loads on
gels are: porcine semitendinosus before Triton, 38
µg; porcine semitendinosus after Triton, 36 µg;
porcine cardiac before Triton, 40 µg; porcine
cardiac after Triton, 40 µg. Gels are courtesy of
Marian McGinley.

inexpensive apparatus and is a rapid procedure with tremendous
potential ability to resolve different polypeptide chains, its use for
study of muscle protein changes that might be related to meat
quality has increased enormously during the last several years (Olson
et al. 1977; Samejima and Wolfe 1976; Wolfe and Samejima 1976).
Although SDS-polyacrylamide gel electrophoresis is a relatively

simple procedure, the very simplicity and ease with which SDS-polyacrylamide gel electrophoresis can be done seems to have led, in some instances, to carelessness in use of this procedure. It is critical when using SDS-polyacrylamide gel electrophoresis to study protein composition of myofibrils that the myofibrils be thoroughly washed and free from membranes and contaminating proteins and that they be completely dissolved in a solution compatible with maximum resolution during electrophoresis. It also is essential that purified myofibrillar proteins be run simultaneously on separate gels for use as markers to identify different bands separated on SDS-polyacrylamide gels of whole myofibrils. Failure to observe these precautions can lead to serious errors in interpretation of experimental results and can completely negate an otherwise well-planned study. Consequently, it is important when evaluating experiments using SDS-polyacrylamide gels of myofibrillar proteins to ensure that gels of marker proteins are included and to examine the procedure used to purify the myofibrils before they were applied to the gels. We have found that sarcoplasmic membrane proteins adhere tenaciously to myofibrils and are not removed even after 14 washes in 100 mM KCl solutions. Electron microscope observations show, however, that washing twice with a 1% Triton X-100 solution (Goll *et al.* 1975) completely eliminates contaminating sarcoplasmic reticular membranes without damaging the myofibrillar apparatus. SDS-polyacrylamide gels of myofibrils before and after washing with Triton X-100 (Fig. 6.1) reveal the presence of the 105,000-dalton polypeptide that makes up most of the protein in sarcoplasmic reticular membranes (MacLennan and Holland 1975). This polypeptide migrates slightly above the 100,000-dalton α-actinin polypeptide (Fig. 6.1). Because α-actinin has a subunit molecular weight of 100,000–105,000 in all animal species studied thus far (Robson, personal communication; Suzuki *et al.* 1976), it seems possible that this sarcoplasmic reticular membrane protein has been identified as α-actinin in some studies where the α-actinin polypeptide has been ascribed a molecular weight of 110,000–120,000. It is evident from the gels in Fig. 6.1, however, that Triton X-100 treatment removes no polypeptide chain from skeletal muscle myofibrils except the 105,000–110,000-dalton sarcoplasmic reticular polypeptide. It should be pointed out explicitly that even the non-Triton-treated skeletal muscle myofibrils in Fig. 6.1 have been washed six times with 100 mM KCl and contain virtually no contaminants other than sarcoplasmic reticular membranes. Sarcoplasmic proteins are somewhat more difficult to remove from cardiac muscle myofibrils, and the gels in Fig. 6.1 show that cardiac myofibrils contain very small amounts of sarcoplasmic protein contaminants after six washes with 100 mM

KCl and before Triton X-100 treatment (faint bands near 50,000, 46,000, and 30,000 daltons in cardiac before Triton treatment, Fig. 6.1). Although these bands are absent after Triton X-100 treatment (Fig. 6.1), other experiments have indicated that removal of these bands is due to the extra washing involved in Triton X-100 treatment and not to the Triton X-100 itself.

SDS-polyacrylamide gels of Triton X-100-washed porcine skeletal muscle myofibrils contain 15 protein bands (Fig. 6.1). With only three exceptions, these bands can be identified as originating from a subunit polypeptide of one of the known myofibrillar proteins. The slowest migrating band (labeled 200,000 daltons in Fig. 6.1) is the heavy chain of the myosin molecule; protein in this band accounts for 42–45% of total protein loaded onto SDS-polyacrylamide gels of myofibrils. Below the 200,000-dalton myosin heavy chain are two fainter protein bands migrating with molecular weights of 170,000 (the upper band, which is the faintest of the two) and 160,000. The lower of these two bands (the 160,000-dalton band) is the M-protein polypeptide (Table 6.3; Stromer *et al.* 1977; Trinick and Lowey 1976). The 170,000-dalton band is often extracted with M-protein, but antibodies to this protein do not bind to the M-line of myofibrils (Trinick and Lowey 1976); hence this band represents one of the unidentified polypeptides in myofibrils. The lower of the two faint bands located immediately below the 160,000-dalton M-protein polypeptide (opposite the 140,000-dalton marker in Fig. 6.1) is the C-protein polypeptide (Table 6.3; Offer *et al.* 1973). The band immediately above the C-protein band is a second unidentified polypeptide. This second unidentified polypeptide and C-protein are not easily discernible in gels of porcine cardiac myofibrils (Fig. 6.1). C-protein has not been isolated from cardiac muscle, and based on the evidence provided by the SDS-polyacrylamide gels shown in Fig. 6.1, it may be predicted that cardiac myofibrils either contain no C-protein or the polypeptide in cardiac C-protein has a different molecular weight than the polypeptide in skeletal muscle C-protein. The 100,000-dalton α-actinin polypeptide has already been discussed. The third unidentified protein in myofibrils is a very faint protein band at 54,000–56,000 daltons (56,000 daltons in the skeletal muscle myofibrils and 54,800 daltons in cardiac myofibrils; Fig. 6.1). This band is barely detectable in the skeletal muscle myofibrils but is more evident in the cardiac myofibrils (Fig. 6.1). Although the nature of the protein in this band is unknown, recent studies in our laboratory indicate that this protein remains un-extracted when thick filaments are completely removed to leave only thin filaments and Z-disks and that it is quickly destroyed by treatments that destroy Z-disks. Other preliminary studies in our laboratory have indicated that intermediate filaments, 10 nm in diameter, may be composed in part of a 55,000–60,000-dalton polypeptide (Huiatt, Schollmeyer, Arakawa, and Robson, unpublished results). These results suggest some intriguing and potentially important possible relationships among Z-disks, intermediate filaments, and the 55,000-dalton unidentified component found in myofibrils. Because Z-disks have an important role in meat quality (Goll *et al.* 1974), the 55,000-dalton polypeptide may have critical significance in meat science.

The 42,000-dalton actin band occurs below the 55,000-dalton band. Just below this actin band are three protein bands migrating so closely in the molecular weight range of 34,000–37,000 daltons that they are barely separated from each other (Fig. 6.1). The bottom two of these bands (at 32,758 and 35,000 daltons) are the two subunits of tropomyosin (Table 6.3). Cummins and Perry (1974) have named these two tropomyosin subunits the α (32,758 daltons) and β (35,000 daltons) subunits and have shown that fast-twitch, glycolytic ("white") fibers contain higher proportions of the α-subunit (4:1 α:β) than slow-twitch, oxidative ("red") fibers (1:1 α:β). Somewhat surprisingly, however, cardiac muscle, which contains oxidative muscle fibers, does not follow this pattern (Cummins and Perry 1974). As shown in the SDS-polyacrylamide gels in Fig. 6.1, myofibrils from porcine cardiac muscle contain no detectable 35,000-dalton tropomyosin subunit but have a prominent protein band at 33,000 daltons [polypeptides with mass less than 100,000 daltons have not migrated as far in the gels of cardiac myofibrils as they have in the gels of porcine skeletal myofibrils in Fig. 6.1 (*cf.* the actin bands); hence, the 33,000 cardiac tropomyosin band appears aligned with the 35,000 β-tropomyosin subunit band in Fig. 6.1]. The upper, most slowly migrating band in the triplet of bands below the actin band in the skeletal sample has been identified as troponin-T (TN-T), the largest of the three subunits of troponin (Table 6.3). Although its amino acid sequence indicates that the actual molecular weight of TN-T is 30,503, this subunit of troponin migrates on SDS-polyacrylamide gels as though it had a molecular weight of 37,000. It is possible that the large number of basic amino acid residues in TN-T (TN-T has an isoelectric point of 8.7) causes extra binding of SDS, which imparts a spuriously high molecular mass to the TN-T subunit in SDS-polyacrylamide gel electrophoresis. Alternatively, the SDS bound to TN-T may be insufficient to completely negate the large positive charge on this polypeptide, and the presence of this residual positive charge may slow the rate of migration toward the anode. The TN-T subunit from cardiac muscle migrates in SDS-polyacrylamide gels as though it had a greater mass yet than skeletal TN-T (Brekke and Greaser 1976), and the TN-T band in cardiac myofibrils is probably partly masked by the actin band (Fig. 6.1).

The four protein bands below the 33,000-dalton subunit of tropomyosin in SDS-polyacrylamide gels of porcine skeletal myofibrils (Fig. 6.1) originate from the alkali-1 light chain of myosin (the 20,700-dalton subunit in Table 6.3; band at 25,000 daltons in Fig. 6.1), the TN-I subunit of troponin (band at 23,000 daltons), the DTNB light chain of myosin (the 19,050-dalton subunit in Table 6.3) and TN-C subunit of troponin comigrating (band at 19,000 daltons), and the alkali-2 light chain of myosin (the 16,500-dalton subunit in Table 6.3; this band is almost undetectable on the 7½% gels shown in Fig. 6.1) in order of increasing distance of migration. Comparison of the molecular weights indicated by SDS-polyacrylamide gel electrophoresis (Fig. 6.1) with actual molecular weights given by amino acid sequence (Table 6.3) shows that the alkali-1 myosin light chain, TN-I, and TN-C all migrate more slowly on SDS-polyacrylamide gels than would be expected for polypeptides of their molecular weights. TN-I has a very high isoelectric point (pH = 9.3), and its slow rate of migration may be due to binding of additional SDS that increases the mass of the polypeptide, or to a residual positive charge that slows its rate of migration, as was described for TN-T. The alkali-1 light chain has been shown to contain a very hydrophobic and basic N-terminal amino acid sequence having the composition $ala_{13} pro_{10} ile_3$ $leu_1 val_1 lys_8 glu_2 asp_1 asn_1 ser_1$ (Frank and Weeds 1974). Because the alkali-1 and alkali-2 light chains of myosin have virtually identical amino acid sequences except for this additional N-terminal, 41 amino acids on alkali-1, and because the alkali-2 light chain migrates normally on SDS-polyacrylamide gels, it is presumed that this 41 amino-acid sequence is responsible for the abnormal migration of the alkali-1 light chain in SDS-polyacrylamide gels. The reason for the slightly slow migration of the TN-C subunit on SDS-polyacrylamide gels is unclear; TN-C has an isoelectric point of 4.1 to 4.4 so it seems unlikely that it should either bind excess SDS or contain a residual positive charge in the presence of SDS as was postulated to account for the abnormal electrophoretic behavior of TN-T and TN-I. Although TN-C and the DTNB light chain of myosin are not resolved on 7½% polyacrylamide gels (Fig. 6.1), they can be separated on 10 or 12% polyacrylamide gels (the TN-C subunit migrates slower than the DTNB light chain on such gels), and 10–12% polyacrylamide gels should be used whenever detailed examination of the myofibrillar polypeptides below 42,000 daltons is needed.

The polypeptide chains below the 33,000-dalton tropomyosin band in cardiac muscle myofibrils very clearly differ from those below the tropomyosin band in skeletal muscle myofibrils (Fig. 6.1). It has been shown that both the light chains of myosin (Lowey and Risby 1971; Sarkar *et al.* 1971; Weeds and Pope 1971) and the troponin subunits (Brekke and Greaser 1976) in cardiac myofibrils have different molecular weights than their counterparts in skeletal myofibrils. Cardiac myosin contains only two different kinds of light chains that migrate on SDS-polyacrylamide gels at a rate corresponding to molecular weights of 27,000 and 18,000–20,000 (Lowey and Risby 1971; Sarkar *et al.* 1971). Cardiac TN-I migrates on SDS-polyacrylamide gels as though it had a molecular weight of 28,000, and cardiac TN-C migrates at a rate corresponding to 20,000 daltons (Brekke and Greaser 1976). Amino acid sequence of the cardiac myosin light chains or of cardiac TN-I and TN-C has not yet been determined, and it is not known whether these polypeptides exhibit the aberrant migration behavior on SDS-polyacrylamide gels that their skeletal muscle counterparts do. Based on these molecular weights for cardiac myosin light chains, TN-I, and TN-C, it seems likely that the two major bands below the 33,000-dalton tropomyosin band in the SDS-polyacrylamide gels of cardiac myofibrils (Fig. 6.1) originate from comigration of the largest myosin light chain with TN-I in the upper band and from the similar migration of the smallest myosin light chain with TN-C in the lower band.

Although cardiac muscle is not as important to the human diet as skeletal muscle, SDS-polyacrylamide gels of cardiac myofibrils have been included in Fig. 6.1, and the differences between these gels and gels of ordinary white skeletal muscle myofibrils have been discussed because cardiac muscle cells are a pure source of oxidative or "red" muscle cells that are found in varying proportions in different muscles and meat cuts. A meat scientist using SDS-polyacrylamide gel electrophoresis to study changes in myofibrillar proteins that may be related to meat quality could encounter samples containing mixtures of myofibrils from fast-twitch, glycolytic ("white") and slow-twitch, oxidative ("red") cells. In such a situation, SDS-polyacrylamide gels could obviously contain all the bands shown in Fig. 6.1 for the skeletal muscle myofibrils as well as all the bands shown for the cardiac muscle myofibrils. This example emphasizes the importance of knowing the purity of myofibrils being subjected to SDS-polyacrylamide gel electrophoresis and of running samples of purified myofibrillar proteins as standards for positive identification of the different protein bands separated by the SDS-polyacrylamide electrophoresis. Standards are especially important for identification of protein bands migrating faster than actin both because a large number of different polypeptides migrate in this area and because the polypeptides that migrate in this area often differ between "red" and "white" muscle cells. Therefore, although SDS-polyacrylamide gel electrophoresis is an experimentally simple technique having exceptional ability to resolve different polypeptides, and although use of this technique by food scientists should be strongly encouraged, care must be taken to insure that proper controls are always run and that the simplicity of the procedure does not lull the investigator into a false sense of security. Physiological function of the different myofibrillar polypeptides will be discussed after a brief review of the structure of the myofibril.

Molecular Architecture in Striated Muscle

The nine or ten different myofibrillar proteins discussed in the preceding section are assembled in a very specific way in skeletal or cardiac muscle cells to produce the interdigitating thick and thin filament array that constitutes a myofibril. Figure 6.2 shows a schematic diagram of the structure of vertebrate skeletal muscle at several different levels of organization ranging from an intact muscle, such as the semitendinosus or psoas, at the top of the figure to molecular details of the structure of thick and thin filaments at the bottom of the figure. Like other mammalian tissues, muscle tissue is composed of cells. Muscle cells (synonomous with muscle fibers) differ from many cells because they are very long and spindle-shaped (Fig. 6.2, second diagram from top) and because the interior of muscle cells is literally packed with longitudinal protein threads called myofibrils (Fig. 6.2, third diagram from top). Myofibrils have no membrane encasing them, and they exist as structural entities simply because they are insoluble at the ionic strengths that exist in muscle cells. Use of stereological techniques to perform quantitative electron microscopy has shown that 80–87% of the interior of skeletal muscle cells and approximately 50% of the interior of rat ventricular cardiac muscle cells are occupied by myofibrils (Eisenberg and Kuda 1976). Because muscle protein concentration is approximately 18–20%, and myofibrillar protein makes up 45–50% of total protein in cardiac muscle cells and 52–56% of total protein in skeletal muscle cells (Table 6.1), protein concentration in myofibrils in skeletal and cardiac muscle cells can be calculated to be approximately 110–160 mg/ml, assuming that sarcoplasmic proteins are excluded from the interior of myofibrils (as will be discussed later in this review, this is a reasonable assumption). This same calculation indicates that sarcoplasmic protein concentration in living skeletal and cardiac muscle cells is a remarkable 270–500 mg/ml; such protein concentrations are 20–30 times higher than those usually used in biochemistry experiments and indicates that sarcoplasm in muscle cells may have physical properties that differ considerably from those encountered in *in vitro* experiments.

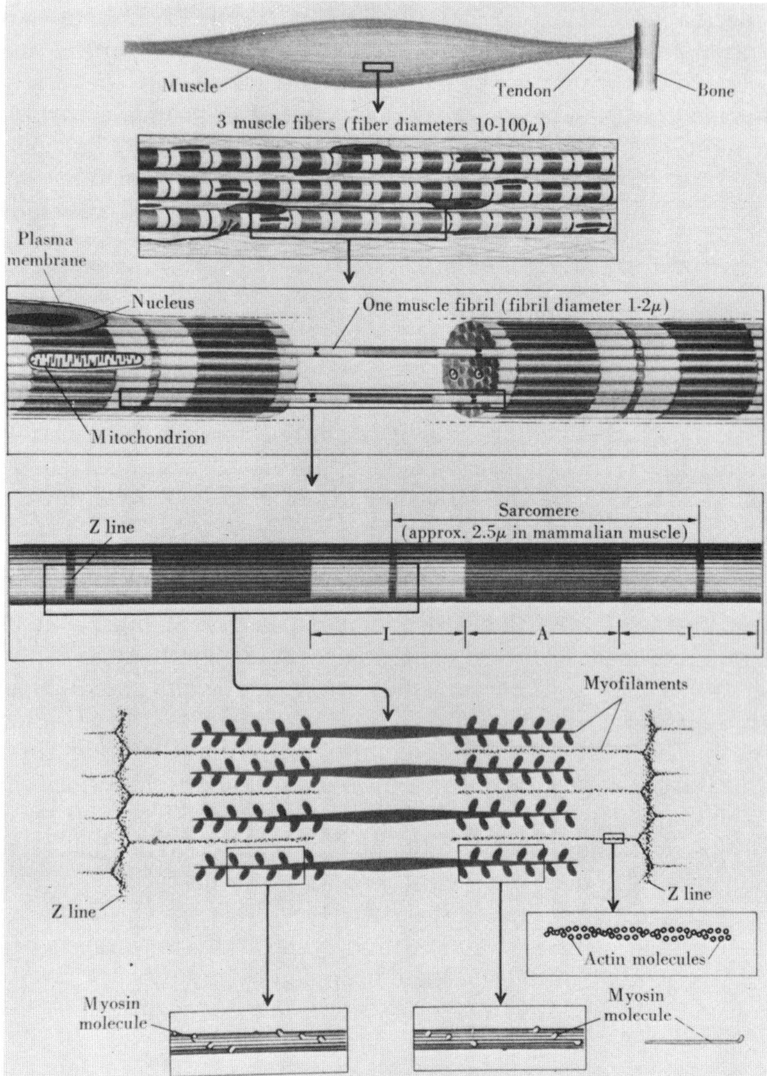

From Novikoff and Holtzman (1970)

FIG. 6.2. SCHEMATIC DIAGRAM SHOWING STRUCTURE OF MATURE VERTEBRATE SKELETAL MUSCLE AT SEVERAL DIFFERENT LEVELS OF ORGANIZATION RANGING FROM AN ENTIRE MUSCLE (TOP OF FIGURE) TO THE MOLECULAR ARCHITECTURE OF THE MYOFILAMENTS (BOTTOM OF FIGURE)

An intact muscle, such as the *biceps femoris* or *semitendinosus*, is shown with tendon attachments to bone at the top of the figure. Muscle tissue is composed of many muscle cells or fibers, three of which are shown in the second panel. As indicated in the diagram, muscle cells are cross-striated and contain nuclei located just under the outer cell membrane. Small lines in the cells in this diagram are intended to represent mitochondria. The third panel portrays a single cell at an intermediate light microscope magnification and shows that the interior of a skeletal or cardiac muscle cell is filled with longitudinal protein threads called myofibrils or muscle fibrils. Myofibrils are cross-striated, and the light and dark bands of adjacent myofibrils lie in register to confer a cross-striated appearance on the entire muscle cell. The fourth panel shows the cross-striated structure of the myofibril at very high light microscope magnification. The dark band is called the A-band, the light band is called the I-band, and the dark line bisecting the light band is called the Z-disk. The contractile unit of the myofibril is called a sarcomere, which is the distance from one Z-disk to the next (as designated). In resting vertebrate skeletal muscle, this distance is 2.3–2.7 μm long and becomes shorter as a muscle contracts. Myofibrils extend completely from one end of the muscle cell to the other. What appears to be single filaments constituting the myofibril at the light microscope level (fourth panel) is seen to be an interdigitating array of two kinds of filaments at a very high electron microscope magnification (fifth panel). The dark band in the light microscope is that area of the myofibril that contains thick filaments, and the light band is that area that contains no thick filaments. Z-disks are transverse structures that link individual sarcomeres in series. As indicated at the bottom of this figure, thin filaments contain two strands of actin molecules wrapped around one another, whereas thick filaments contain myosin molecules packed in a very specific way.

Reproduced by permission of Holt, Rinehart and Winston, Inc.

As shown in Fig. 6.2 (fourth diagram from top), myofibrils in skeletal and cardiac cells are striated and possess alternating light and dark bands when viewed with the phase microscope. Adjacent myofibrils lie with their dark and light bands in register to confer a cross-striated appearance on the entire cell (Fig. 6.2, third diagram from top). For this reason, skeletal and cardiac muscle is sometimes simply called striated muscle. The origin of these alternating light and dark bands was explained simply and directly in 1954 when H. E. Huxley and Jean Hanson (Huxley and Hanson 1954), and independently, A. F. Huxley and Niedergerke (Huxley and Niedergerke 1954) proposed the interdigitating thick and thin filament structure for the myofibril. The essential features of this model are shown in the fifth diagram from the top in Figure 6.2. This interdigitating thick and thin filament array has enormous implications for all phases of meat science, and a strong case can be made for the proposal that discovery of this structure is the single most important finding ever made in meat science.

Briefly, the interdigitating thick and thin filament model states that myofibrils are composed of two sets of interdigitating filaments. One set of filaments, called the thick filaments, are 12 to 16 nm in diameter and are twice the diameter of the second, 6- to 8-nm, set of filaments called the thin filaments. Convincing and varied evidence has shown that 94 to 96% of the protein in thick filaments is myosin and the remaining 5–6% of thick filament protein is C-protein (Offer *et al.* 1973). Thin filaments are composed of actin, tropomyosin, troponin, and β-actinin and are attached at one end (Fig. 6.2, fifth diagram from the top) to a transverse structure called the Z-disk. Z-disks are composed of α-actinin and tropomyosin and possibly also contain actin and the 55,000-dalton polypeptide discussed in the previous section on SDS-polyacrylamide gels of myofibrils (Schollmeyer, Goll, Dayton, Robson, Singh, and Stromer, unpublished results). Current evidence suggests that the amorphous component in Z-disks (shown as a dotted band in Fig. 6.2, fifth panel) contains α-actinin, whereas the Z-filaments are composed of actin and tropomyosin (Schollmeyer, Goll, Robson, and Stromer, unpublished results). The M-line, which is a tranverse structure located at the middle of the thick filaments and which is not shown in Fig. 6.2, is made up of M-protein(s). As shown in Fig. 6.2 (fifth diagram from top), thick filaments contain protrusions or cross-bridges that extend outward at regular intervals from the surface of the filament; it will be indicated subsequently in this review that these cross-bridges are parts of individual myosin molecules.

The interdigitating thick and thin filament structure is easily and directly related to the cross-striated appearance of myofibrils as seen in the phase microscope. The dark bands in the phase microscope are those areas of the myofibril that contain thick filaments, whereas the light bands in the phase microscope are those areas that have no thick filaments (Fig. 6.1, *cf.* fourth and fifth diagrams). Z-disks appear dark in the phase microscope. Fig. 6.3 shows that the dark and light bands seen when myofibrils are viewed in the phase microscope are also seen when myofibrils are viewed in the electron microscope after staining with the substances usually used in electron microscopy. The light or I-band, dark or A-band, two Z-disks and one sarcomere are designated. Fig. 6.3 also shows that the H-zone, which is the light zone in the center of the dark A-band, corresponds to the area where thin filaments do not meet in the center of the sarcomere. Although it is not seen clearly because cross-bridges are not shown in the schematic diagram, the pseudo-H-zone indicated in Fig. 6.3 corresponds to that area in the center of thick filaments that has no cross-bridges (see Fig. 6.2, fifth panel from the top). The M-line is readily seen in the electron micrograph in Fig. 6.3; it is difficult, however, to show the M-line schematically in longitudinal sections, and exact structure of the M-line will become clearer subsequently in this review when cross-sections of the myofibril are examined.

It is extremely difficult to resolve thick and thin filaments in the electron micrograph in Fig. 6.3; this difficulty is not due to lack of magnification but rather to thickness of the section being examined. The thick and thin filaments in the plane immediately under those on the surface of a section do not lie directly under the filaments on the surface but are displaced slightly above or below those filaments. Consequently, if a section is thick enough to include several layers of filaments, the individual thick and thin filaments become difficult to resolve because of interference from adjoining, slightly displaced layers. If sections are sufficiently thin to include only one or two

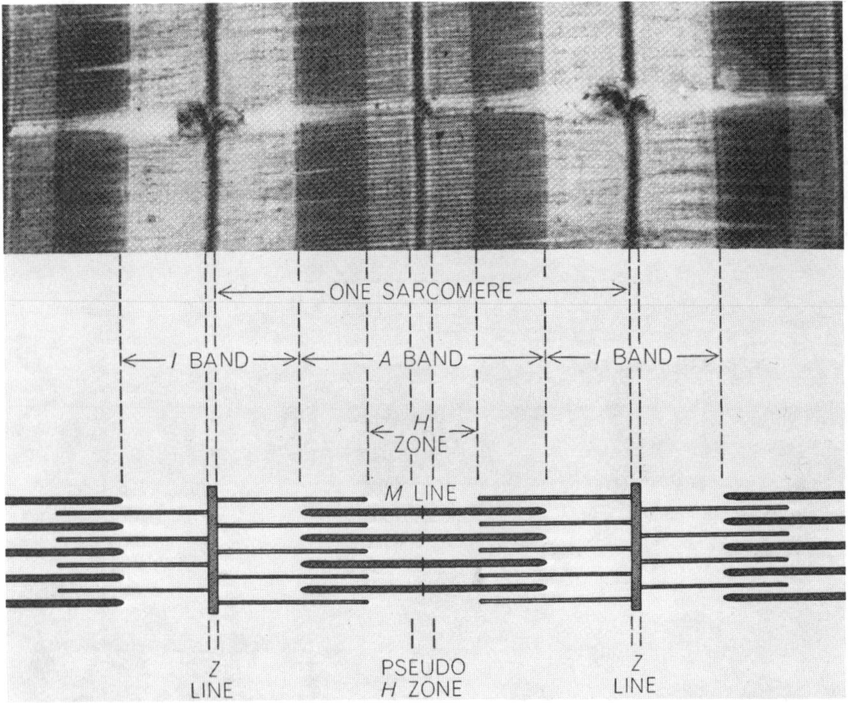

From Huxley (1972)

FIG. 6.3. AN ELECTRON MICROGRAPH OF A LONGITUDINAL SECTION THROUGH A FROG SARTORIUS MUSCLE IS SHOWN AT THE TOP OF THIS FIGURE, AND A SCHEMATIC DIAGRAM OF A LONGITUDINAL VIEW OF THE INTERDIGITATING THICK AND THIN FILAMENT STRUCTURE OF THE MYOFIBRIL IS SHOWN AT THE BOTTOM

The schematic diagram is arranged so that the filaments are aligned with the areas they occupy in the electron micrograph. Sarcomere length is about 2.5 μm.

Reproduced by permission of Academic Press, Inc.

layers of filaments, however, the thick and thin filament structure is readily seen in electron micrographs (Fig. 6.4 a and b). That two thin filaments are interdigitated between each pair of thick filaments in Fig. 6.4a and 6.4b rather than one thin filament between each pair of thick filaments as has been indicated in the schematic diagrams shown thus far (Fig. 6.2 and 6.3) is a consequence of the plane of sectioning. This point will be seen more clearly after the structure of the myofibril in cross-section has been examined.

Structure of Cross-Sections through Myofibrils.—Myofibrils have a highly organized cross-sectional structure in addition to the very characteristic longitudinal pattern discussed in the preceding paragraphs. Cross-sections through the region of overlap of thick and thin filaments in vertebrate skeletal muscle shows that each thick filament is surrounded by six thin filaments and each thin filament is

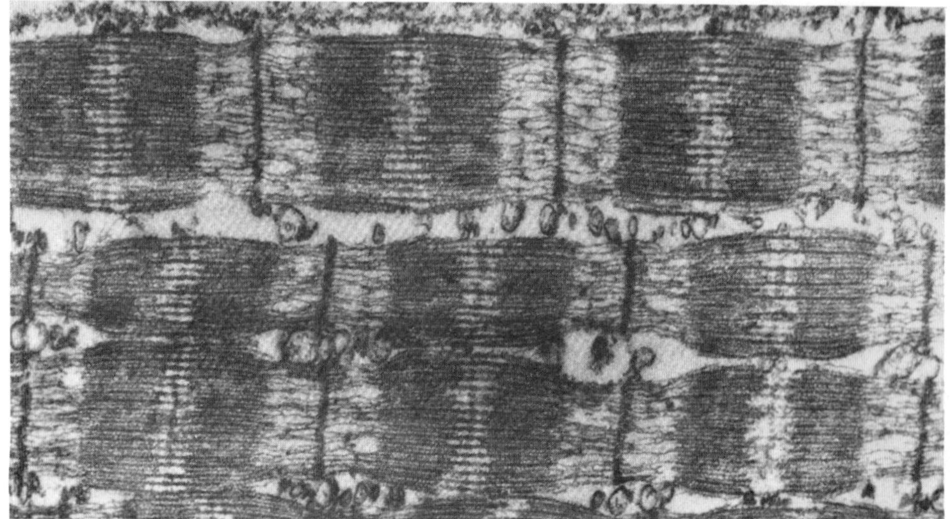

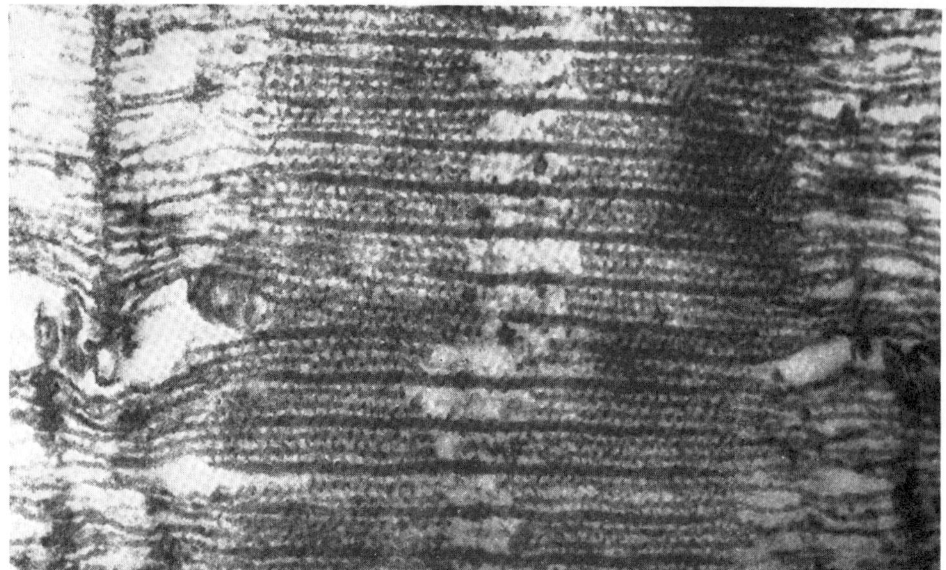

From Huxley (1972)

FIG. 6.4. ELECTRON MICROGRAPHS OF LONGITUDINAL SECTIONS OF GLYCERINATED RABBIT PSOAS MUSCLE CUT SUFFICIENTLY THIN SO THAT ONLY SINGLE LAYERS OF FILAMENTS LIE WITHIN THE PLANE OF THE SECTION

The thick and thin filaments can be clearly seen under these conditions: (a) Lower magnification micrograph, X44,500; (b) High magnification micrograph of the same myofibrils shown in Fig. 6.4a. Only a single sarcomere is seen in this high magnification micrograph, and Z-disks bounding this sarcomere are seen on either edge of the figure. Cross-bridges extending between the thick and thin filaments are visible, X118,000.

Reproduced by permission of Academic Press, Inc.

surrounded by three thick filaments (Fig. 6.5a). Although this cross-sectional structure is observed in all vertebrate striated muscles, invertebrate muscles may contain anywhere from six to twelve thin

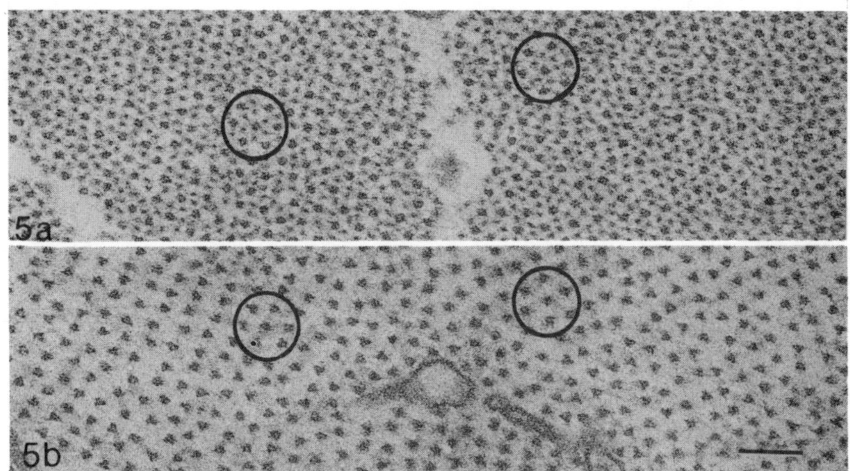

FIG. 6.5. ELECTRON MICROGRAPHS OF CROSS-SECTIONS THROUGH RABBIT PSOAS MYOFIBRILS AT TWO DIFFERENT PLANES

(a) Cross-section through the region of thick and thin filament overlap in the A-band. Both thick and thin filaments are seen in cross-sections. Circles show instances of single thick filaments surrounded by six thin filaments, which are in turn encircled by a triangular array of three thick filaments. (b) Cross-section through the H-zone in the center of the A-band. Only thick filaments are seen in cross-section. The hexagonal array of thick filaments is evident in this cross-section. In some areas, the plane of section passes close to the M-line, and faint outlines of M-bridges can be seen extending from one thick filament to the next (circles). Bar is 100 nm.

filaments surrounding one thick filament, depending on the species and the type of muscle. In vertebrate skeletal muscle, therefore, the ratio of number of thin filaments to number of thick filaments is 2:1, whereas this ratio is 3:1 in insect flight muscle and 6:1 in some types of crustacean muscle. In general, those muscles that contain large thick filaments (25–30 nm in diameter) with paramyosin cores are surrounded by larger numbers of thin filaments. Cross-sections through the H-zone in the middle of the A-band of vertebrate striated muscle show that thick filaments are arranged in a hexagonal lattice in this muscle (Fig. 6.5b). In some instances, where the plane of section approaches the M-line, faint protein bridges can be seen extending from one thick filament to the next (circles in Fig. 6.5b). These bridges are centered in the M-line region of the sarcomere and join thick filaments at their centers to hold them in their hexagonal array. Longitudinal sections of muscle cells reveal arrays of overlapping M-bridges between centers of adjacent thick filaments. These arrays of bridges collectively form the dark M-line seen in the center of the A-band.

The highly organized lattice-structure of myofibrils has special importance to meat science because approximately 65–95% of total water-holding capacity of muscle is due to water trapped in the lattice spacings between the protein filaments of the myofibril. Only

40–80 g out of 290–380 g of water/100 g of protein in vertebrate skeletal muscle is directly bound to proteins, and the remaining 250–300 g water/100 g of protein is found in the thick and thin filament lattice. Of the 40–80 g water/100 g of protein that is bound directly to muscle proteins, 4–10 g/100 g of protein is tightly bound as a monomolecular layer around the thick and thin filaments, another 4–10 g/100 g of protein is immobilized as a second layer just outside the first layer bound to the filaments, and approximately 20–60 g/100 g protein is loosely bound in two to three layers outside the second layer of immobilized water. The tightly bound and immobilized layers of water remain unfrozen when muscle tissue is stored below 0°C and seem unaffected by denaturation of the myofibrillar proteins. The loosely bound third layer, on the other hand, freezes when muscle is stored below 0°C and decreases after denaturation of the myofibrillar proteins. Because most of the water in muscle tissue is simply trapped in the lattice spacings of myofibrils, water-holding capacity of meat is high if these lattice spacings are large, and water-holding capacity decreases if these lattice spacings decrease. Low angle X-ray diffraction measurements of living vertebrate skeletal muscle have shown that thick and thin filaments are 19–31 nm apart center-to-center and that thick filaments are 35–45 nm apart center-to-center, depending on state of contraction (Elliott *et al.* 1963, 1967; Huxley 1968; Huxley and Brown 1967). Electron microscope observations (Franzini-Armstrong 1973) indicate that spacing between actin filaments as they enter the Z-disk varies from 23 nm at a sarcomere length of 2.0 μm to 28 nm at sarcomere lengths near 1.65 μm. Center-to-center distances between thick filaments in rabbit psoas muscle change from 36.8 nm at sarcomere lengths of 2.6 μm to 40.6 nm at 2.3 μm (Huxley 1968). In general, lattice spacing in the thick and thin filament array is proportional to $1/(\text{sarcomere length})^2$ (Elliott *et al.* 1967). These lattice spacings are small enough to exclude almost all sarcoplasmic proteins from the interior of the myofibril lattice (Fig. 6.6). Although the myofibril lattice would have more space in the H-zone, where no thin filaments are present, or in the I-band, where no thick filaments exist, thin and thick filaments ingress into these areas during muscle contraction and would dislodge any sarcoplasmic protein that had diffused into these spaces [indeed, the purported role of creatine kinase as an M-protein (Morimoto and Harrington 1972; Turner *et al.* 1973) may have originated from the fact that this was the only area of the myofibril lattice permeable to the creatine kinase molecule]. Consequently, it seems likely that the myofibril lattice in vertebrate striated muscle contains very little sarcoplasmic protein and is filled largely with water.

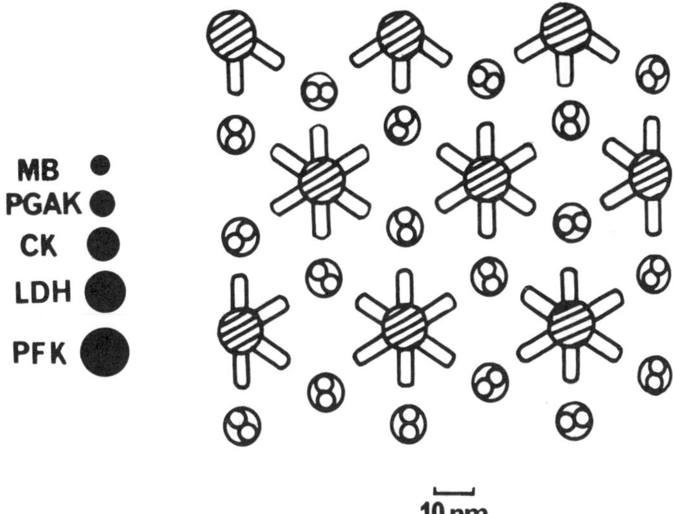

From Scopes (1970)

FIG. 6.6. DIAGRAMMATIC SKETCH OF A CROSS-SECTION THROUGH THE REGION OF THICK AND THIN OVERLAP IN THE A-BAND OF VERTEBRATE STRIATED MUSCLE

Thick filaments with myosin cross-bridges and thin filaments are drawn to scale with proper lattice spacing. Also drawn to scale, assuming a perfectly spherical shape, are some prominent sarcoplasmic proteins. Mb = myoglobin; PGAK = phosphoglycerate kinase; CK = creatine kinase; LDH = lactate dehydrogenase; PFK = phosphofructokinase.

Reproduced by permission of Univ. of Wisconsin Press.

That a number of different factors have similar effects on lattice spacings in myofibrils and water-holding capacity of meat substantiates the conclusion that water-holding capacity of meat is determined largely by size of the lattice spacings between thick and thin filaments in myofibrils. Although myofibrillar lattice spacings increase as a muscle shortens, this increase is accompanied by encroachment of thick filaments into the I-band and of thin filaments into the H-zone. Hence, water-holding capacity decreases. Lattice spacing between thick filaments decreases from 38 nm at pH 7.0 to 33 nm at pH 4.8 (in general, 0.7 pH unit causes a 4% decrease in thick filament lattice spacing in the pH range 4.8–8.0; Rome 1967), and water-holding capacity of meat also decreases in this same range. Thick filament lattice spacing increases markedly from 39.5 nm to 46 nm as ionic strength of the solution bathing the myofibrils is lowered from 0.1 to 0.01, but increasing the ionic strength above 0.2 has little effect on thick filament lattice spacing (Rome 1968). At any given ionic strength, thick filament lattice spacing is approximately 4 nm smaller if Mg^{2+} or Ca^{2+} are present than if Na^+ or K^+ are present (Rome 1968). Recently, April and Wong (1976) have reported that thick filament lattice spacings increase when the sarcolemma is removed or damaged. It is possible, therefore, that part of the slight increase in water-holding capacity of meat during postmortem aging is due to deterioration of the sarcolemma during postmortem aging.

The schematic diagram of cross-section through thick and thin filaments shown in Fig. 6.6 also illustrates why some longitudinal sections through myofibrils will have one thin filament between two adjacent thick filaments, whereas other longitudinal sections will show two thin filaments between two adjacent thick filaments. If a longitudinal section is cut in a horizontal plane through the myofibril shown in Fig. 6.6, one thin filament will be seen between two adjacent thick filaments. If, however, the longitudinal section is cut in a plane 45° to the horizontal, then two thin filaments will be seen between two adjacent thick filaments. Therefore, the appearance of one thin filament between two thick filaments in some longitudinal sections and two thin filaments between two thick filaments in other longitudinal sections is a simple consequence of the three-dimensional structure of the myofibril.

Molecular Arrangement of Myofibrillar Proteins in Thick and Thin Filaments

Myosin and The Molecular Architecture of Thick Filaments.— Myosin has been an extensively studied protein molecule, and much is known about its structure. As indicated in Table 6.3, the myosin molecule from muscle tissue is large and contains six subunit polypeptide chains. An impressive amount of information obtained from direct electron microscope observation (Lowey *et al.* 1969; Slayter and Lowey 1967), selective tryptic and papain digestion

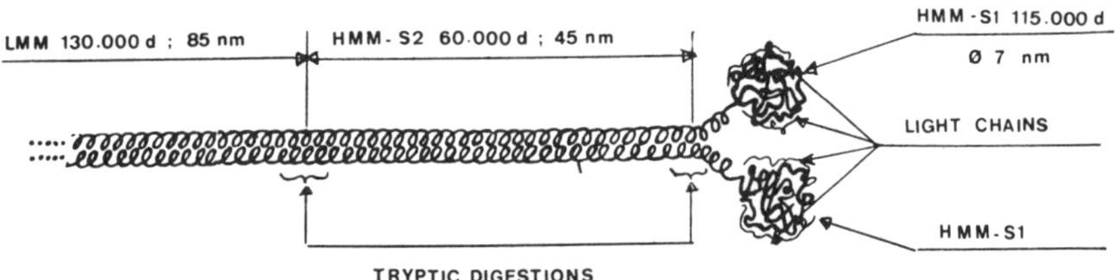

From Morel and Pinset-Härstrom (1975)
FIG. 6.7. SCHEMATIC DIAGRAM OF THE MYOSIN MOLECULE
Part of the long rodlike portion is not shown at the left side of the diagram so that the remainder of the molecule could be shown in proper proportion. The two points at which trypsin cleaves the molecule into light meromyosin (LMM), subfragment 2 (HMM-S2), and subfragment 1 (HMM-S1) are indicated. The α-helical nature of the polypeptide chains in the rod portion of the molecule is shown, and the location of the light polypeptide chains in the head portions of the molecule is indicated. Approximate lengths and molecular weights of the tryptic fragments are shown. Although the diameter of HMM-S1 is given as 7 nm in this diagram, recent evidence shows that HMM-S1 is ellipsoidal with axes of 4 X 15 to 18 nm.
Reproduced by permission of Masson & Cie Press, Paris.

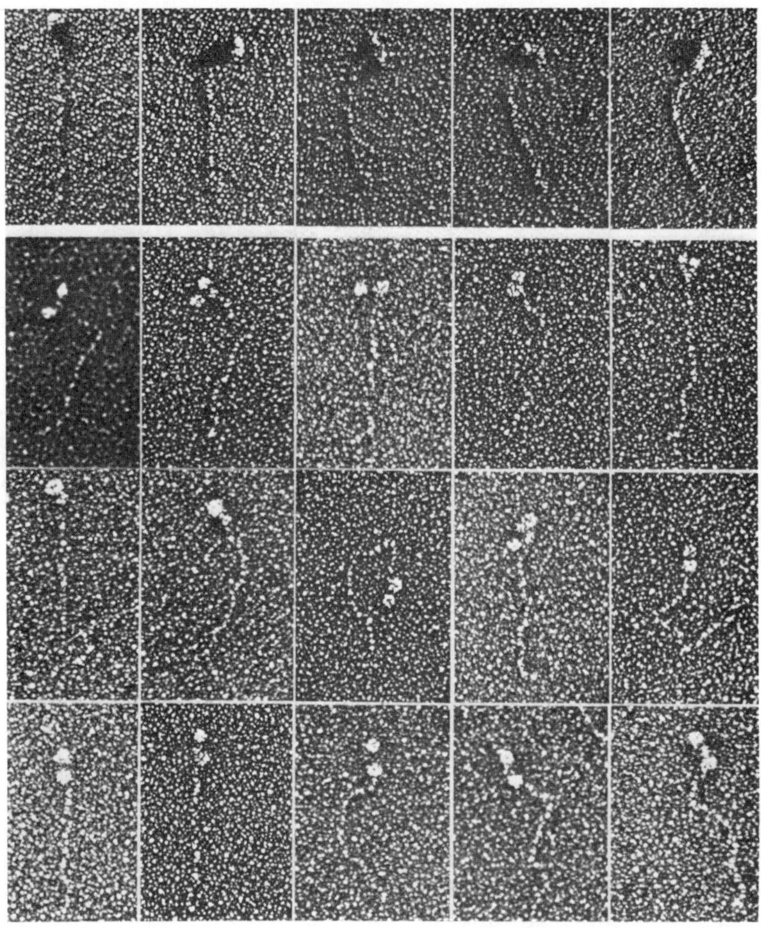

From Slayter and Lowey (1967)

FIG. 6.8. ELECTRON MICROGRAPHS OF SELECTED UNIDIRECTIONALLY
SHADOWED (TOP ROW) AND ROTARY SHADOWED (SECOND, THIRD,
AND FOURTH ROWS) MYOSIN MOLECULES.

The second row contains molecules with the axis of the doublet head
perpendicular to the rod, whereas the third and fourth rows contain molecules
with the axis of the doublet head parallel with the rod, X140,000.

Reproduced by permission of the National Academy of Science.

(Lowey *et al.* 1969), and separation of subunit polypeptide chains
(Lowey and Risby 1971; Weeds and Lowey 1971; Weeds and Pope
1971) has shown that the myosin molecule consists of a long rod
with two globular heads at one end (Fig. 6.7 and 6.8). The rod
portion of the molecule is 135–140 nm long and approximately 1.5
nm in diameter (an axial ratio of 90 to 95). The two globular heads
are actually ellipsoidal and are approximately 15–18 nm long and 4.5
nm in diameter (an axial ratio of 3.3:4.0). The intact myosin
molecule is 155–160 nm long, has 50–55% of its polypeptide chains
in the form of an α-helix, contains 40–42 sulfhydryl groups per

TABLE 6.4
THREE PHYSIOLOGICALLY IMPORTANT PROPERTIES OF MYOSIN

Property	Some Characteristics of the Property
(1) ATPase activity	Without actin, it is activated by Ca^{2+}, inhibited by Mg^{2+} With actin, activated by Ca^{2+}, activated by Mg^{2+}
(2) Binds to actin	$K_{dis} = 10^{-8}$ to 10^{-7} M, actin-myosin complex is specifically dissociated by ATP, pyrophosphate, and a few other polyanions. Mg^{2+} is required for this dissociation. The higher the ionic strength, the less ATP required to dissociate the actin-myosin complex.
(3) Forms filaments	Myosin molecules aggregate head-to-tail to form parallel dimers and these dimers aggregate to form filaments. Aggregation is due to pattern of charged amino acid distribution in the rod portion of myosin molecule and is disrupted by high ionic strength.

molecule, and has no cystine. Extensive biochemical studies of myosin have shown that it has at least three physiologically important properties (Table 6.4).

Much of the detailed information now available on substructure of the myosin molecule has been obtained by using the discovery that trypsin cleaves the myosin molecule into discrete fragments and that the characteristic properties of myosin are segregated among these fragments (Lowey *et al.* 1969). A very brief 5-min incubation with 1 part of trypsin to 300 parts of myosin by weight splits the 160-nm long myosin molecule near its center into two fragments called light meromyosin and heavy meromyosin (Fig. 6.7). Light meromyosin is insoluble in water, soluble at ionic strengths above 0.3, has no ATPase activity or actin-binding ability, is almost 100% α-helical in nature, and under proper conditions will form filaments that resemble thick filaments without cross-bridges (see Fig. 6.2). Heavy meromyosin is soluble in water, has both ATPase activity and actin-binding ability, does not form filaments, is approximately 45% α-helical, and is seen in the electron microscope to consist of two globular heads attached to a short rod (Fig. 6.7). Heavy meromyosin contains all the light chains of the parent myosin molecule. A more extensive 40–60 min incubation with 1 part of trypsin to 50 parts of heavy meromyosin by weight splits heavy meromyosin into two different fragments called subfragment 1 (HMM-S1) and subfragment 2 (HMM-S2; Fig. 6.7). Two moles of subfragment 1 are produced per one mole of subfragment 2. Subfragment 2 is soluble in water, is about 90% α-helical, has no ATPase activity, actin-binding ability, **or** light polypeptide chains, and is a short rod-shaped molecule. Subfragment 1 is soluble in water, has the light chains found in the original myosin molecule (although the extensive tryptic digestion frequently destroys the 19,050-dalton light chain), has ATPase activity and actin-binding ability, and has very little (less than 20%) α-helix. This detailed dissection of the myosin molecule by trypsin [papain can be used to cleave off the two heads of the molecule and leave the long rod intact (Lowey *et al.* 1969)] combined with careful physicochemical examination of the separated fragments show that the long-rod portion of the myosin molecule seen in the electron microscope contains the major part of the two large polypeptide chains of myosin (Table 6.3), each coiled in the form of an α-helix (Fig. 6.7) and then coiled around each other to form a supercoil (not shown in Fig. 6.7). The two ellipsoidal heads of the myosin molecule each contain a small part of one of the two large polypeptide chains, one of the two 19,050-dalton light chains, and one of either the 20,700-dalton or the 16,500-dalton light chains (Table 6.3 and Fig. 6.7). Because the ATPase activity and actin-binding ability of myosin are found only in the subfragment 1 tryptic fragment, the active sites for enzymic activity and actin-binding must be located in the two heads of the myosin molecule. Moreover, because myosin has two binding sites for ATP and because both heads are enzymatically active in subfragment 1 preparations, the myosin molecule has two enzymatically active sites. The 19,050-dalton light chain found in the myosin heads is sometimes referred to as the DTNB-light chain because it can be selectively removed from the myosin molecule by 5,5'-dithio-*bis*(2-nitrobenzoic acid). Removal of the DTNB-light chain has no obvious effect on the ATPase and actin-binding ability of myosin. Current thinking indicates that this light chain may bind Ca^{2+} at Ca^{2+} concentrations of $10^{-6} M$ and that, in some muscles, Ca^{2+}-binding by the DTNB light chain may regulate contractile function of myosin. This proposed physiological function for the DTNB light chain seems vestigial in vertebrate striated muscles, however. The 20,700-dalton and 16,500-dalton light chains of myosin (Table 6.3) are more difficult to remove from the myosin molecule than the DTNB light chain, and strong denaturing solvents, such as 10 M urea, 6 M guanidine HCl, or 1% SDS, or very high concentrations of some chaotropic salts, such as LiCl, or treatment at alkaline pH values above 10 are required to remove these two light chains. Because these two light chains can be removed at alkaline pH, they are sometimes called the alkali-1 (20,700 daltons) and alkali-2 (16,500 daltons) light chains. Removal of the alkali-1 and alkali-2 light chains always is accompanied by complete loss of ATPase activity and actin-binding ability, so it is presumed that these two light chains are in some way necessary for these activities.

Huxley (1963) first demonstrated that it was possible, by incubating myosin solutions at ionic strengths of 0.10–0.20 and in the pH range of 6.5–7.0, to produce thick filaments that were indistinguishable in the electron microscope from native thick filaments isolated directly from vertebrate skeletal muscle [Huxley's myosin preparations were later shown to contain C-protein, but it is now known that the presence of C-protein is unnecessary for formation of filaments resembling thick filaments although the presence of C-protein alters the form of these filaments (Moos *et al.* 1975)]. Observations at various stages of myosin aggregation during formation of thick filaments suggest that the initial aggregation nucleating thick filament formation is a tail-to-tail aggregation of myosin molecules to produce a short filament having a smooth central region flanked on either side by projections (Fig. 6.9A). These projections are the doublet heads of the myosin molecule and represent the cross-bridges observed in electron micrographs of striated muscle cells (Fig. 6.2, 6.4a and b). Hence, each cross-bridge is derived from a single myosin molecule having two heads. That the cross-bridges contain the heads of the myosin molecule and that, as indicated in the preceding paragraph, these heads contain the actin-binding ability of the myosin molecule means that the actin-binding site in this model for thick filament formation is in the cross-bridges where it is free to interact with actin in the thin filament rather than buried in the shaft of the filament.

The myosin molecules themselves evidently contain much of the information necessary for their assembly to form thick filaments. Because thick filament formation is highly dependent on pH and ionic strength, this information must be in the form of patterns of charged amino acids along the rod portion of the myosin molecule. Physicochemical evidence (Burke and Harrington 1972; Harrington and Burke 1972) has shown that dimers of myosin molecules form spontaneously in myosin solutions. These dimers are parallel aggregates formed by tail-to-head aggregation with one molecule being offset by approximately 43 nm from the other. Consequently, it seems likely that thick filaments are formed by successive addition of these parallel dimers rather than by addition of single myosin molecules, and that the nucleating event for thick filament formation is the tail-to-tail aggregation of two parallel dimers rather than tail-to-tail aggregation of the monomers shown in Fig. 6.9A. It is still not clear why some parallel myosin dimers should aggregate

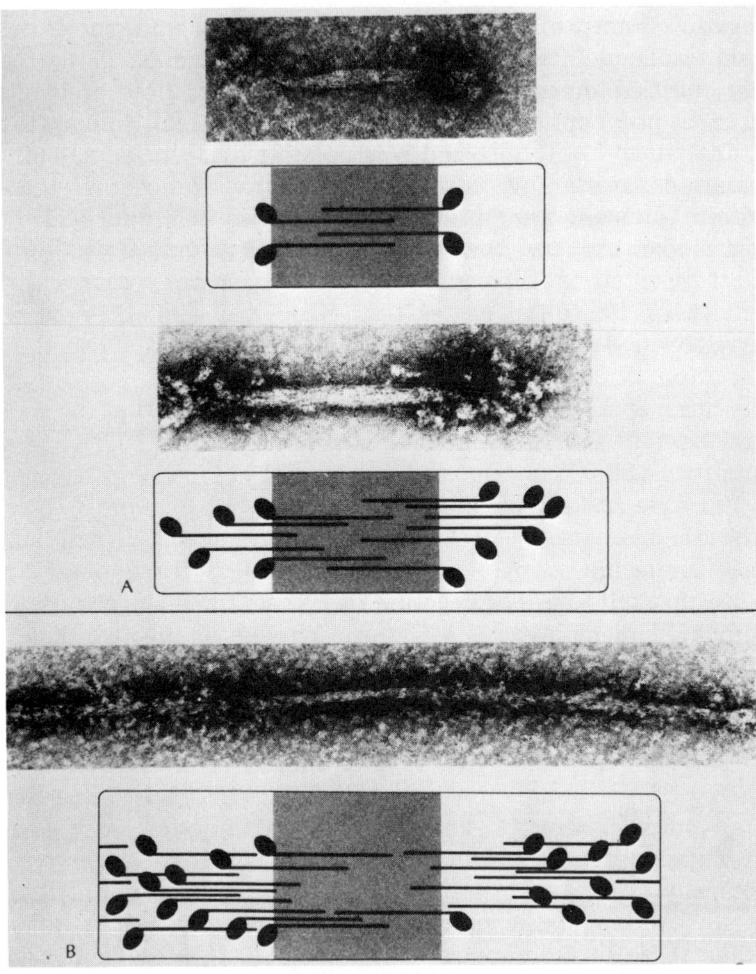

From Loewy and Siekevitz (1969)

FIG. 6.9. ELECTRON MICROGRAPHS AND SCHEMATIC DIAGRAMS SHOW-
ING AGGREGATION OF MYOSIN MOLECULES TO FORM THICK FILA-
MENTS

For simplicity, the double-headed myosin molecule is represented by a single
sphere attached to a rod. Schematic diagrams are intended to portray approximate stage of aggregation shown in the electron micrograph immediately above
the diagram. A. Initial stages of myosin aggregation. Molecules first aggregate
tail-to-tail to form a central region having no heads. Subsequent molecules are
added tail-to-head to produce a regular array of cross-bridges extending from the
surface of the filament. B. A complete thick filament. Only the central part of
this filament is shown in the electron micrograph.

Reproduced by permission of Holt, Rinehart, and Winston, Inc.

tail-to-tail to nucleate thick filament assembly, whereas most of the
myosin dimers must aggregate tail-to-head to produce growth of the
nucleated filament. It is possible that M-protein is involved in the
nucleating tail-to-tail aggregation and that the tail-to-head growth

aggregation is controlled by the profile of charged amino acids on the myosin molecule itself. Indeed, SDS-polyacrylamide gels of even highly purified myosin preparations show that these preparations contain a polypeptide having a molecular weight similar to the 160,000-dalton M-protein, and it is possible that virtually all myosin preparations made by existing purification procedures contain M-protein. Indeed, some preliminary attempts to produce filaments with a myosin that had been specially treated to reduce its M-protein content resulted in filaments that had no smooth central region indicative of tail-to-tail aggregation (Goll and Cullen, unpublished observations). Additional evidence is necessary to establish the role of M-protein in thick filament assembly, however.

Measurements on isolated A-segments have shown that the smooth central region (*i.e.*, the pseudo-H-zone, see Fig. 6.3) in thick filaments is 128.5 nm long (Hanson *et al.* 1971). The M-line itself is 49.5 nm wide and is flanked on either side by a bare zone 39.5 nm in width (Hanson *et al.* 1971). If the subfragment 2 part of the myosin molecule is part of the cross-bridge, these observations indicate that tail-to-tail aggregation of myosin dimers in the center of the thick filament involves a 43.0 nm overlap of the two dimers to produce a 129 nm smooth central region.

Although the general features of the mechanism of myosin aggregation to form thick filaments is now known, at least in the sense that tail-to-head aggregation of dimers is involved, the exact geometry of the completely assembled thick filament in vertebrate striated muscle remains unclear. X-ray diffraction and electron microscope evidence documents that myosin cross-bridges protude periodically every 14.2–14.4 nm along the thick filament (1/3 of the 43.0 nm overlap of the parallel myosin dimer), but it is not clear whether three or four myosin cross-bridges are present at each of these levels. Pedagogically, a model having three myosin cross-bridges at each 14.2 nm interval currently seems favored (Squire 1973). This model (Fig. 6.10) proposes that each set of three cross-bridges is rotated 85° with respect to the sets of cross-bridges on either side, so that the cross-bridges form a helical pattern with a repeat distance of 127.8–129.6 nm along the thick filament; one strand of this helix contains nine cross-bridges in one repeat (Fig. 6.10). Because thick filaments are 1.5–1.6 μm long, it is possible to calculate that each thick filament would contain 300–310 myosin molecules in this three-stranded model for the thick filament.

Antibody studies (Craig and Offer 1976) have shown that C-protein is located in 14 bands that are distributed symmetrically on either side of the pseudo-H-zone of the thick filament and that completely encircle the thick filament. These bands are 43.0 nm

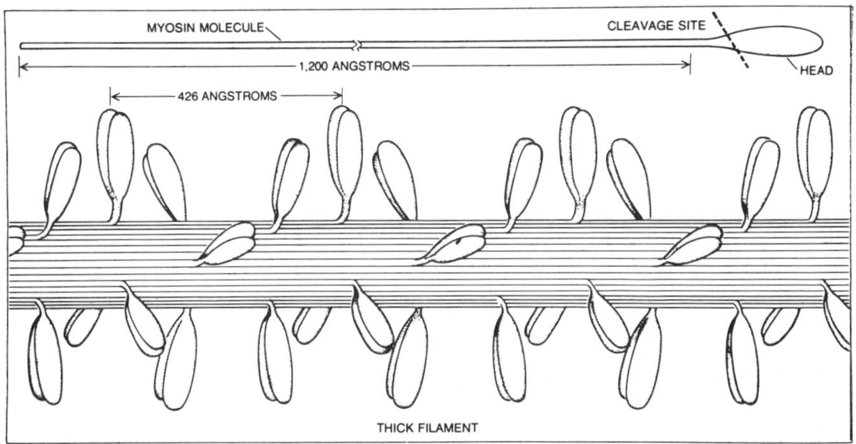

From Murray and Weber (1974)

FIG. 6.10. SCHEMATIC DIAGRAM OF PART OF A THICK FILAMENT SHOWING DETAILS OF A THICK FILAMENT STRUCTURE PROPOSED BY JOHN M. SQUIRE OF IMPERIAL COLLEGE, LONDON

In this model, three sets of double myosin heads (shown as doublets in this diagram) project from the thick filament at intervals spaced 14.2 nm apart along the length of the filament. Each set of these doublet heads is rotated clockwise (to the right) by 60° with respect to the set of three doublet heads just medial (closer to the center of the thick filament) to it, but because the entire filament also has a right-handed twist, the actual rotation between adjacent sites of three doublet heads is 85°. In this model, the helical repeat occurs after two complete rotations and a translational distance of 127.8 nm. Hence, this model suggests that the myosin filament is a three-stranded, 9/1 helix with a repeat distance of 127.8 nm.

Reproduced by permission of Scientific American, Inc.

apart, center-to-center. The innermost two bands are 492 nm apart, and the outermost bands are approximately 300 nm from the tapered ends of the thick filament. Although the function of C-protein is still unclear, Offer *et al.* (1973) have suggested that rings of C-protein may act like staves around a barrel to hold the thick filament together during tension development. Each band of C-protein contains two, three, or four molecules of C-protein (Craig and Offer 1976), but because the amount of myosin in the thick filament is still unclear, it is impossible to ascertain the amount of C-protein in each band within any closer limits. Craig and Offer's antibody studies with C-protein (Craig and Offer 1976) have indicated that thick filaments may contain two other, as yet uncharacterized, proteins that are also located in bands around the thick filament. The significance of these proteins and whether they are indeed components of the myofibril or simply sarcoplasmic contaminants are still unclear.

Actin and the Molecular Architecture of Thin Filaments.—Like myosin, actin is a thoroughly characterized protein molecule. The actin molecule, however, contains only one polypeptide chain and is a much simpler molecule than myosin (Table 6.3). It has been shown

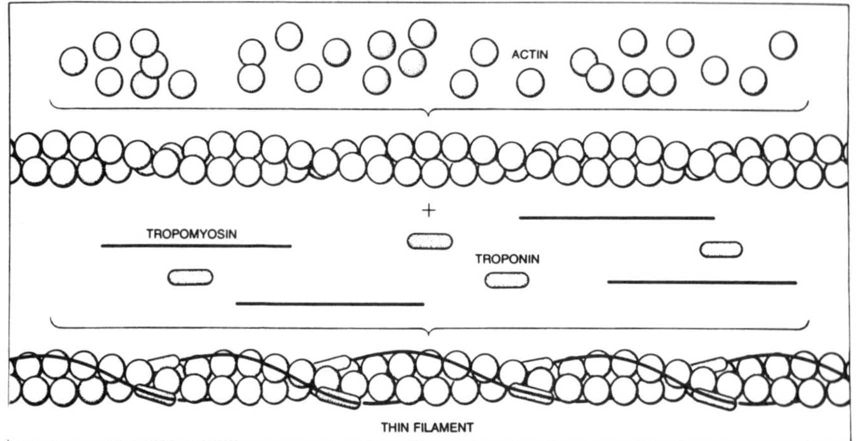

From Murray and Weber (1974)

FIG. 6.11. SCHEMATIC DIAGRAM SHOWING ASSEMBLY OF THE THIN FILAMENT FROM ACTIN, TROPOMYOSIN, AND TROPONIN, AND THE MOLECULAR ARCHI-TECTURE OF THE ASSEMBLED THIN FILAMENT

Actin monomers, shown as spheres, tropomyosin monomers, shown as rods, and troponin monomers, shown as prolate ellipsoids, are protrayed approximately to scale as indicated by physicochemical measurements.

Reproduced by permission of Scientific American, Inc.

that actin contains one molecule of ATP and, as usually prepared, one molecule of Ca^{2+} (Mg^{2+} can be substituted for the Ca^{2+}) per molecule of actin. A large number of different conditions, including monovalent salt concentrations of 50–500 mM or Mg^{2+} or Ca^{2+} concentrations above 1 mM, cause aggregation of monomeric actin to form a double-stranded, helical filament (Fig. 6.11, top). During aggregation, the ATP associated with monomeric actin is hydrolyzed to ADP and inorganic phosphate; the ADP remains associated with the actin monomer after its aggregation into a filament. The physiological significance of the nucleotide and divalent cation bound to actin remains unclear. Actin monomers without nucleotide are very unstable and denature rapidly but, if handled before denaturation, will still polymerize to form double-stranded filaments that are indistinguishable in the electron microscope from ordinary actin filaments. Some very recent work (Cooke 1976) has suggested that actomyosin threads made from actomyosin containing actin filaments that had a nonhydrolyzable ADP analog in place of ADP developed much less tension than actomyosin containing normal actin filaments. It is still unclear, however, whether this effect is due to stabilization of the actin filament by ADP so that the filament will support greater tension, or whether it is due to some direct participation of the actin ADP in tension development. Whatever its physiological role, the bound nucleotide of actin can be used to monitor conformational changes in the actin molecule. For example, the ATP bound to G-actin is freely exchangeable with ATP in the medium, but the ADP bound to actin monomers assembled in a double-stranded filament exchanges very slowly with either ATP or ADP in the medium. Hence, conformation of the actin monomer aggregated in a double stranded filament is different from conformation of unaggregated actin monomers. The complete amino acid sequence of actin is known (Table 6.3). The actin molecule contains 374 amino acids, has five sulfhydryl groups, contains no cystine, has one 3-methyl histidine residue per molecule, and has an intrinsic viscosity, $[\eta]$, near 3.7–4.0 cc/g ($[\eta]$ of a perfect sphere is approximately 3.0–3.3 cc/g); hence, the actin monomer is almost spherical.

As with the myosin aggregates studied by Huxley (1963), *in vitro* aggregation of purified actin elicited by addition of KCl or $MgCl_2$ results in a double stranded filament that, except for heterogeneity of its length, is indistinguishable in the electron microscope (Hanson and Lowy 1963) from the actin backbone of thin filaments (Fig. 6.11). The double-stranded actin filament contains 13 actin monomers per strand per turn, has a helical repeat distance of 35–38 nm, and makes one complete turn every 70–76 nm. Optical diffraction measurements indicate that actin monomers in the double-stranded filament are almost perfectly spherical with a radius of 55 Å (Moore *et al.* 1970). Because native thin filaments from vertebrate striated muscle are uniformly about 1.0 μm long, and because the actin filament has 13 monomers between each cross-over point 35 to 38 nm apart on the double-stranded, helical actin filament, it can be calculated that each thin filament *in vivo* contains 340–380 actin molecules. This is only slightly larger than the 300 to 310 myosin molecules calculated to be present in one thick filament in vertebrate striated myosin (assuming Squire's model for thick filament architecture), but because vertebrate striated muscle has twice as many thin filaments as it does thick filaments, the molar concentration of actin in vertebrate striated muscle is about twice the molar concentration of myosin.

Hanson and Lowy (1963) suggested during the description of their original micrographs of native thin filaments and *in vitro* actin aggregates that strands of tropomyosin might lie in the two grooves of the double-stranded actin filament (Fig. 6.11, bottom). Several years were required, however, before X-ray and optical diffraction evidence (Hanson *et al.* 1973) established that each of the two grooves in the actin filament does indeed contain a strand of tropomyosin (Fig. 6.11). Moreover, optical diffraction evidence (Spudich *et al.* 1972) corroborated with antibody binding studies

(Ohtsuki *et al.* 1967) has shown that troponin is distributed periodically along the thin filament at 38.5 nm intervals (Fig. 6.11). Availability of the complete amino acid sequence of the α-subunit of tropomyosin (Table 6.3) and the knowledge that the tropomyosin molecule contains two polypeptide chains (Table 6.3) and is 100% α-helical in nature has made it possible to calculate that the tropomyosin molecule containing 284 amino acid residues is 42.3 nm long and that tropomyosin molecules aggregated end-to-end in the groove of the double-stranded actin filament are probably overlapped by approximately 1.3 nm or about eight or nine amino acid residues (Johnson and Smillie 1975). Moreover, because the N-terminal region of tropomyosin is relatively rich in lysine residues, whereas the C-terminal end of the tropomyosin molecule contains several glutamic and aspartic acid residues and is devoid of arginine and lysine, the end-to-end aggregation of tropomyosin is probably due to ionic attractions. Other evidence (Stewart 1975) obtained by selective staining of sulfhydryl groups in tropomyosin paracrystals has established that each tropomyosin molecule has one binding site for troponin located at the single cysteine residue of α-tropomyosin. This cysteine residue is 14 nm or 94 amino acid residues from the C-terminal end of the tropomyosin molecule. The site on troponin-T that binds to tropomyosin evidently is about 1/3 of the way in from the N-terminal end of the troponin-T polypeptide chain and is close to one of the three sites on troponin-T that can be phosphorylated (Pearlstone *et al.* 1976). It is clear from this brief summary that molecular structure of the thin filament is known in considerable detail, but paradoxically, the exact helical repeat distance of the double-stranded, helical actin filament has remained elusive.

Before leaving the discussion of actin and the molecular architecture of the thin filament, some of the properties of troponin should be summarized. Greaser and Gergely (1971) first demonstrated that the troponin molecule, molecular weight 69,213, contains three dissimilar subunits, each having a particular physiological role (Table 6.3). These three subunits have been named troponin-T, tropinin-I, and troponin-C, and all three have been completely sequenced (Table 6.3). Troponin-T is the subunit containing the principal binding site that attaches the troponin complex to tropomyosin (T for tropomyosin). Troponin-I binds to both troponin-T and troponin-C and also binds weakly to actin (Hitchcock 1975; Hitchcock *et al.* 1973; Margossian and Cohen 1973). Troponin-I inhibits the ATPase activity of actomyosin in either the absence or presence of Ca^{2+}. Troponin-I also inhibits actomyosin ATPase activity in the absence of troponin-T and tropomyosin, but if these latter two proteins are absent, seven times greater amounts of troponin-I are required to produce complete inhibition. Evidently, one tropomyosin molecule, which spans seven actin monomers in the double-stranded actin-filament, is able to amplify the troponin-I inhibition to all actin monomers it contacts. Troponin-C has four high affinity binding sites (K_b of approximately $5 \times 10^6 \ M^{-1}$) for Ca^{2+} and is therefore the Ca^{2+}-binding subunit of troponin. Troponin-C binds to both troponin-T and troponin-I in the troponin complex (Hitchcock 1975; Hitchcock *et al.* 1973; Margossian and Cohen 1973). The significance of these interactions among actin, tropomyosin, and the troponin subunits will be discussed in the next section on muscle contraction.

MECHANISM OF MUSCLE CONTRACTION

An overwhelming amount of evidence has now been accumulated to show that muscle contraction is accomplished by a sliding together or telescoping of the interdigitating thick and thin filament array (Fig. 6.2 and 6.3) without any detectable shortening of the filaments themselves (see Huxley 1972, for a review). The force causing the thick and thin filaments to slide past one another is generated by the cross-bridges or myosin heads that project outward from the surface of the thick filament (Fig. 6.2, 6.4a and b). Although it is still unclear how the chemical energy in the ATP molecule is converted to the mechanical energy of muscle contraction, a great deal is known about the anatomical and biochemical events that accompany this energy transduction. Indeed, it is possible to divide the events that occur during muscle contraction into two general categories (Table 6.5): (1) those events that occur in the thin filament; and (2) those events that occur in the thick filament. Because the switch for turning muscle contraction on and off is located on the thin filament, this discussion of muscle contraction will start with the processes that occur in thin filaments during muscle contraction.

Changes in Thin Filaments During Muscle Contraction

Although it has been known for almost 10 years that the thin filament was involved in turning muscle contraction on and off, the mechanism of this switching process has been elucidated only recently (Parry and Squire 1973; Spudich *et al.* 1972; Wakabayashi *et al.* 1975). A motor nerve impulse reaching the muscle cell passes along the sarcolemma and into the T-tubules. By a process that is still unclear, passage of an electrical signal along the extracellular T-tubule causes the adjoining, intracellular sarcoplasmic reticular membranes to momentarily lose their ability to retain the calcium that they have accumulated against a concentration gradient. This

TABLE 6.5
TWO BASIC PROCESSES IN MUSCLE CONTRACTION

Process	Description
(1) Turning on the Thin Filament	In resting muscle, the thin filament is in an "off" conformation that prevents myosin from interacting with actin. Binding of Ca^{2+} to troponin on the thin filament turns the thin filament on and allows myosin to bind to actin. Changing from the "on" to "off" conformation is associated with movement of tropomyosin in and out of the groove of the actin helix.
(2) Cycling of Myosin Cross-Bridges	Force for muscle contraction is developed by myosin cross-bridges. In resting muscle, cross-bridges are energized and ready for contraction. Turning the thin filament on (see process 1) permits the cross-bridge to attach to actin and push the actin filament toward the center of the sarcomere. The spent cross-bridge now binds a new molecule of ATP which dissociates the myosin cross-bridge from actin and reorients (energizes) it. The energized cross-bridge splits the ATP so the cross-bridge can now reattach to actin and push. This cycle is repeated until the thin filament is turned off (see process 1).

Ca^{2+} therefore floods into the interior of the cell, and intracellular free Ca^{2+} concentration rises from approximately $10^{-8} M$ in resting muscle to 10^{-5} to $10^{-6} M$. Because troponin-C has a binding constant near $5 \times 10^6\ M^{-1}$ for Ca^{2+} (Potter and Gergely 1975), an intracellular Ca^{2+}-concentration of 10^{-5} to $10^{-6} M$ is sufficiently high to permit binding of Ca^{2+} to troponin-C. Ca^{2+}-binding to troponin-C causes a conformational change (ostensibly, two molecules of Ca^{2+} per molecule of troponin-C are sufficient to cause this change) in the troponin-C subunit (the exact nature of this conformational change is unclear, but it can be detected by changes in tryptophan fluorescence and in circular dichroism spectra), and this conformational change is transmitted to the other two troponin subunits, which then undergo conformational changes of their own. Again, the nature of these conformational changes in troponin-I and troponin-T is unclear, but the overall effect, as measured by *in vitro* biochemical experiments (Hitchcock *et al.* 1973; Margossian and Cohen 1973), is to increase the affinity of the troponin subunits for each other and to decrease the affinity of troponin-I for actin. Hence, the biochemical binding experiments indicate that, in the absence of Ca^{2+} ($10^{-8} M$ or less or so there is no Ca^{2+} on troponin-C), troponin-T binds strongly to tropomyosin, troponin-C binds loosely to troponin-T and troponin-I, and troponin-I binds loosely to troponin-T, but firmly to actin. In the presence of Ca^{2+} (10^{-5} to 10^{-6} M or high enough to permit binding of Ca^{2+} to troponin-C), troponin-T binds strongly to tropomyosin (this linkage is unaffected by Ca^{2+}), troponin-C binds strongly to troponin-T and troponin-I, and troponin-I binds to troponin-T but loses its affinity for actin (it is unclear whether Ca^{2+} affects the troponin-T-troponin-I linkage).

Simultaneously with the biochemical experiments measuring the effects of Ca^{2+} on strength of the interactions among actin, tropomyosin, and the troponin subunits, X-ray and optical diffraction experiments (Parry and Squire 1973; Spudich *et al.* 1972; Wakabayashi *et al.* 1975) indicated that, upon activation of muscle contraction, the tropomyosin strand moves from a radius of 4.4 nm from the center of the double-stranded actin filament inwards to a radius of 2.4 nm (Fig. 6.12). These X-ray and optical diffraction results together with the biochemical binding experiments described

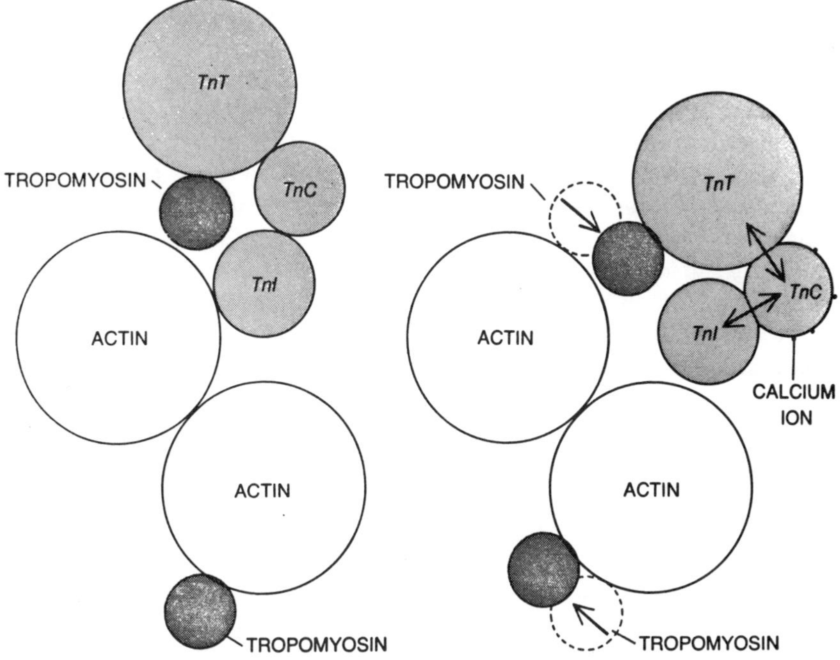

From Cohen (1975)

FIG. 6.12. DIAGRAMMATIC SKETCH OF THIN FILAMENT CROSS-SECTIONS SHOWING HOW INTERACTIONS AMONG THE TROPONIN SUBUNITS, TROPOMYOSIN, AND ACTIN MAY FUNCTION TO REGULATE ABILITY OF THE THIN FILAMENT TO PARTICIPATE IN CONTRACTION

Tropomyosin and actin are indicated; TN-T = troponin T; TN-I = troponin-I; and TN-C = troponin C. Although troponin-I does not contact troponin-T in this schematic diagram, recent evidence (Hitchcock 1975) indicates that troponin-I and troponin-T are in very close proximity in the troponin complex and probably bind to each other.

Reproduced by permission of Scientific American, Inc.

in the preceding paragraph have suggested the following series of events on thin filaments during initiation of muscle contraction. In resting muscle, the tropomyosin strand is located out of the groove of the double-stranded actin filament (left diagram, Fig. 6.12) in a position where it physically blocks or at least sterically interferes with the myosin-binding site of actin (optical diffraction results have shown that location of the tropomyosin strand in resting muscle is very close to the site that subfragment 1 fragments of myosin bind to actin). Troponin is necessary to hold tropomyosin in this out position, and a firm interaction between troponin-I and actin is evidently necessary to maintain this out position of tropomyosin. When troponin-C binds Ca^{2+}, however, the firm linkage between troponin-I and actin is weakened, and troponin can no longer hold the tropomyosin strand in the out or "off" position. The tropomyosin strand rolls back into the groove of the double-stranded actin filament, and the myosin-binding site on actin is exposed (Fig. 6.12). Myosin binds to actin, contraction ensues and continues until the Ca^{2+} is removed from the troponin-C and troponin-I binds to actin and forces the tropomyosin strand back out of the groove of the actin filament to block the myosin-binding site of actin. Because one tropomyosin molecule "covers" or contacts seven actin monomers and because one troponin molecule binds to one tropomyosin molecule, the preceding scheme also provides a simple and direct explanation for why the presence of tropomyosin enhances the inhibitory effects of troponin-I approximately sevenfold. This scheme further provides a mechanism for the ability of troponin-C and Ca^{2+} to derepress the inhibitory effect of troponin-I on actomyosin ATPase activity and muscle contraction. In summary, therefore, turning muscle contraction on and off in vertebrate striated muscle is accomplished by moving the tropomyosin strand in (on) and out (off) of the groove in the double-stranded actin filament (Table 6.5).

Changes in Thick Filaments During Muscle Contraction

X-ray and optical diffraction and electron microscope studies done during the 1960s demonstrated that, during muscle contraction, myosin cross-bridges extend outward, attach to the thin or actin filament, and then swivel or rotate so that the tip of the cross-bridge undergoes approximately a 10 nm translocation, and pushes the actin filament toward the center of the sarcomere (Fig. 6.13). At the end of this stroke, the spent cross-bridge detaches from the thin filament, is reoriented, and is ready to repeat the cycle (Table 6.5 and Fig. 6.13). During a single twitch of a muscle fiber, each myosin cross-bridge may perform many cycles of attaching to actin,

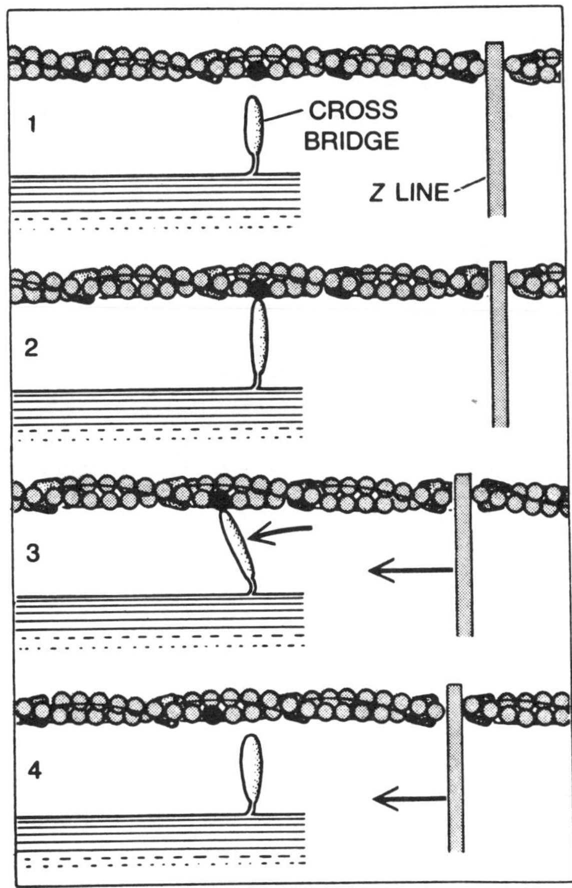

From Murray and Weber (1974)

FIG. 6.13. SCHEMATIC DIAGRAM OF THE CROSS-BRIDGE CYCLE IN STRIATED MUSCLE

In resting muscle, myosin cross-bridges cannot attach to thin filaments because the myosin-binding site on actin is blocked by tropomyosin (1 in diagram). A nerve impulse releases Ca^{2+}, and Ca^{2+} causes a series of conformational changes in the thin filament (see Fig. 6.12) so that the myosin-binding site on actin is unblocked. The myosin cross-bridge attaches to the actin filament (2 in diagram), then swivels or rotates to push the actin filament toward the center of the sarcomere (3 in diagram), and finally, after binding a new molecule of ATP, is detached and reoriented to the resting position ready to repeat the cycle (4 in diagram). In a single contraction, one myosin cross-bridge may go through many such cycles of attachment, swiveling, and detachment.

Reproduced by permission of Scientific American, Inc.

swiveling and then dissociating from the actin filament, but only 10–20% of the total cross-bridges in a single sarcomere are ever attached to thin filaments at any given instant during contraction. The remaining cross-bridges are in various other stages of the cross-bridge cycle. Because cross-bridges interact asynchronously with actin (*i.e.*, they do not all attach to the thin filament simultaneously, then swivel simultaneously, etc.), the thick and thin filaments slide past each other at a smooth, uniform rate during muscle contraction rather than with a "jerky," ratchet-like motion that would be produced by synchronous cross-bridge actin.

Although the mechanics of cross-bridge motion were known eight years ago, only recently has it become possible to relate myosin cross-bridge cycling to actomyosin ATPase activity. Part of the difficulty in establishing this relationship between chemical events and mechanical motion of myosin originated from the fact that ATP has two roles in muscle contraction that until recently have seemed dichotomous. First, ATP hydrolysis is clearly involved in providing energy for muscle contraction (although this involvement may not be as direct as once supposed). Second, convincing evidence has shown that, at physiological concentrations, ATP prevents the actin-myosin interaction and even dissociates the actin-myosin complex necessary for contraction (Table 6.4). Hence, it seemed to early workers that ATP would prevent the very interaction that would lead to its hydrolysis and the release of energy necessary for contraction. Some important recent findings from transient state kinetic studies of the hydrolysis of ATP by myosin (Bagshaw and Trentham 1974; Bagshaw *et al.* 1974; Taylor 1972) have resolved this ostensible dichotomy concerning the roles of ATP in muscle contraction, and apposition of these findings with the earlier results on mechanics of cross-bridge movement during muscle contraction now permits a detailed insight into events in the myosin molecule during muscle contraction.

The transient state kinetic studies showed that, in the presence of Mg^{2+}, myosin binds and splits ATP very rapidly (Steps 1, 2, and 3, Table 6.6). The experimentally determined forward rate constants for the seven steps in the transient state kinetic scheme shown in Table 6.6 are: $k_1 \cong 63 \times 10^7 M^{-1} \text{ sec}^{-1}$, $k_2 = 400 \text{ sec}^{-1}$, $k_3 \cong 160 \text{ sec}^{-1}$, $k_4 = 0.06 \text{ sec}^{-1}$, $k_5 > 50 \text{ sec}^{-1}$, $k_6 = 1.6 \text{ sec}^{-1}$, $k_7 < 2.7 \times 10^3 \text{ sec}^{-1}$. These rate constants clearly show that step 4 is the rate-limiting process during splitting of ATP by myosin in the presence of Mg^{2+} and that, in general, the three steps occurring after step 4 are slower than the three steps occurring before step 4. Step 4 itself involves a conformational change in the myosin molecule because intrinsic fluorescence of a tryptophan residue in the subfragment 1 region of myosin changes during this period. Step 2 in the kinetic mechanism shown in Table 6.6 also involves a conformational change in myosin as measured by intrinsic fluorescence of this same tryptophan residue. Hence, intrinsic fluorescence

TABLE 6.6
TRANSIENT STATE KINETICS OF MYOSIN[1]

$$M+ATP+H^+ \overset{1}{\rightleftharpoons} M{\cdot}ATP \overset{2}{\rightleftharpoons} M^*{\cdot}ATP \overset{3}{\rightleftharpoons} P_i{\cdot}M^{**}{\cdot}ADP \overset{4}{\rightleftharpoons} P_i{\cdot}M^*{\cdot}ADP \overset{5}{\rightleftharpoons} M^*{\cdot}ADP+P_i$$

$$+ \atop H^+$$

$$\downarrow 6$$

$$M{\cdot}ADP$$

$$\downarrow 7$$

$$M+ADP$$

$K_1 = 4.5 \times 10^3 \ M^{-1}$

$K_2 = 3.7 \times 10^9 ; \Delta G = -12{,}900 \ \text{cal/mole}$

$K_3 = 9 ; \Delta G = -1600 \ \text{cal/mole}$

$K_4 = 15 \ (\text{cross-bridge "stroke"})$

$K_5 = 8 \times 10^{-3} \ M$

$K_6 = 3.5 \times 10^{-3}$

$K_7 = 2.7 \times 10^{-4} \ M; \Delta G = +4800 \ \text{cal/mole}$

$K_{\text{overall}} = 1.7 \times 10^7 \ M$

[1] All values are for 100 mM KCl, 6 mM $MgCl_2$, 50 mM Tris·HCl, pH 8.0, 21.0°C. M in the kinetic scheme is myosin subfragment 1. Forward rate constant for Step 4 is 0.06 s^{-1}, and this is the rate-limiting step.

measurements alone show that the subfragment 1 region of the myosin molecule passes through at least five conformational states during hydrolysis of ATP in the presence of Mg^{2+}. One of these conformational states is free myosin, which may be assigned an intrinsic fluorescence of 1.00. Binding of ATP very quickly ($k_1 \cong 63 \times 10^7 \ M^{-1} \ \text{sec}^{-1}$) produces a second conformational state of myosin (M·ATP in Table 6.6) whose intrinsic fluorescence has been estimated at 1.03 relative to the intrinsic fluorescence of free myosin (Bagshaw *et al.* 1974). Immediately after the conformational change resulting from binding of ATP to myosin, the bound ATP induces an additional conformational state ($k_2 = 400 \ \text{sec}^{-1}$) designated as M*·ATP in Table 6.6. This third conformational state has a relative intrinsic fluroescence of 1.10 (Bagshaw *et al.* 1974). Hydrolysis of the bound ATP in this third conformational state occurs at an intermediate rate ($k_3 \cong 160 \ \text{sec}^{-1}$) and produces a fourth conformational form ($P_i{\cdot}M^{**}{\cdot}ADP$ in Table 6.6) that has a relative intrinsic fluorescence of 1.19 (Bagshaw *et al.* 1974). This fourth conformational state decays slowly ($k_4 = 0.06 \ \text{sec}^{-1}$) to a fifth conformational state ($P_i{\cdot}M^*{\cdot}ADP$ in Table 6.6) that has a relative intrinsic fluorescence of 1.06 (Bagshaw and Trentham, 1974). Different intermediate conformational states during hydrolysis of ATP by myosin have also been detected by electron spin resonance spectroscopy (Seidel 1975; Seidel and Gergely 1973) and ultraviolet absorption (Yazawa *et al.* 1973), and at least for the electron spin resonance results, the conformational states detected seem to be the same as those observed by using intrinsic fluorescence (Seidel 1975). Because step 4 is the rate-limiting step in the Mg^{2+}-modified ATPase activity of pure myosin and because actin increases specific activity of the Mg^{2+}-modified ATPase activity of myosin by 100 to 200-fold, the presence of actin must increase the rate constant k_4, and also the rate constant k_6. In the absence of actin, however, the very large rate constants for steps 1, 2, and 3 (Table 6.6) followed by the small rate constant for step 4 results in accumulation of the form $P_i \cdot M^{**} \cdot$ ADP in resting muscle. Indeed, direct chemical measurements have shown that the nucleotide bound to myosin in resting muscle is almost entirely ADP.

Combination of the data from transient state kinetic studies with the X-ray diffraction and electron microscope results on the cross-bridge cycle leads to the following scheme of events in myosin cross-bridges during muscle contraction. In living resting muscle, almost every myosin cross-bridge is "energized" and contains one molecule each of ADP and inorganic phosphate, the hydrolysis products of ATP. "Energized" here refers to some unknown conformational state of myosin in which the energy of ATP is stored; this state is represented by $P_i \cdot M^{**} \cdot$ ADP in Table 6.6. In resting muscle, the energy in energized myosin is dissipated slowly through step 4 of the transient state mechanism shown in Table 6.6. The energized cross-bridge is unable to bind to actin because the thin filament is "turned off," *i.e.*, tropomyosin is blocking the myosin-binding site on actin. As described in the preceding section, muscle contraction is triggered by a release of Ca^{2+} from the sarcoplasmic reticular membranes, and intracellular free Ca^{2+} concentrations rise to approximately 10^{-6} *M*. This enables the troponin-C polypeptide to bind Ca^{2+}, and the thin filament is "switched on" by the series of events described in the preceding section (see also Table 6.5). Because the ATP bound to myosin has already been split to ADP and P_i (the $P_i \cdot M^{**} \cdot$ ADP form), and because neither ADP nor P_i are very effective dissociators of the actin-myosin complex, the myosin cross-bridge interacts with actin in the thin filament immediately after the actin is unblocked in the "switching on" process. Interaction with actin (step 2 in Fig. 6.13) not only accelerates the rate-limiting step (step 4 in Table 6.6) in ATP hydrolysis by myosin but also is accompanied by a swiveling or rotating of the myosin cross-bridge so that the actin filament is pushed (*not* pulled) toward the center of the sarcomere (step 3 in Fig. 6.13). The exact nature of actin's participation in cross-bridge swiveling or force generation is unclear. Actin filaments may simply be rods that myosin cross-bridges push on, or actin may modify myosin, not only to increase the rate of step 4 in the hydrolysis of ATP (Table 6.6) but also to participate actively in swiveling of the cross-bridge and force generation. That is, force for contraction may be generated at the point of the actin-myosin interaction rather than within the myosin cross-bridge itself. Once the myosin cross-bridge has rotated or swiveled, steps 5 through 7, involving release of the products of ATP hydrolysis in the transient state kinetic mechanism for ATP hydrolysis (Table 6.6), follow rapidly.

Only after ADP has been released from the spent cross-bridge (step 7, Table 6.6) can the myosin head bind a new molecule of ATP. Because ATP is a potent dissociator of the actin-myosin complex, binding of a new molecule of ATP immediately (or faster than existing techniques can measure) dissociates the myosin cross-bridge from the actin filament. At this point, however, the myosin cross-bridge is angled with respect to the actin filament, and even it if were to rebind to the actin filament, it could not swivel and push the filament toward the center of the sarcomere. Consequently, binding of ATP and the resulting dissociation of the myosin cross-bridge from actin must be followed by a reorientation of the angled or spent cross-bridge back to the state shown in step 4, Fig. 6.13. Although it was originally assumed that energy for this reorientation of myosin cross-bridges originated from hydrolysis of ATP, determination of the rate constants in the transient state mechanism for ATP hydrolysis by myosin (Table 6.6) has shown that by far the largest free energy change in this mechanism occurs at step 2 immediately after ATP is bound to the myosin head and before it is hydrolyzed to ADP and P_i. Indeed, actual hydrolysis of ATP releases only -1600 cal/mole of free energy (Table 6.6). Therefore, it seems that the energy of ATP is released in the form of a protein conformational change during the binding of ATP to the myosin cross-bridge and that hydrolysis of ATP to ADP and P_i is necessary only to allow the myosin cross-bridge to bind to actin whenever actin is available. Thus, dissociation of the spent cross-bridge (step 3, Fig. 6.13) from actin and reorientation to prepare it for a new cycle are associated with steps 1 and 2 in Table 6.6 and not with the hydrolysis of ATP in step 3.

The cyclic series of events described in the preceding paragraph continues uninterrupted until Ca^{2+} is rebound into the sarcoplasmic reticular membranes and the thin filament is "turned off" (Table 6.5) so that actin is no longer available to bind myosin cross-bridges (this occurs in physiological relaxation) or until muscle ATP is exhausted and no ATP is available to bind to the spent myosin cross-bridge and dissociate it from actin (this occurs during rigor mortis in muscle after death). In the latter situation, all myosin cross-bridges stop attached to actin at an angled position (step 3, Fig. 6.13); this is probably the state of most myosin cross-bridges in meat. As will be mentioned briefly in the next section and as has been discussed more extensively elsewhere (Goll *et al.* 1974, 1977), the nature of the binding of myosin cross-bridges to actin probably has immense effects on meat quality.

Although the events that occur in myosin cross-bridges during contraction of striated muscle can be specified in remarkable detail, several important questions remain completely enigmatic. The uncertain extent to which actin participates in the contractile process and the site at which force for cross-bridge movement is generated have already been mentioned. The significance of the double-headed nature of the myosin cross-bridge is unclear. Binding experiments show that only one subfragment 1 tryptic fragment of myosin binds to one actin monomer in actin filaments, so it seems very unlikely that both heads of the double-headed cross-bridge bind to the same actin monomer during contraction. It is still uncertain whether the two myosin heads act cooperatively during contraction, with one head binding during one cross-bridge cycle and the other head during the successive cross-bridge cycle, etc. Finally, the exact nature of the conformational changes in myosin, and possibly also in actin, during contraction remains completely unclear. Measurements done thus far indicate that the subfragment 1 part of the cross-bridge remains relatively rigid during contraction and functions somewhat like an oar or an impeller. Hence, the conformational change for swiveling or rotation of the cross-bridge must occur in the subfragment 2 part of the myosin molecule or in the proteolytically sensitive regions that link subfragment 1 and subfragment 2 or that link light and heavy meromyosin. It is evident that more information is needed on the changes that occur in myosin cross-bridges during contraction before the molecular nature of these changes and their exact role in meat quality can be specified.

POSSIBLE IMPLICATIONS OF THE MOLECULAR ARCHITECTURE AND BIOCHEMICAL PROPERTIES OF MYOFIBRILLAR PROTEINS IN MEAT QUALITY

As indicated in the introduction of this review and as has been described elsewhere (Goll *et al.* 1970), the myofibrillar proteins are vitally important to many aspects of meat quality. Some possible effects of the myofibrillar proteins on meat quality are summarized in Table 6.7. The role of myofibrillar proteins in meat tenderness has recently been discussed (Goll *et al.* 1974). The existing evidence indicates that myofibrillar proteins can affect meat tenderness in at least two different ways (Goll *et al.* 1974): (1) binding of myosin

TABLE 6.7

IMPLICATIONS OF MOLECULAR ARCHITECTURE AND BIOCHEMICAL PROPERTIES
OF MYOFIBRILLAR PROTEINS TO MEAT SCIENCE

Attribute of Meat	Possible Role of Myofibrillar Architecture and Biochemistry
Tenderness	Much of the variation in tenderness is probably determined by integrity of Z-disks and strength and nature of the actin-myosin interaction.
Water-holding capacity	A function of spacings and integrity of the thick and thin filament lattice and possibly also of the extent of the myosin-actin interaction and the high proportion of charged amino acids in the myofibrillar proteins. Over 90% of water-holding capacity due to myofibrillar proteins.
Emulsifying capacity	Related to solubility and structural integrity of the myofibrillar proteins and possibly also to the high proportion of charged amino acids in the myofibrillar proteins. Myofibrillar proteins are responsible for approximately 90% of the emulsifying capacity of meat.
Nutritive value	Myofibrillar proteins contain relatively high proportions of nutritionally essential amino acids.
Cost	Because myofibrillar proteins constitute 35–45% of total organic material in muscle, cost of meat ultimately depends to a large extent on the efficiency with which ingested nutrients are converted to myofibrillar protein.

cross-bridges to actin filaments as muscle ATP concentrations decline during postmortem storage and become too low to dissociate the actin-myosin complex causes inextensibility and rigidity and increases toughness; and (2) Z-disks are gradually degraded in postmortem muscle; because Z-disk removal leaves myofibrils in the form of short segments of interdigitating thick and thin filaments and because these segments will be only 2.0–2.5 μm long, Z-disk disintegration has enormous potential for increasing meat tenderness. Some recent findings indicate that Z-disk degradation is caused by a specific Ca^{2+}-activated proteolytic enzyme found in muscle (Goll *et al.* 1974). The nature of this enzyme and the possible relation of its activity to meat tenderness have been summarized (Goll *et al.* 1974, 1977). An example of the Z-disk degradation normally observed in postmortem muscle is shown in Fig. 6.14.

The precise role of the actin-myosin interaction in meat tenderness and in the water-holding ability and emulsifying capacities of meat remains unclear, largely because the exact nature of the actin-myosin interaction remains unclear. At least four different properties of the actin-myosin complex change during postmortem storage (Goll *et al.*

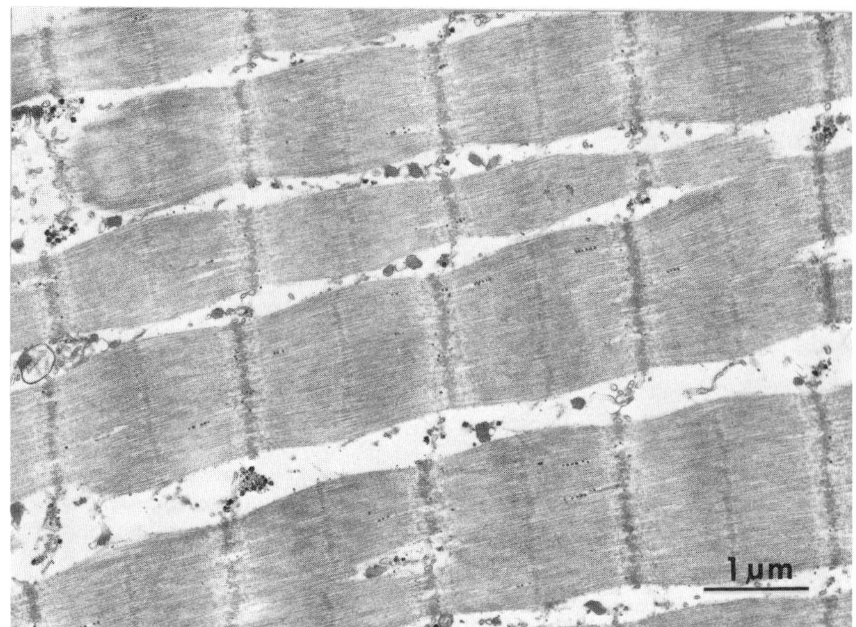

FIG. 6.14. ELECTRON MICROGRAPH OF BOVINE SEMITENDINOSUS MUSCLE
AFTER STORAGE AT 2° C FOR 13 DAYS

The cross-striated appearance of this muscle is still clearly evident after this long period of postmortem storage, and the fibrillar nature of the muscle seems relatively intact. Indeed, an actomyosin having an active ATPase activity and capable of *in vitro* contractile responses can be prepared from this muscle (Goll *et al.* 1970). M-lines are also evident in this micrograph. Z-disks, however, are patchy and have interruptions or holes along their length after this period of postmortem storage. Hence, Z-disk degradation is the major ultrastructural change observed in muscle during postmortem storage. Preservation of the interdigitating thick and thin filament structure after death argues strongly against significant catheptic enzyme destruction of myofibrils in postmortem muscle.

1974), and it is therefore inferred that the actin-myosin interaction may also change during postmortem storage. The nature of this change or changes, however, remains unclear, although circumstantial evidence indicates that the strength of the actin-myosin interaction gradually decreases postmortem. It may be theorized, for example, that, when muscle ATP concentrations fall to a level too low to dissociate myosin cross-bridges from the actin filament, both heads of one myosin cross-bridge attach to adjacent actin monomers in a single actin filament. Because actin monomers are slightly rotated with respect to each other, however, only one of the myosin-binding sites of two adjacent actin monomers in a single actin filament would be in a sterically favorable position to interact with the myosin heads on a single myosin cross-bridge. The binding of the myosin head attached in a sterically unfavorable position may be further weakened during postmortem storage so that this myosin head loses its attachment to its actin monomer. This free myosin head could then compete with its companion head on the same cross-bridge for the sterically more favorably oriented actin monomer and perhaps even displace its companion from this more favorably oriented monomer. This process would account for the slight, gradual lengthening sometimes observed in sarcomeres from postmortem muscle (Goll *et al.* 1974). Detachment of one of the two myosin heads on a cross-bridge could also increase water-holding capacity slightly because the space in the thick and thin filament lattice would increase slightly if the detached head pulled back closer to the surface of the thick filament. Detachment of one of the two heads might also increase myosin solubility slightly and thereby enhance emulsifying capacity of muscle by a small amount.

Obviously, many of the suggestions in the preceding paragraph are simply speculations at the current state of our knowledge of the actin-myosin interaction in postmortem muscle. These suggestions, however, indicate the widespread and far-reaching consequences that could result from additional fundamental information on the molecular nature of the actin-myosin interaction in postmortem muscle. Acquisition of such information will require detailed and highly sophisticated studies involving both biochemical and ultra-structural techniques. Although the eventual impact of such efforts on the meat industry is enormous, the sophistication of the instruments needed for these studies and the inherent difficulty in elucidating molecular details of protein interactions will make these studies both costly and time-consuming. It would seem perspicacious, therefore, to increase emphasis on fundamental studies of muscle contraction as soon as possible.

ACKNOWLEDGMENTS

We gratefully acknowledge the unflagging and skillful efforts of Joan Andersen in preparing this manuscript. We also thank Mary Bremner, Jackie Martin, Darlene Markley, and Mary Edwards for expert technical assistance with those studies done in our laboratory.

Journal Paper No. J-8593 of the Iowa Agriculture and Home Economics Experiment Station, Projects 1795, 1796, 2025, and 2127. The original research discussed in this review that originated in our laboratory was supported in part by N 111 Grants AM-12654 and HL-15679, American Heart Association Grant No. 71-679, grants from the Iowa Heart Association, and the Muscular Dystrophy Association.

BIBLIOGRAPHY

APRIL, E. W., and WONG, D. 1976. Nonisovolumic behavior of the unit cell of skinned, striated muscle fibers. J. Mol. Biol. *101*, 107–114.

BAGSHAW, C. R., ECCLESTON, J. F., ECKSTEIN, F., GOODY, R. S., GUT-FREUND, H., and TRENTHAM, D. R. 1974. The magnesium ion-dependent adenosine triphosphatase of myosin. Two-step processes of adenosine triphosphate association and adenosine diphosphate dissociation. Biochem. J. *141*, 351–364.

BAGSHAW, C. R., and TRENTHAM, D. R. 1974. The characterization of myosin-product complexes and of product-release steps during the magnesium ion-dependent adenosine triphosphatase reaction. Biochem. J. *141*, 331–349.

BENDALL, J. R. 1967. The elastin content of various muscles of beef animals. J. Sci. Food Agric. *18*, 553–558.

BREKKE, C. J., and GREASER, M. L. 1976. Separation and characterization of the troponin components from bovine cardiac muscle. J. Biol. Chem. *251*, 866–871.

BROSTROM, C. O., HUNKELER, F. L., and KREBS, E. G. 1971. The regulation of skeletal muscle phosphorylase kinase by Ca^{2+}. J. Biol. Chem. *246*, 1961–1967.

BUEGE, D. R., and MARSH, B. B. 1975. Mitochondrial calcium and post-mortem muscle shortening. Biochem. Biophys. Res. Commun. *65*, 478–482.

BURKE, M., and HARRINGTON, W. F. 1972. Geometry of the myosin dimer in high-salt media. II. Hydrodynamic studies on macromodels of myosin and its rod segments. Biochemistry *11*, 1456–1462.

COHEN, C. 1975. The protein switch of muscle contraction. Sci. Am. *233*, 36–45.

COOKE, R. 1976. The role of the bound nucleotide of actin in the interaction of actin with myosin. Biophys. J. *16*, 44a.

CRAIG, R., and OFFER, G. 1976. The location of C-protein in rabbit skeletal muscle, Proc. Roy. Soc. London B. *192*, 451–461.

CUMMINS, P., and PERRY, S. V. 1974. Chemical and immunochemical characteristics of tropomyosins from striated and smooth muscle. Biochem. J. *141*, 43–49.

DAWSON, D. M. 1966. Efflux of enzymes from chicken muscle. Biochim. Biophys. Acta *113*, 144–157.

DAYTON, W. R., GOLL, D. E., ZEECE, M. G., ROBSON, R. M., and REVILLE, W. J. 1976A. A Ca^{2+}-activated protease possibly involved in myofibrillar protein turnover. Purification from porcine muscle. Biochemistry *15*, 2150–2158.

DAYTON, W. R., REVILLE, W. J., GOLL, D. E., and STROMER, M. H. 1976B. A Ca^{2+}-activated protease possibly involved in myofibrillar protein turnover. Partial characterization of the purified enzyme. Biochemistry *15*, 2159–2167.

DUBUISSON, M. 1954. Muscular Contraction. Charles C. Thomas, Springfield, Ill.

EASON, B. A. 1969. Purification and properties of skeletal muscle microsomes. Ph. D. Thesis. Iowa State University, Ames, Iowa.

EISENBERG, B. R., and KUDA, A. M. 1976. Discrimination between fiber populations in mammalian skeletal muscle by using ultrastructural parameters. J. Ultrastruct. Res. *54*, 76–88.

ELLIOTT, G. F., LOWY, J., and MILLMAN, B. M. 1967. Low-angle X-ray diffraction studies of living striated muscle during contraction. J. Mol. Biol. *25*, 31–45.

ELLIOTT, G. F., LOWY, J., and WORTHINGTON, C. R. 1963. An X-ray and light-diffraction study of the filament lattice of striated muscle in the living state and in rigor. J. Mol. Biol. *6*, 295–305.

FRANK, G., and WEEDS, A. G. 1974. The amino-acid sequence of the alkali light chains of rabbit skeletal-muscle myosin. Eur. J. Biochem. *44*, 317–334.

FRANZINI-ARMSTRONG, C. 1973. The structure of a simple Z-line. J. Cell Biol. *58*, 630–642.

GOLL, D. E., ARAKAWA, N., STROMER, M. H., BUSCH, W. A., and ROBSON, R. M. 1970. Chemistry of muscle proteins as a food. *In* The Physiology and Biochemistry of Muscle as a Food, E. J. Briskey, R. G. Cassens, and B. B. Marsh (Editors). The University of Wisconsin Press, Madison.

GOLL, D. E., BRAY, R. W., and HOCKSTRA, W. G. 1963. Age-associated changes in muscle composition. The isolation and properties of a collagenous residue from bovine muscle. J. Food Sci. *28*, 503–509.

GOLL, D. E., BRAY, R. W., and HOEKSTRA, W. G. 1964C. Age-associated changes in bovine muscle connective tissue. 3. Rate of solubilization at 100°C. J. Food Sci. *29*, 622–628.

GOLL, D. E., HOEKSTRA, W. G., and BRAY, R. W. 1964A. Age-associated changes in bovine muscle connective tissue. 1. Rate of hydrolysis by collagenase. J. Food Sci. *29*, 608–614.

GOLL, D. E., HOEKSTRA, W. G., and BRAY, R. W. 1964B. Age-associated changes in bovine muscle connective tissue. 2. Exposure to increasing temperature. J. Food Sci. *29*, 615–621.

GOLL, D. E., STROMER, M. H., OLSON, D. G., DAYTON, W. R., SUZUKI, A., and ROBSON, R. M. 1974. The role of myofibrillar proteins in meat tenderness. Proc. Meat Ind. Res. Conf., American Meat Institute Foundation, Arlington, Va.

GOLL, D. E., STROMER, M. H., ROBSON, R. M., and PARRISH, F. C., JR. 1977. Molecular architecture and biochemical properties as bases of quality in muscle foods. J. Food Sci. (in press).

GOLL, D. E., STROMER, M. H., ROBSON, R. M., TEMPLE, J., EASON, B. A., and BUSCH, W. A. 1971. Tryptic digestion of muscle components simulates many of the changes caused by postmortem storage. J. Anim. Sci. *33*, 963–982.

GOLL, D. E., YOUNG, R. B., and STROMER, M. H. 1975. Separation of subcellular organelles by differential and density-gradient centrifugation. Proc. 27th Annual Reciprocal Meats Conf., National Livestock and Meat Board, Chicago, Ill.

GREASER, M. L., CASSENS, R. G., BRISKEY, E. J., and HOESKSTRA, W. G. 1969. Post-mortem changes in subcellular fractions from normal and pale, soft, exudative porcine muscle. 1. Calcium accumulation and adenosine triphosphate activities. J. Food Sci. *34*, 120–124.

GREASER, M. L., and GERGELY, J. 1971. Reconstitution of troponin activity from three protein components. J. Biol. Chem. *246*, 4226–4233.

HANSON, J., LEDNEV, V., O'BRIEN, E. J., and BENNETT, P. M. 1973. Structure of the actin-containing filaments in vertebrate skeletal muscle. Cold Spring Harbor Symp. Quant. Biol. *37*, 311–318.

HANSON, J., and LOWY, J. 1963. The structure of F-actin and of actin filaments isolated from muscle. J. Mol. Biol. *6*, 46–60.

HANSON, J., O'BRIEN, E. J., and BENNETT, P. M. 1971. Structure of the myosin-containing filament assembly (A-segment) separated from frog skeletal muscle. J. Mol. Biol. *58*, 865–871.

HARRINGTON, W. F., and BURKE, M. 1972. Geometry of the myosin dimer in high-salt media. 1. Association behavior of rod segments from myosin. Biochemistry *11*, 1448–1455.

HEILMEYER, L. M. G., MEYER, F., HASCHKE, R. H., and FISCHER, E. H. 1970. Control of phosphorylase activity in a muscle glycogen particle. J. Biol. Chem. *245*, 6649–6656.

HEYWOOD, S. M., and KENNEDY, D. S. 1974. The control of myoglobin synthesis during muscle development. Dev. Biol. *38*, 390–393.

HEYWOOD, S. M., KENNEDY, D. S., and BESTER, A. J. 1974. Separation of specific initiation factors involved in the translation of myosin and myoglobin messenger RNAs and the isolation of a new RNA involved in translation. Proc. Natl. Acad. Sci. USA *71*, 2428–2431.

HITCHCOCK, S. E. 1975. Cross-linking of troponin with dimethylimido esters. Biochemistry *14*, 5162–5167.

HITCHCOCK, S. E., HUXLEY, H. E., and SZENT-GYÖRGYI, A. G. 1973. Calcium sensitive binding of troponin to actin-tropomyosin: a two-site model for troponin action. J. Mol. Biol. *80*, 825–836.

HULTIN, H. O. 1974. Characteristics of immobilized multi-enzymic systems. J. Food Sci. *39*, 647–652.

HUXLEY, A. F., and NIEDERGERKE, R. 1954. Structural changes in muscle during contraction. Interference microscopy of living muscle fibers. Nature *173*, 971–973.

HUXLEY, H. E. 1963. Electron microscope studies on the structure of natural and synthetic protein filaments from striated muscle. J. Mol. Biol. 7, 281–308.

HUXLEY, H. E. 1968. Structural difference between resting and rigor muscle; evidence from intensity changes in the low-angle equatorial X-ray diagram. J. Mol. Biol. *37*, 507–520.

HUXLEY, H. E. 1972. Molecular basis of contraction in cross-striated muscles. *In* The Structure and Function of Muscle, Vol. 1, 2nd Edition, G. H. Bourne (Editor). Academic Press, Inc., New York.

HUXLEY, H. E., and BROWN, W. 1967. The low-angle X-ray diagram of vertebrate striated muscle and its behavior during contraction and rigor. J. Mol. Biol. *30*, 383–434.

HUXLEY, H. E., and HANSON, J. 1954. Changes in the cross-striations of muscle during contraction and stretch and their structural interpretation. Nature *173*, 973–976.

JOHNSON, P., and SMILLIE, L. B. 1975. Rabbit skeletal α-tropomyosin chains are in register. Biochem. Biophys. Res. Commun. *64*, 1316–1322.

LOEWY, A. G., and SIEKEVITZ, P. 1969. Cell Structure and Function, 2nd Edition. Holt, Rinehart, and Winston, Inc., New York.

LOWEY, S., and RISBY, D. 1971. Light chains from fast and slow muscle myosins. Nature *234*, 81–85.

LOWEY, S., SLAYTER, H. S., WEEDS, A. G., and BAKER, H. 1969. Substructure of the myosin molecule. I. Subfragments of myosin by enzymic degradation. J. Mol. Biol. *42*, 1–29.

MacLENNAN, D. H., and HOLLAND, P. C. 1975. Calcium transport in sarcoplasmic reticulum. Ann. Rev. Biophys. Bioeng. *4*, 377–404.

MARGOSSIAN, S. S., and COHEN, C. 1973. Troponin subunit interactions. J. Mol. Biol. *81*, 409–413.

MARSH, B. B. 1972. Postmortem muscle shortening and meat tenderness. Proc. Meat Ind. Res. Conf., American Meat Institute Foundation, Arlington, Va.

MOORE, P. B., HUXLEY, H. E., and deROSIER, D. J. 1970. Three-dimensional reconstruction of F-actin, thin filaments, and decorated thin filaments. J. Mol. Biol. *50*, 279–295.

MOOS, C., OFFER, G., STARR, R., and BENNETT, P. 1975. Interaction of

C-protein with myosin, myosin rod, and light meromyosin. J. Mol. Biol. *97*, 1–9.

MOREL, J. E., and PINSET-HÄRSTRÖM, I. 1975. Ultrastructure of the contractile system of striated skeletal muscle and the processes of muscular contraction. 1. Ultrastructure of the myofibril and source of energy. Biomedicine *22*, 88–96.

MORIMOTO, K., and HARRINGTON, W. F. 1972. Isolation and physico-chemical properties of an M-line protein from skeletal muscle. J. Biol. Chem. *247*, 3052–3061.

MURRAY, J. M., and WEBER, A. 1974. The cooperative action of muscle proteins. Sci. Am. *230*, 58–71.

NOVIKOFF, A. B., and HOLTZMAN, E. 1970. Cells and Organelles. Holt, Rinehart, and Winston, Inc., New York.

OFFER, G., MOOS, C., and STARR, R. 1973. A new protein of the thick filaments of vertebrate skeletal muscle myofibrils. Extraction, purification, and characterization. J. Mol. Biol. *74*, 653–676.

OHTSUKI, I., MASAKI, T., NONOMURA, Y., and EBASHI, S. 1967. Periodic distribution of troponin along the thin filament. J. Biochem. *61*, 817–819.

OLSON, D. G., PARRISH, F. C., JR., DAYTON, W. R., and GOLL, D. E. 1977. Effect of postmortem storage and calcium activated factor on the myofibrillar proteins of bovine skeletal muscle. J. Food Sci. *42*, 117–124.

OSNER, R. C. 1966. Influence of processing procedures on the post-mortem permeability of chicken muscle sarcolemmas to protein. J. Food Sci. *31*, 832–837.

PARRY, D. A. D., and SQUIRE, J. M. 1973. Structural role of tropomyosin in muscle regulation: analysis of the X-ray diffraction patterns from relaxed and contracting muscles. J. Mol. Biol. *75*, 33–55.

PEARLSTONE, J. R., CARPENTER, M. R., JOHNSON, P., and SMILLIE, L. B. 1976. Amino acid sequence of tropomyosin-binding component of rabbit skeletal muscle troponin. Proc. Natl. Acad. Sci. USA *73*, 1902–1906.

POLLARD, T. D., and WEIHING, R. R. 1974. Actin and myosin, and cell movement. Crit. Rev. Biochem. *2*, 1–65.

POTTER, J. D., and GERGELY, J. 1975. The calcium and magnesium binding sites on troponin and their role in the regulation of myofibrillar adenosine triphosphatase. J. Biol. Chem. *250*, 4628–4633.

PURCHAS, R. W. 1972. The relative importance of some determinants of beef tenderness. J. Food Sci. *37*, 341–345.

REED, R., HOUSTON, T. W., and TODD, R. M. 1966. Structure and function of the sarcolemma of skeletal muscle. Nature *211*, 534–536.

REVILLE, W. J., GOLL, D. E., STROMER, M. H., ROBSON, R. M., and DAYTON, W. R. 1976. A Ca^{2+}-activated protease possibly involved in myofibrillar protein turnover. Subcellular localization of the protease in porcine skeletal muscle. J. Cell Biol. *70*, 1–8.

ROME, E. 1967. Light and X-ray diffraction studies of the filament lattice of glycerol-extracted rabbit psoas muscle. J. Mol. Biol. *27*, 591–602.

ROME, E. 1968. X-ray diffraction studies of the filament lattice of striated muscle in various bathing media. J. Mol. Biol. *37*, 331–344.

SAMEJIMA, K., and WOLFE, F. H. 1976. Degradation of myofibrillar protein components during postmortem aging of chicken muscle. J. Food Sci. *41*, 250–254.

SARKAR, S., SRETER, F. A., and GERGELY, J. 1971. Light chains of myosins from white, red, and cardiac muscles. Proc. Natl. Acad. Sci. USA *68*, 946–950.

SCOPES, R. K. 1970. Characterization and study of sarcoplasmic proteins. *In* The Physiology and Biochemistry of Muscle as a Food, E. J. Briskey, R. G. Cassens, and B. B. Marsh (Editors). The University of Wisconsin Press, Madison.

SEIDEL, J. 1975. The effects of ionic conditions, temperature, and chemical modification on the fluorescence of myosin during the steady state of ATP hydrolysis. J. Biol. Chem. *250*, 5681–5687.

SEIDEL, J. C., and GERGELY, J. 1973. Electron spin resonance of myosin spin labeled at the S_1 thiol groups during hydrolysis of adenosine triphosphate. Arch. Biochem. Biophys. *158*, 853–863.

SHERTON, C. C., and WOOL, I. G. 1974. A comparison of the proteins of rat skeletal muscle and liver ribosomes by two-dimensional polyacrylamide gel electrophoresis. Observations on the partition of proteins between ribosomal subunits and a description of two acidic proteins in the large subunit. J. Biol. Chem. *249*, 2258–2267.

SLAYTER, H. S., and LOWEY, S. 1967. Substructure of the myosin molecule as visualized by electron microscopy. Proc. Natl. Acad. Sci. USA *58*, 1611–1618.

SPUDICH, J. A., HUXLEY, H. E., and FINCH, J. T. 1972. Regulation of skeletal muscle contraction. II. Structural studies of the interaction of the tropomyosin-troponin complex with actin. J. Mol. Biol. *72*, 619–632.

SQUIRE, J. M. 1973. General model of myosin filament structure. Molecular packing arrangements in myosin filaments. J. Mol. Biol. *77*, 291–323.

STEWART, M. 1975. The location of the troponin binding site on tropomyosin. Proc. Roy. Soc. London B. *190*, 257–266.

STROMER, M. H., GOLL, D. E., RICHARDSON, F., and ROBSON, R. M. 1977. Extraction and some properties of M-protein. (submitted for publication).

STULL, J. T., and MAYER, S. E. 1971. Regulation of phosphorylase activation in skeletal muscle *in vivo*. J. Biol. Chem. *246*, 5716–5723.

SUZUKI, A., GOLL, D. E., SINGH, I., ALLEN, R. E., ROBSON, R. M., and STROMER, M. H. 1977. Some properties of purified skeletal muscle α-actinin. J. Biol. Chem. (in press).

TAYLOR, E. W. 1972. Chemistry of muscle contraction. Ann. Rev. Biochem. *41*, 577–616.

THOMPSON, W. C., and HEYWOOD, S. M. 1974. Basic aspects of protein synthesis in muscle. J. Anim. Sci. *38*, 1050–1053.

TRINICK, J., and LOWEY, S. 1976. Studies of M-protein from chicken muscle. Biophys. J. *16*, 199a.

TURNER, D. C., WALLIMANN, T., and EPPENBURGER, H. M. 1973. A protein that binds specifically to the M-line of skeletal muscle is identified as the muscle form of creatine kinase. Proc. Natl. Acad. Sci. USA *70*, 702–705.

WAKABAYASHI, T., HUXLEY, H. E., AMOS, L. A., and KLUG, A. 1975. Three-dimensional image reconstruction of actin-tropomyosin complex and actin-tropomyosin-troponin T-troponin I complex. J. Mol. Biol. *93*, 477–497.

WEEDS, A. G., and LOWEY, S. 1971. Substructure of the myosin molecule. II. The light chains of myosin. J. Mol. Biol. *61*, 701–725.

WEEDS, A. G., and POPE, B. 1971. Chemical studies on light chains from cardiac and skeletal muscle myosins. Nature *234*, 85–88.

WOLFE, F. H., and SAMEJIMA, K. 1976. Further studies of postmortem aging effects on chicken actomyosin. J. Food Sci. *41*, 244–249.

YAZAWA, M., MORITA, F., and YAGI, K. 1973. Distinction between two moles of ADP binding to heavy meromyosin in the presence of Mn (II), and its bearing on ATPase action. J. Biochem. *74*, 1107–1117.

7

Milk Proteins

J. Robert Brunner

Although man utilizes milk of various species of domesticated mammals, the Western breeds of dairy cows, *Bos Taurus*, represent the dominant source of milk for the industrialized food industry.

Milk is a multiphasic secretion of the mammary gland containing emulsified fat globules, colloidally dispersed casein micelles and dissolved proteins, lactose and salts with an approximate composition of 4.0% fat, 3.5% protein, 4.8% lactose, and 0.7% ash; water, representing the continuous phase, constitutes the remainder. The fat and protein contents are seasonally variable and reflect breed differences, stage of lactation and feeding regimen.

This treatise presents a discussion of the composition and characteristics of components constituting the protein system of cow's milk. For this purpose, the individual proteins are categorized into three principal classes: caseins; whey proteins, including selected minor proteins and enzymes; and proteins associated with the lipid phase which are usually recognized as components of the milk fat globule membrane. Finally, a brief consideration is given to the mechanism of milk secretion at the cellular level.

CASEINS

Casein is a family of related phosphoproteins comprising about 80% of the total protein content of milk and found in concentrations approximating 2.5–3.2%. Three principal components, α_{s1}-, β-, and κ-casein, and a minor component, γ-casein, represent about 95% of the casein fraction. These caseins as well as the whey proteins, β-lactoglobulin and α-lactalbumin, occur as genetically-controlled compositional variants (polymorphs) which are present in milks of individual cows as single or mixed phenotypes, viz., α_{s1}-casein B/B or B/C. The polymorphic species are products of allelic autosomal genes inherited through simple Mendelian processes. Gene frequencies for polymorphs of the major milk proteins in a random population of the Western dairy breeds are recorded in Table 7.1. From these data, for example, we note that α_{s1}-casein B, β-casein A^1 and A^2, κ-casein A and B, β-lactoglobulin A and B, and α-lactalbumin B are the predominant variants found in Holstein cattle.

TABLE 7.1
GENE FREQUENCIES FOR THE MAJOR MILK PROTEINS
OF FIVE MAJOR DAIRY BREEDS[1]

Protein Variant	Gene Frequency				
	Holstein	Guernsey	Jersey	Ayrshire	Brown Swiss
α_{s1}-Caseins					
Number observed	541	296	287	257	282
A	0.003	0	0	0	0
B	0.94	0.79	0.74	1.00	0.98
C	0.06	0.21	0.26	0	0.02
D	0	0	0	0	0
β-Caseins					
Number observed	526	262	298	202	235
A^1	0.49	0.06	0.09	0.67	0.15
A^2	0.49	0.88	0.54	0.32	0.72
A^3	0.01	0	0	0.01	0
B	0.01	0.01	0.37	0	0.10
C	0	0.05	0.003	0	0.03
D	0	0	0	0	0
κ-Caseins					
Number observed	543	297	297	211	290
A	0.75	0.59	0.12	0.70	0.41
B	0.25	0.41	0.88	0.30	0.59
β-Lactoglobulin					
Number observed	494	278	270	191	259
A	0.50	0.38	0.36	0.17	0.33
B	0.50	0.62	0.62	0.83	0.67
α-Lactalbumin[2]					

[1] Compiled from data of Li and Gaunt (1972).
[2] Western dairy breeds appear to be homozygous for B.

Whole Casein

"Whole casein" is operationally defined as a heterogeneous group of phosphoproteins precipitated from raw skim milk at pH 4.6 and 20°C (Whitney *et al.* 1976). At temperatures below 20°C, precipitation is incomplete. In this form, ionically-associated calcium and colloidal calcium phosphate are essentially ionized and lost to the serum (whey). Intact casein micelles, with associated calcium, are collected as a sedimented pellet by high speed centrifugation at 37°C. However, not all the casein is in micellar form; thus some 5–20% may remain in the supernatant layer as soluble components. Casein can be "salted-out" by the addition of ammonium sulfate (26.4g/100 ml skim milk at 2°C) but may include small amounts of whey components. Casein is modified to an insoluble para-calcium caseinate by the addition of rennin or rennin-like enzymes. Rennin-coagulated casein is the most insoluble form of industrial casein, whereas isoelectric casein is readily solubilized by conversion to its sodium salt by the addition of sodium or ammonium hydroxide to pH 6.7.

Casein is characterized compositionally by relatively high contents of phosphorus (0.85%) and the amino acid, proline. Within recent years, the phosphorus-polypeptide linkage has been established unequivocally as a phosphomonoester, formed principally with the hydroxyamino acid, serine (Ho *et al.* 1969).

Components of Whole Casein

Approximately ten discernible zones are apparent in urea-containing starch or polyacrylamide electropherograms (Fig. 7.1). The α_s-caseins, principally α_{s1}-caseins, contain more anionic phosphate groups than other caseins, and thus migrate the greatest distance in alkaline gels. Under these conditions, the polymorphs β-casein A^1, A^2 and A^3 are not separated. In acid media, variant A separates into three zones, A^1, A^2 and A^3; a ramification of different contents of charged histidine residues (Peterson and Kopfler, 1966). The polymorphs of κ-casein which migrate together as a diffuse zone can be resolved by the addition of 2-mercaptoethanol, a disulfide reducing agent, to the sample solution or the gel media. Recognized components of cow's milk casein and related compositional and physical characteristics are listed in Table 7.2 and 7.3.

α_s-**Caseins.**—This group of calcium-sensitive caseins includes the genetically variable α_{s1}-casein as the principal component and several minor species, α_{s0}-, α_{s2}-, α_{s3}-, α_{s4}- and α_{s5}-caseins, accounting for 50–55% of whole casein. The minor α_s-caseins are apparent in gel electropherograms as minor zones beneath or trailing the predominant α_{s1}-caseins. Apparently, component α_{s5}-casein consists of two polypeptide chains, α_{s3}- and α_{s4}-caseins, joined by a disulfide bond. In the presence of mercaptoethanol, the α_{s5}-zone is not apparent (Hoagland *et al.* 1971).

The amino acid sequence of α_{s1}-casein B is outlined in Fig. 7.2, showing 199 residues with a calculated molecular weight of 23,613 (Mercier *et al.* 1972). Other variants differ from B as indicated. Notably, the rare variant D contains a phosphorylated threonine in place of Ala_{53}, and the equally rare variant A is missing a 13-residue segment from position 14 to 26, inclusive. These modifications affect the net charge of the molecules in such a manner to render them the fastest migrating zones in alkaline gels, viz., A>D>B>C. The sequence reveals several interesting features which help to explain the interacting properties of α_{s1}-casein. The highly charged nonterminal segment (43–79) which contains seven of the eight phosphoseryl residues may predispose the molecule to the activity of calcium ions (Österberg 1964). Hydrophobic areas are encountered on either side of this segment, interrupted only by a short segment of charged residues in the vicinity of the eighth phosphoseryl residue, Ser_{115}.

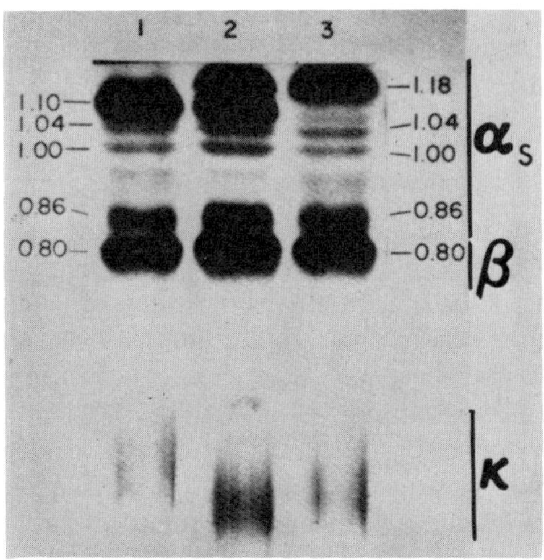

Data of Thompson and Kiddy (1964);
Courtesy of the American
Dairy Science Association

FIG. 7.1. UREA-STARCH GEL (SGE) ELEC-TROPHEROGRAMS OF WHOLE CASEINS FROM INDIVIDUAL COW'S MILK RUN UNDER ALKALINE CONDITIONS AND SHOWING GENETIC POLYMORPHISM IN α_{s1}-CASEIN

Pattern 1, α_{s1}-B; pattern 2, α_{s1}-A/B; pattern 3, α_{s1}-A. All β-caseins are type A (A^1, A^2 and/or A^3).

DISTRIBUTION AND CHARACTERISTICS OF MILK PROTEINS[1]

Component	Approximate Concentration		Genetic Variants[3]	Approximate Molecular Weight	pI	Groups per Mole		
	% of Skim milk Protein	g/liter				P	-S-S-	-SH
Casein	78–85	$(27.2)^2$			4.6			
α_s-Caseins	45–55							
α_{s1}-Casein		13.6	A,$\underline{B}$,$\underline{C}$,D	23,500	5.1	8	0	0
(α_{s0}-, α_{s2}-, α_{s3}-, (α_{s4}-, α_{s5}-, are minor components)								
β-Casein	25–35	8.2	$\underline{A}^1$,$\underline{A}^2$,A^3,$\underline{B}$,C,D	$24,000^5$	5.3	5	0	0
κ-Casein	8–15	4.1	$\underline{A}$,B,	$19,000^4$	3.7–4.2	1	0	0
γ-Casein	3– 7	1.4			5.8			
$\gamma1^5$			A^1,$\underline{A}^2$,A^3,B	20,500		1	0	0
$\gamma2^5$			$A^{1\,or\,2}$,A^3,B	11,800		0	0	0
$\gamma3^5$			$A^{1,2\,or\,3}$,B	11,500		0	0	0
Whey proteins	15–25	(6.8)						
β-Lactoglobulin	7–12	3.6	$\underline{A}$,B,C,D	18,300	5.3	0	2	1
α-Lactalbumin	2– 5	1.7	$\underline{A}$,$\underline{B}$	14,200	5.1	0	4	0
Immunoglobulins	1.5–2.5	0.6			4.6–6.0	0	present and variable	
IgG1[6],IgG2				160,000	(monomers)			
IgM				900,000	(pentamer)			
IgA				400,000	(dimer)			
FSC[7]				70,000				
Serum albumin	0.7–1.3	0.4		69,000	4.7	0	17	1
Proteose-peptone[8]	2.0–4.0	0.7		4,000–40,000	3.7	0.5–2.0	0	0

[1] Compiled partially from data presented by Swaisgood (1973) and Whitney *et al.* (1976).
[2] Values for casein based on 27.2 g/liter of total casein and 55% α_s-, 30% β-, 15% κ-, and 5% γ-caseins.
[3] Underscored variations represent those most frequently found in Western breeds.
[4] Molecular weight of carbohydrate-free species; species containing 0–5 carbohydrate moieties (~650 daltons) exist.
[5] Similar to β-casein segments represented by residue number 29–209, 106–209, and 108–209, respectively.
[6] Principal immunoglobulin in normal milk.
[7] FSC (free secretory component).
[8] Heterogeneous mixture of glycoproteins; usually designated as components 3, 5, 8-fast, and 8-slow in ascending order of electrophoretic mobility.

TABLE 7.3

AMINO ACID COMPOSITION OF REPRESENTATIVE SPECIES OF THE PRINCIPAL MILK PROTEINS[1]

| | Caseins | | | | | | Whey Proteins | | | | |
Residue	α_{s1} - B [23,613 daltons] R/mole[2]	R(wt %)[3]	β - A[1] [24,020 daltons] R/mole	R(wt %)	κ - A [19,037 daltons] R/mole	R(wt %)	β - Lg - A [18,362 daltons] R/mole	R(wt %)	α - La - B [14,174 daltons] R/mole	R(wt %)	Serum albumin[4] $\widetilde{}$ 69,000 daltons] R(wt %)
Asp	15(8)[5]	7.3	9(5)	4.3	12(7)	7.3	16(5)	10.2	21(12)	17.1	9.4
Thr	5	2.1	9	4.0	15	8.0	8	4.5	7	5.0	4.9
Ser	16	6.0	16	5.8	13	6.0	7	3.4	7	4.3	3.5
Glu	39(14)	21.3	39(22)	21.1	27(14)	18.3	25(9)	17.9	13(5)	11.9	14.4
Pro	17	7.0	34	13.8	20	10.2	8	4.3	2	1.4	4.1
Gly	9	2.2	5	1.2	2	0.6	3	1.0	6	2.4	1.4
Ala	9	2.7	5	1.5	14	5.2	14	5.5	3	1.5	5.0
CySH	0	0	0	0	0	0	1	0.6	0	0	5.5
Cys/2	0	0	0	0	2	1.1	4	2.3	8	5.8	
Val	11	4.6	19	7.8	11	5.7	10	5.5	6	4.2	5.0
Met	5	2.8	6	3.3	2	1.4	4	2.9	1	0.9	0.7
Ile	11	5.3	10	4.7	12	7.1	10	6.3	8	6.4	2.2
Leu	17	8.1	22	10.4	8	4.8	22	13.8	13	10.4	10.6
Tyr	10	7.0	4	2.7	9	7.7	4	3.6	4	4.6	4.6
Phe	8	5.0	9	5.5	4	3.1	4	3.3	4	4.2	5.9
Trp	2	1.6	1	0.8	1	1.0	2	2.1	4	5.3	0.5
Lys	14	7.6	11	5.9	9	6.1	15	10.7	12	10.9	11.2
His	5	2.9	6	3.4	3	2.2	2	1.5	3	2.9	3.3
Arg	6	4.0	4	2.6	5	4.1	3	2.6	1	1.1	5.3
Total Residue	199		209		169		162		123		

[1] Compiled from data presented by Swaisgood (1973), Whitney *et al.* (1976) and Gordon and Whittier (1965).
[2] Number of residues per mole derived from primary sequence.
[3] Weight percentage of residue per mole.
[4] Elucidation of the primary structure has not been completed.
[5] Number in brackets represents amide form of residues.

The extensive hydrophobic regions display a rather ubiquitous distribution of proline residues which serve to restrict the formation of secondary structure. α_{s1}-Caseins possess strong endothermic interacting tendencies, indicating that their association occurs through hydrophobic bonds. Thus, these caseins seem to serve as the fundamental units of structure for micelles (Noble and Waugh 1965; Waugh and Noble 1965; Waugh *et al.* 1970).

β-Casein.—β-Caseins account for 30–35% of whole casein. Although sensitive to calcium (II), β-caseins are not precipitated by concentrations sufficient to precipitate α_{s1}-caseins. When sensitized, they form colloidal suspensions rather than the copious precipitates encountered with α_{s1}-caseins. β-Caseins possess temperature-, concentration-, and pH-dependent association-dissociation equilibria. At temperatures below 8°C or at high values of pH, dissociation into monomers occurs. At higher temperatures and near neutrality, association into threadlike polymers takes place. The kinetic association rates are much lower than those for α_{s1}-caseins.

<pre>
                                                                    10   20
H.ARG-PRO-LYS-HIS-PRO-ILE-LYS-HIS-GLN-GLY-LEU-PRO-GLN-[GLU-VAL-LEU-ASN-GLU-ASN-LEU-
                                                                              ABSENT IN VARIANT A

                                              30   40
[LEU-ARG-PHE-PHE-VAL-ALA]-PRO-PHE-PRO-GLN-VAL-PHE-GLY-LYS-GLU-LYS-VAL-ASN-GLU-LEU-

                                          50   60
SER-LYS-ASP-ILE-GLY-SER-GLU-SER-THR-GLU-ASP-GLN-[ALA]-MET-GLU-ASP-ILE-LYS-GLU-MET-
                        P   P   THR-P (VARIANT D)

                                      70   80
GLU-ALA-GLU-SER-ILE-SER-SER-SER-GLU-GLU-ILE-VAL-PRO-ASN-SER-VAL-GLU-GLN-LYS-HIS-
                P   P   P   P   P

                                      90   100
ILE-GLN-LYS-GLU-ASP-VAL-PRO-SER-GLU-ARG-TYR-LEU-GLY-TYR-LEU-GLU-GLN-LEU-LEU-ARG-

                                      110   120
LEU-LYS-LYS-TYR-LYS-VAL-PRO-GLN-LEU-GLU-ILE-VAL-PRO-ASN-SER-ALA-GLU-GLU-ARG-LEU-
                                                                  P

                                    130   140
HIS-SER-MET-LYS-GLN-GLY-ILE-HIS-ALA-GLN-GLN-LYS-GLU-PRO-MET-ILE-GLY-VAL-ASN-GLN-

                                    150   160
GLU-LEU-ALA-TYR-PHE-TYR-PRO-GLU-LEU-PHE-ARG-GLN-PHE-TYR-GLN-LEU-ASP-ALA-TYR-PRO-

                                    170   180
SER-GLY-ALA-TRP-TYR-TYR-VAL-PRO-LEU-GLY-THR-GLN-TYR-THR-ASP-ALA-PRO-SER-PHE-SER-

                                    190   199
ASP-ILE-PRO-ASN-PRO-ILE-GLY-SER-GLU-ASN-SER-[GLU]-LYS-THR-THR-MET-PRO-LEU-TRP.OH
                                                  GLY (VARIANT C)
</pre>

FIG. 7.2. PRIMARY STRUCTURE OF α_{s1}-CASEIN B

Boxed amino acid residues represent positional mutations which differentiate variants A, C and D from B. Adapted from Mercier *et al.* (1972) and Whitney *et al.* (1976).

The primary structure of β-casein A^2 is outlined in Fig. 7.3, showing 209 amino acid residues with a calculated molecular weight of 23,980 (Mercier *et al.* 1972). The sequence reveals a charged N-terminal segment (1–42) in which four phosphoseryl residues occupy a segment (14–21) precisely similar to the highly phosphorylated segment in α_{s1}-casein (63–70). The remaining portion of the molecule (49–209) is strongly hydrophobic, containing a fairly uniform distribution of proline residues. Also to be noted is the presence of a nonphosphorylated serine in position 35 of C variant, concomitant with the substitution of Lys for Glu in position 37. This observation and the discovery that α_{s1}-casein D contained a phosphorylated threonine in position 53 in a sequence of Thr_{53}-Met-Glu_{55} led the French workers to postulate that a sequence of Ser/Thr-X-Glu/Ser P was a minimal requirement for recognition by phosphoryl kinase. All of the phosphorylated seryl and threonyl residues in caseins fit this model.

```
                                        10                              20
H.Arg-Glu-Leu-Glu-Glu-Leu-Asn-Val-Pro-Gly-Glu-Ile-Val-Glu-Ser-Leu-Ser-Ser-Ser-Glu-
                              ┌──────────→ γ₁-CASEINS    P      P   P   P
                              │       30
Glu-Ser-Ile-Thr-Arg-Ile-Asn-Lys┘Lys-Ile-Glu-Lys-Phe-Gln-Ser-Glu-|Glu|-Gln-Gln-Gln-
                            (ABSENT IN VARIANT C)|P|       LYS (VARIANT C)    40
                                        50                              60
Thr-Glu-Asp-Glu-Leu-Gln-Asp-Lys-Ile-His-Pro-Phe-Ala-Gln-Thr-Gln-Ser-Leu-Val-Tyr-

                                        70                              80
Pro-Phe-Pro-Gly-Pro-Ile-|Pro|-Asn-Ser-Leu-Pro-Gln-Asn-Ile-Pro-Pro-Leu-Thr-Gln-Thr-
(VARIANTS C, A¹ AND B)  His

                                        90                             100
Pro-Val-Val-Val-Pro-Pro-Phe-Leu-Gln-Pro-Glu-Val-Met-Gly-Val-Ser-Lys-Val-Lys-Glu-
                              ┌────────────→ γ₃-CASEINS
                              │       110                             120
Ala-Met-Ala-Pro-Lys-|His|-Lys┘Glu-Met-Pro-Phe-Pro-Lys-Tyr-Pro-Val-Gln-Pro-Phe-Thr-
                    |Gln| (VARIANT A³)
                              └────────────→ γ₂-CASEINS
                                      130                             140
Glu-|Ser|-Gln-Ser-Leu-Thr-Leu-Thr-Asp-Val-Glu-Asn-Leu-His-Leu-Pro-Pro-Leu-Leu-Leu-
    Arg (VARIANT B)
                                      150                             160
Gln-Ser-Trp-Met-His-Gln-Pro-His-Gln-Pro-Leu-Pro-Pro-Thr-Val-Met-Phe-Pro-Pro-Gln-

                                      170                             180
Ser-Val-Leu-Ser-Leu-Ser-Gln-Ser-Lys-Val-Leu-Pro-Val-Pro-Glu-Lys-Ala-Val-Pro-Tyr-

                                      190                             200
Pro-Gln-Arg-Asp-Met-Pro-Ile-Gln-Ala-Phe-Leu-Leu-Tyr-Gln-Gln-Pro-Val-Leu-Gly-Pro-

                                      209
Val-Arg-Gly-Pro-Phe-Pro-Ile-Ile-Val-OH
```

FIG. 7.3. PRIMARY STRUCTURE OF β-CASEIN A^2

Boxed amino acid residues represent positional mutations differentiating variants A^1, A^3, B and C from A^2. Segments of the molecule believed to be identical with γ_1-, γ_2- and γ_3-caseins are indicated with arrows. Adapted from Mercier *et al.* (1972) and Whitney *et al.* (1976).

γ**-Caseins.**—γ-Caseins constitute about 5% of whole casein. A relationship between the components of "whole γ-casein" and "β-casein" has been established (Gordon *et al.* 1972; Groves *et al.* 1972). The components previously referred to as γ-, R-, TS-B-, S-, and TS-A^2-caseins have been matched with identical segments of the β-caseins. It should be noted that the γ-caseins detected in a given milk specimen reflect the genetic variants of the β-casein (Fig. 7.3). Thus, the Committee on Milk Protein Nomenclature and Methodology of the American Dairy Science Association, in its most recent report (Whitney *et al.* 1976), recommended the following designations for the γ-caseins: γ_1-casein (β-casein segment 29–209, ~20,500 daltons); γ_2-casein (β-casein segment 106–209, ~11,800 daltons); and γ_3-casein (β-casein segment 108–209, ~11,600 daltons). Probably, the γ-caseins arise from a limited proteolysis of the β-caseins by endogenous casein-associated proteases. Interestingly, the N-terminal phosphopeptides which would result from these cleavages have not been identified in milk.

Being highly hydrophobic in composition, the γ-caseins behave in a manner similar to β-caseins in regard to temperature-, concentration- and pH-dependent association-dissociation equilibria.

κ**-Casein.**—κ-Caseins constitute about 15% of whole casein and were first recognized, *per se*, as the calcium-insensitive fraction in micellar casein (Waugh and Von Hipple 1956). In the presence of calcium ions, κ-caseins interact with calcium-sensitive α_{s1}- and β-caseins to form thermodynamically stable micelles (Waugh 1971). κ-Caseins contain two cysteine residues per mole which participate, at least partially, in intermolecular disulfide linkages, yielding a series of covalent polymers ranging in size from ~60,000–600,000 daltons. This polymerization is apparent as a diffuse zone in urea-containing electrophoretic gels. Upon the addition of mercaptoethanol to the sample or media, the polymers are reduced to monomers which are resolved as a series of discernible zones.

Two genetic variants, A and B, have been identified by electrophoresis in urea- and mercaptoethanol-containing gels. The primary structure of κ-casein B is outlined in Fig. 7.4, showing 169 amino acid residues and a calculated molecular weight of 19,005 (Jollès *et al.* 1972; Mercier *et al.* 1973). Although minor discrepancies exist regarding the precise assignment of a few residues, it is apparent that the substitutions which characterize the variants and the single phosphate residue are located in the C-terminal (glyco) macropeptide. Both cysteine residues are located in the basic N-terminal para-κ-casein portion which is identical in both variants. Urea-gel electropherograms frequently show two zones for para-κ-casein which, apparently, is a ramification of partial carbamylation of lysine residues (Kim *et al.* 1969).

```
                                                      10                                              20
PyroGlu-Glu-Gln-Asn-Gln-Glu-Gln-Pro-Ile-Arg-Cys-Glu-Lys-Asp-Glu-Arg-Phe-Phe-Ser-Asp-
     (Gln)        (Glu)   (Glu)
                                                      30                                              40
Lys-Ile-Ala-Lys-Tyr-Ile-Pro-Ile-Gln-Tyr-Val-Leu-Ser-Arg-Tyr-Pro-Ser-Tyr-Gly-Leu-

                                                      50                                              60
Asn-Tyr-Tyr-Gln-Gln-Lys-Pro-Val-Ala-Leu-Ile-Asn-Asn-Gln-Phe-Leu-Pro-Tyr-Pro-Tyr-

                                                      70                                              80
Tyr-Ala-Lys-Pro-Ala-Ala-Val-Arg-Ser-Pro-Ala-Gln-Ile-Leu-Gln-Trp-Gln-Val-Leu-Ser-

                                                      90                                             100
Asp-Thr-Val-Pro-Ala-Lys-Ser-Cys-Gln-Ala-Gln-Pro-Thr-Thr-Met-Ala-Arg-His-Pro-His-
(Asn)              Rennin                                                                (Pro)
                     ↓                                110                                             120
Pro-His-Leu-Ser-Phe-Met-Ala-Ile-Pro-Pro-Lys-Lys-Asn-Gln-Asp-Lys-Thr-Glu-Ile-Pro-
(His)
                                                     130                                             140
Thr-Ile-Asn-Thr-Ile-Ala-Ser-Gly-Glu-Pro-Thr-Ser-Thr-Pro-Thr-[Ile]-Glu-Ala-Val-Glu-
                                                                      Thr (Variant A)
                                                     150                                             160
Ser-Thr-Val-Ala-Thr-Leu-Glu-[Ala]-Ser-Pro-Glu-Val-Ile-Glu-Ser-Pro-Pro-Glu-Ile-Asn-
     (Variant A) Asp P
                                        169
Thr-Val-Gln-Val-Thr-Ser-Thr-Ala-Val.OH
```

FIG. 7.4. PRIMARY STRUCTURE OF K-CASEIN B

Boxed amino acid residues represent positional mutations differentiating variant A from B. Positions of conflicting assignment with the data of Jollès *et al.* (1972) are indicated with brackets. The arrow indicates site of the primary action of rennin. Adapted from Mercier *et al.* (1973) and Whitney *et al.* (1976).

κ-Caseins exist as mixed species, ranging in carbohydrate moieties ($\sim$657 daltons) from none to five. In gel electropherograms, the slowest migrating zone represents the carbohydrate-free species. Although this zone appears to be the principal zone when stained for protein, it represents only $\sim$25% of the total κ-casein (Pujolle *et al.* 1966). A carbohydrate moiety located on the Thr_{131} residue has been identified as α-*N*-acetylneuraminyl ($2\rightarrow6$) β-galactosyl ($1\rightarrow3$ or 6) *N*-acetylgalactosamine in which neuraminic acid is the moiety terminal residue; *N*-acetylgalactosamine is joined by an o-glycosidic bond to threonine. The precise location of similar moieties is not known although it seems apparent from analysis of the rennin-released macropeptide that essentially all carbohydrate material is located in this portion of the molecule which contains 18 of the 28 serine and threonine residues present in κ-A. Although the functional role of the carbohydrate moieties has not been elucidated, carbohydrate-free κ-caseins stabilize α_{s1}-caseins and are preferentially hydrolyzed by rennin. Indeed, carbohydrate moieties enhance the negative charge and hydrophilicity of the protein. Thus, micellar size and stability could be ramifications of the type of κ-casein incorporated.

The amino acid sequence for κ-casein reveals an overall hydrophobicity somewhere between α_{s1}- and β-casein, with a high concentration of apolar residues in the N-terminal para-κ-casein (segment 1–105). The charged portion of κ-casein exists in the C-terminal macropeptide (106–169). Unlike α_{s1}- and β-caseins, the primary charge contribution is not from phosphate residues which probably accounts for the insensitivity of κ-casein to calcium (II). The negative charge located in the macropeptide is enhanced by the presence of carbohydrate moieties. It is of interest to note the N-terminal pyroglutamic acid. Whether this occurs endogenously or represents process-induced cyclization of the glutamic acid residue has not been verified.

κ-Casein is a specific substrate for the primary action of rennin (chymosin, EC 3.4.4.3.). Cleavage occurs at the Phe_{105}-Met_{106} bond, liberating a soluble C-terminal (glyco) macropeptide of $\sim$6,800 daltons and an insoluble N-terminal para-κ-casein of 12,271 daltons (McDonald and Thomas 1970). Photooxidized κ-casein in which histidine and tryptophan residues are destroyed is a poor substrate for rennin. From this observation, we surmise that His_{102} may be essential to the cleavage mechanism. This was verified by the work of Polzhofer (1972) who demonstrated by the use of synthetic peptides that the minimum chain length for rennin activity is His_{102}-Leu-Ser-Phe-Met-Ala_{107} (see Fig. 7.4).

The high order of hydrophobicity inherent in the insoluble para-κ-casein favors the aggregation of casein micelles following the action of rennin. That rennin-treated milk fails to clot if held cold suggests that hydrophobic interactions do, indeed, play an important role in this process.

Casein Micelles

About 80–90% of the caseins in normal milk are in the form of colloidally-dispersed, nearly spherical micelles ranging in size from 50–300 nm in diameter with particle weights from 10^7–10^9. Micelles are assembled from subunits of uniform size of $\sim$10–20 nm in diameter, containing from 25–30 casein monomers with an average particle weight of $\sim$6$\times$10^5. The remaining portion, viz., 10–20%, constitutes the "soluble" or nonmicellized caseins. The amount and composition of the soluble fraction varies with environmental factors such as temperature, pH and calcium ion concentration. For example, at pH 4.6, the soluble caseins participate concomitantly with micelle dissociation and precipitation. And, at temperatures below 8°C, a portion of the micellar components, particularly β- and κ-caseins, dissociate into the serum, thus increasing the proportion of soluble caseins. Electron micrographs of micelle specimens are shown in Fig. 7.5 (see captions for explanation).

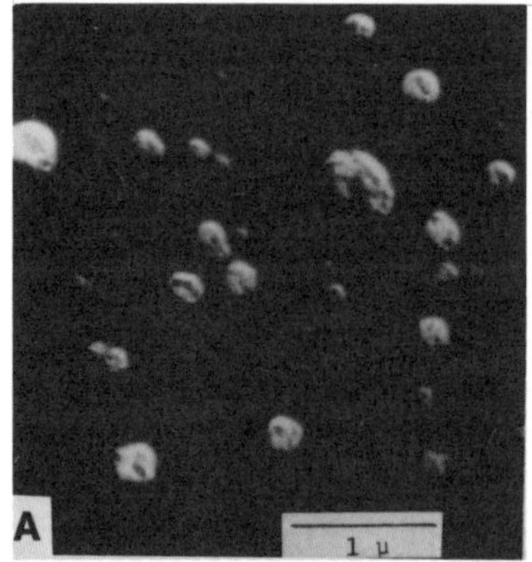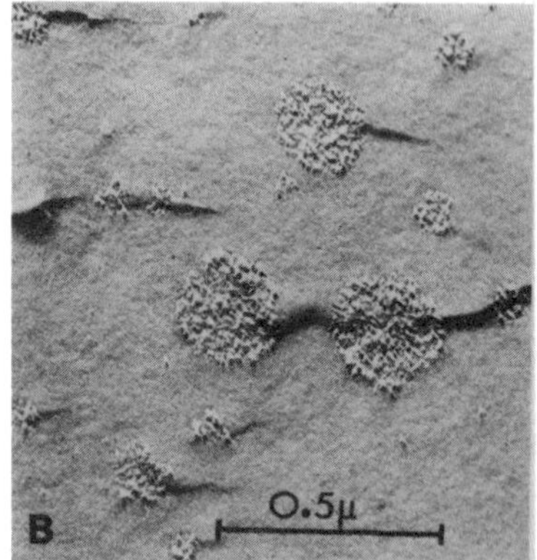

Data of Knoop (1972);
Courtesy of Volkswirt-shaftlicher
Verlag GmbH

FIG. 7.5. ELECTRON MICROGRAPHS OF CASEIN MICELLES

Frame A, specimen fixed with glutaraldehyde and shadowed with platinum (author's data). Frame B, specimen prepared by a freeze-etching technique, showing the open structure in micelles. Numerous individual subunits are apparent.

Although the exact structure of casein subunits and micelles has not been determined, several working models have been proposed which encompass the known compositional and operational characteristics of the micelle and its components. These models fall essentially into two categories. First, a "coat-core" concept in which subunits composed of α_{s1}- and β-caseins assemble in thermodynamically stable core units of uniform size which associate into various sized clusters covered with a peripheral layer of κ-casein (Fig. 7.6A). Second, an open-structured model composed of uniform sized subunits which contain all of the casein components (Fig. 7.6B). In support of the first model is the observation that the size of experimentally generated micelles is influenced by the amount of κ-casein in the system; within limits, micelle size relates inversely with the concentration of κ-casein. Also, most investigators have reported higher κ-/α_{s1}-: β-casein ratios in smaller endogenous micelles. However, other experimental evidence supports the open-structured model. For example, carboxypeptidase A, a molecule with a molecular weight of 34,600, releases C-terminal amino acid residues from all micellar casein components without disrupting the micelle structure, indicating that all parts of the micelle are available to the enzyme (Ribadeau-Dumas and Garnier 1970). The proposition that casein components are distributed throughout the micelle in a fairly uniform manner was tested elegantly by Ashoor *et al.* (1971). A superpolymer of papain, too large to enter micellar interstitial spaces, was prepared and allowed to act against casein micelles. Throughout the course of the reaction, the unhydrolyzed micellar protein contained about the same proportions of α_{s1}-, β- and κ-caseins as originally present in intact micelles. A more complete treatment of the aspects of micelle structure and characteristics can be found in reviews by Rose (1969) and Farrell (1973).

More recently, Slattery and Evard (1973) described an intriguing model which embraces both of the previously discussed concepts and accommodates known characteristics of casein components and endogenous micelles. As visualized in Fig. 7.7, casein micelles are assembled from polymeric subunits containing a mixture of all the casein components in which the apolar portion of each monomer is oriented radially inward, exposing the polar or charged portion on the surface. They developed a rationale for the asymmetric distribution of self-associated κ-casein in the subunit which imparts hydrophilic characteristics to a portion of the surface (grey patch in Fig. 7.7). The remaining surface is composed of α_{s1}- and β-caseins and is available for hydrophobic interactions with similar surfaces on adjacent subunits since the charge on the protein phosphates is decreased by Ca(II) binding. Because β-casein monomers are presumed to be narrower and longer than α_{s1}-casein monomers, they extend beyond the α_{s1}-monomers at the surface. Thus, in its interaction with Ca(II) and other subunits, this portion of the

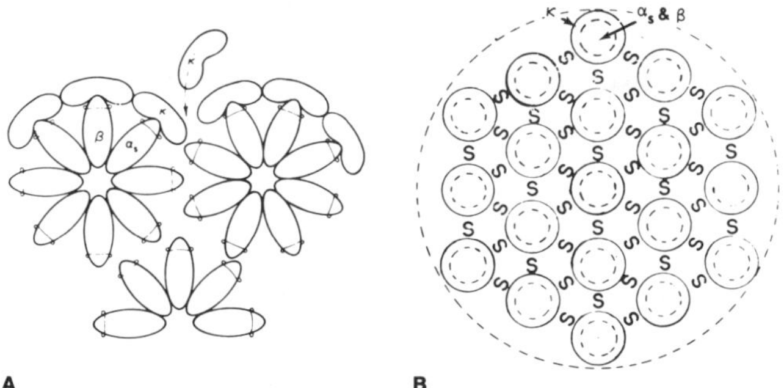

FIG. 7.6. CONCEPTUAL MODELS OF A CASEIN MICELLE

A, "coat-core" concept in which the α_{s1}- and β-caseins are associated radically in a rosette configuration to form the core. κ-Casein is oriented at the peripheral layer forming the "coat" [Reproduced from Waugh (1971); courtesy Academic Press]. B, associated subunits of relatively uniform composition, consisting of α_{s1}-β-casein polymers covered with layers of κ-α_{s1}-casein complex [Adapted from Morr (1967); courtesy of the American Dairy Science Association]. Both models allow for the association of subunits through calcium and colloidal calcium phosphate linkages which are represented by S in B.

surface behaves as though it contained only β-casein. Because each subunit contains a hydrophilic area (associated κ-casein), micelle buildup would not be uniform in all directions. Rather, it would be random directionally, resulting in a porous structure, readily accessible to monomeric caseins, rennin and proteins of similar size.

Limitation of growth and the formation of a noninteracting micelle surface would be natural ramifications of the system. Hydrophobic interactions inward and sideways at the interacting surfaces would align added subunits with the hydrophilic areas radially outward. As the radius of curvature for the micelle increases, less of its surface is available for interaction because of steric factors. Finally, a surface rich in κ-casein would occur, presenting a nonreactive hydrophilic surface of nonassociated subunits, thus, terminating growth of the micelle. Accordingly, micelle size would reflect the relative concentration of κ-casein in constituent subunits. In subunits containing a large concentration of κ-casein, the hydrophilic surface would be correspondingly greater. Then, the number of subunits required to present a non-interacting surface on the micelle would be fewer than for subunits containing relatively less κ-casein.

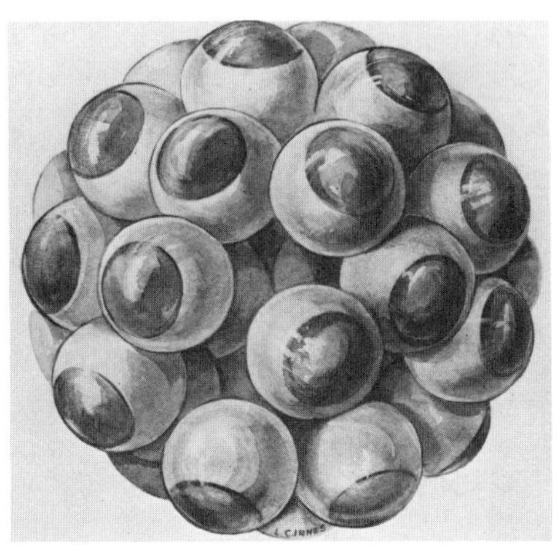

From Slattery and Evard (1973);
Courtesy Elsevier Scientific
Publishing Co., Amsterdam.

FIG. 7.7. CONCEPTUAL MODEL OF A CASEIN MICELLE CONTAINING ABOUT 40 SUBUNITS

The lighter surfaces represent α_{s1}-, and β-casein polymers (hydrophobic area). The darker patches cover about one-fifth of the surface area and represent associated κ-casein polymers (hydrophilic area). The model provides for open channels throughout the micelle. Further growth is impeded by the extensive hydrophilic peripheral surface.

The endogenous casein micelle can be reconciled with this model by the introduction of phosphate and citrate which form, in combination with Ca(II), ion chains between micelle subunits. When these interactions occur predominantly between the subunit α_{s1}-caseins, a three-dimensional network results which resists disruption in the endogenous environment. This concept is supported experimentally by the observation that the micelle framework does not change significantly when a portion of the β-casein is dissociated at low temperatures, viz., $<8°$ C. Other phenomena attributed to micelle behavior are accommodated in the above model.

In a given specimen of milk, the stability of the micellar system is a ramification of calcium (II) activity, temperature, net charge and component distribution. At temperatures below $8°$C, β-casein, in particular, and a small amount of κ-α_{s1}-casein complex move out of the micelle into the serum. Upon warming the milk, these components reassemble in the micelles but it is not known if the endogenous structure is reformed. Similarly, as calcium is removed from the system by chelation or dialysis, micelles dissociate into basic subunits which can be reassembled upon the subsequent addition of calcium (II). Dephosphorylation of α_{s1}-caseins represses the stabilizing interaction of κ-casein, thus rendering the system unstable. The addition of acid to pH 4.6 increases the calcium (II) activity, resulting in gross intermicellar associations and subsequent precipitation. The specific action of rennin on κ-casein destroys its stabilizing function and promotes the hydrophobic association of micelles into the classical gel structure characteristic of casein curd.

Casein micelles, on average, contain α_{s1}-, β-, and κ-caseins in a ratio of about 3:2:1, respectively. However, not all micelles are of similar composition; smaller micelles contain higher concentrations of κ-casein. Typical micelle composition is 93% protein, 2.8% calcium, 2.3% organic phosphorus, 2.9% inorganic phosphorus, 0.4% citrate, and low levels of Mg, Na and K. The organic PO_3^- groups on α_{s1}- and β-caseins, in particular, bind ionic calcium at the subunit level. Inorganic phosphorus exists essentially in complex with calcium and citrate as a colloidal moiety uniformly dispersed through the micellar phase where it plays a role in stabilizing the intermicellar structure.

The reader should consult the excellent papers by Swaisgood (1973) and Bloomfield and Mead (1975) for additional information on the chemistry and structure of the caseins.

WHEY PROTEINS

When casein is removed from skim milk, the remaining aqueous phase is designated as whey or milk serum. A further refinement in terminology reflects the method by which the casein is removed from skim milk. If precipitated with acid at pH 4.6, the aqueous phase is "acid" whey. If removed by the action of rennin or rennin-like enzymes, as in the production of Cheddar cheese, the aqueous phase is "rennin" or "sweet" whey. Wheys differ slightly in composition, reflecting their origin. Acid whey, although free of casein components, contains more calcium, whereas sweet whey, in addition to the usual complement of whey protein, has a greater content of free amino acids and casein fragments resulting from enzymatic cleavage.

The proteins of whey represent about 20% of total milk proteins. The two principal components, β-lactoglobulin and α-lactalbumin, both products of mammary synthesis, account for 70–80% of the protein content (see PAG electropherogram, Fig. 7.8). Readily identifiable amounts of blood serum albumin, immunoglobulins and the proteose-peptones account for the remaining principal proteins of whey. Additionally, numerous enzyme proteins and proteins with specific metabolic functions, *viz.*, lactoferrin, have been identified and in many instances isolated from whey. However, it should be

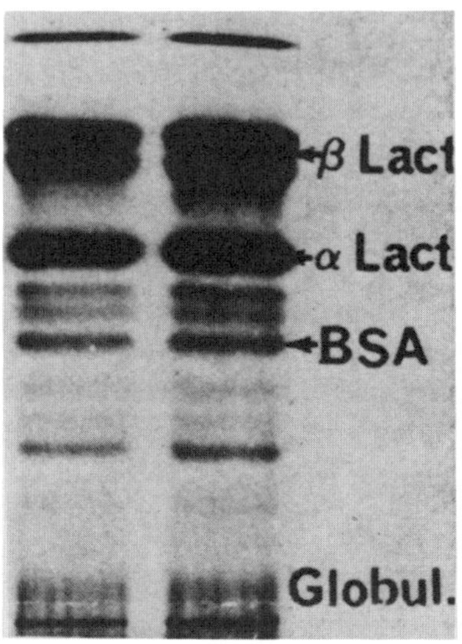

FIG. 7.8. POLYACRYLAMIDE GEL ELECTROPHEROGRAMS OF WHEY PROTEINS OBTAINED WITH A DISCONTINUOUS BUFFER SYSTEM UNDER ALKALINE CONDITIONS AND STAINED WITH COOMASSIE BLUE

Adapted from Melachouris (1969), courtesy of the American Dairy Science Association.

pointed out that not all enzymes found in milk exist as soluble components in the aqueous phase. Many are particulate- or membrane-associated species associated principally with the milk fat globule membrane or casein complex. Selected compositional and physical characteristics of the whey proteins are listed in Tables 7.2 and 7.3.

β-Lactoglobulin

This protein was the first milk protein crystallized (Palmer 1934) and derived its "β" designation by virtue of being the second boundary in the schlieren pattern of ultracentrifuged whey (Pedersen 1936). Consequently, it has been studied extensively as a model system for physicochemical phenomena. Also, the observation by Aschaffenburg and Drewry (1955) that this protein exhibited genetically controlled compositional differences marked the first recognition of genetic polymorphism in milk proteins.

The primary structure of β-lactoglobulin has been established by Frank and Braunitzer (1967) and Braunitizer *et al.* (1972, 1973), revealing 162 residues and calculated molecular weights of 18,362 for A and 18,276 for B (Fig. 7.9). Monomeric β-lactoglobulins are further characterized by the presence of one free sulfhydryl group and two disulfide groups. By applying the rationale of McKenzie *et al.* (1972) to the most recent sequence proposed by Braunitzer *et al.* (1972, 1973), Whitney *et al.* (1976) deduced the positions of the two disulfide bonds. One is always found between the residues Cys_{66} and Cys_{160}; the other is present in equal distribution between Cys_{106} and Cys_{119} or between Cys_{106} and Cys_{121}. Thus, the thiol group, too, is equally distributed between Cys_{119} and Cys_{121}. This is an unusual situation in protein structure and could impart a unique binding property to β-lactoglobulin. As yet, no specific biological function has been assigned to this milk protein.

β-Lactoglobulin has served as a model system for extensive thermodynamic studies (Timasheff and Townsend 1968). In the isoelectric area (pH 5.1–5.6) and probably in normal milk (pH 6.7), monomeric subunits form a stable dimer, consisting of two spheres, 17.9 Å in diameter. Below pH 3, dimers dissociate into monomers, as they do above pH 8. In the pH range just below the isoelectric point, *viz.*, pH 3.8–5.1, β-lactoglobulin tends to undergo a specific association to an octamer at low temperatures and high protein concentration. In this interaction, A variant associates strongly, B variant weakly. The octamerization phenomenon is attributed to the ionization of carboxyl groups at the site of interaction which, apparently, is in the region of A/B amino acid substitution $(Asp/Gly)_{64}$. Thus, the A variant possesses an additional carboxyl group at the association interface, rendering it more reactive.

H.Leu-Ile-Val-Thr-Cln-Thr-Met-Lys-Gly-Leu(10)-Asp-Ile-Gln-Lys-Val-Ala-Gly-Thr-Trp-Tyr(20)-

Ser-Leu-Ala-Met-Ala-Ala-Ser-Asp-Ile-Ser(30)-Leu-Leu-Asp-Ala-Gln-Ser-Ala-Pro-Leu-Arg(40)-

Val-Tyr-Val-Glu-Glu-Leu-Lys-Pro-Thr-Pro(50)-Glu-Gly-Asp-Leu-Glu-Ile-Leu-Leu-[Gln]-Lys(60)-
(variant C) His

Trp-Glu-Asn-[Asp]-Glu-Cys-Ala-Gln-Lys-Lys(70)-Ile-Ile-Ala-Glu-Lys-Thr-Lys-Ile-Pro-Ala(80)-
Gly (variant B and C)

Val-Phe-Lys-Leu-Asp-Ala-Ile-Asn-Glu-Asn(90)-Lys-Val-Leu-Val-Leu-Asp-Thr-Asp-Tyr-Lys(100)-

Lys-Tyr-Leu-Leu-Phe-Cys-Met-Glu-Asn-Ser(110)-Ala-Glu-Pro-Glu-Gln-Ser-Leu-[Val]-Cys[SH](120)-Gln-
(variant B and C) Ala

[SH]Cys-Leu-Val-Arg-Thr-Pro-Glu-Val-Asp-Asp(130)-Glu-Ala-Leu-Gly-Lys-Phe-Asp-Lys-Ala-Leu(140)-

Lys-Ala-Leu-Pro-Met-His-Ile-Arg-Leu-Ser(150)-Phe-Asn-Pro-Thr-Leu-Gln-Glu-Glu-Gln-Cys(160)-

His-Ile(162).OH

FIG. 7.9. PRIMARY STRUCTURE OF β-LACTOGLOBULIN A

Boxed amino acid residues represent positional mutations differentiating variants B and C from A. Variant D differs from A by a $(Gln/Glu)_{51}$ substitution. The SH-group is believed to exist in an equal distribution between Cys_{119} and Cys_{121}. Thus, there is also an equal distribution of disulfide bridges between Cys_{106}-Cys_{119} and Cys_{106}-Cys_{121}; the Cys_{66}-Cys_{160} bridge is not variable. Adapted from Braunitzer et al. (1973) and Whitney et al. (1976).

Like other whey proteins, β-lactoglobulin undergoes time- and temperature-dependent, thermo-denaturation at temperatures in excess of 65°C (Fig. 7.10). Extensive conformational transitions occur, exposing highly reactive nucleophilic groups, viz., -SH, ϵ-NH_2-, and hydrophobic areas. In the absence of casein micelles, as in whey, they participate in homogeneous and heterogeneous, irreversible association interactions, culminating in aggregation and precipitation. This is the case especially in the isoelectric pH range. In the presence of casein, as in skim milk, the interactions are somewhat more specific. A classical example is the heat-induced interaction of β-lactoglobulin and κ-casein which proceeds through the mechanism of disulfide interchange involving the exposed β-Lg thiol group and the disulfides of κ-casein (McKenzie et al. 1971). It is of interest to note that β-lactoglobulin B exhibits a higher rate of thermo-denaturation than the A variant and, hence, interacts more rapidly with κ-casein. The presence of κ-casein during the thermo-denaturation retards the complete aggregation of β-lactoglobulin. Also, the association of β-lactoglobulin with κ-casein increases the heat stability and rennin clotting time of colloidial caseinate.

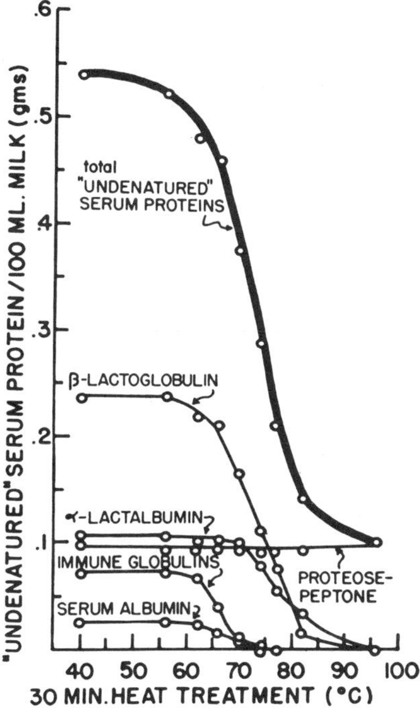

*Data of Larson
and Rolleri (1955);
Courtesy of the American
Dairy Science Association*

FIG. 7.10. THE DENATURATION OF WHEY PROTEIN COMPONENTS IN MILK HEATED FOR 30 MIN AT TEMPERATURES RANGING FROM 40–96°C

Quantitative values were estimated from Tiselius patterns of the undenatured proteins.

α-Lactalbumin

This protein derived its α-designation from the observation that it was the first peak in the schlieren pattern of ultracentrifuged whey (Pedersen 1936). Its most characteristic physical property is the tendency to undergo time-dependent associations at pH values below its isoelectric region. At the pH of milk (~6.6) and above, α-lactalbumin exists essentially as a monomer.

α-Lactalbumin exerts a significant biological function by participating as a "modifier" protein in the lactose synthetase (UDP-galactose: D-glucose 4-β-galactosyl transferase, EC.2.4.1.22) system which is responsible for the synthesis of lactose (Brodbeck *et al.* 1967). The milks of all mammals in which lactose is the principal sugar contain α-lactalbumin. In its absence, the enzyme galactosyl transferase preferentially transfers galactose from UDP-galactose to N-acetylglucosaminyl-glycoprotein, even in the presence of glucose, since the transfer of galactose to glucose is slow (K_m = 1400 mM). In the presence of α-lactalbumin, the transfer of galactose to glucose is rapid (K_m = 5 mM) and glucose becomes the preferred substrate. These reactions are illustrated as follows (Ebner and McKenzie, 1972):

(a) UDP-galactose + N-acetylglucosaminyl-glycoprotein

$$\xrightarrow[\text{Mn (II)}]{\text{galactosyl transferase}} \text{galactose-}\beta\text{- } (1\rightarrow 4)$$

$$N\text{-acetylglucosaminyl-glycoprotein} + \text{UDP}$$

(b) UDP-galactose + glucose $\xrightarrow[\alpha\text{-lactalbumin; Mn(II)}]{\text{galactosyl transferase}}$ lactose + UDP

Two genetic variants, A and B, have been identified but only the B variant has been observed in the milk of Western breeds. Minor forms of bovine α-lactalbumin have been reported which are distinguished by the presence of carbohydrate moieties. One species contains three disulfide bonds instead of the usual four. All species of the protein exhibit similar modifier activity in the lactose synthetase system.

The primary structure of α-lactalbumin was established by Brew *et al.* (1970), yielding 123 amino acid residues with a calculated molecular weight of 14,176 (Fig. 7.11). Of some interest is the apparent similarity of α-lactalbumin and hen's egg-white lysozyme when the former is adapted to the crystallographic coordinates of the latter (Vanaman *et al.* 1970). Despite differences in amino acid composition between the two proteins, there exists 51 identical positional residues, including the four disulfide bonds, and 24 additional residues with similar structures. Although this model suggests extensive structural homology between the two proteins, there is evidence of conformational differences. α-Lactalbumin does not act upon the substrates of lysozyme nor does lysozyme participate in the synthesis of lactose. Yet, α-lactalbumin plays a role in the synthesis and lysozyme in the cleavage of the same $\beta 1 \rightarrow 4$ glucopyranosyl linkage—the former in lactose and the latter in mucoprotein. Also, the two proteins are immunologically different. Despite these differences in composition and function, many authorities contend that lysozyme and α-lactalbumin may have evolved from the same ancestral genes by the process of gene duplication.

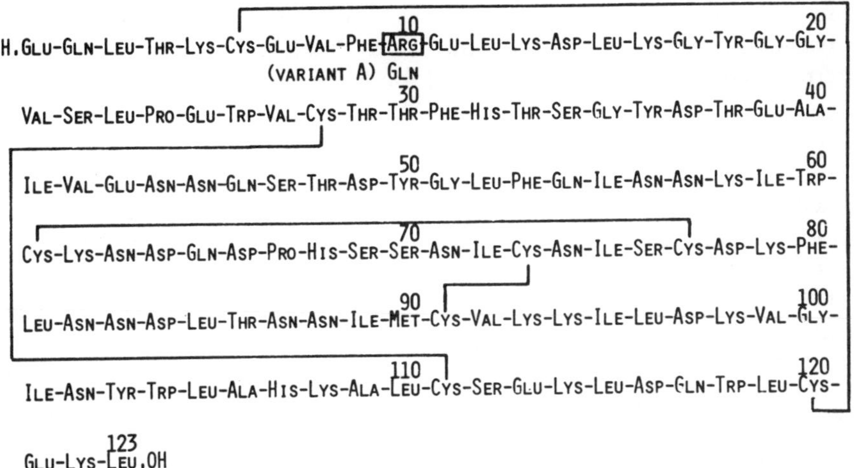

FIG. 7.11. PRIMARY STRUCTURE OF α-LACTALBUMIN B

The boxed amino acid residue represents the positional mutations differentiating variant A and B. Adapted from Brew *et al.* (1970) and Whitney *et al.* (1976).

Bovine Serum Albumin

This protein has been crystallized from milk and was shown to be compositionally and physically similar to bovine blood serum albumin (Polis *et al.* 1950). Although the entire amino acid sequence has not been elucidated, present evidence suggests that there exists some degree of microheterogeneity (Spencer and King 1971). Two features of its structure include a free sulfhydryl group at position 34 in its N-terminal peptide and 17 intramolecular disulfide bonds. At low values of pH, intermolecular associations occur which are apparent as multizoned patterns in gel electropherograms.

Bovine serum albumin enters milk from the vascular system, possibly by a route similar to serum immunoglobulins and is present at elevated concentrations in mastitic milk.

Immunoglobulins

The term immunoglobulins describes a family of high molecular weight proteins that share common physical and chemical characteristics and antigenic determinants (Butler 1969). Immunoglobulins occur in serum and other body fluids. In cow's colostrum they serve to transfer passive immunity to the calf, protecting it against disease until its own immune defenses are activated.

All immunoglobulins are monomers or polymers of a four-chain molecule consisting of two light chain polypeptides (~20,000 daltons) and two heavy chain polypeptides (50,000–70,000 daltons) which are cross-linked by disulfide bonds. Additionally, the structure is characterized by a variable number of intra-chain disulfide bonds. Three classes of bovine immunoglobulins have been identified, Ig G (Ig G1 and Ig G2), Ig A, and Ig M. All occur in bovine serum and milk.

Bovine Ig G1 and Ig G2.—In bovine serum these two subclasses exist in nearly equal proportions and at relatively high concentrations compared to normal milk. Ig G1 is, by a considerable margin, the dominant species in milk and appears to be selectively transported from the serum. In colostrum, the concentration of Ig G1 is unusually high, amounting to about one-half of the colostral-whey proteins. Ig G2 is present in both colostrum and milk in much lower concentrations. These lacteal immunoglobulins exist essentially as monomers (~160,000 daltons) and contain 2-4% carbohydrate.

Bovine Ig A.—This immunoglobulin is antigenically distinct from the Ig G and Ig M classes. Lacteal Ig A differs from its serum counterpart by virtue of a bound glycoprotein recognized as the free secretory component (FSC). The FSC exists also in a free state in milk and was first detected by Groves and Gordon (1967) who gave it the designation, glycoprotein-a. The identity of FSC and glycoprotein-a was established by Butler *et al.* (1968) and Butler (1971). Ig A appears to be a dimer, possessing a molecular weight of ~400,000 and contains 8-9% carbohydrate.

Bovine Ig M.—This macroglobulin occurs in milk in relatively low concentrations and is probably homologous to Ig M of other species. It exists as a pentamer (~900,000 daltons) and contains ~12% carbohydrate. The primary immune response of the cow occurs almost exclusively in the Ig M class of immunoglobulins.

Lacteal immunoglobulins have long been recognized as participants in the "creaming" of milk. When freshly drawn milk is submitted to quiescent cooling, fat globules "cluster" and rise to form a distinct layer of cream. Temperatures in excess of 62°C for 30 min (or equivalent) and homogenization destroy this property. Payens *et al.* (1965) demonstrated that Ig M is a cryoglobulin, viz., aggregates at low temperatures, and that it promotes the clustering of fat globules through nonspecific interactions with the fat globule surface. Heat denatures the macroglobulin and homogenization alters the fat globule surface. Both treatments impair cluster formation (Koops *et al.* 1966).

Proteose-Peptone Fraction

The proteose-peptone fraction of skim milk is conveniently defined as that portion of the milk protein system not precipitated by heating to 95°C for 20 min and subsequent acidification to pH

4.7, but precipitated by 12% (w/v) trichloroacetic acid (Rowland 1938). This protein fraction accounts for about 4% of the milk protein system and exhibits three distinct boundaries in Tiselius electrophoretic patterns which are designated as components "3," "5" and "8" in ascending order of mobility in alkaline buffers (Larson and Rolleri, 1955). Zonal electrophoresis in polyacrylamide gels, employing both discontinuous and continuous alkaline buffer systems, reveal a greater degree of heterogeneity. This is especially apparent for component 8 which resolves into two multi-zoned areas, designated "8-fast" and "8-slow."

Compositional and physical characteristics of the major components indicate that they are low to intermediate molecular-weight phosphoglycoproteins (4,000–40,000 daltons). Their amino acid profiles are characterized by low concentrations of aromatic residues and relatively high concentrations of glutamic and aspartic acids. Low concentrations of methionine account for all the sulfur-containing amino acids. Component 3 is found only in the whey fraction, whereas components 5 and 8 are constituents of whey and can be extracted from casein micelles. Indeed, components 5 and 8 may be minor caseins or casein-derived fragments. The proteose-peptone have been isolated from both heated and unheated milk, thus refuting the idea that they represent heat-induced artifacts. As a family of proteins, they are quite surface active and probably account for the relatively low surface tension of whey. For additional information regarding these proteins, the reader is referred to publications by Ng *et al.* (1970) and Kolar and Brunner (1969, 1970).

MINOR PROTEINS OF WHEY

Whey contains several unique proteins which are present in lower concentrations than the components already identified. Usually, these "minor" proteins escape detection when the whole whey protein system is examined by gel electrophoresis. The reader should consult the review article by Groves (1971) for a more extensive coverage of these milk proteins.

Lactoferrin

Lactoferrin has been identified in both internal and external secretions as well as in neutrophilic leucocytes. However, it was discovered first in milk, hence, the prefix "lacto." Lactoferrin is a metalloprotein, tenaciously binding two moles of iron per mole of protein (86,100 daltons), yielding a salmon-red color in solution. When isolated in an iron-free state, the apolactoferrin solution is colorless. Lactoferrin is a basic protein with an isoelectric point near pH 8. It demonstrates heterogeneous migration in both alkaline and acid electrophoretic gels, reflecting a strong tendency to form interaction products or, possibly, genetic polymorphis. With bound iron it resists heat, chemical and enzymatic denaturation.

The biological function of lactoferrin involves the fixation of iron from its serum homologue, transferrin. Because of this ability to chelate iron, the protein exerts a bacteriostatic activity against iron-dependent pathogenic organisms in milk and, eventually, in the gut of animals consuming milk.

Lactollin

This basic protein (43,000 daltons) is found in association with lactoferrin and is present in colostrum in higher concentrations than in normal milk. The amino acid profile differs markedly from lactoferrin and is characterized by high concentrations of aromatic residues and no methionine. Lactollin participates in pH-dependent association-dissociation phenomena.

Serum Transferrin

A small amount of transferrin which is identical to serum transferrin is found in milk; larger amounts are present in colostrum. It exhibits genetically controlled polymorphism similar to its serum counterpart. Transferrin is differentiated from lactoferrin by physicochemical, immunological and electrophoretic properties. As in the case of human serum transferrin, bovine serum transferrin may be composed of two identical polypeptide chains, each binding one mole of iron. Like lactoferrin, transferrin exhibits bacteriostatic activity due to its chelating property.

M-1 Glycoproteins

Bovine milk and colostrum contain a small amount of M-1 acid glycoproteins which, apparently, are derived from serum. They represent a family of related molecular species, ranging in molecular weight from 7,200–12,000 and containing 28.4% and 39.0% carbohydrates, respectively. These glycoproteins contain high concentrations of glutamic acid, threonine and proline, but no cystine/2 or tryptophan.

LIPID PHASE—ASSOCIATED PROTEINS

Thus far, we have identified the protein complement of the aqueous phase of milk, viz., components of the casein complex and whey fraction. An additional system of proteins exists as components of the milk fat globule membrane (MFGM) which surrounds the fat droplets in freshly secreted milk. The MFGM is composed of a complex mixture of neutral and polar lipids in association with lipid-compatible proteins. It is derived from the mammary secretory cell plasma membrane in the process of fat secretion and may contain small amounts of entrapped cytoplasmic substance. Numerous enzymes have been detected as constituents of the membrane material and a few have been isolated in significant quantities (Table 7.4). Presumably, some of these enzymes and, possibly, additional protein constituents contribute to the structural integrity of the membrane. However, no evidence exists which would support the presence of a strictly structural component. Indeed, the only proteins isolated from the MFGM in relatively pure state exhibit enzymatic activities. Thus, the membrane enzymes may be serving both as functional and structural components.

Prior to the introduction of the sodium dodecyl sulfate (SDS)-polyacrylamide gel (PAG) electrophoretic techniques by Weber and Osborn (1969), only limited progress was made toward the resolution of this complex protein system. Because of inherent hydrophobicity and the attendant tendency to form protein-protein interaction complexes, the membrane protein system resists solution in aqueous buffer systems normally employed for solution analyses. The SDS-PAG system seems to destabilize membrane proteins *in situ*, thus permitting their dispersion and resolution. Data reported from several laboratories reveal from 10 to more than 20 protein-stained electrophoretic zones, ranging in molecular weight from about 11,000 to >200,000. At least six zones react with periodate-Schiff base (PAS), several of which are not stained with the protein dye, indicating a relatively high population of glycoproteins.

Despite the resolution of membrane proteins in the presence of SDS and 2-mercaptoethanol, limited data are available on the physicochemical properties of these proteins. In this regard, it is paradoxical that the agent employed to effect the elegant resolution of the protein system binds tenaciously to the denatured molecules, thus negating most characterization studies. On the other hand, agent-free specimens tend to reassociate into water-insoluble complexes. However, two or three of the water-extractable glycoproteins have been partially characterized, yielding molecular weights of from 20,000–150,000. Amino acid profiles of total MFGM and fractions isolated therefrom reveal concentrations of hydrophobic residues somewhat lower than found in the caseins. Thus, until the primary structures of these proteins are elucidated, little understanding of their organization in the membrane complex will be forthcoming.

For additional information on MFGM proteins, the reader is referred to recent reviews by Anderson and Cawston (1975) and by Patton and Keenan (1975).

TABLE 7.4
A PARTIAL LIST OF ENZYMES IDENTIFIED IN MILK

Enzyme[1]	Classification[2] Number (E.C.)	Reported Distribution[3] in Milk	
Acid phosphatase	3.1.3.2.	FGM,	SM
Alkaline phosphatase	3.1.3.1.	FGM,	SM
Aldolase	4.1.2.13.	FGM	
α-Amylase	3.2.1.1.		SM
β-Amylase	3.2.1.2.		SM
Mg^{2+}-activated ATPase	3.6.1.3.	FGM	
Catalase	1.11.1.6.	FGM,	SM
Choline esterase	3.1.1.8.	FGM	
Diaphorase	1.6.4.3.	FGM	
A-Esterase	3.1.1.2.		SM
B-Esterase	3.1.1.1.		SM
C-Esterase	3.1.1.8.		SM
β-Galactosidase	3.2.1.23.	FGM	
β-Glucosidase	3.2.1.21.	FGM	
Glucose-6-phosphatase	3.1.3.9.	FGM	
Glycoprotein β-galactosyl-transferase	2.4.1.38.	FGM,	SM
D-Glutamyl transferase	2.3.2.1.	FGM	
Inorganic pyrophosphatase	3.6.1.1.	FGM	
Lactic acid dehydrogenase	1.1.1.27.	FGM	
Lipase	3.1.1.3.		SM
Lipoprotein lipase	3.1.1.34.		SM
Lysozyme	3.2.1.17.		SM
α-Mannosidase	3.2.1.24.		SM
NADH-Dehydrogenase	1.6.99.3.	FGM	
Nucleotide pyrophosphatase	3.6.1.9.	FGM	
5' Nucleotidase	3.1.3.5.	FGM	
Peroxidase (Lacto)	1.11.1.7.		SM
Phosphoprotein phosphatase	3.1.3.16.	FGM,	SM
Phosphodiesterase	3.1.4.1.	FGM	
Protease	3.4.4.-.		SM
Protein disulfide isomerase	5.3.4.1.	FGM	
Ribonuclease	2.7.7.16.		SM
Sulfhydryl oxidase	—		SM
Thiol oxidase	1.8.3.2.	FGM	
UDP-Glucosyl hydrolase	—	FGM	
UDP-Galactosyl hydrolase	—	FGM	
UDP-Galactosyl transferase	2.4.1.22.		SM
Xanthine oxidase	1.2.3.2.	FGM,	SM

[1] Compiled from Groves (1971), Shahani *et al.* (1973), Anderson and Cawston (1975).
[2] According to the Committee on Enzymes of the International Union of Biochemists.
[3] Notation: FGM—fat globule membrane: SM—skim milk phase.

ENZYMES

Milk contains a myriad of enzymes, reflecting the complex biological nature of the synthetic and secretory processes (Table 7.4). As might be expected, these enzymes possess specialized functions and are associated with the membranous material or casein complex. Several exist freely dispersed in the aqueous phase and a few are distributed throughout the total system. Being proteins, they exhibit the usual environmentally induced association-dissociation phenomena and conformational changes ascribed to protein systems. Many of the enzymes are glycoproteins and a number exist in the form of lipoprotein particulates.

Despite the array of enzymes present in milk, only a few are of interest to the food technologist. For a more comprehensive presentation on the subject of the enzyme complement of milk, the reader is referred to reviews by Groves (1971) and Shahani *et al.* (1973).

Alkaline Phosphatase

This enzyme exists as a lipoprotein complex and is distributed between the MFGM and the aqueous phase. Its specificity involves a Mg(II) catalyzed hydrolysis of phosphomonoesters, with a maximum activity around pH 9. Under appropriate conditions, it will dephosphorylate the caseins, although at a slow rate. Its sensitivity to heat as usually employed for the pasteurization of milk, viz., 71.5°C for 16 sec or more, is fortuitous. Thus, its detection in processed dairy products usually signifies inadequate pasteurization. However, in some instances the enzyme is reactivated in processed milk following storage at temperatures in excess of 5°C.

Lipases

Milk contains lipases which under appropriate conditions partially hydrolyze the milk fat, producing a characteristic bitter or "rancid" flavor defect. Although lipolysis occurs at the normal pH of milk, viz., 6.6–6.7, lipase exerts its maximum activity around pH 9. There is some question relative to the number of lipases found in milk. Several investigators have noted that lipase is associated with casein micelles and that most of it can be dissociated into the serum by the addition of sodium chloride (Downey and Andrews 1965A, B). When the casein-free serum was passed over a Sephadex G-200 column, three lipolytically active zones were observed, ranging in molecular weight from 62,000–112,000. Other investigators reported a molecular weight of 7,000 for a preparation obtained from the sediment collected during the clarification of raw milk (Chandan and Shahani 1963). Lipases are quite sensitive to heavy metals, heat and light and are inactivated by pasteurization.

Protease

A proteolytic enzyme, apparently trypsin-like and of endogenous origin, is found in milk in association with its substrate, casein. It has been extracted from acid-precipitated caseins and seems to be preferentially associated with κ-casein. The activity of this protease exhibits a wide pH optimum from 6.5–9. The enzyme is sometimes referred to as "caseinase" because of its apparent preference for caseins as substrates, β-casein being most susceptible, followed by α_{s1}- and κ-caseins. Presumably, this endogenous enzymatic activity is the mechanism involved in the creation of the γ-caseins from β-caseins. In the free state, the enzyme is inactivated by heat at 80°C for 10 min, but could be more heat resistant when present in milk. Thus, its survival in processed milk products could play an important role in product stability. Also, its activity during the cheese curing process may be of significance in flavor and texture development.

Xanthine Oxidase

This enzyme catalyzes the oxidation of purines, hypoxanthine and xanthine to uric acid. It is found in relatively high concentrations in the MFGM comprising about 10% of the protein mass. The enzyme has a molecular weight of ~300,000 and firmly binds the cofactor flavin adenine dinucleotide (FAD), molybdenum and iron in a ratio of 2:2:8. More recent studies indicate that the 300,000 dalton species is a dimer consisting of two identical subunits (Waud *et al.* 1975). Because xanthine oxidase is not a component of secretory cell plasmalemma, its presence in MFGM preparations raises the question of its origin and the manner by which it became associated with fat globules.

Sulfhydryl Oxidase

This enzyme catalyzes the oxidation of sulfhydryl groups to corresponding disulfides. In milk, the enzyme may catalyze the formation of disulfide bonds in proteins. When isolated from the whey fraction, it exhibits concentration-dependent association-dissociation. One zone was revealed by gel electrophoresis in the presence of dodecyl sulfate with an estimated subunit weight of 89,000 (Janolino and Swaisgood 1975). Currently, the Swaisgood group is investigating the use of this enzyme, immobilized in a flow-through reactor, to eliminate the "heated" flavor in ultra high temperature processed milk.

Peroxidase

Lactoperoxidase accounts for about 0.5–1.0% of the total whey protein, being the most abundant enzyme in milk. It catalyzes the decomposition of hydrogen peroxide in the presence of a hydrogen donor or an oxidizable component. It is a haemprotein, possessing a molecular weight of about 77,000. The enzyme functions as a bacterial inhibitor in milk, acting against *Salmonella* and pathogenic *Streptococcus* in the presence of thiocyanate and peroxide, both of which are present in milk.

MECHANISM OF MILK SECRETION

As food scientists we are interested in milk more as a food or as a source of components for food systems than as a biological secretion. Yet, we would be remiss if we neglected to recognize the progress being made toward a better understanding of the mechanism of milk secretion. Briefly summarized, secretion involves intracellular processes in which fat droplets are extruded from the apical end of mammary epithelial cells into the alveolar lumen of the udder. In the process of exocytosis, the fat droplets are enshrouded with plasma membrane and small amounts of cytoplasmic material. These fat globules, with their membrane coat (MFGM), range in size from 1 to ~10 μm in diameter. One cubic centimeter of milk contains up to 3 × 10^9 globules, possessing a total surface area of about 600 cm^2.

Concomitant with the process of fat secretion is the replenishment of the plasmalemma with Golgi vesicle membrane. Proteins synthesized on the rough endoplasmic reticulum (ER) enter the cisternae and migrate to Golgi vesicles for final assembly. Product-filled, mature Golgi vacuoles move away from the Golgi apparatus toward the apical end of the cell. Here they contact the apical plasmalemma and discharge their contents to the lumen. These events are illustrated in Fig. 7.12. Implicit to this process is the "flow" of membrane material from the endoplasmic reticulum through the organelle chain to the Golgi vacuoles. Electron micrographs indicate the vacuoles, upon reaching the inner surface of the plasmalemma, "blend" with the plasmalemma in such a manner that their inner surface becomes the outer surface of the plasma membrane and, consequently, the outer surface of the MFGM.

Presumably, all of the components of milk which are synthesized in the secretory cells, including the aqueous phase, are discharged to the lumen by either of the processes described. The process by which the serum-derived components, viz., serum albumin and immunoglobulins, traverse the secretory cell and enter the lumen requires further study. Contemporary reviews by Patton and Keenan (1975) and Patton and Jensen (1975) should be consulted for additional details.

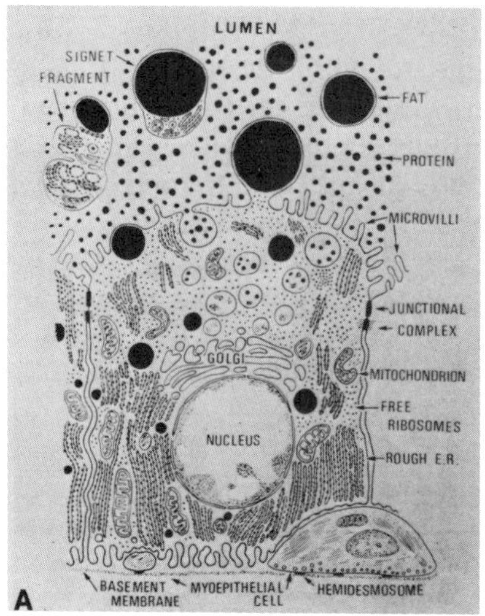

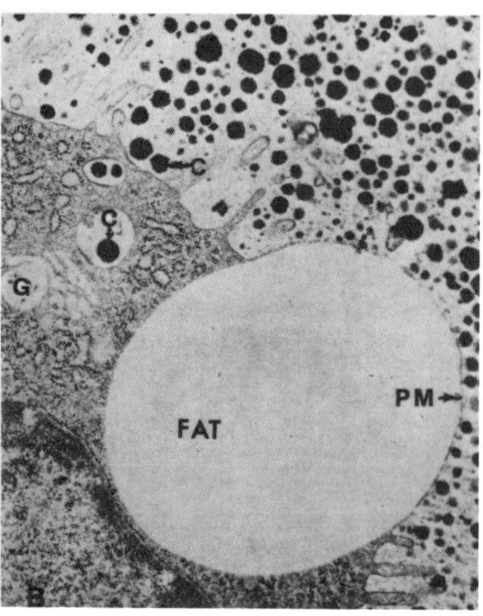

FIG. 7.12. THE MECHANISM OF MILK SECRETION

(A) A schematic representation of the secretory processes, showing the secretion of plasma membrane-covered fat globules into the mammary lumen and a concomitant release of proteins from Golgi vacuoles. Note that the vacuolar membrane becomes part of the plasmalemma [Reprinted from Linzell and Peaker (1971), by courtesy of the American Physiological Society]. (B) An electron micrograph of a section from a secretory cell, showing the same process: G designates an intracellular Golgi vacuole; C designates casein micelles in the intervacuolar space and in the lumen; and PM identifies the plasmalemma. [Reprinted from Toben *et al.* (1972), by courtesy of J. B. Lippincott Co.].

BIBLIOGRAPHY

ANDERSON, M., and CAWSTON, T. E. 1975. Reviews of the progress of dairy science. The milk-fat globule membrane. J. Dairy Res. 42, 459–483.

ASCHAFFENBURG, R., and DREWRY, J. 1955. Occurrence of different beta-lactoglobulins in cow's milk. Nature *176*, 218–219.

ASHOOR, S. H., SAIR, R. A., OLSON, N. F., and RICHARDSON, T. 1971. Use of a papain superpolymer to elucidate the structure of bovine casein micelles. Biochim. Biophys. Acta *229*, 423–430.

BLOOMFIELD, V. A., and MEAD, R. J. 1975. Structure and stability of casein micelles. J. Dairy Sci. *58*, 592–601.

BRAUNITZER, G., CHEN, R., SCHRANK, B., and STANGL, A. 1972. Automatic sequential analysis of the proteins β-lactoglobulin AB. Hoppe-Seyler's Z. Physiol. Chem. *353*, 832–834. (German)

BRAUNITZER, G., CHEN, R., SCHRANK, B., and STANGL, A. 1973. The sequential analysis of β-lactoglobulins. Hoppe-Seyler's Z. Physiol. Chem. *354*, 867–878. (German)

BREW, K., CASTELLINO, F. J., VANAMAN, T. C., and HILL, R. L. 1970. The complete amino-acid sequence of bovine α-lactalbumin. J. Biol. Chem. *245*, 4570–4582.

BRODBECK, U., PENTON, W. L., TANAHASHI, N., and EBNER, K. E. 1967. The isolation and identification of the B protein of lactose synthetase as α-lactalbumin. J. Biol. Chem. *242*, 1391–1397.

BUTLER, J. E. 1969. Bovine immunoglobulins: A review. J. Dairy Sci. *52*, 1895–1909.

BUTLER, J. E. 1971. Physicochemical and immunochemical studies of bovine Ig A and glycoprotein-a. Biochim. Biophys. Acta *251*, 435–448.

BUTLER, J. E., COULSON, E. J., and GROVES, M. L. 1968. Identification of glycoprotein-a as a probable fragment of bovine Ig A. Federation Proc. *27*, 617.

CHANDAN, R. C., and SHAHANI, K. 1963. Purification and characterization of milk lipase. I. Purification. J. Dairy Sci. *46*, 275–283.

DOWNEY, W. K., and ANDREWS, P. 1965A. Gel filtration applied to the study of lipases and other esterases. Biochem. J. *94*, 642–650.

DOWNEY, W. K., and ANDREWS, P. 1965B. Milk lipases and the association of lipases with casein micelles studied by gel-filtration. Biochem. J. *94*, 33.

EBNER, K. E., and McKENZIE, L. M. 1972. α-Lactalbumin and galactosyltransferase in rat serum and their relationship to milk secretion. Biochem. Biophys. Res. Commun. *49*, 1624–1630.

FARRELL, H. M. 1973. Models for casein micelle formation. J. Dairy Sci. *56*, 1195–1206.

FRANK, G., and BRAUNITZER, G. 1967. The primary structure of β-lactoglobulins (German). Hoppe-Seyler's Z. Physiol. Chem. *348*, 1691–1692.

GORDON, W. G., GROVES, M. L., GREENBERG, R., JONES, S. B., and KALAN, E. B. 1972. Probable identification of γ-, TS-, R-, and S-caseins as fragments of β-casein. J. Dairy Sci. *55*, 261–263.

GORDON, W. G., and WHITTIER, E. O. 1965. Proteins of milk. *In* Fundamentals of Dairy Chemistry. B. H. Webb and A. H. Johnson (Editors). AVI Publishing Co., Westport, Conn.

GROVES, M. L. 1971. Minor milk proteins and enzymes. *In* Milk Proteins, Chemistry and Molecular Biology, Vol. 2. A. A. McKenzie (Editor). Academic Press, N.Y.

GROVES, M. L., and GORDON, W. G. 1967. Isolation of a new glycoprotein-a and a γ G globulin from individual cow milks. Biochemistry *6*, 2388–2394.

GROVES, M. L., GORDON, W. G., KALAN, E. B., and JONES, S. B. 1972. Composition of bovine γ-caseins A^1 and A^3 and further evidence for a relationship in biosynthesis of γ- and β-caseins. J. Dairy Sci. *55*, 1041–1049.

HO, C., MAGNUSSON, J. A., WILSON, J. B., and KURLAND, R. J. 1969. Phosphorus nuclear magnetic resonance studies of phosphoproteins and phosphorylated molecules. 2. Chemical nature of phosphorus atoms in α_s-casein B and phosvitin. Biochemistry *8*, 2074–2082.

HOAGLAND, P. D., THOMPSON, M. P., and KALAN, E. B. 1971. Amino acid composition of α_{s3}-, α_{s4}-, and α_{s5}-caseins. J. Dairy Sci. *54*, 1103–1110.

JANOLINO, V. G., and SWAISGOOD, H. E. 1975. Isolation and characterization of sulfhydryl oxidase. J. Biol. Chem. *250*, 2532–2538.

JOLLÈS, J., SCHOENTGEN, F., ALAIS, C., and JOLLÈS, P. 1972. Studies on the primary structure of cow κ-casein: The primary structure of cow para-κ-casein. Chimia *26*, 645–646.

KIM, Y. K., YAGUCHI, M., and ROSE, D. 1969. Isolation and amino acid composition of para-κ-casein. J. Dairy Sci. *52*, 316–320.

KNOOP, E. 1972. Electron microscopic studies on the structure of milk fat and protein. Milchwissenschaft *27*, 364–373. (German)

KOLAR, C. W., and BRUNNER, J. R. 1969. Proteose-peptone fraction of bovine milk: Distribution in the protein system. J. Dairy Sci. *52*, 1541–1556.

KOLAR, C. W., and BRUNNER, J. R. 1970. Proteose-peptone fraction of bovine milk: Lacteal serum components 5 and 8—Casein-associated glycoproteins. J. Dairy Sci. *53*, 997–1008.

KOOPS, J., PAYENS, T. A. J., and MAGOT, M. F. K. 1966. The effect of homogenization on the spontaneous creaming of milk. Neth. Milk Dairy J. *20*, 296–299.

LARSON, B. L., and ROLLERI, G. D. 1955. Heat denaturation of specific serum proteins of milk. J. Dairy Sci. *38*, 351–360.

LI, F. H. F., and GAUNT, S. N. 1972. A study of genetic polymorphisms of milk β-lactoglobulin, α_{s1}-casein, β-casein, and κ-casein in five dairy breeds. Biochem. Genet. *6*, 9–19.

LINZELL, J. L., and PEAKER, M. 1971. Mechanism of milk secretion. Physiol. Rev. *51*, 564–597.

MAC DONALD, C. A., and THOMAS, M. A. W. 1970. The rennin-sensitive bond of bovine κ-casein. Biochim. Biophys. Acta *207*, 139–142.

MC KENZIE, G. H., NORTON, R. S., and SAWYER, W. H. 1971. Heat-induced interaction of β-lactoglobulin and κ-casein. J. Dairy Res. *38*, 343–351.

MC KENZIE, H. A., RALSTON, G. B., and SHAW, D. C. 1972. Location of sulfhydryl and disulfide groups in bovine β-lactoglobulin and effect of urea. Biochemistry *11*, 4539–4547.

MELACHOURIS, N. 1969. Discontinuous gel electrophoresis of whey proteins, casein and clotting enzymes. J. Dairy Sci. *52*, 456–459.

MERCIER, J. C., BRIGNON, G., and RIBADEAU-DUMAS, B. 1973. Primary structure of bovine κ-casein B. Complete sequence. Eur. J. Biochem. *35*, 222–235. (French)

MERCIER, J. D., GROSCLAUDE, F., and RIBADEAU-DUMAS, B. 1972. Primary structure of bovine caseins. A review. Milchwissenschaft *27*, 402–408.

MORR, C. V. 1967. Effect of oxalate and urea on the ultracentrifugation properties of raw and heated skim milk micelles. J. Dairy Sci. *50*, 1744–1751.

NG, W. C., BRUNNER, J. R., and RHEE, K. C. 1970. Proteose-peptone fraction of bovine milk: Lacteal serum component 3—A whey glycoprotein. J. Dairy Sci. *53*, 987–996.

NOBLE, R. W., and WAUGH, D. F. 1965. Casein micelles formation and structure. 1. J. Am. Chem. Soc. *87*, 2236–2245.

ÖSTERBERG, R. 1964. Enzymic degradation of a phosphopeptide obtained by trypsin hydrolysis of α-casein. A partial structural formula. Acta Chem. Scand. *18*, 795–804.

PALMER, A. H. 1934. The preparation of a crystalline globulin from the albumin in fraction of cows' milk. J. Biol. Chem. *104*, 359–371.

PATTON, S., and JENSEN, R. G. 1975. Lipid metabolism and membrane functions of the mammary gland. *In* Progress in the Chemistry of Fats and Other Lipids, Vol. 14, R. T. Holman (Editor). Pergamon Press, Oxford, England.

PATTON, S., and KEENAN, T. W. 1975. The milk fat globule membrane. Biochim. Biophys. Acta *415*, 273–309.

PAYENS, T. A. J., KOOPS, J., and MOGOT, M. F. K. 1965. Adsorption of euglobulin on agglutinating milk fat globules. Biochim. Biophys. Acta *94*, 576–578.

PEDERSEN, K. O. 1936. Ultracentrifugal and electrophoretic studies on the milk proteins. 1. Introduction and preliminary studies with fractions from skim milk. Biochem. J. *30*, 948–960.

PETERSON, R. F., and KOPFLER, F. C. 1966. Detection of new types of β-casein by polyacrylamide gel electrophoresis at acid pH: A proposed nomenclature. Biochem. Biophys. Res. Commun. 22, 388–392.

POLIS, B. D., SHMUKLER, H. W., and CUSTER, J. H. 1950. Isolation of a crystalline albumin from milk. J. Biol. Chem. 187, 349–354.

POLZHOFER, K. P. 1972. Synthesis of the rennin-sensitive pentadecapeptide of bovine-κ-casein. Tetrahedron 28, 855–864. (German)

PUJOLLE, J., RIBADEAU-DUMAS, B., GARNIER, J., and PION, R. 1966. A study of κ-casein components. Biochem. Biophys. Res. Commun. 25, 285–290.

RIBADEAU-DUMAS, B., and GARNIER, J. 1970. Structure of casein micelle. The accessibility of the subunits to various reagents. J. Dairy Res. 37, 269–278.

ROSE, D. 1969. A proposed model of micelle structure in bovine milk. J. Dairy Res. 31, 171–175.

ROWLAND, S. J. 1938. The precipitation of the proteins in milk. 1. Casein; 2. Total proteins; 3. Globulin; 4. Albumin, and proteose-peptone. J. Dairy Res. 9, 30–41.

SHAHANI, K. M., HARPER, W. J., JENSEN, R. G., PERRY, R. M., and ZITTLE, C. A. 1973. Enzymes in bovine milk: A review. J. Dairy Sci. 56, 531–543.

SLATTERY, C. W., and EVARD, R. 1973. A model for the formation and structure of casein micelles from subunits of variable composition. Biochim. Biophys. Acta 317, 529–538.

SPENCER, E. M., and KING, T. P. 1971. Isoelectric heterogeneity of bovine plasma albumin. J. Biol. Chem. 246, 201–207.

SWAISGOOD, H. E. 1973. The caseins. CRC Critical Rev. Food Tech. 3, No. 4, 375–414.

THOMPSON, M. P., and KIDDY, C. 1964. Genetic polymorphism in caseins of cow's milk. 3. Isolation and properties of α_{s1}-caseins A, B and C. J. Dairy Sci. 47, 626–632.

TIMASHEFF, S. N., and TOWNSEND, R. 1968. β-Lactoglobulin as a model of subunit enzymes. In Protides of the Biological Fluids, Proc. 16th Colloquium, Bruges. H. Peeters (Editor). Pergamon Press, Oxford, England.

TOBEN, H., JOSIMOVICH, J. B., and SALAZAR, H. 1972. The ultrastructure of the mammary gland during prolactin—induced lactogenesis in the rabbit. Endocrinology 90, 1569–1577.

VANAMAN, T. C., BREW, K., and HILL, R. L. 1970. The disulfide bonds of bovine α-lactalbumin. J. Biol. Chem. 245, 4583–4590.

WAUD, W. R., BRADY, F. O., WILEY, R. D., and RAJAGOPALAN, K. V. 1975. A new purification procedure for bovine xanthine oxidase: Effect of proteolysis on the subunit structure. Arch. Biochem. Biophys. 169, 695–701.

WAUGH, D. F. 1971. Formation and structure of casein micelles. In Milk Proteins, Chemistry and Molecular Biology, Vol. 2, H. A. McKenzie (Editor). Academic Press, N.Y.

WAUGH, D. F., CREAMER, L. K., SLATTERY, C. W., and DRESDEN, G. W. 1970. Core polymers of casein micelles. Biochemistry 9, 786–795.

WAUGH, D. F., and NOBLE, R. W. 1965. Casein micelles. Formation and structure. 2. J. Am. Chem. Soc. 87, 2246–2257.

WAUGH, D. F., and von HIPPEL, P. H. 1956. κ-Caseins and the stabilizing of casein micelles. J. Am. Chem. Soc. 78, 4576–4582.

WEBER, K., and OSBORN, M. 1969. The reliability of molecular weight determinations by dodecyl sulfate polyacrylamide gel electrophoresis. J. Biol. Chem. *244*, 4406–4412.

WHITNEY, R. McL., BRUNNER, J. R., EBNER, K. E., FARRELL, H. M., JOSEPHSON, R. V., MOOR, C. V. and SWAISGOOD, H. E. 1976. Nomenclature of the proteins of cow's milk: Fourth revision. J. Dairy Sci. *59*, 785–815.

Additional General References

LYSTER, R. L. J. 1972. Reviews of the progress of dairy science. Section C. Chemistry of milk proteins. J. Dairy Res. *39*, 279–318.

McKENZIE, H. A. 1970. Milk Proteins, Chemistry and Molecular Biology, Vol. 1. Academic Press, New York.

McKENZIE, H. A. 1971. Milk Proteins, Chemistry and Molecular Biology, Vol. 2. Academic Press, New York.

TABORSKY, G. 1974. Phosphoproteins. *In* Advances in Protein Chemistry. C. B. Anfinsen, J. T. Edsall, and F. M. Richards (Editors). Academic Press, New York.

WEBB, B. H., and JOHNSON, A. H. 1965. Fundamentals of Dairy Chemistry. Avi Publishing Co., Westport, Conn.

WEBB, B. H., JOHNSON, A. H., and ALFORD, J. A. 1974. Fundamentals of Dairy Chemistry, 2nd Edition. Avi Publishing Co., Westport, Conn.

Nomenclature Reports, American Dairy Science Association

JENNESS, R. *et al.* 1956. Nomenclature of the proteins of bovine milk. J. Dairy Sci. *39*, 536–541.

BRUNNER, J. R. *et al.* 1960. Nomenclature of the proteins of bovine milk—First revision. J. Dairy Sci. *43*, 901–911.

THOMPSON, M. P. *et al.* 1965. Nomenclature of the proteins of cow's milk—Second revision. J. Dairy Sci. *48*, 159–169.

ROSE, D. *et al.* 1970. Nomenclature of the proteins of cow's milk—Third revision. J. Dairy Sci. *53*, 1–17.

WHITNEY, R. McL. *et al.* 1977. Nomenclature of the proteins of cow's milk—Fourth revision. J. Dairy Sci. *59*, (in press).

8

Egg Proteins

David T. Osuga
Robert E. Feeney

Egg proteins have been investigated for many years by many researchers from different fields of endeavor. Egg proteins are studied by the embryologist, poultry scientist, and consumer-orientated researcher, and researchers from many disciplines use the egg as a source of typical or atypical proteins. This, of course, is due to the unique properties and structures of the individual protein components in the egg. As a consequence of this diverse interest in the egg proteins, the literature on egg proteins is enormous and a detailed review of the subject would be voluminous. Thus, the intent of this article is to serve as a summary of the current state of the knowledge of the egg proteins from the domestic chicken. Unless otherwise mentioned, the egg proteins will be those from the chicken, *Gallus gallus*.

The following recent references will provide extensive background information on avian eggs: Carter (1968), Feeney and Allison (1969), Bell and Freeman (1971), Stadelman and Cotterill (1973), Vadehra and Nath (1973), Board and Fuller (1974), and Osuga and Feeney (1974).

Most of the proteins present in the egg are glycoproteins in which the carbohydrates are either glycosidically linked to the seryl or threonyl residues or attached by an amide linkage to the asparaginyl residues. Some proteins contain phosphorus and sulfur linked to the seryl or threonyl residues. Lipids are also present to further complicate the structure and composition of proteins. Moreover, any combination of these may be attached to the protein at various locations on the polypeptide chain. These macromolecules can exist as a basic subunit or structure which can complex to form a more intricate macromolecular system or morphological entity. Nonetheless, there are present several proteins which are easily isolated and crystallized. These have served as important cornerstones in the development of the basic concepts of protein chemistry.

THE WHOLE EGG

The whole egg consists of 11.8% protein which can be divided into three distinct parts: shell, yolk, and white, which contain approximately 3, 17.5 and 11% protein, respectively. Table 8.1 summarizes

TABLE 8.1
DISTRIBUTION OF EGG COMPONENTS

Constituents	Shell %	White %	Yolk %
Water	1.0	88.5	47.5
Protein	4.0 (2)[1]	10.5	17.4
Lipid	—	0.02[2]	33.0
Carbohydrate[3]	—	0.5[4]	0.2
Inorganic ions	95.0 (98)[1]	0.5	1.1
Other	—	—	0.8

Source: Gilbert (1971), except where noted.

[1] Simkiss and Taylor (1973).
[2] Sato *et al.* (1973).
[3] Small amount in shell protein.
[4] Tunmann and Silberzahn (1961) reported that 98% of this 0.5% is glucose.

the gross content and Fig. 8.1 is a diagram of the egg. The yolk and egg white proteins will be discussed in detail in the latter part of this article.

Enclosing the whole egg is the thin (10 μm) outermost layer, the cuticle, which may exist as two layers. It is composed of 86% protein including from 3.5–6.5% water and 1% KCl soluble proteins, 4% carbohydrates, and 3% lipid (Wedral *et al.* 1974).

The shell is some thirtyfold thicker than the cuticle and composed of only 2% proteinaceous material and 98% crystalline calcium carbonate (calcite). The proteins (70% of the proteinaceous material is protein) and carbohydrates (11%) are present in two layers—the

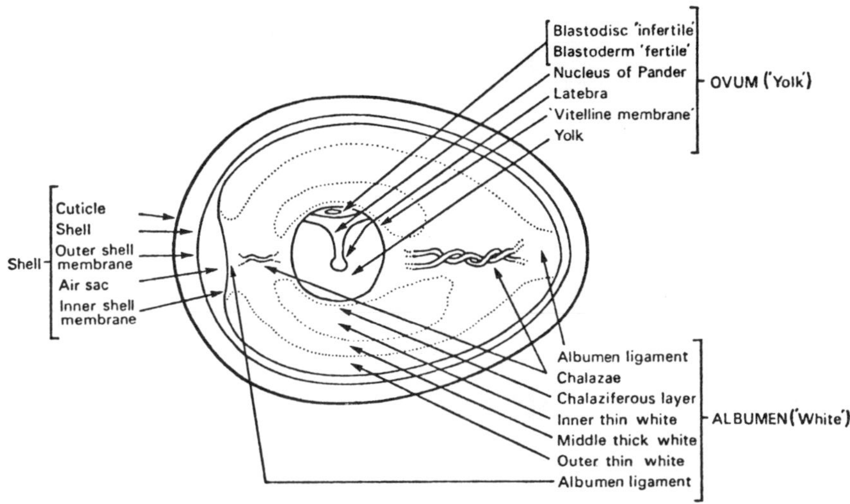

Courtesy of Gilbert (1971)

FIG. 8.1. DIAGRAM OF THE HEN'S EGG

spongy and mammillary layers. The proteins in the latter layer are fibrous.

Below the shell is the shell membrane which consists of two tough fibrous-like proteins. The inner and outer layers appear by electron micrographs to be composed of possibly three and six layers, respectively (Mosshoff and Stolpmann 1961; Simons and Wiertz 1963). The membranes contain approximately 90% protein (including 1.0–4.5% soluble proteins), 3% lipids, and 2% carbohydrates (Wedral *et al.* 1974), and they are layered at right angles to each other. The presence of hydroxylysine (Candlish and Scougall 1969) and hydroxyproline (Balch and Cooke 1970) suggests a collagen-like protein. Picard *et al.* (1973) isolated and characterized two sulfate glycoproteins (A and B) by enzymatic (Pronase and trypsin) treatment of the membranes. Glycopeptide B had approximately 40% of the amino acid residues O-glycosidically linked between the carbohyrates and the seryl and threonyl residues. [See Candlish (1972) for further details on the shell membranes.]

The vitelline membrane which separates the yolk from the white is composed of four layers of proteins (Gilbert 1971). The layers consist of 87% protein (Britten 1973) with 3% lipid and 10% carbohydrates (Bellairs 1964). The absence of hydroxylysine indicates that collagen was not present (Bellairs *et al.* 1963). Recently Kido *et al.* (1975) reported that the membranes can be dissolved in sodium dodecyl sulfate (SDS) and that three bands were produced on electrophoresis. The molecular weights by SDS-electrophoresis indicated 32,000 and 260,000 daltons for two of the three bands.

Three enzymes have been recently reported to be present in the shell and/or the membranous material in the egg. β-N-Acetygluco-saminidase activity was reportedly present in the shell and the vitelline membranes (Winn and Ball 1975). Small amounts of lysozyme activity were found in the cuticle and shell, but much more was found in the shell membranes (Vadehra *et al.* 1972). ATP phosphohydrolases were isolated from the vitelline membrane, and the effects of ions and pH on the enzymatic reaction were investigated by Haaland and associates (Haaland *et al.* 1971; Etheredge *et al.* 1971).

YOLK

Egg yolk proteins are composed of a number of complex types of macromolecules (glycoproteins, phosphoglycoproteins, lipoproteins, and phosphoglycolipoproteins) which are soluble or insoluble in an aqueous medium. They can be purified by such techniques as centrifugation, extraction, electrophoresis, and chromatography. These proteins can interact in a distinct fashion and exhibit a definite morphological organization (Bellairs *et al.* 1972) due to their unique composition and properties.

The constituents of yolk are diagrammed in Fig. 8.2 and the gross

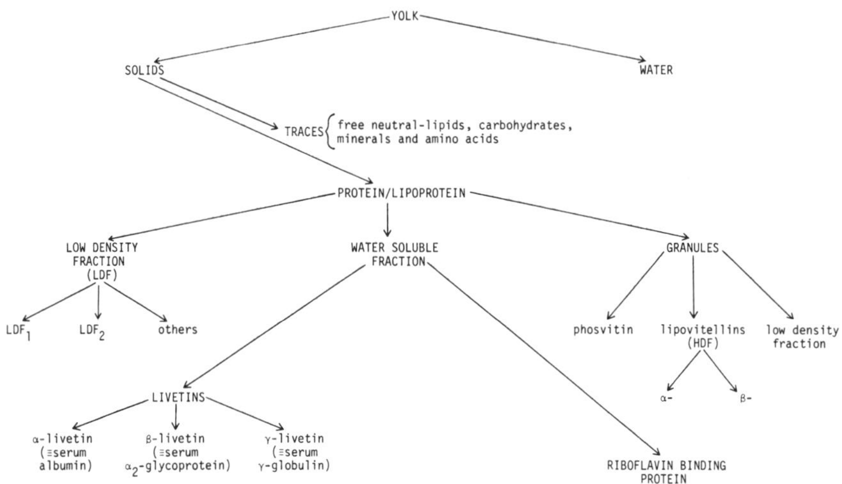

Modified from Gilbert (1971)

FIG. 8.2. SCHEMATIC OF THE CHARACTERIZED EGG YOLK CONSTITUENTS

composition listed in Table 8.2. A general separation of the constituents can be achieved with relative ease since the low density fraction (LDF) has a density less than unity. Thus, upon centrifugation the LDF will layer on top, and the heavier granules will settle to

TABLE 8.2
PROTEIN COMPOSITION OF YOLK

| | | | % of Total | | | |
| | | | | | Phosphorous | |
Protein	Weight	Solid	Proteins	Lipid	Lipid	Protein
Low density fraction	33.7	65.0	22.0	93.0	19.0	—
Livetin	5.3	10.0	30.0	—	—	—
Phosvitin	2.1	4.0	12.0	—	—	69.0
Lipovitellin	8.3	16.0	36.0	7.0	←12.0→	
Others	2.6	5.0	—	—	—	—
Riboflavin binding protein (YRBP)	—	—	0.4	—	—	some[1]

Source: Modified from Gilbert (1971).

[1] 0.2% P or 2 M P/M YRBP.

the bottom, with the water soluble fraction in between. The granules contain 4% LDF. Although the gross separation of the macromolecules appears quite simple, obtaining knowledge of their interaction, composition, and structure presents a formidable task for the investigator.

Extensive general reviews on the yolk proteins have been recently presented by Cook (1968), Shenstone (1968), McIndoe (1971), Powrie (1973), and Vadehra and Nath (1973); structure of the LDF by Schneider et al. (1973); yolk lipoproteins by Cook and Martin (1969); the chemical and structural aspects of lipoproteins by Burley (1971); lipovitellin and phosvitin by Taborsky (1974).

Low Density Lipoproteins

The LDF appears to have a micelle type of structure which exists as a phospholipoprotein with a lipid-rich core and with the phospholipid and protein moieties radiating towards the surface. This type of structure was supported by the susceptibilities of the LDF to phosphatidase C (Burley and Kushner 1963), and the proteolytic enzymes, papain (Saari et al. 1964), trypsin and Pronase (Augustyniak 1968; Steer et al. 1968). The major forces in this very complex system of lipids, phospholipids, and glycoproteins appear to be hydrophobic (Burley 1971).

Nicholes et al. (1954) first isolated LDF by preparative centrifugation and established that the lipoprotein was 89% lipid. Turner and Cook (1958) reported that the lipovitellenin, originally isolated by Fevold and Lausten (1946), was derived from the LDF. The LDF is basically composed of two large lipoproteins, LDF$_1$ and LDF$_2$. The LDF$_1$ makes up 20% of the LDF and is two to three times larger than LDF_2. The molecular weight of LDF_1 is 10.3×10^6 daltons and that of LDF_2 is 3.3×10^6 daltons (Martin *et al.* 1964). There is 3–4% more lipid in LDF_1 than in LDF_2, and it appears to be more heterogenous (Martin *et al.* 1964). Augustyniak *et al.* (1964) reported that the LDF_1 and LDF_2 will dissociate into five similar subunits.

LDF is not only a lipoprotein but also a glycolipoprotein, since it contains 3% carbohydrates glycosidically linked to an asparaginyl residue (Augustyniak and Martin 1968). It may be composed of at least two polypeptide chains, since two N-terminal groups, arginine and lysine, have been reported with the same C-terminal tyrosine (Neelin and Cook 1961). Hillyard *et al.* (1972) confirmed the presence of lysine and tyrosine as terminal groups.

The evidence suggests that the LDF proteins are composed of lipid-free 10,000 daltons subunits (Hillyard *et al.* 1972; Burley 1975). Burley (1975) isolated and analyzed two subunits of different molecular weights, 10,000 and 20,000 daltons. The smaller component was composed of two forms which were essentially identical in size and composition except for the amide groups. The larger one had an amino acid composition different from that of the small one. The subunit described by Burley (1975) is different in composition when compared to the subunit reported by Hillyard *et al.* (1972).

High Density Fraction

The high density fraction (HDF or lipovitellin) is found in a complex association with phosvitin in the granules (Fig. 8.2). This association is probably due to the acidic and very atypical properties of phosvitin. Electrophoretic analysis of the HDF showed the presence of two forms, the α- and β-lipovitellin (Bernardi and Cook 1960). Amino acid, N, S, and P analyses of the lipid-free protein (vitellin) indicated that the difference was due to the phosphorus content, 0.53% and 0.30% for α- and β-vitellin, respectively (Cook *et al.* 1962). This was supported by Martin *et al.* (1963) who showed that the lipid compositions of the α- and β-lipovitellins were essentially identical. The phosphorus was presumably attached to seryl residues (Belitz 1963). Small amounts of carbohydrates (0.75%) also were reported to be present (Ito and Fujii 1962).

The molecular weights (4×10^5 daltons) were identical for the HDF and the two separate forms (Bernardi and Cook 1960). The two forms were composed of similarly sized subunits which were approximately half the size of the parent macromolecule. Furthermore, they may be composed of at least two polypeptide chains with N-terminal arginines and lysines (Neelin and Cook 1961). Although they are similar in composition, there appear to be distinct differences in their macromolecular binding forces and sites in the subunit structure. This was demonstrated by the differences in the rates of, and condition for, dissociation of the dimer to monomer subunits (Burley and Cook 1962A) and the fact that they will not hybridize (Radomski and Cook 1964).

The HDF appears to have similarities to globular proteins since it has definite structural integrity. The association forces appear to be hydrophobic. The following reports favor this interpretation: (1) Cook and Tsai (1967) showed that the lipid content was necessary for structural stability; (2) dissociation of the dimers decreased with increasing temperatures, ionic strength, and lipoprotein concentration (Burley and Cook 1962A); (3) sulfhydryl and phosphorus groups were not involved in the dimerization (Burley and Cook 1962B). Unlike the LDF, most of the phosphorus was not available for enzymatic dephosphorylation (Burley and Kushner 1963).

Phosvitin

Phosvitin, a phosphoglycoprotein, was first purified and named by Mecham and Olcott (1949). This protein contains 69% of the total yolk phosphorus as a monoesterified orthophosphate, constitutes 12% of the total yolk protein (Table 8.2), and probably exists as a complex with HDF in the yolk granules (Fig. 8.2). All the phosphorus appears to be linked as the O-phosphorylserine (Ho *et al.* 1969), with the exception of one O-phosphorylthreonine (Allerton and Perlmann 1965).

The molecular weight is 35,500 daltons. It has an unusually low partial specific volume, 0.545 ml/g (a normal value for typical proteins is 0.7 ml/g) (Taborsky 1974). This atypical nature of phosvitin is due to its unique composition and the sequence of its amino acids. It is composed of an unusually high amount of serine residues (54%) (122 residues out of 225) (Allerton and Perlmann 1965); 120 phosphate/mole (Allerton and Perlmann 1965); 6.5% carbohydrates, attached as a single heteropolysaccharide by a glycosidic linkage to asparagine (Shainkin and Perlmann 1971A); no cystine and possibly only a single residue each of tyrosine and tryptophan (Clark 1970; Clark and Joubert 1971). Clark (1970, 1973) isolated two phosvitins, one with a methionine residue and a second without one.

A partial sequence of the phosphoglycoprotein has been reported by Clark (1973). The N-terminal was alanine, and the C-terminal was tyrosine (Clark 1970). Clark (1973) found a 37 amino acid sequence of the N-terminal polypeptide chain from a purified fraction of phosvitin after CNBr cleavage. The 63-residue sequence of amino acids by Belitz (1965) on the unfractionated preparation and the glycopeptide sequence of Shainkin and Perlmann (1971A, B) are probably present in the C-terminal portion reported by Clark (1973). These workers and Williams and Sanger (1959) demonstrated the presence of up to six and eight seryl residues in sequence which were usually terminated with a basic amino acid.

Due to its composition and sequence, phosvitin behaves similarly to an acidic polyelectrolyte. It readily binds a number of substances: cations — Ca^{2+} and Mg^{2+} (McIndoe 1961; Grizzuti and Perlmann 1973A) in addition to Mn^{2+}, Co^{2+}, and Sr^{2+} (Grizzuti and Perlmann 1973B), Fe^{2+} (1 Fe:2 P) (Taborsky 1963), Fe^{3+} (Webb *et al.* 1973), poly-*L*-lysine and protamine (Perlmann and Grizzuti 1971), and cytochrome c (20 cytochrome:1 phosvitin) (Taborsky 1970). Taborsky (1974) extensively reviews this topic and the effects of the media on the structural configuration of the protein.

Livetin

Plimmer in 1908 named a water soluble fraction from egg yolk livetin (Fig. 8.2 and Table 8.2). Then in 1957, Martin *et al.* isolated three acidic proteins with an isoelectric pH of 4.8–5.0, which they named α-, β-, and γ-livetin. Although there are mainly three livetins, a multitude and variable number have been reported by electrophoretic and chromatographic analyses. This has been well reviewed recently by Powrie (1973).

The α-, β-, and γ-livetins were reported to be present in a ratio of 2:3:5 (Shepard and Hottle 1949) and as 2:5:3 (Bernardi and Cook 1960) with molecular weights of 80,000, 45,000, and 150,000 daltons, respectively (Martin *et al.* 1957; Martin and Cook 1958). Williams (1962A) immunologically identified the α-, β-, and γ-livetins as homologous to blood serum albumin, α_2-glycoprotein, and γ-globulin, respectively.

Yolk Riboflavin Binding Protein (YRBP)

The YRBP is a minor (0.4% of the total yolk proteins) water soluble globular yolk phosphoglycoprotein which binds riboflavin much more strongly than FMN or FAD (Zak *et al.* 1972). It forms a 1:1 complex (Zak and Ostrowski 1963) with a very low dissociation constant, $K_{dissoc} = 2.65 \times 10^{-9}$ M (Ostrowski and Krawczyk 1963). This protein was first isolated by Ostrowski *et al.* (1962) and was reported to contain 13% N and 0.2% phosphorus (2 moles per mole YRBP). This acidic protein (pI 4.1–4.2) has been characterized to have the following properties by Ostrowski *et al.* (1968): stable between pH 3.8–8.5, riboflavin dissociates at pH 3.0, molecular weight of 36,000 daltons, partial specific volume of 0.699 ml/g, amino acid analysis (total 297 residues), 12% carbohydrate (6 glucosamine, 3 glucose, 5 galactose, 6 mannose, and 1 fucose per mole YRBP), histidine as the N-terminal, and possibly composed of two subunits.

Recently Steczko and Ostrowski (1975) further demonstrated the importance of the tryptophan residue in the formation of the riboflavin-protein complex and also the importance of hydrophobic interactions upon binding.

The yolk RBP has many similarities to the egg-white RBP, and Farrell *et al.* (1970) reported that these two RBP proteins and the one in the blood serum were immunologically identical.

EGG WHITE

In marked contrast to the other components of the egg, the egg white is very low in lipids (0.02%) and inorganic ions (0.5%) and is mainly water (88.5%) and protein (10.5%), with a little sugar (0.5%) (Table 8.1). Most of the proteins are therefore present in soluble form and can be easily isolated and characterized. In addition, most of the proteins have very characteristic biological and chemical activities. Unlike the yolk proteins, the egg white proteins are not lipoproteins, but are more "typical" globular type proteins, many of them being glycoproteins. As a consequence of these attributes, the egg white proteins have been extensively studied and exploited by scientists.

The small amount (2%) of insoluble material in the white consists of the chalezae and the chalaziferous layer (Fig. 8.1), and they are probably composed of ovomucin fibers. Apparently the only difference between the thin and the thick egg white (besides their obvious physical differences) is due to the amount of ovomucin present (Forsythe and Foster 1949; Lanni *et al.* 1949). The thick egg white contains approximately four times as much ovomucin as does the thin egg white (Feeney *et al.* 1952).

The egg white contains as many as 40 different proteins (Gilbert 1971; Vadehra and Nath 1973) and most of these have not been isolated or characterized. The proteins mentioned in this section will be described because of their interesting properties and their presence in relatively large amounts. Tables 8.3, 8.4, and 8.5 present the composition, physical, and biological properties of the proteins. Further information on egg white proteins can be obtained from the following reviews: Baker (1968), Feeney and Allison (1969), Gilbert (1971), Robinson (1972), Powrie (1973), Vadehra and Nath (1973), and Osuga and Feeney (1974).

TABLE 8.3
COMPOSITION OF EGG-WHITE PROTEINS

Constituents	Ovalbumin[1] (Residues per 45,000 g)	Ovotransferrin (Residues per 76,000 g)	Ovomucoid (Residues per 28,000 g)	Ovoinhibitor (Residues per 49,000 g)	Ficin Inhibitor[2] (Residues per 12,700 g)	Lysozyme (Residues per 14,307 g)	Ovomucin[3] (Residues per 10,000 g)	Ovoflavo-protein (Residues per 32,000 g)	Ovomacro-globulin[4] (Residues per 10,000 g)	Avidin (Residues per subunit)
Alanine	34	52	11.7	20.1	8.50	12	4.18	13.9	3.95	5
Arginine	19	33	6.3	20.5	7.18	11	2.80	5.6	2.60	8
Aspartic acid	31	79	31.9	47.3	11.3	8(13)[5]	7.49	20.0	6.50	15
Cystine/2	6[6]	22	17.5	34.7	3.35	8	4.72	15.9	1.15	2
Glutamic acid	50	69	14.9	40.7	18.0	2(3)[5]	8.60	36.4	7.82	10
Glycine	18	58	16.1	32.3	6.48	12	4.70	8.3	3.49	11
Histidine	8	13	4.3	12.9	1.25	1	1.63	9.2	1.25	1
Isoleucine	24	24	3.2	17.3	5.69	6	3.58	7.1	4.42	8
Leucine	32	48	12.2	22.4	10.0	8	5.38	14.8	6.25	7
Lysine	20	62	13.6	24.2	7.21	6	4.68	17.2	4.11	9
Methionine	15	11	1.9	3.6	2.05	2	1.50	8.2	1.43	2
Phenylalanine	20	25	5.3	6.4	3.07	3	3.15	7.0	3.45	7

Proline	16	31	7.7	17.2	2.48	2	4.19	9.7	3.65	2
Serine	36	42	12.5	26.0	10.9	10	6.14	28.8	5.05	9
Threonine	15	35	14.6	28.0	3.53	7	5.41[7]	7.7	4.55	21
Tryptophan	3	18	0	1.0		6	1.11[7]	8.8	0.5	4
Tyrosine	9	20	6.7	15.1	4.79	3	3.33	9.5	2.68	1
Valine	30	44	16.0	25.9	9.51	6	4.99	5.7	5.41	7
Sialic acid	0	0	0.3				0.61(1.38)[7]	0.5	<0.01	0
Hexose			12–16.5	5–10	0		2.84(4.57)[7]	3.09[9]	2.2	
Mannose	5[8]	4	10–13.4				2.72[7]	2[9]		5
Galactose			0.8–6.3				1.36[7]	1[9]		
Glucosamine	3[8]	6	14.8–27.7	7–15			2.79(3.64)[7]	4.41[9]	3.21	4
Galactosamine			0	0			0.75(0.99)[7]			0
N-Terminal	Acetylglycine[8]	Alanine	Alanine			Lysine				Alanine
C-Terminal	Proline		Phenylalanine[10]			Leucine		Serine and glycine		Glutamic acid

Source: Osuga and Feeney (1974) unless otherwise noted.

[1] Fothergill and Fothergill (1970).
[2] Sen and Whitaker (1973).
[3] Osuga and Feeney (1968).
[4] Miller and Feeney (1966).
[5] Values in parentheses are amides.
[6] Sulfhydryls-4; disulfide-1.
[7] See Donovan *et al.* (1970).
[8] See Marshall and Neuberger (1972).
[9] Miller and Clagett (1973).
[10] Penasse *et al.* (1952).

TABLE 8.4
PHYSICAL PROPERTIES OF EGG-WHITE PROTEINS

Protein	Amount in Egg White %	pI	Mol Wt g	$S_{20,w}$	$D_{20,w} \times 10^7$ $cm^2\ sec^{-1}$	$\overline{V}$ cm^3/g	$E\lambda^{1\%}$
Ovalbumin	54	4.5	45,000	3.27	7.67	0.750	$E_{280}^{1\%} = 7.50$
Ovotransferrin	12	6.05	76,600	5.05	5.72(Fe)	0.732	$E_{280}^{1\%} = 11.6$
Ovomucoid	11	4.1	28,000	2.62	7.7	0.685	$E_{280}^{1\%} = 4.55$
Ovoinhibitor	1.5	5.1	49,000	—	nd	0.693	$E_{278}^{1\%} = 7.4$
Ficin inhibitor[1]	0.05	~5.1	12,700	—	nd	—	$E_{278}^{1\%} = 8.88$
Ovomucin	3.5	4.5–5.0	110,000[2]	6.4(10%)[2]	nd	nd	$E_{277.5}^{1\%} = 9.3$
Lysozyme	3.4	10.7	14,307	1.91	11.2	0.703	$E_{280}^{1\%} = 26.35$
Ovoglycoprotein	1.0	3.9	24,400	2.47	nd	nd	$E_{280}^{1\%} = 3.8$
Ovoflavoprotein	0.8	4.0	32,000	2.76	6.4	0.70	$E_{280}^{1\%} = 15.3$
Ovomacroglobulin	0.5	4.5	900,000 760,000	15.1	1.98	0.745	—
Avidin	0.5	10	68,300	4.55	5.98	0.73	$E_{280}^{1\%} = 15.7$

Source: Osuga and Feeney (1974), unless otherwise noted.

[1] Fossum and Whitaker (1968) and Sen and Whitaker (1973).
[2] Reducing conditions used; three peaks during ultracentrifugation: 6.4S (10%), 2.9S (5%), and 1.4S (85%).

TABLE 8.5
BIOLOGICAL PROPERTIES OF EGG WHITE PROTEINS

Constituents	Characteristic Properties
Ovalbumin	Denatures easily; has four poorly reactive sulfhydryls
Ovotransferrin	Complexes iron (K_D = $10^{-29}M$) and other metals; homologous to serum transferrin; antimicrobial
Ovomucoid	Specific trypsin inhibitor − 1:1 complex (K_D = 1.5 X $10^{-7}M$)
Ovoinhibitor	Inhibitor of serine proteinases: two trypsins and two chymotrypsins simultaneously (K_D = 4 X $10^{-8}M$, trypsin), subtilisin competes with chymotrypsin; alkaline proteinase competes with elastase. Homologous to blood serum α_2-proteinase inhibitor
Ficin inhibitor	Inhibitor of thiol proteinases: binding site for cathepsin C and a second site for cathepsin B_1, bromelain, papain, and ficin (K_D = 1.47 X $10^{-8}M$, ficin)
Ovomucin	Viscous; high in sialic acid; inhibitor of virus hemagglutination and clotting of κ-casein by rennin; important in egg deterioration
Lysozyme	Cleaves polysaccharides; antimicrobial; homologous with human lysozyme and milk α-lactalbumin
Ovoflavoprotein	Binds riboflavin $\gg$ FMN > FAD in 1:1 complex (K_A = 7.9 X $10^8 M^{-1}$, riboflavin); weakly antimicrobial; homologous with egg yolk and blood serum riboflavin binding protein
Ovomacroglobulin	Strongly antigenic and shows extensive immunological cross-reactivity with ovomacroglobulin of other species
Avidin	Binds four biotins per mole of avidin; antimicrobial

Source: Adapted from Feeney and Osuga (1976).

Ovalbumin

Ovalbumin, first crystallized by Hofmeister in 1889, is probably one of the most widely investigated proteins, second only to lysozyme.

This monomeric, nearly spherical globular phosphoglycoprotein can be easily crystallized. It exists in three distinct forms in the approximate ratio of 85:12:3, and they differ only in phosphorus content. Perlmann (1952) showed that ovalbumin existed in three forms, the A_1, A_2, and A_3 with two, one and zero phosphorus atoms per molecule, respectively. The diphosphorylated (A_1) species is the most abundant one. Phosphorylated seryl peptides have been isolated and characterized by Flavin (1954) and Milstein (1968). Ovalbumin has an acetylated glycine N-terminal (Narita 1961), a C-terminal proline (Niu and Fraenkel-Conrat 1955), and a single branched chain carbohydrate moiety composed of D-mannose (2%) and N-acetyl-glucosamine (1.2%) linked N-glycosidically to an asparagine (Cunningham *et al.* 1957). The first polysaccharide moiety to be clearly defined from a glycoprotein was found in ovalbumin by Neuberger in 1938. Since then, the carbohydrate on the protein has attracted numerous investigators. This has been reviewed recently by Spiro (1973), Marshall and Neuberger (1972), and Taborsky (1974). The latter two reports also present an excellent detailed general review on ovalbumin.

Ovalbumin is very susceptible to denaturation. The protein is not very highly cross-linked, containing only one disulfide bond per molecule of 45,000 daltons (Fothergill and Fothergill 1970). Some 50% of the residues are hydrophobic (Table 8.3) and it contains four sulfhydryls (MacDonnell *et al.* 1951B). These sulfhydryl groups have variable reactivities with different reagents (Fernandez Diez *et al.* 1964). One to three may react with certain reagents, but all four groups are only reactive upon the denaturation of ovalbumin. This "masked" nature is not only restricted to the reactivity of the sulfhydryls, but also other groups have shown variable susceptibilities (phenolic hydroxyl, Crammer and Neuberger 1943; lysine residue, Steven and Tristram 1958; and carboxyl groups, Atassi and Rosenthal 1969).

Two interesting products from ovalbumin have been reported: plakalbumin, due to its plate-like crystalline characteristics, and S-ovalbumin, due to its inherent increased stability to denaturation. The former is the result of the removal of a heptapeptide from the C-terminal end by subtilopeptidase A from *B. subtilis* (Linderstrom-Lang and Ottesen 1949; Ottesen 1958). This phenomenon of "limited proteolysis" is not only caused by the action of this particular enzyme but also by other enzymes (see Vadehra and Nath 1973; Taborsky 1974). S-Ovalbumin, first reported by Smith (1964), is found in small amounts in egg white and increases to 81% after six months of cold storage. Smith and associates have extensively studied this yet unexplained formation (see Vadehra and Nath 1973). The conversion of ovalbumin to S-ovalbumin does not involve the phosphoseryl residues or sulfhydryl groups, and there is no change in solubility, electrophoretic migration or ultraviolet spectrum. However, changes in optical rotatory dispersion, viscosity, and $S_{20,w}$ were reported (Smith and Black 1965, 1968).

Ovotransferrin (OT)

Osborne in 1899 named an egg-white protein conalbumin because it fractionated with the albumin. Schade and Caroline (1944) reported on an inhibitory effect of egg white on the bacterial growth that could be overcome by the addition of iron. Then, in 1946, the iron-binding protein was purified by Alderton *et al.* (1946). Its iron-binding sites were studied by chemical modification in 1950 (Fraenkel-Conrat 1950; Fraenkel-Conrat and Feeney 1950), and it was first crystallized a year later by Warner and Weber (1951). Although the name conalbumin is still used by some investigators, ovotransferrin has been the name in more general use for many years (Feeney and Komatsu 1966).

This egg-white protein is unique in that it can be classified in a group of homologous metal binding proteins, the transferrins, found in various fluids of higher (vertebrate) animals (Feeney and Komatsu 1966; Feeney and Allison 1969). The reviews of Feeney and Allison (1969) and Bezkorovainy and Zschocke (1974) report that the proteins have similar physical and chemical properties. All transferrins bind iron, forming a red color with an absorption maximum around 465 nm, contain carbohydrates, and have no sulfhydryl groups or prosthetic groups. The iron-binding sites in the different transferrins all appear to be very similar, and the binding ligands are the side chains of similar amino acids. The properties of OT are given in Tables 8.3, 8.4, and 8.5.

Williams (1962B) reported that the chicken serum transferrin is identical to OT with the exception of the difference in the carbohydrate moiety, and he subsequently confirmed this in 1968 (Williams 1968). This similarity of OT to serum transferrin and the importance of the latter in the transportation and storage of Fe^{3+} with its implication of involvement in certain illnesses, has caused an avalanche of recent reviews and publications (Aisen 1973; Morgan 1974; Bezkorovainy and Zschocke 1974; Zschocke and Bezkorovainy 1974; Aisen *et al.* 1974; Crichton 1975).

This iron-binding protein complexes with two moles of Fe^{3+} per mole of OT and requires HCO_3^- (Schade *et al.* 1949). Other metals early reported bound were Cu^{2+}, with an absorption maximum at 440 nm, and Zn^{2+}, forming a colorless complex (Fraenkel-Conrat and Feeney 1950). The Fe^{3+} is bound extremely tightly with an association constant of 1×10^{29} M^{-1} (Warner and Weber 1953). Stoichiometrically, two metals ions are bound in two specific sites on the polypeptide chain with a relative strength of $Fe^{3+} > Cu^{2+} > Zn^{2+}$ (Fraenkel-Conrat and Feeney 1950; Warner and Weber 1953). Tan and Woodworth (1969) indicated the following relative binding stabilities: $Fe^{3+} > Cr^{3+}$, $Cu^{2+} > Mn^{2+}$, Co^{2+}, $Cd^{2+} > Zn^{2+} > Ni^{2+}$. Other metals will also bind, *i.e.*, Eu^{2+} (Prados *et al.* 1975), Al^{3+} (Cunningham and Lineweaver 1965; Donovan and Ross 1975A), and Tb^{3+} and Ho^{3+} (Gafni and Steinberg 1974).

Upon binding Fe^{3+}, the OT was reported to be a more stable protein. It was more resistant to denaturation by heat, pressure, proteolytic enzymes and denaturating reagents (Fraenkel-Conrat and Feeney 1950; Azari and Feeney 1958, 1961; Tan and Woodworth 1970; Donovan and Ross 1975A).

OT has a single polypeptide chain (Bezkorovainy *et al.* 1968; Greene and Feeney 1968), and it can exist in equilibrium as three forms with different Fe^{3+} contents (zero, one, or two Fe^{3+} per molecule of OT) (Aisen *et al.* 1970; Williams *et al.* 1970) and not as a dimer (Woodworth 1967; Woodworth *et al.* 1969).

The kinetic study of the binding of Fe^{3+} and Cu^{2+} as chelates in the presence of HCO_3^- was recently investigated by Phelps and Antonini (1975). They concluded that the chelate and the metals were first bound to the OT and the chelating agent was then subsequently displaced by the HCO_3^- resulting in the chromophore. The specific binding area of the metal ions, its exact ligands, and the equivalence or nonequivalence of the two binding sites have been under much debate for many years. Windle *et al.* (1963), Aasa *et al.* (1963), Aisen *et al.* (1966), Gafni and Steinberg (1974), and Prados *et al.* (1975) favor equivalent or nearly similar sites. Contrary to this, numerous reports of nonequivalent sites have been reported (Luk 1971; Aasa 1972; Price and Gibson 1972; Tomimatsu and Vickery 1972; Aisen *et al.* 1973; Tomimatsu *et al.* 1973; Evans and Holbrook 1975; Williams 1975A, B; Donovan and Ross 1975B; Woodworth *et al.* 1975; Butterworth *et al.* 1975).

Although the structure of the complex is still under investigation, it might well be an octahedral complex as first proposed by Windle *et al.* (1963) or very similar to it (Fig. 8.3). This complex involves two histidines, two or three tyrosines and a bicarbonate. The latter is involved in a specific anion binding site on the molecule (Aisen 1973). Detailed reviews on this topic have been presented in the previously mentioned review articles and further substantiations of the involvement of tyrosine (Phelp and Antonini 1975) and histidine (Rogers *et al.* 1976) have been recently presented.

The primary sequence of OT has not been completed, but the indications thus far do not support the presence of two identical half structures (Bezkorovainy and Zschocke 1974). This was further supported recently by Williams (1975A, B) who isolated a 35,000 daltons C-terminal Fe^{3+} binding fraction which contained all of the carbohydrate of the intact OT. However, two homologous halves

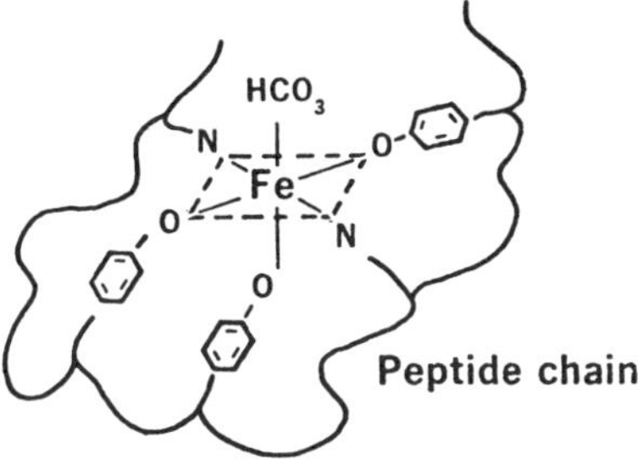

Iron binding

Courtesy of
Windle et al. (1963)

FIG. 8.3. A SCHEMATIC DIAGRAM INDICATING THE POSSIBLE BINDING OF IRON IN EITHER OVOTRANSFERRIN OR TRANSFERRIN SHOWING THE SYMMETRY AS INFERRED FROM THE EPR RESULTS

may exist (Greene and Feeney 1968), since "each hypothetical half molecule could have undergone mutation changes, or the original gene fusion might have been unequal" (Osuga and Feeney 1974). Homology between the two domains of human serum transferrin, first suggested by Greene and Feeney (1968), has recently been demonstrated by determination of primary sequence (MacGillivray and Brew 1975).

The strong binding capabilities of OT for iron may indeed serve a purpose in the egg white as an antimicrobial agent. Garibaldi and Bayne (1962) demonstrated that there was a direct correlation between the amount of iron present in the water used to wash the egg shell free of unsightly debris and the rate that the egg spoiled. Higher iron contents in the wash water resulted in an increase in microbial spoilage.

Inhibitors of Proteolytic Enzymes

Three proteinase inhibitors have been well characterized in the egg white. Although they are all inhibitors, they are far from similar in most other respects. They differ in composition, size, and inhibitory specificities (Tables 8.3, 8.4, and 8.5). Some similarities do exist in that they will form complexes with catalytically inactive enzymes and they are quite resistant to denaturation. The addition of trypsin enhances the stability of ovomucoid to denaturation (Donovan and Beardslee 1975). Another common similarity with these inhibitors is that their complexes are stabilized primarily by hydrophobic forces. The importance and involvement of these types of forces were recently reviewed by Means *et al.* (1974).

Ovomucoid and ovoinhibitor inhibit serine proteinases, while ficin inhibitor inhibits thiol proteinases. Ovoinhibitor and ficin inhibitor can inhibit more than one enzyme simultaneously whereas ovomucoid is specific for (bovine) trypsin. Ovomucoids from different avian species have different inhibitory specificities and stoichiometry (Feeney and Allison 1969). Chicken ovomucoid will not inhibit a major component of human trypsin (Buck *et al.* 1962), although quail ovomucoid can (Feeney *et al.* 1969).

The inhibitors have been extensively reviewed recently by a number of investigators: Vogel *et al.* (1969), Feeney and Allison (1969), Kassell (1970), Feeney (1971A, B), Laskowski and Sealock (1971), Lin and Feeney (1972), Means *et al.* (1974), and Osuga and Feeney (1974).

The ficin inhibitor in the egg white was first isolated and characterized by Fossum and Whitaker (1968) and further characterized by Sen and Whitaker (1973). They reported that this inhibitor was distinctly different from ovomucoid and ovoinhibitor (see Tables 8.3, 8.4, and 8.5). This inhibitor has a different amino acid composition (Table 8.3) and inhibitory specificity (Table 8.5), a complete absence of carbohydrate (Table 8.3), a low molecular weight (Table 8.4), and a greater heat stability than ovomucoid (Sen and Whitaker 1973). It inhibits ficin and papain but gives only a slight inhibition against bromelain (Fossum and Whitaker 1968). Subsequent work demonstrated a 1:1 complex formation with a K_{dissoc} of $1.47 \pm 0.68 \times 10^{-8}$ M for ficin, hydrophobic nature of the inhibitor-enzyme complex, and the nonessentiality of an active sulfhydryl group on the enzyme to form a complex (Sen and Whitaker 1973). Recently Keilova and Tomasek (1974, 1975) further increased the inhibitory spectrum of this inhibitor. They reported that the enzymes with chemically modified sulfhydryl groups (catalytically inactive) and the active cathepsin C_1 and cathepsin B_1 also formed complexes or were inhibited by this inhibitor. Chloride ions were necessary for the complex formation with cathepsin C_1. These investigators found that two independent sites exist, one for cathepsin C_1, and a second for ficin, papain, cathepsin B_1, and bromelain.

Lineweaver and Murray (1947) showed that a protein named ovomucoid by Morner in 1894 was responsible for the tryptic inhibitory effect in egg white. Matsushima in 1958 isolated an inhibitor different from ovomucoid and named it ovoinhibitor. Matsushima demonstrated that this protein inhibits trypsin, bacterial proteinase from *B. subtilis* var. *biotecus*, and a fungal proteinase from an unidentified species of Aspergillus (Matsushima 1958A, B). Since this initital report, the following enzymes have been reported to be inhibited by ovoinhibitor: bovine α-chymotrypsin, fungal proteinase from *A. oryzae* (Rhodes *et al.* 1960B), bacterial proteinase (Nagarse) (Feeney *et al.* 1963B), turkey trypsin and chicken chymotrypsin (Ryan and Clary 1964; Ryan *et al.* 1965), Pronase from *Streptomyces* (Haynes and Feeney 1967), porcine elastase (Gertler and Feinstein 1971), and porcine trypsin (Zahnley 1974, 1975). Feeney *et al.* (1963B) demonstrated that contamination with ovoinhibitor was responsible for the weak chymotrypsin inhibitory activity found by various investigators in some preparations of ovomucoid (Weil and Timasheff 1960).

Since Longsworth *et al.* (1940) first noted the heterogeneity of ovomucoid, several workers have isolated and investigated a number of isoproteins from ovomucoids, i.e., seven ovomucoids (Jakubezak and Montreuil 1970) and five ovomucoids (Beeley 1971). They reported that the difference was due to the carbohydrate composition, thus substantiating previous reports (Beeley and Jevons 1965; Feeney *et al.* 1967).

Ovomucoid contains 20–25% carbohydrates (Table 8.3) composed of some five different saccharides linked by a glycosidic bond between *N*-acetylglucosamine and asparagine as three polysaccharide units (Montgomery and Wu 1963). Although there are a large number of carbohydrates, it apparently is not essential for the configuration of the polypeptide (Christensen *et al.* 1974A) but is necessary for the stability of ovomucoid (Christensen *et al.* 1974B). Christensen and Johansen (1974) reported that 80% of the inhibitory activity was still retained after the removal of 50% of the carbohydrates. Recently Waheed and Salahuddin (1975) described the atypical globular nature of this protein.

Amino acid sequence of the single chain molecule has not yet been completed. Murthy *et al.* (1973) reported on the N-terminal sequence of 40 of the 160 residues present in ovomucoid. Kato *et al.* (1974B) reported that there were three similar domains present but only one was active. Beeley (1972) isolated two fragments, I with a molecular weight of 15,000 and II, 10,000 daltons, from CNBr-treated ovomucoid. Only fragment I was still active as an inhibitor.

Tomimatsu *et al.* (1966) reported on the existence of three forms of ovoinhibitor. Davis *et al.* (1969) isolated five forms similar to ovomucoids and showed that the differences were due to carbohydrate content. Another common feature of ovoinhibitors and ovomucoids is that they both have an essential arginine (Ozawa and Laskowski 1966; Liu *et al.* 1968, 1971).

One molecule of ovomucoid will strongly inhibit one molecule of trypsin, but ovoinhibitor is much more complex in that one molecule of ovoinhibitor will complex two trypsin and two chymotrypsin simultaneously and independently (Rhodes *et al.* 1960B; Tomimatsu *et al.* 1966). In addition to this, subtilisin will compete for the same chymotrypsin sites (Tomimatsu *et al.* 1966; Osuga *et al.* 1974). Subtilisin was reported to be more strongly complexed with ovoinhibitor than chymotrypsin (Osuga *et al.* 1974). Porcine elastase will form a 1:1 complex with ovoinhibitor, and this complex will not inhibit alkaline proteinase from *A. sojae*, but it will inhibit bovine chymotrypsin. Therefore three binding sites were reportedly present, one for chymotrypsin, a second for trypsin, and a third for alkaline proteinase or porcine elastase (Gertler and Feinstein 1971).

The dissociations of the complexes of trypsin with ovomucoid and ovoinhibitor were found to be different (Zahnley and Davis 1970). The dissociation of trypsin from the latter was faster than the former. Zahnley (1974, 1975) demonstrated that the two binding sites for bovine and porcine trypsin on the ovoinhibitor were not equivalent and they can form 1:1 and 2:1 complexes at two preferentially different binding sites.

Oegema and Jourdian (1974) reported the isolation and characterization of a protein which was very similar to ovomucoid in molecular weight and amino acid composition, but nonetheless somewhat different in isoelectric pH (4.8), solubility, and CD spectrum, and was very high in glucosamine (29 moles/mole protein). Kanamori and Kawabata (1969) described still another trypsin inhibitor which was different from previously reported inhibitors.

Barrett (1974) found an α_2-proteinase inhibitor in the chicken serum (1% of the serum proteins). The α_2-proteinase inhibitor was identical to ovoinhibitor except for the presence of sialic acid.

Lysozyme

A lytic activity in egg white was observed in 1909 by Laschtschenko, and in 1922 Fleming reported the lytic action of an egg white constituent on micrococcus cells. He proposed the name lysozyme for the enzyme and *Micrococcus lysodeikticus* for the organism serving as the substrate (Fleming 1922).

Since the first demonstration by Alderton and Fevold (1946) that large quantities of easily crystallizable lysozyme could be easily obtained, this low molecular weight protein has been one of the most intensively investigated and characterized enzymes. Two comprehensive reports have recently appeared, one by Imoto *et al.* (1972) and a second by Osserman *et al.* (1974). In the latter text, Moore and Osserman (1974) compiled a lysozyme bibliography listing 2600 entries which were published from 1922 to 1972.

Osuga and Feeney (1974) list some of the other recent reviews on lysozyme including some of the major roles that this protein has played in the fundamental investigations of proteins and enzymes: (1) development of the hypothesis that the conformation of a molecule is dictated by its primary structure (Isemura *et al.* 1961); (2) proof that the direction of peptide chain synthesis is from the NH_2-terminal toward the COOH-terminal end (Canfield and Anfinsen 1963); (3) status as the first enzyme and the third protein to have its tertiary structure determined by X-ray crystallography (Blake *et al.* 1965); (4) status as the first enzyme in which a reasonable reaction mechanism was proposed entirely by structure analysis of crystals (Blake *et al.* 1967).

Lysozyme appears to have the form of a prolate sphere with the dimensions 45Å × 30Å × 30Å (Imoto *et al.* 1972) and exists as a dimer between pH 5 to 9 (Sophianopoulos and Van Holde 1964). Its amino acid sequence was reported simultaneously by two independent groups in 1963 (Canfield 1963; Jolles *et al.* 1963). The properties and composition of this protein are summarized in Tables 8.3, 8.4, and 8.5, the structure of the α-carbon backbone is shown in Fig. 8.4, and the active site in Fig. 8.5.

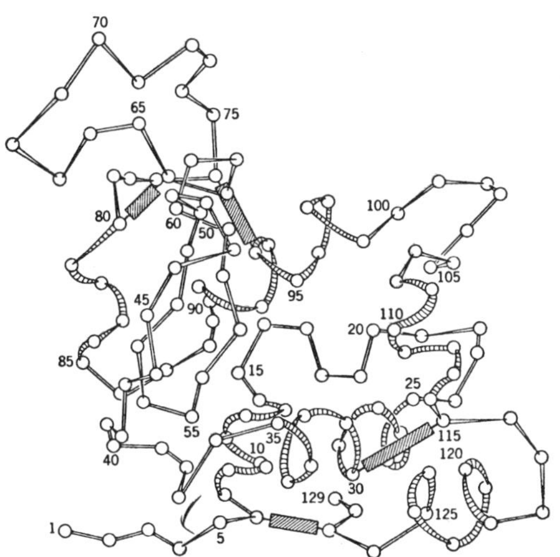

Courtesy of Blake et al. (1965)

FIG. 8.4. SCHEMATIC DRAWING OF THE MAIN CHAIN CONFORMATION OF LYSOZYME

Due to charge interactions, this basic protein binds with several of the egg white proteins: ovomucin (see ovomucin section), ovotransferrin (Ehrenpreis and Warner 1956), and ovalbumin (Nakai and Kason 1974).

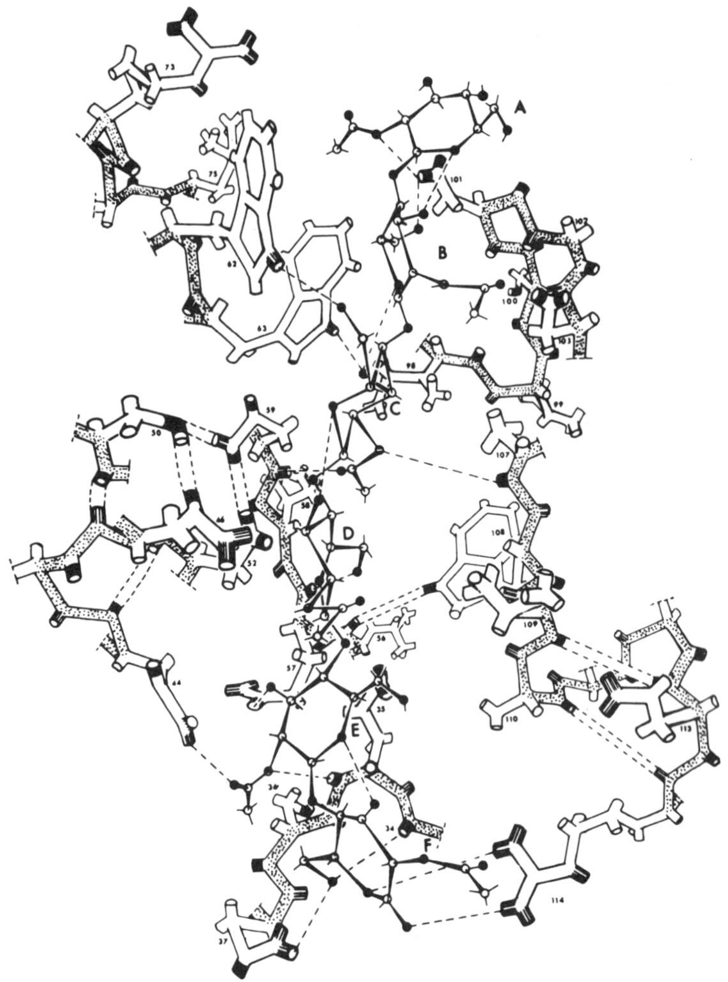

Courtesy of Johnson et al. (1969)

FIG. 8.5. ATOMIC ARRANGEMENT IN THE LYSOZYME MOLECULE IN
THE NEIGHBORHOOD OF THE CLEFT WITH A HEXA-N-ACETYL-
CHITOHEXAOSE SHOWN BOUND TO THE ENZYME

The main polypeptide chain is shown speckled, and NH and CO are indicated
by line and full shading respectively. Sugar residues A, B, and C are as
observed in the binding of tri-N-acetylchitotriose. Residues D, E, and F
occupy positions inferred from model building. It is suggested that the linkage
hydrolyzed by the action of the enzyme is between residues D and E.

Feeney *et al.* (1956) reported an unusually high stability of the
enzyme. When lysozyme was stored at room temperature and
exposed to laboratory lighting (daylight and artificial) for 6 years at
pH 3.4–9.1, it still retained enzymatic activity. Solutions containing
100 μg/ml lysozyme at pH 4–5 retained more than 75% activity after
the six year period. Lysozyme was stable for 1–2 min at pH 4.5 and

100°C. At alkaline pH values, it was inactivated by Cu^{2+} either in the presence or absence of air (Feeney *et al.* 1956). Little structural change has been noted in 9 *M* urea (Warren and Gordon 1970), but it is extensively effected in 6 *M* guanidine hydrochloride (Aune and Tanford 1969). This stability might be due to the compactness of this protein, since it has four disulfide cross-linkages and the presence of three molecules of water per molecule of lysozyme (Imoto *et al.* 1972; Rao and Bryan 1975).

Lysozyme cleaves the β-(1,4)-glycosidic linkages between *N*-acetylated glucosamine and muramic acid found in bacterial cell walls and chitin or other oligomers. In addition, the enzyme reportedly has other activities, including transglycosylation reaction (Chipman and Sharon 1969) and esterase activity (Pirzkiewiez and Bruice 1969). Thus numerous names have been proposed for this enzyme. The Commission on Enzyme Nomenclature in 1964 recommended "E. C. 3.2.1.17. Systematic name: mucopeptide *N*-acetylmuramyl hydrolase; trivial name: mucopeptide glucohydrolase, lysozyme; not recommended: muramidase."

The active site of the enzyme is a cleft in the molecule with six binding areas for the polysaccharide units (Fig. 8.5). The fourth site, position D, imposes a strain on the polysaccharide moiety and thus causes the cleavage of the molecule. Within the cleft are located three aspartyl, three tryptophanyl and one glutamyl residue which are necessary for enzymatic reaction (Osserman *et al.* 1974) (Fig. 8.5). The enzyme can be competitively inhibited by various small *N*-acetylhexosamines which fit into the cleft, and it is irreversibly inhibited by 2′, 3′-epoxypropyl-β-glucosides of *N*-acetylhexosamine oligomers (Sharon and Eshdat 1974).

Recent studies offer another example of the role of lysozyme in pointing the way towards new and interesting areas of protein chemistry. Jolles and Berthou (1973) studied the crystal growth of lysozyme at different temperatures and reported that lysozyme crystals went through a phase transition at 25°C. Below 25°C and from 30–55°C two different types of crystals were formed, tetragonal and orthorhombic forms, respectively. Tetragonal crystals were changed to orthorhombic at 40°C.

Ovomucin

Ovomucin was first isolated and described by Eichlolz (1898). This protein is different from the other egg white proteins in that it is an extremely large protein, contains sulfate esters, galactosamine, large amounts of cystine which interconnect the subunits by intermolecular linkages, and some 50% of the total sialic acid in the egg white (Tables 8.3 and 8.4). This polydispersed glycosulfoprotein, which is filamentous and fiber-like (Sharp *et al.* 1950; Robinson and Monsey 1975), exhibits various unique biological properties (Table 8.5) and appears to be directly involved in the deterioration of shell eggs. During the ensuing deterioration of the egg, the egg white becomes thinner (less rigid), lower in Haugh units, and the yolk membrane tends to rupture easily. Soon after the egg is laid, there is an increase in pH from approximately 6.5 to more than 9.5, due to the loss of carbon dioxide (Feeney and Allison 1969). The analyses of the structure, size, and composition of ovomucin are greatly influenced by the method and conditions of analysis and the mechanism of deterioration. The structure of this complex protein is still in a very perplexing state.

Several mechanisms have been proposed for the role of ovomucin in the deterioration of the egg: (1) Hawthorne (1950) proposed that the interaction of lysozyme with ovomucin causes thinning. (2) Cotterill and Winter (1955) and Rhodes and Feeney (1957) reported that less ovomucin-lysozyme complex was formed as the pH

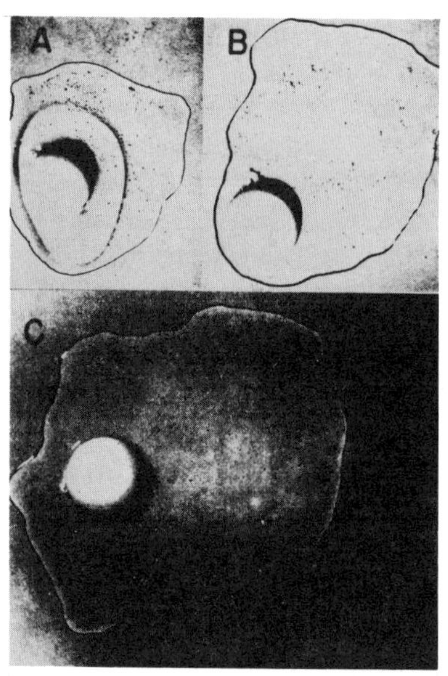

Courtesy of
MacDonnell et al. (1951A)

FIG. 8.6. EFFECT OF ADDING DILUTE THIOGLYCOL TO BROKEN-OUT EGGS

(A) Control after 4 hr at 25°C; (B) 30 ppm thioglycol after 4 hr at 25°C; (C) 300 ppm thioglycol after 4 hr at 25°C. Essentially the same results were obtained by a 2-hr treatment with the same amounts of reducing agent.

was raised above seven and suggested that the dissociation of the ovomucin-lysozyme complex, rather than its formation, might cause thinning. Others (Robinson and Monsey 1972B; Kato *et al.* 1974A) have also claimed that lysozyme is involved in the stability of the thick egg white. (3) Thinning is caused by reductive scission of disulfide bonds (Hoover 1940; MacDonnell *et al.* 1951A) (see Fig. 8.6). (4) Kato *et al.* (1971) proposed that thinning was due to the loss of carbohydrates from ovomucin. (5) The solubility of ovalbumin, which represents over half the protein of egg white, changes during incubation (Smith and Back 1962). (6) The interaction of glucose with protein directly or indirectly causes thinning (Feeney *et al.* 1964). (7) Protein-bound sialic acid is important in the complexes changing during deteriorations (Feeney *et al.* 1960B). (8) The alkaline conditions cause hydrolysis of disulfide bonds of ovomucin (Feeney *et al.* 1952; Tomimatsu and Donovan 1972).

Many arguments have been raised for and against each of these hypotheses. For example, several investigators have supported the existence of an interaction between ovomucin and lysozyme that results in an insoluble complex and causes thinning (Feeney *et al.* 1952; Cotterill and Winter 1955; Garibaldi *et al.* 1968). Duck egg white contains about one-third the amount of lysozyme, and it is reportedly much more stable than chicken egg white. However, penguin egg white which contains very small amounts of lysozyme (0.05%) deteriorated faster than the duck (see Feeney and Allison 1969).

Reducing agents cause the thinning of egg white (mechanism 3) (Hoover 1940; MacDonnell *et al.* 1951A), and rupture of the yolk membrane (MacDonnell *et al.* 1951A) (Fig. 8.6). However, there are not enough sulfhydryl groups present in the egg to cause thinning (Balls and Hoover 1940; Ducay *et al.* 1960), and also thinning occurs even with the absence of this reactive group (Feeney *et al.* 1951). Denaturation of less than 1% of the ovalbumin could, however, provide enough reducing power to cause thinning (MacDonnell *et al.* 1951B).

The importance of the lysozyme-ovomucin complex (mechanism 2) was supported by the interpretation of the results obtained by Robinson and Monsey (1972B) in which sodium chloride and magnesium salts inhibited the natural thinning process. Magnesium was 15 times more effective than sodium ions. The effect of these ions was also reported recently when it was found that mature hens' eggs which contain less magnesium deteriorate faster than those from the young hen which contains more of these ions (Monsey and Robinson 1974). In addition, there is a decrease in these ions during the thinning process of stored eggs (Sauveur 1971, 1976).

Dam (1971) supported a mechanism that the changes in the ovomucin itself were the cause of thinning, and that the lysozyme-ovomucin complex does exist, but is not necessarily related to the changes in the process of thinning. Also, sialic acid may not function in complex formation. He reported that disulfide crosslinks between the lysozyme and the ovomucin do not exist, thus substantiating previously reported results by Feeney *et al.* (1952). Sulfhydryls or disulfide exchange were also reportedly not involved, since complexes could be formed after reduction and alkylation of either or both lysozyme and ovomucin. Recently Kato *et al.* (1975) stated that the ϵ-amino group of lysine and the sialic acid on the ovomucins were essential for complex formation, since chemically modifying the lysines and removing the sialic acid by neuraminidase decreased complex formation.

Donovan *et al.* (1972) reported that thinning was first-order in the ovomucin concentration and hydrogen ion activity and also noted that there was no relationship with lysozyme concentration on the thinning process. The pH of their reaction mixture (pH 11.4) was much higher than encountered in natural thinning (pH 9.7). Thus Donovan and associates proposed that thinning was caused by disulfide cleavage (mechanism 8) by the alkaline hydrolysis of the cystine (Donovan *et al.* 1972; Tomimatsu and Donovan 1972).

Beveridge and Nakai (1975), by blocking sulfhydryl groups and then investigating the effects by viscosity and ultracentrifugation measurements, support mechanism 3 of thinning.

The composition of ovomucin first reported by Osuga and Feeney (1968) (Table 8.3) has since been shown to be generally similar to that reported by others (Donovan *et al.* 1970; Robinson and Monsey 1971; Young and Gardner 1972; Smith *et al.* 1974). The presence of sulfate esters was reported by Astudillo *et al.* (1957) and confirmed by Donovan *et al.* (1970). Lower values were reported for half cystines and methionine by Donovan *et al.* (1970) and Young and Gardner (1972), respectively. Other amino acid contents have been reported, but most of the preparations were made by different techniques and treatments of the ovomucin fractions. Unlike the general similarity of values for amino acids, very different values for carbohydrate have been found.

Kato *et al.* (1973) analyzed their ovomucin fractions for carbohydrate content after trypsin, Pronase, and papain hydrolysis. Three types of carbohydrate moieties were described in the following one-to-one ratios: (A) galactose:galactosamine:sialic acid:sulfate; (B) galactose: glucose; and (C) mannose:glucosamine. In moiety (A) the carbohydrate was reportedly attached to the peptide by an *O*-glycosidic linkage to serine or threonine.

Smith *et al.* (1974) found that the composition of the carbo-hydrate moieties varied with different preparations. They reduced the ovomucin, digested it with papain and separated two peaks, designated A and B, on Sephadex G-25. Type A, which was resistant to Pronase and mild acid digestion, had a molecular weight of 20,000 daltons, with 50% serine and threonine, 40% proline, and most of the galactose, sialic acid, and sulfates of the ovomucin. Type B had a molecular weight of 5,000 daltons, with 25% aspartic acid, 25% hydroxylamino acids, less than 3% cystine, and most of the glucosamine and mannose present in ovomucin. The type A carbo-hydrate moiety was *O*-glycosidically linked from *N*-acetylgalacto-samine to serine or threonine, and the Type B moiety was attached by an amide linkage from N-acetylglucosamine to asparagine. The carbohydrate moieties may exist on Type A as groups of 2 galactosamine, 3 glucosamine, 5 galactose, and 1-2 sialic acid residues, and on Type B as 6 glucosamine with 3 mannose residues.

Variable molecular weight values have been reported. This undoubtedly is due to the nature of ovomucin itself, in addition to the method of preparation, conditions of analysis, and the conse-quences of the analytical technique. Lanni *et al.* (1949) reported a molecular weight of 7.6×10^6 daltons by sedimentation and viscosity measurements; Tomimatsu and Donovan (1972) by light-scattering, 240×10^6, 40×10^6 and 23×10^6 daltons at pH 6.2, 7.9, and with 6.5 M guanidine hydrochloride, respectively; Young and Gardner (1972) by molecular exclusion on Sepharose 4B, greater than 3×10^6 daltons. The largest value at pH 6.2 was considered to be due to aggregation. In the presence of 6 M guanidine hydro-chloride and 0.2 M mercaptoethanol, an average molecular weight from 1.1×10^5 to 1.5×10^5 daltons was also reported by Donovan *et al.* (1970). They also found that the ovomucins have little α-helical content.

Two groups of investigators have isolated ovomucin by slightly different techniques. Kato and Sato (1971) reported on the amino acid and carbohydrate analysis of two pure components isolated from ovomucin treated with 0.01 M mercaptoethanol (gel B). Gel B was the precipitate prepared by centrifugation of the thick egg white for 60 min at 59,000 g, and the supernatant was referred to as the sol B. Both gel B and sol B were washed clean of other egg white contaminants. Two peaks were isolated by density gradient electrophoresis, and purity was demonstrated by free boundary electrophoresis and ultracentrifugation. Peak F, the first peak to be eluted, was richer in carbohydrates than Peak S. Peaks F and S represented 50 and 15% of the dry weight, respectively. The percent compositions of Peaks S and F were reported as follows: hexose,

18.4 and 6.8; hexosamine, 18.3 and 6.7; sialic acid, 11.4 and 0.8; sulfate, 1.18 and 0.06; total nitrogen, 8.1 and 13.7, respectively (Kato and Sato 1971). They further characterized the composition of gel B, Peaks F and S, and demonstrated that during storage of eggs there was a decrease in carbohydrate and sulfate content in Peak F and the loss of affinity of Peak F to bind lysozyme (Kato *et al.* 1971; Kato and Sato 1972). This loss in carbohydrate was diminished by storage of the eggs at $30°C$ for 15 days in an atmosphere of carbon dioxide (Kato *et al.* 1972). Kato *et al.* (1973) reported that Peaks F and S had carbohydrate contents similar to that reported by Smith *et al.* (1974) on Types A and C, respectively. Peak F, which contains more polysaccharides than Peak S, interacts with lysozyme to a greater degree than Peak S and is also a better inhibitor for the aggregation of κ-casein by rennin (Kato and Sato 1972; Kato *et al.* 1974A, 1975). They also showed that the decrease in sialic acid, and the increase in pH and ionic strength did favor less protein-lysozyme complex formation (Kato *et al.* 1974A, 1975).

Robinson and Monsey (1971) isolated two fractions from ovomucin, a "pure" α-ovomucin and a heterogeneous β-ovomucin. Their method of isolation was by precipitation and sequential washing with water, 2% potassium chloride and 5 M guanidine hydrochloride. The reduced β-ovomucin was analyzed and found to be similar to Peak F reported by Kato and associates. Similarly, β-ovomucin was shown to decrease during the thinning process (Robinson and Monsey 1972A, 1972B). Subsequently, Robinson and Monsey (1975) described the properties of a homogeneous S-carboxy-methyl-β-ovomucin with a molecular weight of 720,000 daltons by sedimentation equilibrium. When this was degraded by cyanogen bromide or trypsin, a "subunit" of 112,300 or 103,000 daltons was obtained. The lower value was determined by electron microscopy of the subunit. These investigators reported that the w/w percentage composition of this large fraction is as follows: ester sulfate (4.24%) and carbohydrate (60%) which was composed of large amounts of galactose (22%), galactosamine (8.9%) and sialic acid (10.6%). Extraction of the crude ovomucin preparation with 5 M guanidinium chloride produced a large soluble fibrous component (11.5 $\times$ 10^6 daltons) which contained approximately 70% β-ovomucin (Robinson and Monsey 1975).

Ovoflavoprotein

Over 40 years ago the amount of riboflavin in the egg white was reported to be influenced by the amount of riboflavin in the diet (Heiman 1935; Norris and Bauernfeind 1940), but this dietary effect is not ubiquitously applicable to all avian species. The very low contents of riboflavin in the egg whites of the duck, goose, and penguin were not increased by increasing the intake of riboflavin (Rhodes *et al.* 1959; Feeney *et al.* 1968). Maw (1954) described a lethal hereditary malfunction which was cured by the addition of riboflavin into the egg white of developing chick embryos. The effect of this lethal gene caused a lack of the ovoflavoprotein or riboflavin binding protein (RBP) in the egg and blood serum (Winter *et al.* 1967). The RBP in the egg white, egg yolk, and serum of the laying hen were serologically identical (Farrell *et al.* 1970). Although the amount of riboflavin complexed with the RBP in the egg white was influenced by the diet, the total amount of RBP in the egg from a normal hen was not influenced by the diet (Rhodes *et al.* 1959). Ovoflavoprotein may function as a storage and transport system of riboflavin for the developing embryo (Winter *et al.* 1967).

Rhodes *et al.* (1958, 1959) were first to isolate and characterize RBP in the egg white. They isolated this acidic protein, with an approximate isoelectric pH of 4.2, as the only protein which would bind riboflavin in egg white. At pHs more acidic than 4.2, the riboflavin dissociated; it reformed above pH 4.3.

Rhodes *et al.* (1959) described the various chemical, physical, and biological activities of this phosphoglycoprotein (Tables 8.3, 8.4, and 8.5) and its existence in the riboflavin-free (apoproteins) and bound forms. The RBP contains 13.4% nitrogen, mannose, no sulfhydryl groups, seven to eight phosphates (all linked to serine, Osuga and Feeney 1976), has no enzymatic activity, a weak antimicrobial activity, and binds one mole of riboflavin per mole of RBP with a molecular weight of 32,000 daltons. RBP loses its binding capacity by treatment with trypsin or chymotrypsin and in 8 M urea. It is unaffected by heating at $100°C$ for 15 min at pH 7.0 in 0.05 M Tris. Although pure by ultracentrifugation, the RBP preparations isolated at pH 4.3 and 4.5 were apparently different in phosphate content, containing eight and seven atoms per molecule, respectively.

Riboflavin was bound more tenaciously than riboflavin mononucleotide (FMN) or flavin adenosine dinucleotide (FAD) and other isoalloxazine derivatives (Rhodes *et al.* 1959). Becvar (1973) investigated the interaction of numerous isoalloxazine derivatives with RBP and reported the association constants for riboflavin, FMN and FAD as 7.9×10^8 M^{-1}, 7.3×10^5 M^{-1}, and 7×10^4 M^{-1}, respectively. These results on the binding characteristics of the various isoalloxazine derivatives have been utilized by Blankenhorn *et al.* (1975) to isolate pure RBP by affinity chromatography. They prepared an affinity column by covalently linking 3-carboxymethyl-FMN to aminoalkyl substituted agarose.

Investigations by Becvar (1973) with circular dichroism (CD) and

Nishikimi and Yagi (1969) indicated that the RBP contains approximately 25% α-helix, and that the polypeptide backbone is not greatly affected by the binding. The riboflavin is reportedly buried in an hydrophobic region of the RBP (Nishikimi and Yagi 1969), and tryptophan has been suggested to be involved in the binding site. This has been corroborated by the following investigators: Farrell *et al.* (1969) by oxidation with H_2O_2, Phillips (1969) by CD, Kawabata and Taguchi (1973) by modification with 2-hydroxy-5-nitrobenzyl bromide, and Murthy *et al.* (1976) by *N*-bromosuccinimide. The following other side chains have also been implicated: carboxyl (Murthy *et al.* 1976), carboxyl and histidine (Farrell *et al.* 1969; Kawabata and Taguchi 1973), and disulfide (Phillips 1969; Kawabata and Taguchi 1973; Murthy *et al.* 1976). Not all the disulfide bonds appear to be necessary for activity since a fully active 24,000 daltons subunit and a smaller inactive 8,000 daltons subunit were isolated by selective cleavage of two disulfide bonds (Cotner and Clagett 1972). Phosphate groups (Rhodes *et al.* 1959), tyrosine (Cotner and Clagett 1972) and amino groups (Kawabata and Taguchi 1973) were reported not to be involved in binding.

The presence of subunits has been claimed by Clagett and associates. Farrell *et al.* (1968) suggested the possible existence of two subunits from C-terminal analysis, and then Phillips (1969) and Phillips *et al.* (1969) reported that serine and glycine residues were the C-terminals. The N-terminal residues could not be determined by the usual techniques of analysis. Phillips reported that all the carbohydrates and three-fourths of the 16 half-cystines were on the larger subunit (24,000 daltons). The RBP residues contained 5% hexose (2 mannose:1 galactose) and 9% *N*-acetylglucosamine (Miller and Clagett 1973). The latter investigators also reported that, after Pronase digestion of the large subunit, they isolated one glycopeptide in which the carbohydrate would not β-eliminate. Thus an amide carbohydrate linkage to the protein is implicated. In addition, cyanide cleavage of the RBP produced two glycoproteins; one was similar to the one isolated by Pronase digestion. Becvar (1973) could not find any evidence for the existence of the subunits based on SDS-electrophoresis. Thus he concluded that the RBP is a single polypeptide chain. Our laboratory has also not found the presence of subunits (Nashef *et al.* 1977).

Ovomacroglobulin

Lush (1961) observed the presence of a protein in egg white by starch gel electrophoresis and named it component 18 or C-18 due to its relative position of migration compared to the other egg white proteins. This component was isolated, characterized (Tables 8.3,

8.4, and 8.5) and named ovomacroglobulin by Miller and Feeney (1966). It is the second largest protein in the egg white, second only to ovomucin. It migrates relatively slowly in gel electrophoresis (Miller and Feeney 1966). During the earlier studies of Miller and Feeney (1964) on comparing egg white from various avian species, they reported that this protein had one of the highest degrees of immunological reactivity. Then in 1966, Miller and Feeney reported that this protein had the widest spectrum of immunological cross-reactivity among egg white proteins. In addition, it was not present in all avian species, nor in the egg yolk or serum of chickens (Miller and Feeney 1966).

Donovan *et al.* (1969) confirmed and extended the characterization of this large glycoprotein. They found that it had no inhibitory activity against trypsin or chymotrypsin, was a nearly spherical protein, was denatured in 6 *M* guanidine hydrochloride but not in 8 *M* urea by interpretation of phenolic titration results (although subunits were formed by both reagents), had a transition temperature of 62–64°C at pH 7, and had little α-helix. In addition, it dissociated at pH 2 into two equal halves with the same frictional ratio of 1.6 as the native protein. Two indoles and 24 phenol chromaphores were exposed on each subunit.

Avidin

Avidin, a minor constituent in the egg white (0.05%) which binds biotin, is reportedly the major toxic constituent in the egg (see Osuga and Feeney 1974). A nutritional syndrome from consuming egg white was reported by Boas (1924). The causative agent was named the "Chinese dried egg white injury factor" due to the presence of the injury factor in Chinese fermented dried egg white (Parsons and Kelly 1933A, B). The protein was first named avidalbumin (Gyorgy *et al.* 1941) and later avidin (Eakin *et al.* 1941). It was nearly ten years after the discovery of avidin that biotin deficiency in both animals and microorganisms maintained on an egg white diet was found to be a result of its being made unavailable by the formation of a complex with avidin (Eakin *et al.* 1941). Unfortunately, the name "Chinese egg white injury factor" resulted in some people believing that there was something particularly injurious in Chinese egg whites but not in other whites.

Thus far, avidin is the second and the only other egg protein to have its complete amino acid sequence determined. This alkaline protein has distinctive characteristic properties (Tables 8.3, 8.4, and 8.5). Avidin has been reviewed by Osuga and Feeney (1974) and Green (1975). It contains little or no α-helix and is composed of four polypeptide subunits wherein each subunit binds one molecule of biotin (Green 1964A; Green and Toms 1973). Each subunit contains 128 amino acid residues, including 13 amide groups, half residues each of isoleucine and threonine at residue number 34, one disulfide bond, a carbohydrate moiety (four of N-acetylglucosamine and five of mannose) linked to the amino sugar by an amide bond at asparaginyl residue (number 17), a C-terminal glutamic acid and an N-terminal alanine (DeLange 1970; DeLange and Huang 1971; Huang and DeLange 1971).

Avidin was crystallized by Pennington *et al.* in 1942, and more recently by Green and Toms (1970) in three crystalline forms: square platelets, prisms and as needle-like crystals. Further work on the crystals by Green and Joynson (1970) indicated that the size of the unit cell of the crystal was 62Å × 107Å × 43Å. Green *et al.* (1971), by using bisbiotinylpolymethylenediamine with various numbers of methylene groups, determined the physical distances between binding sites in the tetramer. They reported that the tetramer was composed of two pairs of subunits with binding sites at opposite ends of biotin and carboxyl groups 9Å below the surface of the binding sites. The avidin molecular dimensions were calculated as 55Å × 55Å × 41Å with 222 symmetry. Chignell *et al.* (1975) attached various spin labels to the carboxy group and by electron spin resonance deduced that the binding site was a hydrophobic depression in the molecule and proposed the quaternary structure of avidin based on their results and those of Green and associates (Green *et al.* 1971) (Fig. 8.7). The involvement of hydrogen bonds during complexing was interpreted by Suurkuusk and Wadso (1972) from the high negative change in heat capacity at constant pressure (237 cal/°C/biotin) as a result of the loss of hydrophobic surface during binding.

The subunits are reportedly not linked by covalent bonds since the protein dissociates above 3.5 M guanidine hydrochloride, to single polypeptide chains at 6 M guanidine hydrochloride, and it re-associates when the guanidine hydrochloride concentration is less than 2 M (Green 1963B). However, avidin is stable in 8 M urea (Fraenkel-Conrat *et al.* 1952A). This hysteresis, a peculiar binding characteristic in guanidine hydrochloride, and the irreversible effects by guanidine hydrochloride were recently reviewed by Green (1975). Green and Toms (1973) reported that the single subunit can bind biotin under some conditions.

Avidin binds biotin strongly, with a dissociation constant of 10^{-15} M and a free energy change of 20 kcal/mole of biotin bound. The rate constant of formation was 7×10^7 sec^{-1} and the rate constant of dissociation was 9×10^{-8} sec^{-1} (Green 1963B). This binding results in an unusually resistant complex. Donovan and

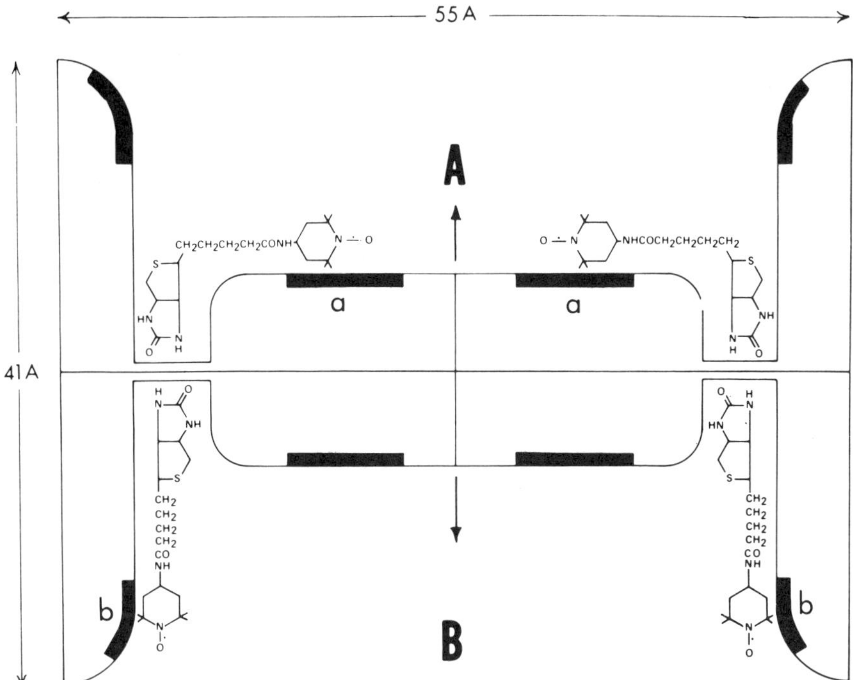

Courtesy of Chignell et al. (1975)

FIG. 8.7. A MODEL FOR THE BIOTIN BINDING SITES OF AVIDIN

The convention of Green *et al.* (1971) was followed and ascribed a 222 symmetry to the avidin tetramer. The figure represents a diagrammatic projection of the biotin binding sites onto a plane containing the twofold axis indicated by the arrows. A, spin label (4-biotinamido-2, 2, 6, 6-tetramethyl-1-piperidinyloxy) bound to adjacent accessory sites (a). B, spin label bound to nonadjacent accessory sites (b).

Ross (1973) reported that the thermal transitions in 0.1 M KCl-0.02 M in potassium phosphate at pH 6.62 of avidin and its complex were 85°C and 132°C per subunit, respectively. Gyorgy and Rose (1943) showed that the complex was resistant to denaturation and proteolysis (trypsin, chymotrypsin, pancreatin, and Pronase). The formation of the complex and its dissociation were influenced by the presence of various ions (Pai and Lichstein 1964) and by the various biotin derivatives (Green 1975). In biotin binding, the imidazolidone ring is much more important than the thiophane ring, while the carboxyl group has little effect.

The random binding of biotin at sites other than the binding sites (Green 1964B) does not cause any gross changes in the structure of avidin which can be detected by optical rotation, optical rotatory dispersion, or polarization of fluorescence. The crystalline structure of avidin and the complex were isomorphous (see Osuga and Feeney 1974; Green 1975).

The binding site does not involve reactive amino, phenolic, imidazole, carboxyl, or disulfide groups (Fraenkel-Conrat *et al.* 1952B). Tryptophan, however, is directly involved as shown by difference spectra (Green 1962) and by chemical modification with *N*-bromosuccinamide (Green 1963A). Two out of the four tryptophans are involved in the binding site (Green 1963A; Huang 1971). Green (1975) recently reported that a second group may be involved in the binding site. Chemical modification of one amino group per subunit with 1-fluoro-2,4-dinitrobenzene caused inactivation, but, in contrast, no inactivation was found on modification with trinitrobenzene sulfonic acid or dansyl chloride (Green 1963B).

Some Minor or Least Characterized Constituents

The proteins in this section are present in small or variable amounts or are less stable or lack the biochemically distinctive properties of the previously described egg proteins.

Longsworth *et al.* (1940) demonstrated the presence of three globulins: G_1, G_2, and G_3. G_1 was identified as lysozyme by Alderton *et al.* (1945). G_2 and G_3, which constitute some 4% of the egg white proteins, show large genetic variabilities (see Feeney *et al.* 1963A; Baker 1968). G_2 and G_3 have isoelectric pHs of 5.5 and 4.8, respectively (Baker 1968), and G_2 has a molecular weight between 30,000–45,000 daltons (Feeney *et al.* 1963A).

Ketterer (1962 and 1965) named and characterized a protein, ovoglycoprotein (Table 8.3), which constitutes approximately 0.5% of the egg white proteins. This acidic glycoprotein (pI = 3.9) with a molecular weight of 24,400 daltons, was characterized to contain the following: 13.6% mannose plus galactose (2:1 ratio, respectively); 13.8% glucosamine; 3% sialic acid; 11.6% nitrogen; an N-terminal threonine.

Lineweaver *et al.* (1948) analyzed the egg white and egg yolk for enzymatic activities. Amylase was present in the egg yolk but not in the egg white, and tributyrinase, peptidase, phosphatase and catalase were also present. Some of these enzymes have been described in more detail recently. Brooks (1962) and Shrimption *et al.* (1962) investigated amylase activities and applied them to test the adequacy of pasteurization of the egg. Catalase activity was investigated by Ball and Cotterill (1971A, B). The pH optimum was near pH 8.0, the optimum reaction temperature was at 20°C, and it was inactivated above 50°C and by NaCN and NaN_3. Baker and Manwell (1972) showed the presence of alanyl and glutamyl peptidases by substrate staining on starch gel electrophorograms. Moors and Stockx (1966, 1968) described pH optima and some kinetic data of two phosphatases—phosphodiesterase and phosphomonoesterase. The phosphodiesterase had two pH optima (pH 4.6 and 8.0) in both the egg white and yolk. The phosphomonoesterase also had two pH optima (pH 4.4 and 9.6) in the egg white. The egg yolk phosphomonoesterase, however, had three pH optima (pH 4.4, 5.4 and 9.6). There are also complexities in isolating and investigating these phosphoesterases (Moors and Stockx 1972A, B). Most of the phosphomonoesterases have similar molecular weights composed of basic 11,000–14,000 daltons subunits. The activities were based upon the state of aggregation-dissociation and the subunits (Moors and Stockx 1972B).

Neumann and Sela (1960) reported that the egg white digested polymers and copolymers of several amino acids. This digestive activity was lost when heated at 60°C or by 24 hr incubation at 37°C.

β-N-Acetylglucosaminidase and α-mannosidase activities were reportedly present, but there was no β-galactosidase activity in the egg white (Lush and Conchie 1966). Some kinetic, stability, and heat inactivation parameters of β-N-acetylglucosaminidase were investigated by Donovan and Hansen (1971A, B). The inactivation of this enzyme in the various areas of the stored eggs (egg white layers, vitelline membrane, and shell) was reported by Winn and Ball (1975).

De Moor and Stockx (1971) found similarities in crystalline RNAase from the egg yolk and white. Approximately 20 times more activity was found in the yolk than in the white.

COMPARATIVE BIOCHEMISTRY

Areas of Comparative Biochemistry

Comparative biochemistry is a phrase used to cover a broad area related to chemical comparisons of different organisms, systems, tissues, and molecules. At the organismal level, comparative biochemistry may be used to determine gross differences in metabolic pathways. At the molecular level, comparative biochemistry may use the differences in molecular structure to interpret how molecules function (Feeney and Allison 1969; Feeney and Osuga 1976). Comparative molecular studies may include examination of the properties of homologous proteins from the same type of organ or fluid in different species, *e.g.*, lysozymes from egg whites of different birds, or homologous proteins from different organs or fluids, *e.g.*, transferrin from egg white and blood serum.

Salient Properties of Individual Proteins

All the homologous egg white proteins show interesting differences, but some have been particularly useful. Excellent examples of variable characteristic properties due to species variations are the

TABLE 8.6
INHIBITION OF FIVE PROTEINASES BY OVOMUCOIDS
AND OVOINHIBITORS

	Enzymes Inhibited[1]				
	Human Trypsin[2]	Bovine Trypsin	Bovine α-Chymotrypsin	Subtilisin	Fungal Proteinase[3]
Ovomucoids					
Chicken	—	+++	—	—	—
Quail	++	+++	—	—	—
Cassowary	—	++	—	—	—
Ostrich	—	+	—	—	—
Emu	—	++	+	+	—
Rhea	—	++	+	+	—
Pheasant[4]	—	—	++++	+++	—
Duck	—	++++	++++	+++	—
Turkey[5]	—	++++	++++	++	—
Tinamou	—	+	++++	++	—
Penguin	—	+	++	++++	—
Ovoinhibitors					
Chicken[5]	—	++++	++++	++++	++++
Quail	?	++++	++++	++++	++++
Turkey	?	++++	++++	++++	++++
Penguin	?	++++	—	++++	++++

Source: Adapted from Feeney (1971B).

[1] Enzymatic activity determined by the casein digestion assay. The degree of inhibition is indicated as (—) for extremely weak or inactive ($K_D > 10^{-5}$ M) and +, ++, +++, and ++++, for varying degrees, progressing from weak ($K_D \sim 10^{-5}$ M) to strong ($K_D \sim < 10^{-9}$ M).

[2] There are two human trypsins which have been found. This one is not inhibited by chicken and most other ovomucoids. The other one is inhibited.

[3] Alkaline proteinase from *Aspergillus oryzae*.

[4] Golden pheasant.

[5] Gertler and Feinstein (1971) reported that elastase was inhibited by turkey ovomucoid and chicken ovoinhibitor; elastase competed with α-chymotrypsin and fungal proteinase (from *A. sojae*) for ovomucoid and ovoinhibitor, respectively.

inhibitors—ovomucoids and ovoinhibitors. These are summarized in Table 8.6. Although they have distinctly different inhibitory properties, their general protein properties are very similar to those of chicken (Tables 8.3 and 8.4), *i.e.*, acidic isoelectric pH, size, and general composition. Chicken ovomucoid is a specific inhibitor of trypsin, but golden pheasant ovomucoid inhibits chymotrypsin. Turkey and duck ovomucoids can inhibit both trypsin and chymotrypsin simultaneously in a 1:1 and 2:1 molecular ratio, respectively. They are thus double or triple headed (Rhodes *et al.* 1960B). The inhibitory characteristics of these four ovomucoids are diagrammed in Fig. 8.8. Recently three similar domains have been reported to be present in various ovomucoids (Kato 1976). Kato (1976) reported on the isolation of the third domain and that this was inactive from

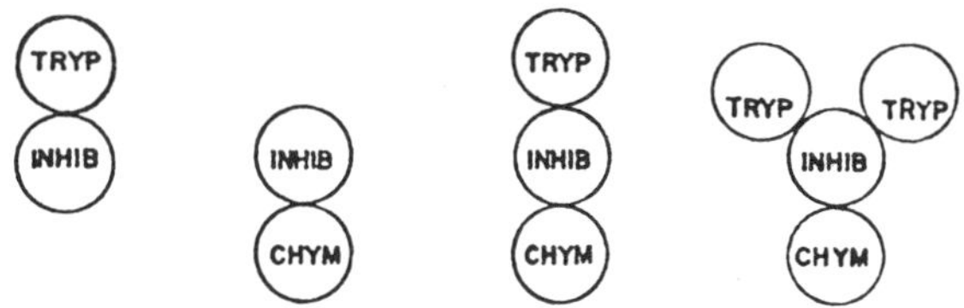

Courtesy of Rhodes et al. (1960B)

FIG. 8.8. DIAGRAMMATIC REPRESENTATION OF THE ENZYME-INHIBITOR COMPLEXES

Each circle represents a molecule of protein: INHIB is the described ovomucoid at the bottom of the figure. TRYP and CHYM are trypsin and chymotrypsin, respectively.

chicken; the one from turkey inhibited chymotrypsin, subtilisin, and elastase; the one from quail inhibited trypsin. Both the degree of complex formation and the inhibitory specificities are different (see Table 8.6). Quail ovomucoid, a trypsin inhibitor, is the only egg white inhibitor which will significantly inhibit both forms of human trypsin (Feeney *et al.* 1969). Penguin inhibitors are notably unusual in inhibitory activities in that the complex between ovomucoid and subtilisin is the strongest among the ovomucoids analyzed (Feeney 1971B; Osuga *et al.* 1974), and the ovoinhibitor does not inhibit chymotrypsin (Table 8.6). Specific amino acid residues on the inhibitors are necessary for activity. All ovomucoids appear to require lysine for trypsin inhibitory activity except the chicken ovomucoid. Chicken ovomucoid and ovoinhibitors from various species require arginine (Feeney 1971B).

Lysozymes from various avian species have had their amino acid sequence determined, and polymorphism exists in some species (see Osuga and Feeney 1974). There appears to be two types of avian egg white lysozymes, that which is similar to chicken and that from the goose, and these two lysozymes do not appear to interact immunologically. Arnheim and Steller (1970) demonstrated the presence of these two types of lysozyme in the black swan. Osuga and Feeney (1974) have recently summarized the comparative aspects of lysozyme.

Most, if not all, ovalbumin from various avian species contains two to three sulfhydryl groups except the chicken and the red jungle fowl which contain four sulfhydryl groups after denaturation (Fernandez Diez *et al.* 1964; Feeney and Allison 1969).

Variations in composition of the penalbumin and the ovotransferrin have been recently investigated in studying the phylogenetic relationship of the penguins (Ho 1975; Ho *et al.* 1976) and the evaluation of the flightless land birds (Prager *et al.* 1976).

Sialic acid, which is part of the carbohydrate side chain of some of the proteins, is one of the most variable constituents among egg whites (see Feeney and Allison 1969). The sialic acid varies from a low of 0.1% in the duck to over 5% in the penguin. No reason for this difference and no function for the sialic acid in penguin white has yet been found.

Applications of Comparative Biochemistry

Our laboratories have used comparative biochemistry for the study of egg white proteins for at least three different purposes:

(1) Determining taxonomic relationships among birds.

(a) Analyzing the relative amounts (quantitative differences) of different proteins in the whites of the eggs of different species (Feeney *et al.* 1960A). This approach relies heavily on direct biochemical and chemical tests on the whites. The amount of homologous proteins varies greatly from species to species. The variations can be large or small, or a protein may appear to be completely absent in some species. Table 8.7 shows the composition of four species selected to demonstrate these variabilities. Penalbumin, a recently isolated and characterized protein, is completely absent or present in very low amounts in chicken and turkey

TABLE 8.7

COMPOSITIONS OF DIFFERENT AVIAN EGG WHITES

Proteins	Chicken %	Turkey %	Duck %	Penguin %
Ovalbumin	54	40	40	25
Penalbumin	0	0	0.1[1]	10
Ovotransferrin	12	11	2	4.5
Ovomucoid	11	15	10[1]	10
Ovomucin	3.5	3[1]	3[1]	5[1]
Lysozyme	3.4	3.1	1.2	0.05
Ovoinhibitor	1.5	0.5	nd[2]	0.05
Ovomacroglobulin	0.5	0	1.0[1]	1.0[1]
Ovoflavoprotein	0.8	0.4	0.3	0.3
Avidin	0.05	0.06	0.03	0.02
Others[3]	15	27	42	44

Source: Adapted from Feeney and Osuga (1976) and Feeney and Allison (1969).
[1] Estimated by starch gel patterns, assays, or isolation.
[2] nd = Not determined.
[3] Calculated to give a total of 100% for each column.

egg white (Ho 1975; Ho *et al.* 1976). This protein is larger than ovalbumin and has sulfhydryl groups which will react with most sulfhydryl reagents without the presence of a denaturating reagent. Ovomacroglobulin is completely absent in turkey egg white but present in the other three species. Also notable are the variable amounts of each of the constituents from species to species. Particularly variable is the lysozyme; turkey and chicken contain approximately 65 times more lysozyme than the penguin.

(b) Using structural changes and immunological interactions between homologous proteins (Ho *et al.* 1976; Prager *et al.* 1976). Similar extensive studies using gel electrophoresis have also been done by Sibley (1970) and Sibley and Ahlquist (1972).

(2) As a tool for studying the fundamental properties of proteins.

(a) Using structural differences to help understand how a particular protein functions, *i.e.*, amino acid differences to show the essentiality or nonessentiality of a particular amino acid in a particular function, *e.g.*, the roles of arginine and lysine in the binding site of ovomucoids for trypsin (Liu *et al.* 1968).

(b) Using structural differences to help develop chemical methods for modifying or analyzing amino acids in a protein, *e.g.*, developing methods for arginine by using ovomucoids containing 0, 1, 2, 3, or 6 arginines per molecule (Liu *et al.* 1968).

(3) To study problems related to food uses. This approach uses differences in both composition, *i.e.*, relative amounts quantitative differences) of constituents, and differences in properties (qualitative differences) of constituents to understand the gross structural properties of the egg and the use of egg white in cookery.

(a) Chicken, turkey, and Adelie penguin eggs deteriorate much more rapidly than goose or duck eggs (Feeney and Allison 1969). The thinning of the whites of these species can not be easily correlated to the composition of their egg whites (see Table 8.7). In fact, the numerous causes of thinning discussed in the ovomucin section do not appear to be applicable to all species. Lysozyme was reported important in the thinning rate of chicken egg whites, but chicken egg white contains some 68 times more than penguin. Penguin egg white has the highest amount of reactive sulfhydryl groups (penalbumin), whereas chicken and turkey have little or no penalbumin. Nonetheless, one of the egg whites more resistant to thinning, the duck, still contains 0.1% penalbumin. Rhodes and Feeney (1957), investigating the deterioration of chicken and duck eggs, reported that when a duck yolk was incubated in chicken egg white and when a chicken yolk was incubated in duck egg white, the deteriorations progressed at rates characteristic of the species providing the yolk. Thus, the presence of the white from the other species had no influence, indicating the absence of a diffusible agent.

(b) Duck egg white was found to perform poorly in the preparation of angel-food cakes (Rhodes *et al.* 1960A). It whipped poorly and the cake volume was very poor. A simple procedure for making duck egg white perform well resulted from comparisons of pH titration curves of chicken and duck egg whites. It was found that duck egg white coagulated better when acidified. The solution was simply the addition of a small amount of acetic acid to the batter before whipping.

CONCLUSIONS

Egg proteins have been and still are a challenge to the scientists working either in the basically oriented fields or in the applied fields of food product research and development. The egg remains a unique isolated system having all the materials necessary for the development of the embryo. Yet most of the egg proteins have not been well characterized. Lysozyme is the only one that has its sequence and structure "completely" elucidated. There are many proteins present in the egg which have no known function, and other proteins which have been shown to possess unique biochemical or chemical properties, but in most cases the question still remains: Why are they present?

BIBLIOGRAPHY

AASA, R. 1972. Re-interpretation of the electron paramagnetic resonance spectra of transferrin. Biochem. Biophys. Res. Commun. *49*, 806–812.

AASA, R., MALMSTROM, B. G., SALTMAN, P., and VANNGARD, T. 1963. The specific binding of iron (III) and copper to transferrin and conalbumin. Biochim. Biophys. Acta *75*, 203–222.

AISEN, P. 1973. The transferrins (siderophilins). *In* Inorganic Biochemistry, G. L. Eichorn (Editor). Vol. 1. Scientific Publishing Co., New York.

AISEN, P., KOENIG, S. H., SCHILLINGER, W. E., SCHEINBERG, I. H., MANN, K. G., and FISK, W. 1970. Absence of dimers and nature of iron binding in transferrin solutions. Nature *226*, 859–861.

AISEN, P., LANG, G., and WOODWORTH, R. C. 1973. Spectroscopic evidence for a difference between the iron-binding sites of conalbumin. J. Biol. Chem. *248*, 649–653.

AISEN, P., LEIBMAN, A., and PINKOWITZ, R. A. 1974. The anion-binding functions of transferrin. *In* Protein-Metal Interactions, M. Friedman (Editor). Plenum Publishing Corp., New York.

AISEN, P., LEIBMAN, A., and REICH, H. A. 1966. Studies on the binding of iron to transferrin and conalbumin. J. Biol. Chem. *241*, 1666–1671.

ALDERTON, G., and FEVOLD, H. L. 1946. Direct crystallization of lysozyme from egg white and some crystalline salts of lysozyme. J. Biol. Chem. *164*, 1–5.

ALDERTON, G., WARD, W. H., and FEVOLD, H. L. 1945. Isolation of lysozyme from egg white. J. Biol. Chem. *157*, 43–58.

ALDERTON, G., WARD, W. H., and FEVOLD, H. L. 1946. Identification of the bacteria-inhibiting iron-binding protein of egg white as conalbumin. Arch. Biochem. *11*, 9–13.

ALLERTON, S. E., and PERLMANN, G. E. 1965. Chemical characterization of the phosphoprotein, phosvitin. J. Biol. Chem. *240*, 3892–3898.

ARNHEIM, N., and STELLER, R. 1970. Multiple genes for lysozyme in birds. Arch. Biochem. Biophys. *141*, 656–661.

ASTUDILLO, M. D., FERNANDEZ, V., and SANZ, F. 1957. Determination of the different chemical forms of sulfur found in ovomucin. Rev. Espan. Fisiol. *13*, 95–105. (Spanish)

ATASSI, M. Z., and ROSENTHAL, A. F. 1969. Specific reduction of carboxyl groups in peptides and proteins by diborane. Biochem. J. *111*, 593–601.

AUGUSTYNIAK, J. 1968. Proteolysis of low-density lipoprotein fraction (LDF) of hen's egg yolk and preparation of glycopeptides from its protein moiety. J. Acta Biochim. Polon. *15*, 67–76.

AUGUSTYNIAK, J. Z., and MARTIN, W. G. 1968. Characterization of glycopeptides derived from the low-density lipoprotein of hen's egg yolk. Can. J. Biochem. *46*, 983–988.

AUGUSTYNIAK, J., MARTIN, W. G., and COOK, W. H. 1964. Characterization of lipovitellenin components and their relation to low-density lipoprotein structure. Biochim. Biophys. Acta *84*, 721–728.

AUNE, K.,C., and TANFORD, C. 1969. Thermodynamics of the denaturation of lysozyme by guanidine hydrochloride. I. Dependence on pH at 25°. Biochemistry *8*, 4579–4585.

AZARI, P. R., and FEENEY, R. E. 1958. Resistance of metal complexes of conalbumin and transferrin to proteolysis and to thermal denaturation. J. Biol. Chem. *232*, 293–302.

AZARI, P. R., and FEENEY, R. E. 1961. The resistances of conalbumin and its iron complex to physical and chemical treatments. Arch. Biochem. Biophys. *92*, 44–52.

BAKER, C. M. A. 1968. The proteins of egg white. *In* Egg Quality, A Study of Hen's Eggs, T. C. Carter (Editor). Oliver and Boyd, Edinburgh.

BAKER, C. M. A., and MANWELL, C. 1972. Molecular genetics of avian proteins. 11. Egg proteins of *Gallus gallus*, *G. sonnerati* and hybrids. Anim. Blood Groups Biochem. Genet. *3*, 101–107.

BALCH, D. -A., and COOKE, R. A. 1970. A study of the composition of hen's egg-shell membranes. Ann. Biol. Anim. Biochem. Biophys. *10*, 13–25.

BALL, H. R., JR., and COTTERILL, O. J. 1971A. Egg white catalase: 1. Catalatic reaction. Poultry Sci. *50*, 435–446.

BALL, H. R., JR., and COTTERILL, O. J. 1971B. Egg white catalase: 2. Active component. Poultry Sci. *50*, 446–452.

BALLS, A. K., and HOOVER, S. R. 1940. Behavior of ovomucin in the liquefaction of egg white. Ind. Eng. Chem. *32*, 594–596.

BARRETT, A. J. 1974. Chicken α_2-proteinase inhibitor: A serum protein homologous with ovoinhibitor of egg white. Biochim. Biophys. Acta *371*, 52–62.

BECVAR, J. E. 1973. Studies on the binding of flavins to the riboflavin binding protein from leghorn egg white. Ph.D. Thesis, Univ. of Michigan, Ann Arbor.

BEELEY, J. G. 1971. The isolation of ovomucoid variants differing in carbohydrate composition. Biochem. J. *123*, 399–405.

BEELEY, J. G. 1972. Cyanogen bromide cleavage of ovomucoid. Biochem. J. *128*, 119–120.

BEELEY, J. G. and JEVONS, F. R. 1965. Heterogeneity of ovomucoid. Biochim. Biophys. Acta *101*, 133–135.

BELITZ, H. D. 1963. Egg yolk proteins and their cleaved products. 3. Phosphopeptides from α- and β-vitellin. Z. Lebensm.-Unters. Forsch. *119*, 492–497. (German)

BELITZ, H. D. 1965. Egg yolk proteins and their cleaved products. 4. Amino acid sequence of phosvitin. Z. Lebensm.-Unters. Forsch. *127*, 341–352. (German)

BELL, D. J., and FREEMAN, B. M. 1971. Physiology and Biochemistry of the Domestic Fowl. Vol. 3. Academic Press, New York.

BELLAIRS, R. 1964. Biological aspects of the yolk of the hen's egg. Adv. Morphogenes's *4*, 217–272.

BELLAIRS, R., BACKHOUSE, M., and EVANS, R. J. 1972. A correlated chemical and morphological study of egg yolk and its constituents. Micron *3*, 328–346.

BELLAIRS, R., HARKNESS, M., and HARKNESS, R. D. 1963. The vitelline membrane of the hen's egg: A chemical and electron microscopical study. J. Ultrastructure Res. *8*, 339–359.

BERNARDI, G., and COOK, W. H. 1960. An electrophoretic and ultracentrifugal study on the proteins of the high density fraction of egg yolk. Biochim. Biophys. Acta *44*, 86–96.

BEVERIDGE, T., and NAKAI, S. 1975. Effects of sulphydryl blocking on the thinning of egg white. J. Food Sci. *40*, 864–868.

BEZKOROVAINY, A., GROHLICH, D., and GERBECK, C. M. 1968. Some physical-chemical properties of reduced-alkylated and sulphitolysed human serum transferrins and hen's egg conalbumin. Biochem. J. *111*, 765–770.

BEZKOROVAINY, A., and ZSCHOCKE, R. H. 1974. Structure and function of transferrins. 1. Physical, chemical, and iron-binding properties. Arzneim.-Forsch. *24*, 476–485.

BLAKE, C. C. F., KOENING, D. F., MAIR, G. A., NORTH, A. C. T., PHILLIPS, D. C., and SARMA, V. R. 1965. Structure of hen egg-white lysozyme. A three-dimensional Fourier synthesis at two angstrom resolution. Nature *206*, 757–761.

BLAKE, C. C. F., MAIR, G. A., NORTH, A. C. T., PHILLIPS, D. C., and SARMA, V. R. 1967. On the conformation of the hen egg-white lysozyme molecule. Proc. Roy. Soc. London, Ser. B. *167*, 365–377.

BLANKENHORN, G., OSUGA, D. T., Lee, H., and FEENEY, R. E. 1975. Synthesis of immobilized flavin derivatives and their use in purification of chicken egg-white ovoflavoprotein. Biochim. Biophys. Acta *386*, 470–478.

BOARD, R. G., and FULLER, R. 1974. Non-specific antimicrobial defenses of the avian egg, embryo and neonate. Biol. Rev. *49*, 15–49.

BOAS, M. A. 1924. An observation on the value of egg-white as the sole source of nitrogen for young growing rats. Biochem. J. *18*, 422–424.

BRITTEN, W. M. 1973. Vitelline membrane chemical composition in natural and induced yolk mottling. Poultry Sci. *52*, 459–464.

BROOKS, J. 1962. α-Amylase in whole egg and its sensitivity to pasteurization temperature. J. Hyg. (Cambridge) *60*, 145–151.

BUCK, F. F., BIER, M., and NORD, F. F. 1962. Some properties of human trypsin. Arch. Biochem. Biophys. *98*, 528–530.

BURLEY, R. W. 1971. Lipoproteins. *In* Biochemistry and Methodology of Lipids, A. R. Johnson and J. B. Davenport (Editors). John Wiley and Sons, New York.

BURLEY, R. W. 1975. Studies on the apoproteins of the major lipoprotein of the yolk of hen's eggs. 1. Isolation and properties of the low-molecular-weight apoproteins. Aust. J. Biol. Sci. *28*, 121-132.

BURLEY, R. W., and COOK, W. H. 1962A. The dissociation of α- and β-lipovitellin in aqueous solution. 1. Effect of pH, temperature, and other factors. Can. J. Biochem. Physiol. *40*, 363-372.

BURLEY, R. W., and COOK, W. H. 1962B. The dissociation of α- and β-lipovitellin in aqueous solution. 2. Influence of protein phosphate groups, sulfhydryl groups and related factors. Can. J. Biochem. Physiol. *40*, 373-379.

BURLEY, K. W., and KUSHNER, D. J. 1963. The action of *Clostridium perfringens* phosphatidase on the lipovitellins and other egg yolk constituents. Can. J. Biochem. Physiol. *41*, 409-416.

BUTTERWORTH, R. M., GIBSON, J. F., and WILLIAMS, J. 1975. Electron-paramagnetic-resonance spectroscopy of iron-binding fragments of hen ovotransferrins. Biochem. J. *149*, 559-563.

CANDLISH, J. K. 1972. The role of the shell membranes in the functional integrity of the egg. *In* Egg Formation and Production, B. M. Freeman and P. E. Lake (Editors). British Poultry Science Ltd., Edinburgh.

CANDLISH, J. K., and SCOUGALL, P. K. 1969. L-5-Hydroxylysine as a constituent of the shell membranes of the hen's egg. Int. J. Protein Res. *1*, 299-302.

CANFIELD, R. E. 1963. The amino acid sequence of egg white lysozyme. J. Biol. Chem. *238*, 2698-2707.

CANFIELD, R. E., and ANFINSEN, C. B. 1963. Nonuniform labeling of egg white lysozyme. Biochemistry *2*, 1073-1075.

CARTER, T. C. 1968. Egg Quality: A Study of the Hen's Egg. Oliver and Boyd, Edinburgh.

CHIGNELL, C. F., STARKWEATHER, D. K., and SINHA, B. K. 1975. A spin label study of egg white avidin. J. Biol. Chem. *250*, 5622-5630.

CHIPMAN, D. M., and SHARON, N. 1969. Mechanism of lysozyme action. Science *165*, 454-465.

CHRISTENSEN, T. B., HANSEN, E., and JOHANSEN, A. 1974B. Carbohydrate constituents: Effect on thermal stability of glycoproteins. IRCS (Biochem.; Biophys.) *2*, 1031.

CHRISTENSEN, T. B., and JOHANSEN, A. 1974. Carbohydrate removal from glycoproteins by Smith degradation: Optimalisation of conditions of ovomucoid. IRCS (Biochem.; Biophys.) *2*, 1350.

CHRISTENSEN, T. B., ROLAND, I. J., and SVENDSEN, G. M. 1974A. Carbohydrate constituents of globular glycoproteins: Lack of effect of removal on polypeptide conformation. IRCS (Biochem.; Biophys.) *2*, 1030.

CLARK, R. C. 1970. The isolation and composition of two phosphoproteins from hen's egg. Biochem. J. *118*, 537-542.

CLARK, R. C. 1973. Amino acid sequence of a cyanogen bromide cleavage peptide from hen's egg phosvitin. Biochim. Biophys. Acta *310*, 174-187.

CLARK, R. C., and JOUBERT, F. J. 1971. The composition of cyanogen bromide cleavage products from hen's egg phosvitin. FEBS Letters *13*, 225-228.

COOK, W. H. 1968. Macromolecular components of egg yolk. *In* Egg Quality, A Study of the Hen's Egg, T. C. Carter (Editor). Oliver and Boyd, Edinburgh.

COOK, W. H., BURLEY, R. W., MARTIN, W. G., and HOPKINS, J. W. 1962. Amino acid compositions of the egg-yolk lipoproteins and a statistical comparison of their amino acid ratios. Biochim. Biophys. Acta *60*, 98-103.

COOK, W. H., and MARTIN, W. G. 1969. Egg lipoproteins. *In* Structural and

Functional Aspects of Lipoproteins in Living Systems, E. Tria and A. M. Scanu (Editors). Academic Press, New York.

COOK, W. H., and TSAI, C. S. 1967. Dissociative behavior of a lipoprotein-β-lipovitellin. 7th Int. Congr. Biochem., Tokyo. (Abstr.)

COTNER, R. C., and CLAGETT, C. O. 1972. Isolation of a fully active subunit from riboflavin-binding protein of egg white of White Leghorn. Fed. Proc. *31*, 851.

COTTERILL, O. J., and WINTER, A. R. 1955. Egg white lysozyme. 3. The effect of pH on the lysozyme-ovomucin interaction. Poultry Sci. *34*, 679-686.

CRAMMER, J. L., and NEUBERGER, A. 1943. The state of tyrosine in egg albumin and in insulin as determined by spectrophotometric titration. Biochem. J. *37*, 302-310.

CRICHTON, R. R. 1975. Proteins of Iron Storage and Transportation in Biochemistry and Medicine. North-Holland Publishing Co., Amsterdam.

CUNNINGHAM, F. E., and LINEWEAVER, H. 1965. Stabilization of egg white proteins to pasteurization temperatures above 60°C. Food Technol. *19*, 136-141.

CUNNINGHAM, L. W. NUENKE, B. J., and NUENKE, R. B. 1957. Preparation of glycopeptides from ovalbumin. Biochim. Biophys. Acta *26*, 660-661.

DAM, R. 1971. *In vitro* studies on the lysozyme-ovomucin complex. Poultry Sci. *50*, 1824-1831.

DAVIS, J. G., ZAHNLEY, J. C., and DONOVAN, J. W. 1969. Separation and characterization of ovoinhibitors from chicken egg white. Biochemistry *8*, 2044-2053.

DeLANGE, R. J. 1970. Egg white avidin. 1. Amino acid composition; sequence of the amino- and carboxyl-terminal cyanogen bromide peptides. J. Biol. Chem. *245*, 907-916.

DeLANGE, R. J. and HUANG, T.-S. 1971. Egg white avidin. 3. Sequence of the 78-residue middle cyanogen bromide peptide. Complete amino acid sequence of the protein subunit. J. Biol. Chem. *246*, 698-709.

DeMOOR, M., and STOCKX, J. 1971. Purification and crystallization of the chicken egg ribonuclease. Arch. Intern. Physiol. Biochem. *79*, 630-631.

DONOVAN, J. W., and BEARDSLEE, R. A. 1975. Heat stabilization produced by protein-protein association. A differential scanning calorimetric study of the heat denaturation of the trypsin-soybean trypsin inhibitor and trypsin-ovomucoid complexes. J. Biol. Chem. *250*, 1966-1971.

DONOVAN, J. W., DAVIS, J. G., and WHITE, L. M. 1970. Chemical and physical characterization of ovomucin, a sulfated glycoprotein complex from chicken eggs. Biochim. Biophys. Acta *207*, 190-201.

DONOVAN, J. W., DAVIS, J. G., and WIELE, M. B. 1972. Viscosimetric studies of alkaline degradation of ovomucin. J. Ag. Food Chem. *20*, 223-228.

DONOVAN, J. W., and HANSEN, L. U. 1971A. The β-N-acetylglucosaminidase activity of egg white. 1. Kinetics of the reaction and determination of the factors affecting the stability of the enzyme in egg white. J. Food Sci. *36*, 174-177.

DONOVAN, J. W., and HANSEN, L. U. 1971B. The β-N-acetylglucosaminidase activity of egg white. 2. Heat inactivation of the enzyme in egg white and whole egg. J. Food Sci. *36*, 178-181.

DONOVAN, J. W., MAPES, C. J., DAVIS, J. G., and HAMBURG, R. D. 1969. Dissociation of chicken egg-white macroglobulin into subunits in acid, hydrodynamic, spectrophotometric, and optical rotatory measurements. Biochemistry *8*, 4190-4199.

DONOVAN, J. W., and ROSS, K. D. 1973. Increase in the stability of avidin produced by binding of biotin. A differential scanning calorimetric study of denaturation by heat. Biochemistry *12*, 512–517.

DONOVAN, J. W., and ROSS, K. D. 1975A. Nonequivalence of the metal binding sites of conalbumin: Calorimetric and spectrophotometric studies of aluminum binding. J. Biol. Chem. *250*, 6022–6025.

DONOVAN, J. W., and ROSS, K. D. 1975B. Iron binding to conalbumin: Calorimetric evidence for two distinct species with one bound iron atom. J. Biol. Chem. *250*, 6026–6031.

DUCAY, E. D., KLINE, L., and MANDELES, S. 1960. Free amino acid content of infertile chicken eggs. Poultry Sci. *39*, 831–835.

EAKIN, R. E., SNELL, E. E., and WILLIAMS, R. J. 1941. The concentration and assay of avidin, the injury-producing protein in raw egg white. J. Biol. Chem. *140*, 535–543.

EHRENPREIS, S., and WARNER, R. C. 1956. The interaction of conalbumin and lysozyme. Arch. Biochem. Biophys. *61*, 38–50.

EICHLOLZ, A. 1898. The hydrolysis of proteins. J. Physiol. *23*, 163–177.

ETHEREDGE, E., HAALAND, J. E., and ROSENBERG, M. D. 1971. The functional properties of ATPase bound to and solubilized from the membrane complex of the hen's egg. Biochim. Biophys. Acta *233*, 145–154.

EVANS, R. W., and HOLBROOK, J. J. 1975. Differences in the protein fluorescence of the two iron (III)-binding sites of ovotransferrin. Biochem. J. *145*, 201–207.

FARRELL, H. M., BUSS, E. G., and CLAGETT, C. O. 1970. The nature of the biochemical lesion in avian renal riboflavinuria. 5. Elucidation of riboflavin transport in the laying hen. Int. J. Biochem. *1*, 168–172.

FARRELL, H. M., MALLETTE, M. F., BUSS, E. G., and CLAGETT, C. O. 1969. The nature of the biochemical lesion in avian renal riboflavinuria. 3. The isolation and characterization of the riboflavin-binding protein from egg albumin. Biochim. Biophys. Acta *194*, 433–442.

FARRELL, H. M., MALLETTE, M. F., and CLAGETT, C. O. 1968. Isolation of flavoprotein of egg white from genetic variants of White Leghorns. Fed. Proc. *27*, 456.

FEENEY, R. E. 1971A. The non-bond splitting mechanism of action of inhibitors of proteolytic enzymes—the conservative interpretation. *In* Proceedings of the International Research Conference on Protein as Inhibitors, H. Fritz and H. Tschesche (Editors). Walter de Gruyter and Co., Berlin.

FEENEY, R. E. 1971B. Comparative biochemistry of avian egg white ovomucoids and ovoinhibitors. *In* Proceedings of the International Research Conference on Protein as Inhibitors, H. Fritz and H. Tschesche (Editors). Walter de Gruyter and Co., Berlin

FEENEY, R. E., ABPLANALP, H., CLARY, J. J., EDWARDS, D. L., and CLARK, J. R. 1963A. A genetically varying minor proteins constituent of chicken egg white. J. Biol. Chem. *238*, 1732–1736.

FEENEY, R. E., and ALLISON, R. G. 1969. Evolutionary Biochemistry of Proteins. Homologous and analogous proteins from avian egg whites, blood sera, milk, and other substances. John Wiley and Sons, New York.

FEENEY, R. E., ALLISON, R. G., OSUGA, D. T., BIGLER, J. C., and MILLER, H. T. 1968. Biochemistry of the Adelie penguin: Studies on egg and blood serum proteins. *In* Antarctic Research Series, O. L. Austin Jr. (Editor). Vol. 12. Amer. Geophys. Union, Washington, D.C.

FEENEY, R. E., ANDERSON, J. S., AZARI, P. R., BENNETT, N., and

RHODES, M. B. 1960A. The comparative biochemistry of avian egg white proteins. J. Biol. Chem. *235*, 2307–2311.

FEENEY, R. E., CLARY, J. J., and CLARK, J. R. 1964. A reaction between glucose and egg white proteins in incubated eggs. Nature *201*, 192–193.

FEENEY, R. E., DUCAY, E. D., SILVA, R. B., and MacDONNELL, L. R. 1952. Chemistry of shell egg deteriorations: The egg white proteins. Poultry Sci. *31*, 639–647.

FEENEY, R. E., and KOMATSU, S. K. 1966. The transferrins. *In* Structure and Bonding, C. K. Jørgensen, J. B. Neilands, R. S. Nyholm, D. Reinen, and R. J. P. Williams (Editors). Springer-Verlag, Berlin.

FEENEY, R. E., MacDONNELL, L. R., and DUCAY, E. D. 1956. Irreversible inactivation of lysozyme by copper. Arch. Biochem. Biophys. *61*, 72–83.

FEENEY, R. E., MEANS, G. E., and BIGLER, J. C. 1969. Inhibition of human trypsin, plasmin, and thrombin by naturally occurring inhibitors of proteolytic enzymes. J. Biol. Chem. *244*, 1957–1960.

FEENEY, R. E., and OSUGA, D. T. 1976. Comparative biochemistry of Antarctic proteins. Comp. Biochem. Physiol. *54A*, 281–286.

FEENEY, R. E., OSUGA, D. T., and MAEDA, H. 1967. Heterogeneity of avian ovomucoids. Arch. Biochem. Biophys. *119*, 124–132.

FEENEY, R. E., RHODES, M. B., and ANDERSON, J. S. 1960B. The distribution and role of sialic acid in chicken egg white. J. Biol. Chem. *235*, 2633–2637.

FEENEY, R. E., SILVA, R. B., and MacDONNELL, L. R. 1951. Chemistry of shell egg deterioration: The deterioration of separated components. Poultry Sci. *30*, 645–650.

FEENEY, R. E., STEVENS, F. C., and OSUGA, D. T. 1963B. The specifications of chicken ovomucoid and ovoinhibitor. J. Biol. Chem. *238*, 1415–1418.

FERNANDEZ DIEZ, M. J., OSUGA, D. T., and FEENEY, R. E. 1964. The sulfhydryls of avian ovalbumins, bovine β-lactoglobulin, and bovine serum albumin. Arch. Biochem. Biophys. *107*, 448–458.

FEVOLD, H. L., and LAUSTEN, A. 1946. Isolation of a new lipoprotein, lipovitellin, from egg yolk. Arch. Biochem. *11*, 1–7.

FLAVIN, M. 1954. The linkage of phosphate to protein in pepsin and ovalbumin. J. Biol. Chem. *210*, 771–784.

FLEMING, A. 1922. On a remarkable bacteriolytic element found in tissues and secretions. Proc. Roy. Soc. London, Ser. B. *93*, 306–317.

FORSYTHE, R. H., and FOSTER, J. F. 1949. Note on the electrophoretic composition of egg white. Arch. Biochem. *20*, 161–163.

FOSSUM, K., and WHITAKER, J. R. 1968. Ficin and papain inhibitor from chicken egg white. Arch. Biochem. Biophys. *125*, 367–375.

FOTHERGILL, L. A., and FOTHERGILL, J. E. 1970. Thiol and disulphide contents of hen ovalbumin: C-terminal sequence and location of disulphide bond. Biochem. J. *116*, 555–561.

FRAENKEL-CONRAT, H. 1950. Comparison of iron-binding activities of conalbumin and of hydroxylamidoproteins. Arch. Biochem. *28*, 452–463.

FRAENKEL-CONRAT, H., and FEENEY, R. E. 1950. The metal-binding activity of conalbumin. Arch. Biochem. *29*, 101–113.

FRAENKEL-CONRAT, H., SNELL, N. S., and DUCAY, E. D. 1952A. Avidin. 1. Isolation and characterization of the protein and nucleic acid. Arch. Biochem. Biophys. *39*, 80–96.

FRAENKEL-CONRAT, H., SNELL, N. S., and DUCAY, E. D., 1952B. Avidin.

2. Composition and mode of action of avidin A. Arch. Biochem. Biophys. *39*, 97–107.

GAFNI, A., and STEINBERG, I. Z. 1974. Optical activity of terbium ions bound to transferrin and conalbumin studied by circular polarization of luminescence. Biochemistry *13*, 800–803.

GARIBALDI, J. A., and BAYNE, H. G. 1962. Iron and the bacterial spoilage of shell eggs. J. Food Sci. *27*, 57–59.

GARIBALDI, J. A., DONOVAN, J. W., DAVIS, J. G., and CIMINO, S. L. 1968. Heat denaturation of the ovomucin-lysozyme electrostatic complex; a source of damage to the whipping properties of pasteurized egg white. J. Food Sci. *33*, 514–524.

GERTLER, A., and FEINSTEIN, G. 1971. Inhibition of porcine elastase by turkey ovomucoid and chicken ovoinhibitor. Eur. J. Biochem. *20*, 547–552.

GILBERT, A. B. 1971. The egg: its physical and chemical aspects. *In* Physiology and Biochemistry of the Domestic Fowl, D. J. Bell and B. M. Freeman (Editors). Vol. 3. Academic Press, New York.

GREEN, N. M. 1962. Spectroscopic evidence for the participation of tryptophan residues in the binding of biotin by avidin. Biochim. Biophys. Acta *59*, 244–246.

GREEN, N. M. 1963A. Avidin. 2. Purification and composition. Biochem. J. *89*, 592–599.

GREEN, N. M. 1963B. Avidin. 3. The nature of the biotin-binding site. Biochem. J. *89*, 599–609.

GREEN, N. M. 1964A. The molecular weight of avidin. Biochem. J. *92*, 16c–17c.

GREEN, N. M. 1964B. Quenching of fluorescence by dinitrophenyl groups. Biochem. J. *90*, 564–568.

GREEN, N. M. 1975. Avidin. Adv. Prot. Chem. *29*, 85–133.

GREEN, N. M., and JOYNSON, M. A. 1970. A preliminary crystallographic investigation of avidin. Biochem. J. *118*, 71–72.

GREEN, N. M., KONIECZNY, L., TOMS, E. J., and VALENTINE, R. C. 1971. The use of bifunctional biotinyl compounds to determine the arrangements of subunits in avidin. Biochem. J. *125*, 781–791.

GREEN, N. M., and TOMS, E. J. 1970. Purification and crystallization of avidin. Biochem. J. *118*, 67–70.

GREEN, N. M., and TOMS, E. J. 1973. The properties of subunits of avidin coupled to sepharose. Biochem. J. *133*, 687–700.

GREENE, F. C., and FEENEY, R. E. 1968. Physical evidence for transferrins as single polypeptide chain. Biochemistry 7, 1366–1370.

GRIZZUTI, K., and PERLMANN, G. E. 1973A. Binding of magnesium and calcium ions to the phosphoglycoprotein phosvitin. Biochemistry *12*, 4399–4403.

GRIZZUTI, K., and PERLMANN, G. E. 1973B. The binding of divalent ions to the phosphoglycoprotein phosvitin. *In* Liquid Crystals and Ordered Fluids, J. F. Johnson and R. S. Porter (Editors). Plenum Publishing Corp., New York.

GYORGY, P., and ROSE, C. S. 1943. The liberation of biotin from the avidin-biotin complex (AB). Proc. Soc. Exp. Biol. Med. *53*, 55–57.

GYORGY, P., ROSE, C. S., EAKIN, R. E., SNELL, E. E. and WILLIAMS, R. J. 1941. Egg-white injury as the result of non-absorption or inactivation of biotin. Science *93*, 477–478.

HAALAND, J. E., ETHEREDGE, E., and ROSENBERG, M. D. 1971. Isolation of an ATPase from the membrane complex of the hen's egg. Biochim.

Biophys. Acta *223*, 137–144.

HAWTHORNE, J. R. 1950. The action of egg white lysozyme on ovomucoid and ovomucin. Biochim. Biophys. Acta *6*, 28–35.

HAYNES, R., and FEENEY, R. E. 1967. Fractionation and properties of trypsin and chymotrypsin inhibitors from lima beans. J. Biol. Chem. *242*, 5378–5385.

HEIMAN, V. 1935. The relative vitamin G content of dried whey and dried skimmilk. Poultry Sci. *14*, 137–146.

HILLYARD, L. A., WHITE, H. M., and PANGBURN, S. A. 1972. Characterization of apolipoproteins in chicken serum and egg yolk. Biochemistry *11*, 511–518.

HO, C. Y. -K. 1975. Biochemistry and immunochemistry of penguin egg-white proteins, Ph.D. Thesis, Univ. of California, Davis.

HO, C., MAGNUSON, J. A., WILSON, J. B., MAGNUSON, N. S., and KURLAND, R. J. 1969. Phosphorus nuclear magnetic resonance studies of phosphoproteins and phosphorylated molecules. 2. Chemical nature of phosphorus atoms in α_s-casein B and phosvitin. Biochemistry *8*, 2074–2082.

HO, C. Y.-K, PRAGER, E. M., WILSON, A. C., OSUGA, D. T., and FEENEY, R. E. 1976. Penguin evolution: protein comparisons demonstrate phylogenetic relationship to flying aquatic birds. J. Molec. Evol., *8*, 271–282.

HOFMEISTER, F. 1889. The formation of crystalline egg albumin and the crystallizability of colloidal material. Z. Physiol. Chem. *14*, 165–172. (German)

HOOVER, S. R. 1940. A physical and chemical study of ovomucin. Ph.D. Thesis. Georgetown Univ., Washington, D.C.

HUANG, T.-S. 1971. Structural studies on egg white avidin. Ph.D. Thesis. Univ. of California, Los Angeles.

HUANG, T.-S., and DeLANGE, R. J. 1971. Egg white avidin. 2. Isolation, composition, and amino acid sequences of the tryptic peptides. J. Biol. Chem. *246*, 686–697.

IMOTO, T., JOHNSON, L. N., NORTH, A. C. T., PHILLIPS, D. C., and RUPLEY, J. A. 1972. Vertebrate lysozymes. *In* The Enzymes, 3rd Edition, P. D. Boyer (Editor). Vol. 7. Academic Press, New York.

ISEMURA, T., TAKAGI, T., MAEDA, Y., and IMAI, K. 1961. Recovery of enzymatic activity of reduced Taka-amylase A and reduced lysozyme by air-oxidation. Biochem. Biophys. Res. Commun. *5*, 373–377.

ITO, Y., and FUJII, T. 1962. Chemical composition of the egg-yolk lipoproteins. J. Biochem. *52*, 221–222.

JAKUBEZAK, E., and MONTREUIL, J. 1970. Study of the glycopeptides. Heterogeneity of chicken egg ovomucoid. Physico-chemical study of 7 variant "alloglycoproteins". C. R. Acad. Sci. Paris, Ser. D, 537–540. (French)

JOHNSON, L. N., PHILLIPS, D. C., and RUPLEY, J. A. 1969. The activity of lysozyme: an interim review of crystallographic and chemical evidence. Brookhaven Symp. Biol. No. 21 *1*, 122–138.

JOLLES, P., and BERTHOU, J. 1973. Studies on lysozyme crystals prepared at high temperature. Abstr. 2u26 presented at the 9th Int. Congr. Biochem., Stockholm, July 1–7.

JOLLES, J., JAUREQUI-ADELL, J., BERNIER, I., and JOLLES, P. 1963. The chemical structure of chicken egg lysozyme: Detailed study. Biochim. Biophys. Acta *78*, 668–689. (French)

KANAMORI, M., and KAWABATA, M. 1969. Studies on egg white protein. Heterogeneity of ovomucoid. Agric. Biol. Chem. *33*, 75–79.

KASSELL, B. 1970. Proteinase inhibitors from egg white. *In* Methods in Enzymology, G. E. Perlmann and L. Lorand (Editors). Vol. 19. Academic Press, New York.

KATO, I. 1976. Amino acid sequences of third domains of turkey, chicken and Japanese quail ovomucoids. Fed. Proc. *35*, 1333.

KATO, A., FUJINAGA, K., and YAGISHITA, K. 1973. Nature of the carbohydrate side chains and their linkage to the protein in chicken egg white ovomucin. Agric. Biol. Chem. *37*, 2479-2485.

KATO, A., HAYASHI, H., and YAGISHITA, K. 1974A. The inhibitory properties of chicken egg white ovomucin against aggregation of κ-casein by rennin. Agric. Biol. Chem. *38*, 1137-1140.

KATO, A., IMOTO, T., and YAGISHITA, K. 1975. The binding groups in ovomucin-lysozyme interaction. Agric. Biol. Chem. *39*, 541-544.

KATO, A., NAKAMURA, R., and SATO, Y. 1971. Studies on changes in stored shell eggs. 7. Changes in the physicochemical properties of ovomucin solubilized by treatment with mercaptoethanol during storage. Agric. Biol. Chem. *35*, 351-356.

KATO, A., NAKAMURA, R., and SATO, Y. 1972. Effects of the storage in an atmosphere of carbon dioxide on ovomucin. Agric. Biol. Chem. *36*, 947-950.

KATO, A., and SATO, Y. 1971. The separation and characterization of carbohydrate rich component from ovomucin in chicken eggs. Agric. Biol. Chem. *35*, 439-440.

KATO, A., and SATO, Y. 1972. The release of carbohydrate rich component from ovomucin gel during storage. Agric. Biol. Chem. *36*, 831-836.

KATO, I., SCHRODE, J., and LASKOWSKI, M., JR. 1974B. Evidence for the presence of three homologous regions within avian ovomucoids. Fed. Proc. *33*, No. 5, 544.

KAWABATA, M., and TAGUCHI, K. 1973. Flavoprotein. Chemical modification of egg white flavoprotein. Sci. Rep. Kyoto. Pref. Univ. (Nat. Sci. and Liv. Sci.) *24B*, 7-10.

KEILOVA, H., and TOMASEK, V. 1974. Effect of papain inhibitor from chicken egg white on cathepsin B_1. Biochim. Biophys. Acta *334*, 179-186.

KEILOVA, H., and TOMASEK, V. 1975. Inhibition of cathepsin C by papain inhibitor from chicken egg white and by complex of this inhibitor with cathepsin B_1. Collection Czechoslov. Chem. Commun. *40*, 218-224.

KETTERER, B. 1962. A glycoprotein component of hen egg white. Life Sci. *1*, 163-165.

KETTERER, B. 1965. Ovoglycoprotein, a protein of hen's-egg white. Biochem. J. *96*, 372-376.

KIDO, S., JANADO, M., and NUNOURA, H. 1975. Macromolecular components of the vitellin membrane of hen's egg. I. Membrane structure and its deterioration with age. J. Biochem. *78*, 261-268.

LANNI, F., SHARP, D. G., ECKERT, E. A., DILLON, E. S., BEARD, D., and BEARD, J. W. 1949. The egg-white inhibitor of influenza virus hemagglutination. I. Preparation and properties of semi-purified inhibitor. J. Biol. Chem. *179*, 1275-1287.

LASCHTSCHENKO, P. 1909. The antiseptic and growth inhibitory activity of hen white. Z. Hyg. Infektionskrankh. *64*, 419-427. (German)

LASKOWSKI, M., JR., and SEALOCK, R. W. 1971. Protein proteinase inhibitors—molecular aspects. *In* The Enzymes, 3rd Edition, Vol. 3, P. D. Boyer (Editor). Academic Press, New York.

LIN, Y., and FEENEY, R. E. 1972. Ovomucoids and ovoinhibitors. *In*

Glycoproteins, A. Gottschalk (Editor). Vol. 5B. Elsevier Publishing Co., Amsterdam.

LINDERSTROM-LANG, K., and OTTESEN, M. 1949. Formation of plakalbumin from ovalbumin. C. R. Trav. Lab. Carlsberg, Ser. Chim. *26*, 403–442.

LINEWEAVER, H., MORRIS, H. J., KLINE, L., and BEAN, R. S. 1948. Enzymes of fresh hen's eggs. Arch. Biochem. *16*, 443–472.

LINEWEAVER, H., and MURRAY, C. W. 1947. Identification of the trypsin inhibitor of egg white with ovomucoid. J. Biol. Chem. *171*, 565–581.

LIU, W.-H., FEINSTEIN, G., OSUGA, D. T., HAYNES, R., and FEENEY, R. E. 1968. Modification of arginines in trypsin inhibitors by 1,2-cyclohexanedione. Biochemistry 7, 2886–2892.

LIU, W.-H., MEANS, G. E., and FEENEY, R. E. 1971. The inhibitory properties of avian ovoinhibitors against proteolytic enzymes. Biochim. Biophys. Acta *229*, 176–185.

LONGSWORTH, L. G., CANNON, R. K., and MacINNES, D. A. 1940. Electrophoretic study of the proteins of egg white. J. Amer. Chem. Soc. *62*, 2580–2590.

LUK, C. K. 1971. Study of the nature of the metal-binding sites and estimate of the distance between the metal-binding sites in transferrin using trivalent lanthanide ions as fluorescent probes. Biochemistry *10*, 2838–2843.

LUSH, I. E. 1961. Genetic polymorphisms in the egg albumen proteins of the domestic fowl. Nature *189*, 981–984.

LUSH, I. E., and CONCHIE, J. 1966. Glycosidases in the egg albumin of the hen, the turkey and the Japanese quail. Biochim. Biophys. Acta *130*, 81–86.

MacDONNELL, L. R., LINEWEAVER, H., and FEENEY, R. E. 1951A. Chemistry of shell egg deterioration: Effect of reducing agents. Poultry Sci. *30*, 856–863.

MacDONNELL, L. R., SILVA, R. B., and FEENEY, R. E. 1951B. The sulfhydryl groups of ovalbumin. Arch. Biochem. Biophys. *32*, 288–289.

MacGILLIVRAY, R. T. A., and BREW, K. 1975. Transferrin. Internal homology in the amino acid sequence. Science *190*, 1306–1307.

MARSHALL, R. D., and NEUBERGER, A. 1972. Hen's egg albumin. *In* Glycoproteins, A. Gottschalk (Editor). Vol. 5B. Elsevier Scientific Publishing Co., New York

MARTIN, W. G., AUGUSTYNIAK, J., and COOK, W. H. 1964. Fractionation and characterization of the low-density lipoproteins of hen's egg yolk. Biochim. Biophys. Acta *84*, 714–720.

MARTIN, W. G., and COOK, W. H. 1958. Preparation and molecular weight of γ-livetin from egg yolk. Can. J. Biochem. Physiol. *36*, 153–160.

MARTIN, W. G., TATTRIE, W. G., and COOK, W. H. 1963. Lipid extraction and distribution studies of egg yolk lipoproteins. Can. J. Biochem. Physiol. *37*, 1197–1207.

MARTIN, W. G., VANDEGAER, J. E., and COOK, W. H. 1957. Fractionation of livetin and the molecular weights of the α- and β-components. Can. J. Biochem. Physiol. *35*, 241–250.

MATSUSHIMA, K. 1958A. An undescribed trypsin inhibitor in egg white. Science *127*, 1178–1179.

MATSUSHIMA, K. 1958B. On the naturally occurring inhibitors for *Aspergillus* protease. 3. Ovoinhibitor. J. Agric. Chem. Soc. (Japan) *32*, 211–215.

MAW, A. J. G. 1954. Inherited riboflavin deficiency in chicken eggs. Poultry Sci. *33*, 216–217.

MC INDOE, W. M. 1961. The interaction of phosvitin with cations. Biochem. J. *80*, 41.

MC INDOE, W. M. 1971. Yolk synthesis. *In* Physiology and Biochemistry of the Domestic Fowl, D. J. Bell and B. M. Freeman (Editors). Vol. 3. Academic Press, New York.

MEANS, G. E., RYAN, D. S., and FEENEY, R. E. 1974. Protein inhibitors of proteolytic enzymes. Accounts Chem. Res. 7, 315-320.

MECHAM, D. K., and OLCOTT, H. S. 1949. Phosvitin, the principal phosphoprotein of egg yolk. J. Amer. Chem. Soc. *71*, 3670-3679.

MILLER, H. T., and FEENEY, R. E. 1964. Immunochemical relationships of proteins of avian egg whites. Arch. Biochem. Biophys. *108*, 117-124.

MILLER, H. T., and FEENEY, R. E. 1966. The physical and chemical properties of an immunologically cross-reacting protein from avian egg whites. Biochemistry 5, 952-958.

MILLER, M. S., and CLAGETT, C. O. 1973. Characterization of the glycopeptides of the riboflavin-binding protein. Fed. Proc. *32*, 624.

MILSTEIN, C. 1968. An application of diagonal electrophoresis to the selective purification of serine phosphate peptides. Biochem. J. *110*, 127-134.

MONSEY, J. B., and ROBINSON, D. S. 1974. The relationship between the concentration of metals and the rate of liquefaction of thick egg white. Br. Poultry Sci. *15*, 369-373.

MONTGOMERY, R., and WU, Y. C. 1963. The carbohydrate of ovomucoid. Isolation of glycopeptides and the carbohydrate-protein linkage. J. Biol. Chem. *238*, 3547-3554.

MOORE, B. R., and OSSERMAN, E. F. 1974. Lysozyme bibliography 1922-1972. *In* Lysozyme, E. F. Osserman, R. E. Canfield, and S. Beychok (Editors). Academic Press, New York.

MOORS, A., and STOCKX, J. 1966. Alkaline and acid phosphomonoesterase and phosphodiesterase activities in chicken eggs. Arch. Intern. Physiol. Biochim. *74*, 728-729.

MOORS, A., and STOCKX, J. 1968. Alkaline and acid phosphomonoesterase and phosphodiesterase activities in chicken egg. 2. Arch. Intern. Physiol. Biochim. *76*, 195-197.

MOORS, A., and STOCKX, J. 1972A. Phosphoesterase activities in the hen's egg. 1. Purification and characterization. Arch. Intern. Physiol. Biochim. *80*, 717-732.

MOORS, A., and STOCKX, J. 1972B. Phosphoesterase activities in the hen's egg. 2. Phosphomonoesterases. Arch. Intern. Physiol. Biochim. *80*, 923-934.

MORGAN, E. H. 1974. Transferrin and transferrin iron. *In* Iron in Biochemistry and Medicine, A. Jacobs and M. Worwood (Editors). Academic Press, New York.

MORNER, C. T. 1894. A mucin present in high amounts in hen egg white. Z. Physiol. Chem. *18*, 525-532. (German)

MOSSHOFF, W., and STOLPMANN, H. J. 1961. Light and electron microscopic study of yolk and shell membranes of hen eggs. Z. Zellforsch. Mikrosk. Anat. *55*, 818-832. (German)

MURTHY, G. S., GILARDEAU, C., and CHRETIEN, M. 1973. Partial sequence of trypsin inhibitor from ovomucoid. Can. J. Biochem. *51*, 1548-1551.

MURTHY, U. S., PODDER, S. K., and ADIGA, P. R. 1976. The interaction of riboflavin with a protein isolated from hen's egg white: a spectrofluorimetric study. Biochim. Biophys. Acta *434*, 69-81.

NAKAI, S., and KASON, C. M. 1974. A fluorescence study of the interaction between κ- and α_{s1}-casein and between lysozyme and ovalbumin. Biochim. Biophys. Acta *351*, 21–27.

NARITA, K. 1961. N-terminal group of ovalbumin. Biochem. Biophys. Res. Commun. *5*, 160–164.

NASHEF, A. S., LEE, H. S., OSUGA, D. T., and FEENEY, R. E. 1977. Chemical and physical properties of avian egg white riboflavin binding protein. Manuscript in preparation.

NEELIN, J. M., and COOK, W. H. 1961. Terminal amino acids of egg yolk lipoproteins. Can. J. Biochem. Physiol. *39*, 1075–1084.

NEUBERGER, A. 1938. Carbohydrates in proteins. 1. The carbohydrate component of crystalline egg albumin. Biochem. J. *32*, 1435–1451.

NEUMANN, H., and SELA, M. 1960. Proteolytic activity in egg white. Bull. Res. Coun. Israel *9A*, 103–104.

NICHOLES, A. V., RUBIN, L., and LINDGREN, F. T. 1954. Interaction of heparin active factor and egg yolk lipoprotein. Proc. Soc. Exp. Biol. Med. *85*, 352–355.

NISHIKIMI, M., and YAGI, K. 1969. Flavin-protein interaction in egg white flavoprotein. J. Biochem. *66*, 427–429.

NIU, C. I., and FRAENKEL-CONRAT, H. 1955. Determination of C-terminal amino acids and peptides by hydrazinolysis. J. Am. Chem. Soc. *77*, 5882–5885.

NORRIS, L. C., and BAUERNFEIND, J. C. 1940. Effect of level of dietary riboflavin upon quantity stored in eggs and rate of storage. Food Res. *5*, 521–532.

OEGEMA, T. R., JR., and JOURDIAN, G. W. 1974. The physical and chemical properties of a chicken egg white glycoprotein purified by nondenaturing methodology. Arch. Biochem. Biophys. *160*, 26–39.

OSBORNE, T. B. 1899. Egg albumin. J. Am. Chem. Soc. *21*, 477–485.

OSSERMAN, E. F., CANFIELD, R. E., and BEYCHOK, S. 1974. Lysozyme. Academic Press, New York.

OSTROWSKI, W., and KRAWCZYK, A. 1963. The riboflavin flavoprotein from egg yolk. Acta Chem. Scand. *17*, S241–S245.

OSTROWSKI, W., SKARZYNSKI, B., and ZAK, Z. 1962. Isolation and properties of flavoprotein from egg yolk. Biochim. Biophys. Acta *59*, 515–517.

OSTROWSKI, W., ZAK, A., and KRAWCZYK, A. 1968. Riboflavin flavoprotein from egg yolk. Analytical and biophysical data. Acta Biochim. Polon. *15*, 241–259.

OSUGA, D. T., BIGLER, J. C., UY, R. L., SJOBERG, L., and FEENEY, R. E. 1974. Comparative biochemistry of penguin egg-white proteins. 1. Ovomucoids: composition and inhibitory activities for trypsin, α-chymotrypsin and subtilisin. Comp. Biochem. Physiol. *48B*, 519–533.

OSUGA, D. T., and FEENEY, R. E. 1968. Biochemistry of the egg-white proteins of the ratite group. Arch. Biochem. Biophys. *124*, 560–574.

OSUGA, D. T., and FEENEY, R. E. 1974. Avian egg whites. *In* Toxic Constituents of Animal Foodstuffs, I. E. Liener (Editor). Academic Press, New York.

OSUGA, D. T., and FEENEY, R. E. 1976. Unpublished data.

OTTESEN, M. 1958. The transformation of ovalbumin into plakalbumin. A case of limited proteolysis. C. R. Trav. Lab. Carlsberg *30*, 211–270.

OZAWA, K., and LASKOWSKI, M., JR. 1966. The reactive site of trypsin inhibitors. J. Biol. Chem. *241*, 3955–3961.

PAI, C. H., and LICHSTEIN, H. C. 1964. Observations on the use of avidin in bacteriological media. Proc. Soc. Exp. Biol. Med. *116*, 197–200.

PARSONS, H. T., and KELLY, E. 1933A. The effect of heating egg white on certain characteristic pellagra-like manifestations produced in rats by its dietary use. Am. J. Physiol. *104*, 150–164.

PARSONS, H. T., and KELLY, E. 1933B. The character of the dermatitis-producing factor in dietary egg white as shown by certain chemical treatments. J. Biol. Chem. *100*, 645–652.

PENASSE, L., JUTISZ, M., FROMAGEOT, C., and FRAENKEL-CONRAT, H. 1952. The determination of carboxy groups of proteins. 2. Carboxy terminal groups of ovomucoids. Biochim. Biophys. Acta *9*, 551–556. (French)

PENNINGTON, D., SNELL, E. E., and EAKIN, R. E. 1942. Crystalline avidin. J. Am. Chem. Soc. *64*, 469.

PERLMANN, G. E. 1952. Enzymic dephosphorylation of ovalbumin and plakalbumin. J. Gen. Physiol. *25*, 711–726.

PERLMANN, G. E., and GRIZZUTI, K. 1971. Complexes of phosvitin with poly-L-lysine and protamine. Conformational analysis. Biochemistry *10*, 4168–4176.

PHELPS, C. F., and ANTONINI, E. 1975. A study of the kinetics of iron and copper binding to hen ovotransferrin. Biochem. J. *147*, 385–391.

PHILLIPS, J. W. 1969. Physical and chemical properties of riboflavin-binding protein. Ph.D. Thesis. Pennsylvania State Univ., University Park.

PHILLIPS, J. W., MALLETTE, M. F., and CLAGETT, C. O. 1969. Subunit isolation from riboflavin-binding protein of egg white of White Leghorns. Fed. Proc. *28*, 888.

PICARD, J., PAUL-GARDAIS, A., and VEDEL, M. 1973. Glycoprotein sulfates of the membranes of chicken eggs and from the oviduct. Isolation and characterization of glycopeptide sulfates. Biochim. Biophys. Acta *320*, 427–441. (French)

PIRZKIEWIEZ, D., and BRUICE, T. C. 1969. Interaction of cellodextrins with lysozyme: the necessity of the 2-acetamido group for binding and hydrolysis. Arch. Biochem. Biophys. *129*, 317–320.

PLIMMER, R. H. A. 1908. The proteins of egg yolk. J. Chem. Soc. *93*, 1500–1506.

POWRIE, W. D. 1973. Chemistry of eggs and egg products. *In* Egg Science and Technology, W. J. Stadelman and O. J. Cotterill (Editors). Avi Publishing Co., Westport, Conn.

PRADOS, R., BOGGESS, R. K., MARTIN, R. B., and WOODWORTH, R. C. 1975. Fe(III) and Cu(II) conalbumin visible circular dichroism spectra. Bioinorg. Chem. *4*, 135–142.

PRAGER, E. M., WILSON, A. C., OSUGA, D. T., and FEENEY, R. E. 1976. Evolution of flightless land birds on southern continents: transferrin comparison shows monophyletic origin of ratites, J. Molec. Evol. *8*, 283–294.

PRICE, E. M., and GIBSON, J. F. 1972. Electron paramagnetic resonance evidence for a distinction between the two iron-binding sites in transferrin and in conalbumin. J. Biol. Chem. *247*, 8031–8035.

RADOMSKI, M. W., and COOK, W. H. 1964. Chromatographic separation of phosvitin, α- and β-lipovitellin of egg yolk granules on TEAE-cellulose. Can. J. Biochem. *42*, 1203–1215.

RAO, P. B., and BRYAN, W. P. 1975. Measurement of strongly held water of lysozyme. J. Mol. Biol. *97*, 119–122.

RHODES, M. B., ADAMS, J. L., BENNETT, N., and FEENEY, R. E. 1960A.

Properties and food uses of duck eggs. Poultry Sci. *39*, 1473–1478.

RHODES, M. B., AZARI, P. R., and FEENEY, R. E. 1958. Analysis, fractionation and purification of egg white proteins with cellulose cation exchanger. J. Biol. Chem. *230*, 399–408.

RHODES, M. B., BENNETT, N., and FEENEY, R. E. 1959. The flavoprotein-apoprotein system of egg white. J. Biol. Chem. *234*, 2054–2060.

RHODES, M. B., BENNETT, N., and FEENEY, R. E. 1960B. The trypsin and chymotrypsin inhibitors from avian egg whites. J. Biol. Chem. *235*, 1686–1693.

RHODES, M. B., and FEENEY, R. E. 1957. Mechanisms of shell egg deterioration: comparisons of chicken and duck eggs. Poultry Sci. *36*, 891–897.

ROBINSON, D. S. 1972. Egg white glycoproteins and the physical properties of egg white. *In* Egg Formation and Production, B. M. Freeman and P. E. Lake (Editors). British Poultry Sci. Ltd., Edinburgh.

ROBINSON, D. S., and MONSEY, J. B. 1971. Studies on the composition of egg-white ovomucin. Biochem. J. *121*, 537–547.

ROBINSON, D. S., and MONSEY, J. B. 1972A. Changes in the composition of ovomucin during liquefaction of thick egg white. J. Sci. Food Ag. *23*, 29–38.

ROBINSON, D. S., and MONSEY, J. B. 1972B. Changes in the composition of ovomucin during liquefaction of thick egg white: The effect of ionic strength and magnesium salts. J. Sci. Food Ag. *23*, 893–904.

ROBINSON, D. S., and MONSEY, J. B. 1975. The composition and proposed subunit structure of egg-white β-ovomucin: The isolation of an unreduced soluble ovomucin. Biochem. J. *147*, 55–62.

ROGERS, T. B., GOLD, R. A., and FEENEY, R. E. 1976. Chemical modification of the histidines in ovotransferrin. Abstr. 79 presented at the 172nd National Meeting of the Am. Chem. Soc., Div. Biol. Chem., San Francisco, Aug. 29–Sept. 3.

RYAN, C. A., and CLARY, J. J. 1964. Some reactions of chicken chymotrypsin and turkey trypsin with substrates and naturally occurring protease inhibitors. Arch. Biochem. Biophys. *108*, 169–171.

RYAN, C. A., CLARY, J. J., and TOMIMATSU, Y. 1965. Chicken chymotrypsin and turkey trypsin. 2. Physical and enzymatic properties. Arch. Biochem. Biophys. *110*, 175–183.

SAARI, A., POWRIE, W. D., and FENNEMA, O. 1964. Isolation and characterization of low density lipoproteins in native egg yolk plasma. J. Food Sci. *29*, 307–315.

SATO, Y., WATANABE, K., and TAKAHASHI, T. 1973. Lipids in egg white. Poultry Sci. *52*, 1564–1570.

SAUVEUR, B. 1971. Repartition of electrolytes and glucose in eggs preserved by carbon dioxide. Ann. Biol. Anim. Bioch. Biophys. *11*, 625–643. (French)

SAUVEUR, B. 1976. Delayed thinning of thick egg white during storage in eggs produced by acidotic hens. Ann. Biol. Anim. Biochem. Biophys. *16*, 145–153.

SCHADE, A. L., and CAROLINE, L. 1944. Raw hen egg white and the role of iron inhibition of *Shigella dysenteriae, Staphylococcus aureus, Escherichia coli* and *Saccharomyces cerevisiae*. Science *100*, 14–15.

SCHADE, A. L., REINHART, R. W., and LEVY, H. 1949. Carbon dioxide and oxygen in complex formation with iron and siderophilin, the iron-binding component of human plasma. Arch. Biochem. *20*, 170–172.

SCHNEIDER, H., MORROD, R. S., COLVIN, J. R., and TATTRIE, N. H. 1973. The lipid core model of lipoproteins. Chem. Phys. Lipids *10*, 328–353.

SEN, L. C., and WHITAKER, J. R. 1973. Some properties of a ficin-papain inhibitor from avian egg white. Arch. Biochem. Biophys. *158*, 623–632.

SHAINKIN, R., and PERLMANN, G. E. 1971A. Phosvitin, a phosphoglycoprotein. 1. Isolation and characterization of a glycopeptide from phosvitin. J. Biol. Chem. *246*, 2278–2284.

SHAINKIN, R., and PERLMANN, G. E. 1971B. Phosvitin, a phosphoglycoprotein: composition and partial structure of carbohydrate moiety. Arch. Biochem. Biophys. *145*, 693–700.

SHARON, N., and ESHDAT, Y. 1974. Affinity labeling of lysozyme. *In* Lysozyme, E. F. Osserman, R. E. Canfield, and S. Beychok (Editors). Academic Press, New York.

SHARP, D. G., LANNI, F., and BEARD, J. W. 1950. The egg white inhibitor of influenza virus hemagglutination. 2. Electron microscopy of the inhibitor. J. Biol. Chem. *185*, 681–688.

SHENSTONE, F. S. 1968. The gross composition, chemistry and physicochemical basis of organization of the yolk and the white. *In* Egg Quality. A Study of the Hen's Egg, T. C. Carter (Editor). Oliver and Boyd, Edinburgh.

SHEPARD, C. C., and HOTTLE, G. A. 1949. Studies of the composition of the livetin fraction of the yolk of hen eggs with the use of electrophoretic analysis. J. Biol. Chem. *179*, 349–357.

SHRIMPTION, D. H., MONSEY, J. B., HOBBS, B. C., and SMITH, M. E. 1962. A laboratory determination of the destruction of α-amylase and salmonellae in whole egg by heat pasteurization. J. Hyg. (Cambridge) *60*, 153–162.

SIBLEY, C. G. 1970. A comparative study of the egg white proteins of passerine birds. Peabody Mus. Nat. Hist. Bull (New Haven: Yale Univ.) *32*.

SIBLEY, C. G., and AHLQUIST, J. E. 1972. A comparative study of the egg white proteins of nonpasserine birds. Peabody Mus. Nat. Hist. Bull. (New Haven: Yale Univ.) *39*.

SIMKISS, K., and TAYLOR, T. G. 1973. Shell formation. *In* Physiology and Biochemistry of the Domestic Fowl, D. J. Bell and B. M. Freeman (Editors). Vol. 3. Academic Press, New York.

SIMONS, P. C. M., and WIERTZ, G. 1963. Notes on the structure of membranes and shell in hen's egg. An electron microscopical study. J. Zellforsch. mikrosk. Anat. *59*, 555–567.

SMITH, M. B. 1964. Studies on ovalbumin. 1. Denaturation by heat and the heterogeneity of ovalbumin. Aust. J. Biol. Sci. *17*, 261–270.

SMITH, M. B., and BACK, J. F. 1962. Modification of ovalbumin in stored eggs detected by heat denaturation. Nature *193*, 878–879.

SMITH, M. B., and BACK, J. F. 1965. Studies on ovalbumin. 2. The formation and properties of S-ovalbumin, a more stable form of ovalbumin. Aust. J. Biol. Sci. *18*, 365–377.

SMITH, M. B., and BACK, J. F. 1968. Studies on ovalbumin. 3. Denaturation of ovalbumin and S-ovalbumin. Aust. J. Biol. Sci. *21*, 539–548.

SMITH, M. B., REYNOLDS, T. M., BUCKINGHAM, C. P., and BACK, J. F. 1974. Studies on the carbohydrate of egg-white ovomucin. Aust. J. Biol. Sci. *27*, 349–360.

SOPHIANOPOULOS, A. J., and VAN HOLDE, K. E. 1964. Physical studies of muramidase (lysozyme). 2. pH-dependent dimerization. J. Biol. Chem. *239*, 2516–2524.

SPIRO, R. G. 1973. Glycoproteins. Adv. Protein Chem. *27*, 349–467.

STADELMAN, W. J., and COTTERILL, O. J. 1973. Egg Science and Technology. Avi Publishing Co., Westport, Conn.

STECZKO, J., and OSTROWSKI, W. 1975. The role of tryptophan residues and hydrophobic interaction in the binding of riboflavin in egg-yolk flavoprotein. Biochim. Biophys. Acta *393*, 253–266.

STEER, D. C., MARTIN, G., and COOK, W. H. 1968. Structural investigation of the low-density lipoprotein of hen's egg yolk using proteolysis. Biochemistry *7*, 3309–3315.

STEVEN, F. C. and TRISTRAM, G. R. 1958. The reactivity of free amino groups in native and denatured ovalbumin towards fluorodinitrobenzene. Biochem. J. *70*, 179–182.

SUURKUUSK, J., and WADSO, I. 1972. Thermochemistry of the avidin-biotin reaction. Eur. J. Biochem. *28*, 438–441.

TABORSKY, G. 1963. Interaction between phosvitin and iron and its effects on a rearrangement of phosvitin structure. Biochemistry *2*, 266–271.

TABORSKY, G. 1970. Interaction of cytochrome c and the phosphoprotein phosvitin. Formation of a complex with an intact 695 mμ absorption band. Biochemistry *9*, 3768–3774.

TABORSKY, G. 1974. Phosphoproteins. Adv. Prot. Chem. *28*, 1–210.

TAN, A. T., and WOODWORTH, R. C. 1969. Ultraviolet difference spectral studies of conalbumin complexes with transition metal ions. Biochemistry *8*, 3711–3716.

TAN, A. T., and WOODWORTH, R. C. 1970. Differences in absorption and emission properties of conalbumin and metal-saturated conalbumin. J. Polymer Sci. *30C*, 599–606.

TOMIMATSU, Y., CLARY, J. J., and BARTULOVICH, J. J. 1966. Physical characterization of ovoinhibitor, a trypsin and chymotrypsin inhibitor from chicken egg white. Arch. Biochem. Biophys. *115*, 536–544.

TOMIMATSU, Y., and DONOVAN, J. W. 1972. Light scattering study of ovomucin. J. Ag. Food Chem. *20*, 1067–1073.

TOMIMATSU, Y., KINT, S., and SCHERER, J. R. 1973. Resonance raman spectroscopy of iron (III)-ovotransferrin and iron (III)-human serum transferrin. Biochem. Biophys. Res. Commun. *51*, 1067–1074.

TOMIMATSU, Y., and VICKERY, L. E. 1972. Circular dichroism studies of human serum transferrin and chicken ovotransferrin and their copper complexes. Biochim. Biophys. Acta *285*, 72–83.

TUNMANN, P., and SILBERZAHN, H. 1961. The carbohydrate in hens eggs. 1. Free carbohydrate. Z. Lebensmittelunters. U.-Forsch. *115*, 121–128. (German)

TURNER, K. J., and COOK, W. H. 1958. Molecular weight and physical properties of a lipoprotein from the floating fraction of egg yolk. Can. J. Biochem. Physiol. *36*, 937–949.

VADEHRA, D. V., BAKER, R. C., and NAYLOR, H. B. 1972. Distribution of lysozyme activity in the exteriors of eggs from *Gallus gallus*. Comp. Biochem. Physiol. *43B*, 503–508.

VADEHRA, D. V., and NATH, K. R. 1973. Eggs as a source of protein. CRC Crit. Rev. Food Tech. *4*, 193–309.

VOGEL, R., TRAUTSCHOLD, I., and WERLE, E. 1969. Natural Proteinase Inhibitors. Academic Press, New York.

WAHEED, A., and SALAHUDDIN, A. 1975. Isolation and characterization of a variant of ovomucoid. Biochem. J. *147*, 139–144.

WARNER, R. C., and WEBER, I. 1951. The preparation of crystalline conalbumin. J. Biol. Chem. *191*, 173–180.

WARNER, R. C., and WEBER, I. 1953. The metal combining properties of conalbumin. J. Am. Chem. Soc. *75*, 5094–5101.

WARREN, J. R., and GORDON, J. A. 1970. Denaturation of globular proteins. 2. The interaction of urea with lysozyme. J. Biol. Chem. *245*, 4097-4104.

WEBB, J., MULTANI, J. S., SALTMAN, P., BEACH, N. A., and GRAY, H. B. 1973. Spectroscopic and magnetic studies of iron (III) phosvitins. Biochemistry *12*, 1797-1802.

WEDRAL, E. M., VEDEHRA, D. V., and BAKER, R. C. 1974. Chemical composition of the cuticle, and the inner and outer shell membranes from eggs of *Gallus gallus*. Comp. Biochem. Physiol. *47B*, 631-640.

WEIL, L., and TIMASHEFF, S. N. 1960. On the inhibition of chymotrypsin by ovomucoid. Arch. Biochem. Biophys. *87*, 134-136.

WILLIAMS, J. 1962A. Serum proteins and the livetins of hen's egg yolk. Biochem. J. *83*, 346-355.

WILLIAMS, J. 1962B. Comparison of conalbumin and transferrin in the domestic fowl. Biochem. J. *83*, 355-364.

WILLIAMS, J. 1968. A comparison of glycopeptides from the ovotransferrin and serum transferrin of the hen. Biochem. J. *108*, 57-67.

WILLIAMS, J. 1975A. Iron-binding fragments of ovotransferrin. *In* Proteins of Iron Storage and Transport in Biochemistry and Medicine, R. R. Crichton (Editor). North-Holland Publishing Co., Amsterdam.

WILLIAMS, J. 1975B. Iron-binding fragments from the carboxylterminal region of hen ovotransferrin. Biochem. J. *149*, 237-244.

WILLIAMS, J., PHELPS, C. F., and LOWE, J. M. 1970. Ovotransferrin with one iron atom. Nature *226*, 858-859.

WILLIAMS, J., and SANGER, F. 1959. The grouping of serine phosphate in phosvitin and casein. Biochim. Biophys. Acta *33*, 294-296.

WINDLE, J. J., WIERSMA, A. K., CLARK, J. R., and FEENEY, R. E. 1963. Investigation of the iron and copper complexes of avian conalbumins and human transferrins by electron paramagnetic resonance. Biochemistry *2*, 1341-1345.

WINN, S. E., and BALL, H. R. JR. 1975. β-N-Acetylglucosaminidase activity of the albumin layers and membranes of the chicken egg. Poultry Sci. *54*, 799-805.

WINTER, W. P., BUSS, E. G., CLAGETT, C. O., and BOUCHER, R. V. 1967. The nature of the biochemical lesion in avian renal riboflavinuria. 2. The inherited change of a riboflavin-binding protein from blood and eggs. Comp. Biochem. Physiol. *22*, 897-906.

WOODWORTH, R. C. 1967. The mechanism of metal binding to conalbumin and siderophilin. Peptides Biol. Fluids *14*, 37-44.

WOODWORTH, R. C., TAN, A. T., and VIRKAITIS, L. R. 1969. Concerning the proposed existence of a one-iron protein species of conalbumin or siderophilin. Nature *223*, 833-834.

WOODWORTH, R. C., VIRKAITIS, L. M., WOODBURY, R. G., and FAVA, R. A. 1975. Anion binding to conalbumin and transferrin. *In* Proteins of Iron Storage and Transportation in Biochemistry and Medicine, R. R. Crichton (Editor). North-Holland Publishing Co., Amsterdam.

YOUNG, L. L., and GARDNER, F. A. 1972. Preparation of egg white ovomucin by gel filtration. J. Food Sci. *37*, 8-11.

ZAHNLEY, J. C. 1974. Evidence that the two binding sites for trypsin on chicken ovoinhibitor are not equivalent. J. Biol. Chem. *249*, 4282-4285.

ZAHNLEY, J. C. 1975. Preferred binding of bovine and porcine trypsins at two different sites on chicken ovoinhibitor. J. Biol. Chem. *250*, 7879-7884.

ZAHNLEY, J. C., and DAVIS, J. G. 1970. Determination of trypsin-inhibitor complex dissociation by use of the active site titrant, p-nitrophenyl-p -guanidinobenzoate. Biochemistry *9*, 1428–1433.

ZAK, Z., and OSTROWSKI, W. 1963. Preparation and properties of riboflavin flavoprotein of soluble fraction of egg yolk. Acta Biochim. Polon. *10*, 427–441.

ZAK, Z., OSTROWSKI, W., STECZKO, J., WEBER, M., GIZLER, M., and MORAWIECKI, A. 1972. Riboflavin carrier protein from egg yolk. Spectral and other properties observed upon binding of flavin to apoprotein. Acta Biochim. Polon. *19*, 307–323.

ZSCHOCKE, R. H., and BEZKOROVAINY, A. 1974. Structure and function of transferrins. 2. Transferrin and iron metabolism. Arzneim.-Forsch. *24*, 726–737.

9

Cereal Proteins

G. E. Inglett

Cereals are seed grains grown worldwide principally as a food or feed source. The major cereals are wheat, maize, rice, oats, barley, rye, grain sorghum, millet, and triticale. Civilization could not exist today without man's careful perpetuation of these seeds. All cereal grains contain starch as the principal component with protein being the second highest constituent. Average protein composition values of the major cereals (Inglett 1975) vary from 7.5% for rice to 14.2% for oats.

Wheat Proteins

Wheat flour, the principal refined product of wheat milling, is the major ingredient in almost all breads, rolls, chapatties, crackers, cookies, biscuits, cakes, doughnuts, muffins, pancakes, waffles, noodles, macaroni, and spaghetti. Flour composition and functionality vary greatly depending upon the milled wheat's heredity and the environmental conditions of its culture and harvest (Inglett 1974).

The protein content of wheat is usually around 12%, but heredity and environment exert a strong influence on that level. Since the discovery of high-lysine corn (Mertz *et al.* 1964), greater emphasis has been placed on increasing the protein quality and quantity of wheat (Johnson *et al.* 1972). In the World Wheat Collection, 15,000 hexaploid and tetraploid wheats have been screened for protein and lysine contents. Mean protein value was nearly 13%, ranging between 7 and 22%. The lysine content of wheat protein varied between 2.2 and 4.2%, with a mean value of approximately 3.0%.

Wheat endosperm proteins contain primarily gliadin and glutenin; together, these proteins constitute gluten. Glutenin is the wheat flour protein that is primarily responsible for giving strength and elasticity to dough (Bietz *et al.* 1973). The technique of sodium dodecyl sulfate-polyacrylamide gel electrophoresis (SDS-PAGE) was helpful in understanding the nature of glutenin (Bietz and Wall 1972). This technique is illustrated in Fig. 9.1. The glutenin disulfide bonds are

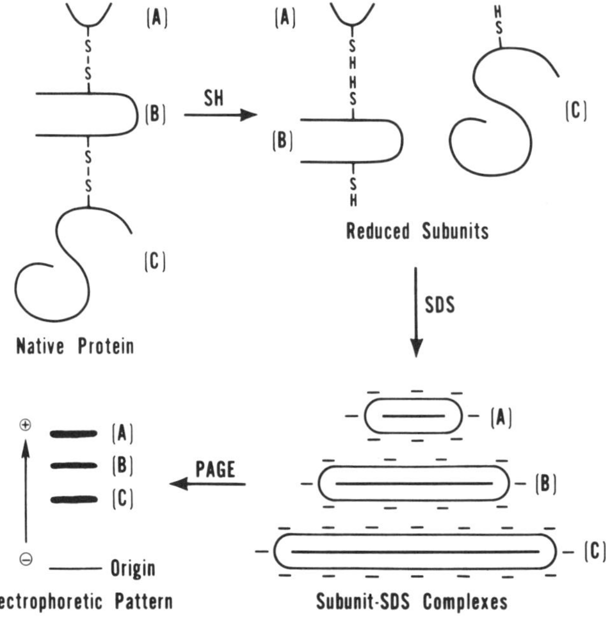

FIG. 9.1. PRINCIPLES OF SODIUM DODECYL SULFATE-POLYACRYLAMIDE GEL ELECTROPHORESIS (SDS-PAGE) ANALYSIS OF PROTEINS

cleaved with 2-mercaptoethanol, and the reduced subunits are complexed with SDS. The rodlike particles migrate at a rate determined by the ease with which they penetrate the gel during electrophoresis. Large molecular weight subunits migrate slowly, small ones quickly. Glutenin is seen to consist of many polypeptides of differing MW which in the native molecule were held together by disulfide bonds. Some of the subunits closely resemble gliadin (Bietz and Wall 1973), but most subunits are unique to glutenin (Fig. 9.2).

Wheat gluten is the major cereal protein product used in foods in

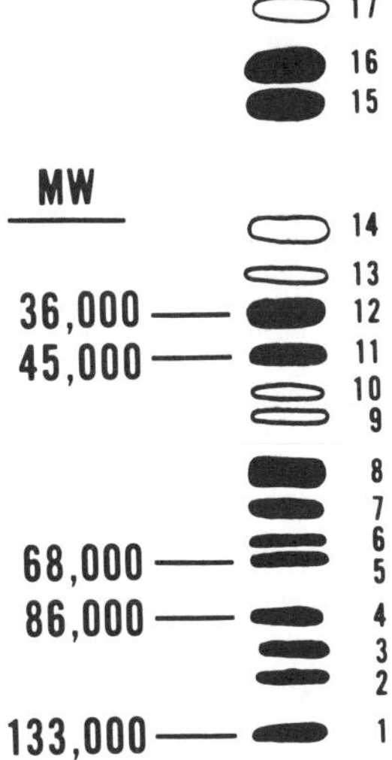

FIG. 9.2. SDS-PAGE OF GLUTENIN FROM THE COMMON WHEAT VARIETY CHINESE SPRING Numbering system and representative molecular weights (MW) of subunits are indicated.

the United States, and studies continue on the preparation of gluten and other protein products from wheat and its milled products (Saunders *et al.* 1975A; Anderson *et al.* 1965). Wheat proteins have the property of producing an elastic, cohesive mass that can be separated from starch by various procedures. A process, called the "batter process," involves making a flour-water batter, breaking up the batter in the presence of additional water, and washing the starch away from the fine curds of gluten (Hilbert *et al.* 1944). This process was used in several manufacturing plants during World War II. A continuous batter process was developed that incorporated a number of improvements on gluten separation from low-grade flours (Anderson *et al.* 1958, 1960, 1965). Flour and water are continuously mixed in the proper proportions followed by breaking up the batter mechanically in a pump in the presence of more water. Suspended gluten is separated from starch by screening the slurry.

The Fesca process described by Eynon and Lane (1928) has been examined by researchers at the Western Regional Research Center (Fellers *et al.* 1969). Centrifugal separation of starch from a thick flour slurry gives a gluten concentrate containing 20–40% protein depending upon processing variables and the type of flour (Johnston and Fellers 1971A, B).

Alkaline-extraction procedures similar to those applied to wheat flour have been used to yield protein and starch from whole wheat (Wu *et al.* 1974) and mill feeds (Saunders *et al.* 1975A). In the milling of whole wheat, Wu and his co-workers used an alkaline solution at pH 10.8 on ground wheat to extract the protein. By adjusting pH to 6 with hydrochloric acid, a concentrate that had a protein content of 64% was obtained in 17% yield.

Protein concentrates were also prepared from wheat shorts and millrun by alkaline extraction (Saunders *et al.* 1975A; Woerman and Satterlee 1974). The concentrates were collected after heat coagulation or acid precipitation of the extract. Without starch removal, the concentrates contained about 30–40% protein, 36–60% starch, and 7–11% fat, and were produced in yields of 15–25%. If starch were removed before protein precipitation, concentrates contained from 63–79% protein.

No morphological variations have been observed in the electron microscope which would indicate that wheat protein is a complex mixture of various molecular species. Buttrose (1963) and Jennings *et al.* (1963) found in developing wheat kernels that all the protein was deposited in distinct bodies and, on maturity, these protein bodies fuse to form the matrix protein of the endosperm which is without any structural features of separate bodies (Adams *et al.* 1976).

Denaturation of wheat proteins related to their functionality and food applications was reviewed by Wu and Inglett (1974). The heat denaturation of gluten as measured by baking-test and solubility correlates well with its functionality (Pence *et al.* 1953).

Wheat gluten has been known to contribute to film-forming properties important to breadmaking for a long time. Sandstedt *et al.* (1954), in a classic study of bread dough and crumb microstructure, showed that proteinaceous films containing starch granules formed around gas bubbles. Starch granules were found to take on various shapes to fit around the gas bubbles, which accounts for the desirable crumb grain and texture. More recent studies on the structure of doughs and breads by scanning and transmission microscopy were made by Khoo *et al.* (1975). Scanning electron micrographs of fermented dough and bread crumb are shown in Fig. 9.3. The gas cells in the dough after fermentation (Fig. 9.3A) are seen to be expanded (FS). The protein coating on the surface of the starch granule, due mainly to the increase in the size of the air cells, stretches and rolls up into fibrils. After kneading and proofing, these fibrils aggregate and form longer and larger fibrils (Fig. 9.3B). In the bread crumb (Fig. 9.3C), a small area of an air cell (AC1) dramatically shows stretched protein identifiable but not as clearly stranded as seen in fermented doughs. Veil-like protein and under-

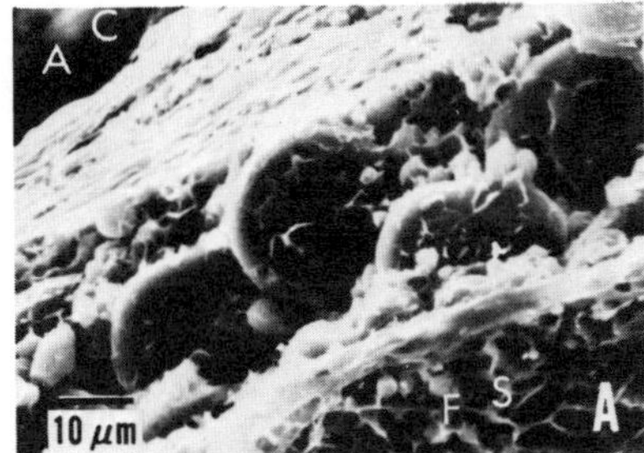

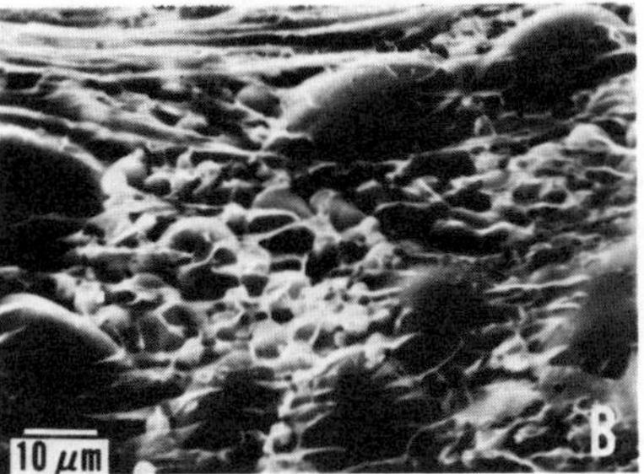

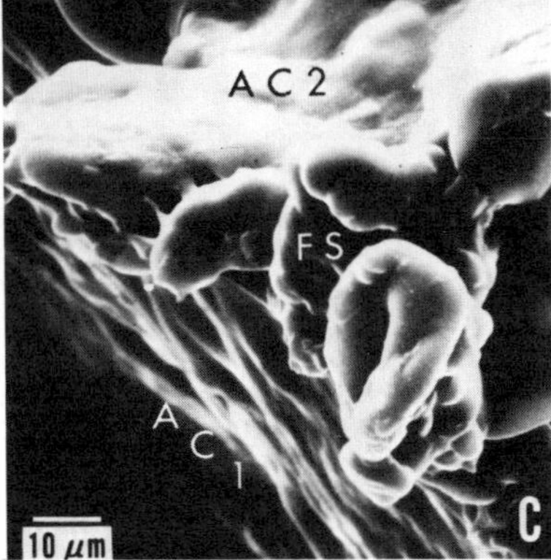

FIG. 9.3. SCANNING ELECTRON MICROGRAPHS OF FERMENTED DOUGH AND BREAD CRUMB (MAGNIFICATION = 500X)

(A) Dough after fermentation. Gas cells in protein matrix within dough (FS) have expanded. Part of large air cell is seen at AC. (B) Dough after fermentation and proofing. (C) Micrograph of bread crumb showing small areas of air cells (AC) and a fractured surface (FS).

lying starch are seen as a cohesive mass at AC2 and a fractured area at FS.

Transmission electron microscopy of flour, dough, and bread was used by Khoo *et al.* (1975) for finding additional information. After fermentation of the dough, some starch granules begin to show partial internal degradation without distortion. However, after proofing, some starch granules show a degree of enzymatic digestion which probably improves the granules' capacity to stretch and distort during oven spring. In bread crumb, starch granules in various forms of distortion and degradation appear in a protein matrix (Fig. 9.4A and B) which remains unchanged in structure except for its more intense strain. Denaturation of gluten during the baking may be attributed to its stainability, but baking does not seem to change

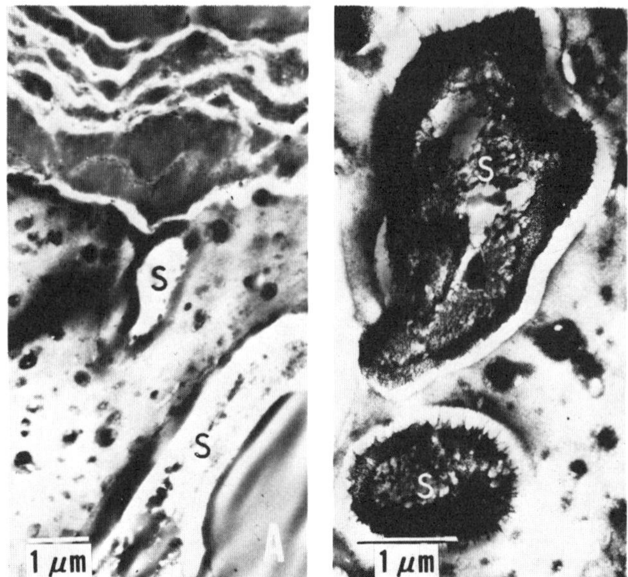

FIG. 9.4. TRANSMISSION ELECTRON MICROGRAPHS OF
BREAD CRUMB SHOWING STARCH GRANULES (S) IN VARI-
OUS STAGES OF DISTORTION AND DEGRADATION
(A) shows alignment of cell walls during mixing. [Magnification: (A)
= 5,000X; (B) = 8,350X].

protein ultrastructure. Some starch granules were observed to be covered with intact membranes. All dough and bread samples show cell walls separated from their contents and neatly aligned in the protein matrix.

Maize, Sorghum, and Millet Proteins

Maize, also called corn, is the number one cereal in U.S. agriculture. In 1973, 5.65 billion bu were produced having a farm value of $14.4 billion (USDA 1975). Approximately 10% of the corn crop in the United States is processed by wet milling, dry milling, and distilling industries (Inglett 1970). Of the whole corn kernels, about 10% consists of proteins. More than 75% of these proteins is contained in endosperm tissue (Table 9.1). The embryo is twice as high in protein as the endosperm (19% vs 9%) but it is only 10% of the kernel. Four classes of corn proteins are known based on their solubility: albumins (water soluble), globulins (saline solution soluble), prolamine (70% ethanol soluble), and glutelin (soluble in 0.1 M NaOH). Most of the corn embryo consists of albumins and globulins, which are less than 10% of the protein in the endosperm.

Endosperm protein in ordinary maize contains about 40–50% zein and from 30–40% glutelin. A marked deviation from these

TABLE 9.1
PROTEIN DISTRIBUTION IN CORN AND CORN FRACTIONS

Component	Whole Grain %	Endosperm %	Germ %	Bran and Tip Cap %
Kernel	100	84	10	6
Kernel protein	100	76	20	3
Protein in fraction	10	9	19	5
Part protein:				
Albumins	8	4	30	—
Globulins	9	4	30	—
Zein	39	47	5	—
Glutelin	40	39	25	—

Source: Reiners *et al.* 1973.

proportions was noted by Mertz and his co-workers (1964) in high-lysine maize (*opaque-2*). In *opaque-2*, the proportions of zein and glutelin were reversed: zein was down to 15% of the protein and glutelin rose to more than 40%. The nutritional importance of this finding becomes obvious from a study of the amino acid composition of maize glutelin and zein, particularly as related to lysine. Zein contains essentially no lysine, whereas maize glutelin contains nearly 4.7 gm of lysine per 100 g of protein. Therefore, any increase in glutelin content at the expense of zein increases the nutritional value of the maize protein, since lysine is the first limiting amino acid in ordinary maize.

Almost all seeds during development deposit protein granules, which are also referred to as aleurone grains or protein bodies. As is commonly recognized, protein granules vary widely in size, quantity, and composition in different seeds (Inglett 1972; Adams *et al.* 1976). Studies have been reported on the granules found in barley, corn, sorghum, millet, rice, cottonseed, peanut, soybean, pea, lima beans, hempseed, and shepherd's purse.

Protein granules of corn, sorghum, and millet are prolamines having poor nutritive value in contrast to the protein bodies of soybeans, which are composed of proteins having an excellent balance of essential amino acids. The nutrition of seeds and their processed products are determined in varying degrees by these protein bodies.

The subcellular structure of proteins of ordinary and high-lysine maize was carefully examined by Khoo and Wolf (1970).

In ordinary dent corn, protein granules averaging about 2 μm in diameter can be seen in endosperm sections that have been

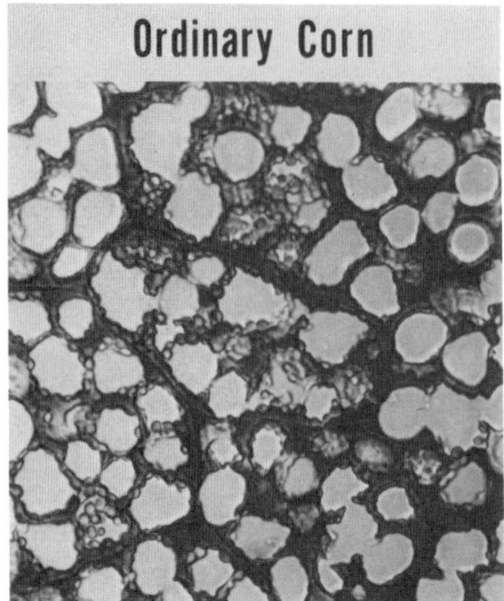

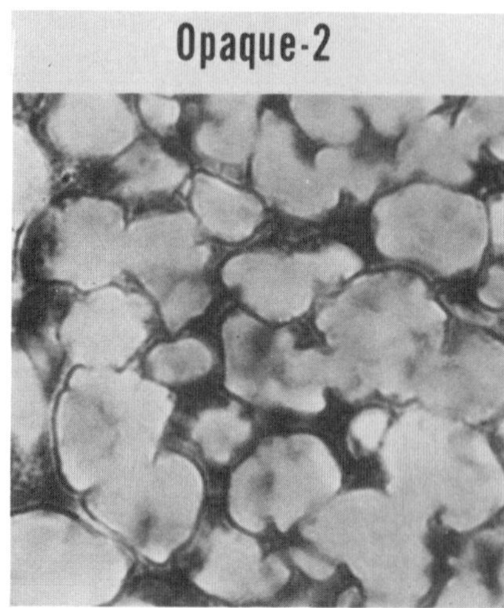

FIG. 9.5. COMPARISON OF SUBCELLULAR PROTEIN STRUCTURE OF ORDINARY CORN (W64A) AND OPAQUE-2 MUTANT (W64AO$_2$)

Ordinary corn endosperm protein has two components: granular bodies and matrix protein (left). *Opaque-2* has no protein bodies visible in light microscopy (right).

enzymatically destarched (Fig. 9.5). A sharp contrast is shown between the ordinary corn section and the high-lysine mutant, *opaque-2*. Duvick (1961) suggested that these granules could be sites of zein deposition. Although examination of the subcellular structure of *opaque-2* endosperm protein under a light microscope revealed no protein granules, electron microscopy revealed granules having diameters about 0.1 μm (Fig. 9.6). The electron photomicrograph also reveals the matrix protein, glutelin. A vivid illustration of protein bodies in ordinary corn and their apparent absence in high-lysine corn can be seen in a scanning electron micrograph (Fig. 9.7).

Protein subcellular particulates were isolated by zone sedimentation (Christianson *et al.* 1969) from homogenates of immature endosperm cells. Insoluble components of each of the zones were collected by ultracentrifugation and examined directly by scanning electron microscopy (Fig. 9.8). The free protein bodies are clearly distinct from the clumped bodies probably held together with matrix protein, glutelin. The mobility of the free bodies protein on starch gels reveals a close similarity to zein.

Adams *et al.* (1976) isolated protein bodies from mature maize by centrifugation and found a considerable amount of fine structure adhering to the outside edge of the granules.

Proteins have been prepared from whole corn or from fractions after dry or wet milling the corn. A large source of protein is available from this grain if suitable products can be prepared and utilized. Corn products containing 73% protein or higher can be

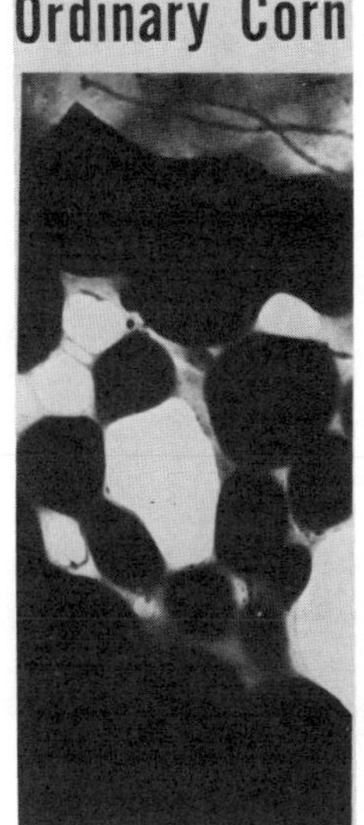

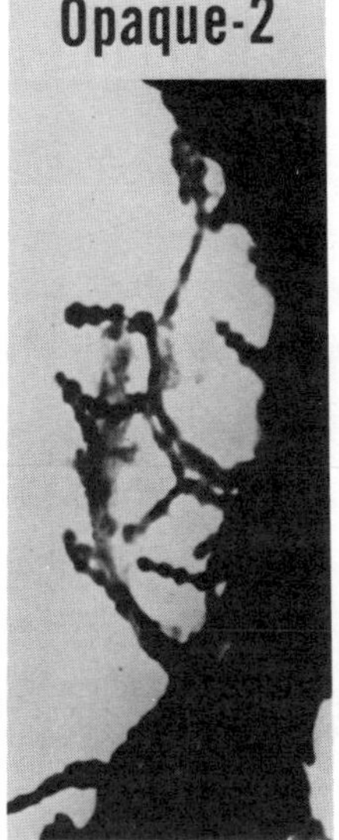

FIG. 9.6. ELECTRON MICROGRAPHS OF SUB-
CELLULAR ENDOSPERM PROTEIN IN OR-
DINARY CORN (LEFT) AND OPAQUE-2 MU-
TANT (RIGHT)
Protein bodies of ordinary corn are around 2 μm
compared to 0.1 μm for *opaque-2* granules.

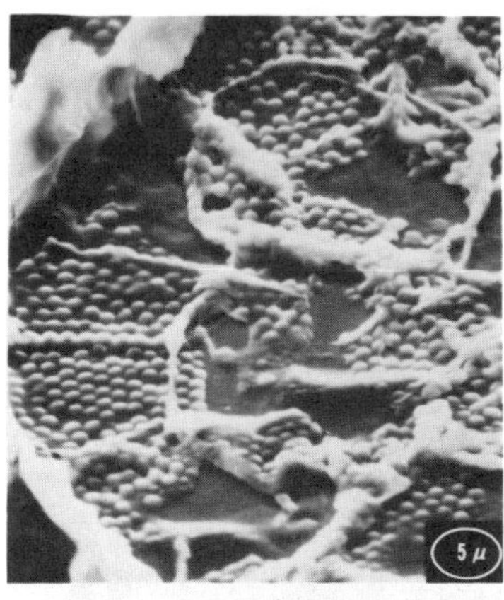

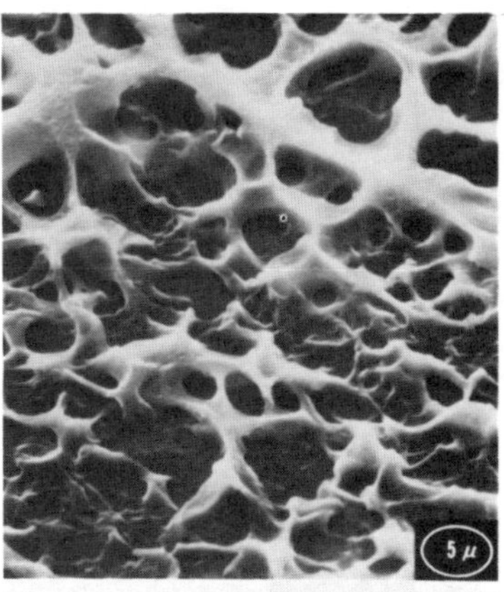

FIG. 9.7. SCANNING ELECTRON MICROGRAPH OF SUBCELLULAR ENDO-
SPERM PROTEIN IN ORDINARY CORN (LEFT) AND OPAQUE-2 MUTANT
(RIGHT)

obtained by using wet processing techniques on the defatted
dry-milled germ. An alkaline extraction of this germ meal at pH 8.7
extracted 50% of the total germ nitrogen (Nielson *et al.* 1973). Acid

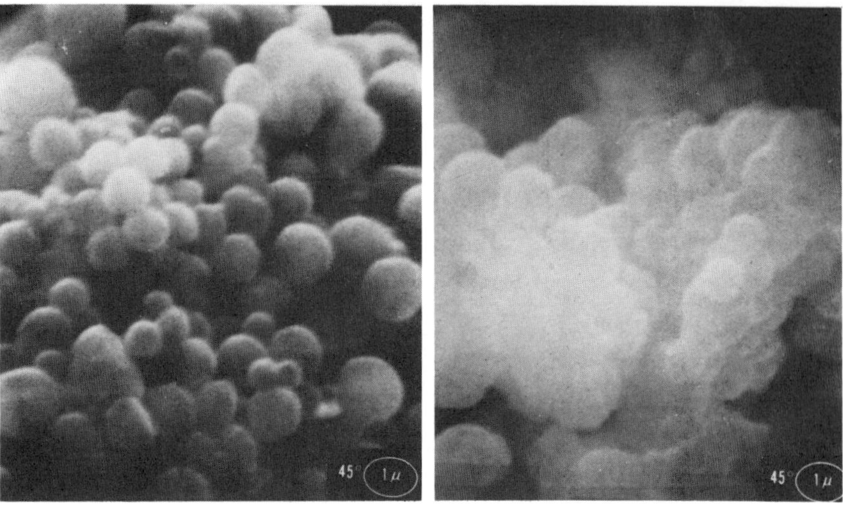

FIG. 9.8. SCANNING ELECTRON MICROGRAPHS OF FREE PROTEIN GRANULES
(LEFT) AND CLUMPED GRANULES (RIGHT) IN DENT CORN

precipitation of the extract gave a product containing more than 73%
protein.

Corn Germ Flour.—The commercial germ fraction from corn dry
milling contains fragments of bran, grits, and tip cap. Additional
screening provides an enhanced embryo-rich material ready for more
processing. Suitable corn germ is passed through a No. 6 screen (U.S.
Standard series), collected on a No. 10, and aspirated. The No. 10
fraction is flaked, defatted with hexane, and ground to a flour. Final
screening on a 9XX silk gave a flour (Blessin *et al.* 1972, 1974). The
composition of this flour shows a protein content of 25% with the
remaining parts being composed of starch, pentosans, fiber, vitamins,
and minerals (Table 9.2). This product was examined as a nutrient
fortifier for bread (Tsen *et al.* 1974) and in meat and baked goods
(Blessin *et al.* 1973). Corn germ flour became a product of commerce
in 1976.

Grain Sorghum.—Generally, the protein content of grain sorghum
varies between 8 and 12%. These proteins are distributed unevenly
throughout the grain and even among the three main types of
endosperm cells: subaleurone, vitreous, and floury. Subaleurone cells
are principally protein; vitreous and floury cells are typical starchy
endosperm cells, but differ in density, texture, and protein content.
Vitreous cells contain about twice as much protein as floury cells
(Shoup *et al.* 1969). Sorghum products containing vitreous endosperm are nutritionally inferior to products containing floury
endosperm. The vitreous endosperm products have higher prolamine

TABLE 9.2
COMPOSITION OF CORN GERM FLOUR

Constituent	%[1]
Protein	25.3
Lysine	1.49
Methionine	0.54
Tryptophan	0.35
Fiber	4.2
Starch	24.7
Sugars	13.8
Pentosans	11.7
Fat	0.5
Ash	10.3
Phosphorus	2.74
Potassium	2.36
Magnesium	1.02
Calcium	0.015
Sodium	0.014
Iron	0.020

Source: Blessin *et al.* 1972.

[1] Moisture-free basis.

content, which results in lower lysine values, than obtained from floury endosperm products (Virupaksha and Sastry 1968). A high prolamine content is undesirable because this protein has the lowest lysine value of all the protein components (Jones and Beckwith 1970).

Microscopy of Grain Sorghum Proteins.—Light microscopy reveals much of the general protein pattern over a large area of the kernel (Fig. 9.9). The starch was removed by amylase digestion leaving clearly definable cell walls and the protein network (Seckinger and Wolf 1973). The subaleurone cells are almost completely filled with protein. The endosperm cells below the subaleurone gradually decrease in protein content toward the center of the endosperm.

At higher magnification, the protein network has a beaded appearance (Fig. 9.10). The subaleurone protein appears as a mass of granules held together by a matrix. Protein granules are clearly evident by electron microscopy. Some granules appear as spherical bodies with a smooth outer surface while others appear coated with matrix protein (Fig. 9.11).

The internal structure of protein granules is shown in Fig. 9.12. In this transmission electron micrograph of a section through the peripheral part of the kernel, almost all the granules have a darkly stained nucleus. Around the nucleus some granules exhibit stained

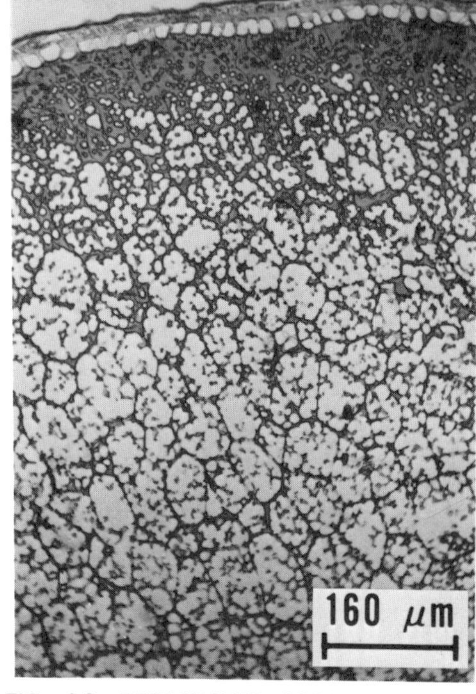

FIG. 9.9. DESTARCHED SORGHUM ENDO-SPERM SHOWING CELL WALLS AND PROTEIN NETWORK (63X)

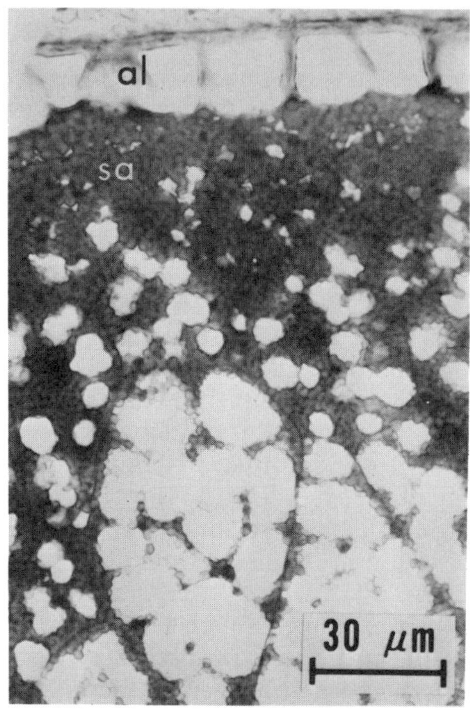

FIG. 9.10. DESTARCHED PERIPHERAL AREA OF A SORGHUM KERNEL SHOWING SUB-ALEURONE (sa) PROTEINACEOUS CELLS, PROTEIN GRANULES, AND THE ALEURONE LAYER (al); 325X

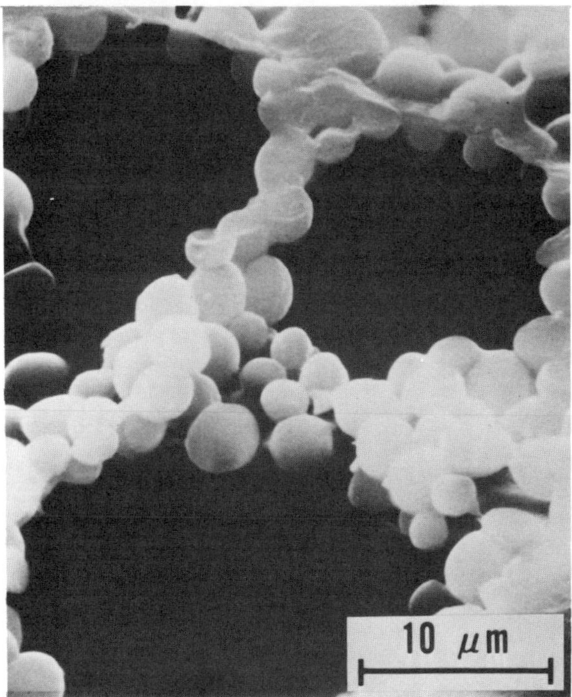

FIG. 9.11. TRANSMISSION ELECTRON MICROGRAPH OF DESTARCHED SORGHUM ENDOSPERM SHOWING PROTEIN GRANULES (1,500X)

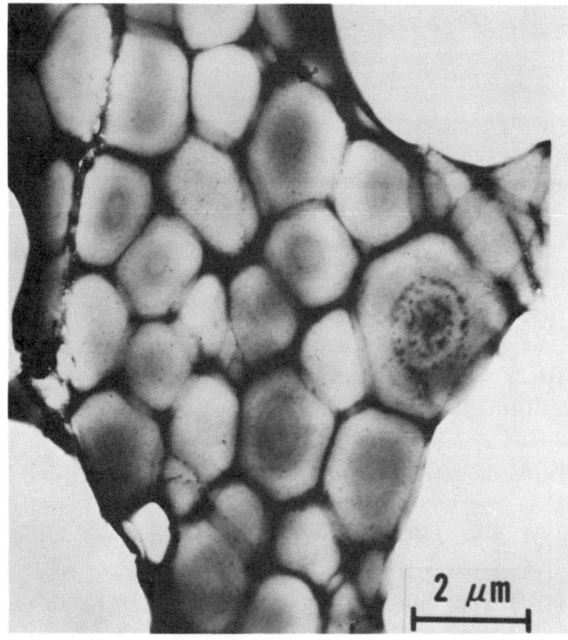

FIG. 9.12. TRANSMISSION ELECTRON MICRO-GRAPH OF PROTEIN GRANULES IN GRAIN SORGHUM; THE MATRIX HAS BEEN STAINED WITH URANYL ACETATE (5,000X)

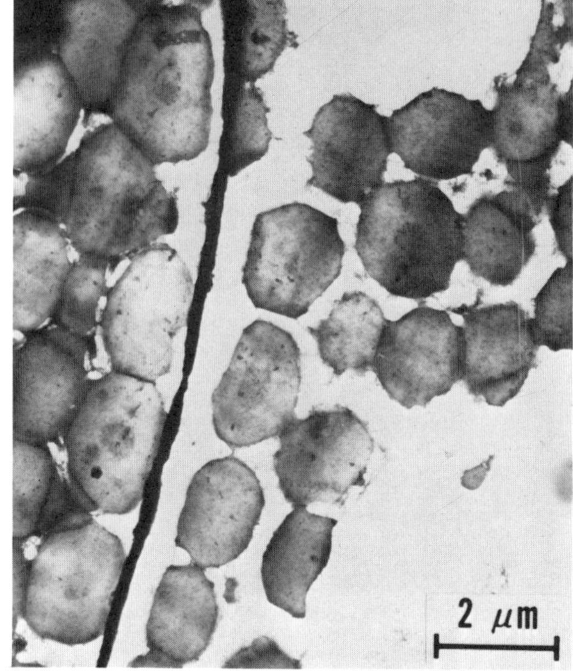

FIG. 9.13. SCANNING ELECTRON MICROGRAPH OF SORGHUM PROTEIN AFTER A 2-HR PRONASE® DIGESTION
Matrix protein digested; protein granules intact (5,000X).

concentric rings, which indicate a spherical laminar structure. The prolamine nature of the granules was established on their solubility in 60% *tert*-butanol and hot ethanol (Jones and Beckwith 1970). The proteolytic enzyme, Pronase®, degraded the granules and matrix proteins at different rates. After 2 hr, this enzyme degraded the matrix protein leaving the prolamine granules (Fig. 9.13).

Sorghum protein bodies were isolated from the mature grain by Adams *et al.* (1976). They were mostly circular in section and varied in size from 0.5–3.5 μm. Each granule had a discrete border which may be membrane bounded. The isolated bodies were very similar to those observed *in situ* by Seckinger and Wolf (1973).

Millet.—Millet is a minor cereal crop in the United State, but elsewhere it is grown in large quantities for food. Microscopic examination of destarched proso endosperm sections showed large deposits of spherical protein granules measuring up to 2.5 μm in diameter (Fig. 9.14). The granules are embedded in matrix protein in the outer endosperm cells, but farther into the endosperm center there are small quantities of matrix protein and mainly protein granules. An electron micrograph of the outer endosperm shows

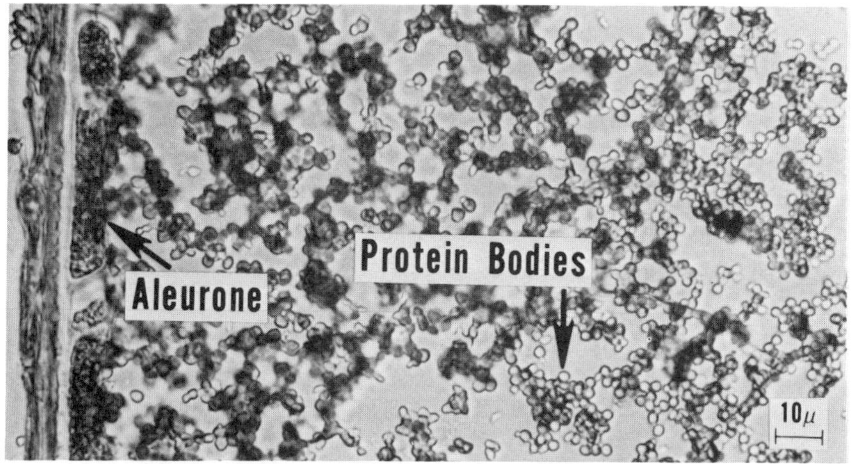

FIG. 9.14. AMYLASE-TREATED SECTION OF WHITE PROSO MILLET

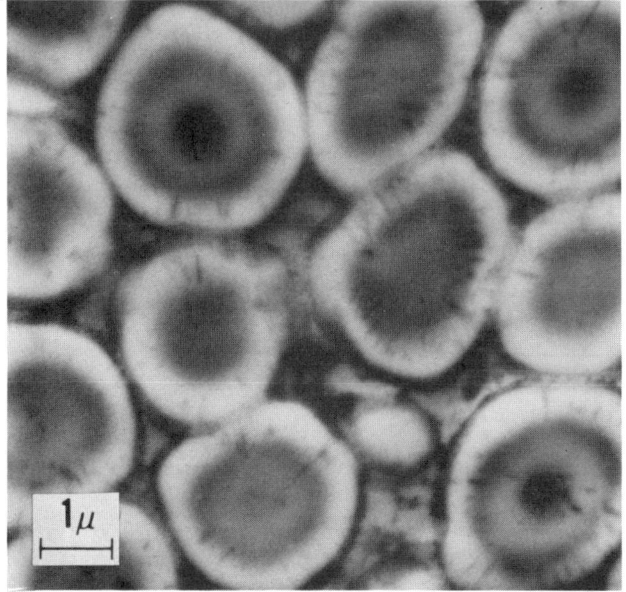

FIG. 9.15. ELECTRON MICROGRAPH OF PROSO MILLET SHOWING PROTEIN BODIES AND MATRIX PROTEIN (4,250X)

protein bodies having a central core surrounded by two or more bands (Fig. 9.15). The matrix protein appears heterogeneous, as indicated by different staining intensities (Jones *et al.* 1970).

Protein bodies isolated from mature millet seeds by Adams *et al.* (1976) appeared as discrete organelles possibly surrounded by a membrane.

Oat Proteins

The nutritional quality of oat protein is good; rolled oats have a protein efficiency ratio (PER) of 2.2 compared to casein that has a PER of 2.5. Feeding studies of seven pure oat varieties—Garland, Clintland, Bonkee, Newton, Beedee, Lodi, and Newaha, which had PER values between 2.25 and 2.38—indicated that the small variations in amino acids observed in these oat samples were not great enough to influence growth response (Clark and Potter 1972; Hischke *et al.* 1968). The protein content of oat groats being used for food has a value between 11 and 15%. However, some oats found in the Near East have protein contents varying between 14 and 25% in the groats.

Practically no research had been done on individual oat protein fractions until the pioneering work by Wu *et al.* (1972). Protein concentrates have been prepared from oat flours of ordinary and high-protein varieties by air classification (Wu and Stringfellow 1973) and by wet-milling procedures (Wu *et al.* 1973; Cluskey *et al.* 1973).

Alkaline extraction of defatted ordinary oat groats (15% protein) gave a concentrate with 67% protein obtained in a 16% yield. A higher yield of concentrate, 19%, was obtained from a high-protein oat variety (Garland groat, 19% protein). The concentrate had a protein content of 68%.

Oat flour from a high-protein variety, Garland, was air-classified to yield fractions with protein contents as high as 88% (Table 9.3).

TABLE 9.3

AIR CLASSIFICATION OF GARLAND FIRST OAT FLOUR
(DRY BASIS, DEFATTED)

	Garland Variety			
Fraction[1]	Yield %	Protein %	Protein % of Total	Ash %
First flour ground				
3X 14,000 rpm		15.8		0.9
1A	2	88.4	14	2.5
1B	27	26.9	48	1.1
2	20	11.7	15	0.9
3	33	7.6	16	0.8
4	13	5.7	5	0.7
5	5	5.7	2	0.8

Source: Wu and Stringfellow (1973).

[1] Fraction 1A was collected as an ultrafine fraction in an air filter bag during the first pass through the classifier, and fraction 1B is the normal fine fraction collected during the first pass.

Such a high-protein fraction (83–88%) has not been obtained from wheat, rye, corn, sorghum, or triticale flours. This unique oat protein fraction (2–5% by weight of the flour) accounted for 14–16% of the total protein in first and second flours. The high protein variety of oats (Garland) gave better results than ordinary oats (Wu and Stringfellow 1973). Additional data show these flours to be of good amino acid composition and to possess interesting properties for food applications.

Rice Proteins

The major constituent of milled rice is starch, and it is most concentrated in the endosperm portion of the kernel. Protein is the second most abundant constituent of rice grain and is unique among the cereal proteins because it contains at least 80% glutelin (alkali-soluble protein). Glutelin has the closest amino acid composition to milled rice protein, probably because it is the major protein

TABLE 9.4
LEVELS OF ESSENTIAL AMINO ACIDS OF PROTEIN FRACTIONS
AND PROTEIN OF MILLED RICE (g/16.8 g N)

Amino Acid	Protein Fraction				Milled Rice Protein, %
	Albumin	Globulin	Prolamin	Glutelin	
Isoleucine	4.05	3.03	4.68	5.27	4.13
Leucine	7.89	6.56	11.3	8.19	8.24
Lysine	4.92	2.56	0.51	3.47	3.80
Methionine	2.54	2.27	0.50	2.61	3.37
Methionine + cystine	5.40	2.27	0.80	4.09	4.97
Phenylalanine	2.97	3.32	6.26	5.42	6.02
Threonine	4.65	4.55	2.86	3.92	4.34
Tryptophan	1.88	1.34	0.94	1.16	1.21
Valine	8.72	6.18	6.97	7.31	7.21

Source: Juliano (1972).

fraction (Table 9.4). The protein content of rice of any variety can vary considerably even when grown at the same location (Cagampang *et al.* 1966; Tecson *et al.* 1971). For example, protein content of the high-protein variety BPI-76-1 may range from 8–14% and the low protein variety, Intan, from 5–11% protein. The effect of differences in protein contents of milled rice on their nutritive quality (Juliano 1972) shows that protein quality tended to decrease as protein contents increased; however, the decrease in quality was less than proportional to the increase in protein content.

Protein concentrates can be prepared from rice bran by wet processing techniques (Saunders *et al.* 1975B; Lew *et al.* 1975). A high-protein flour (18%) can be separated from rice bran in about 50% yield by wet grinding and wet particle size classification (Barber *et al.* 1974). Air classification of rice bran gave some fractions with slightly increased protein content (Houston and Mohammad 1966); however, Wrenshall (1974) found that pin-milling and air classification were useful in preparing high-protein products from rice bran.

Triticale Proteins

Present commercial varieties of triticale are crosses of durum wheat and rye. Triticale flour using standard dough processing and baking conditions produces bread with poor loaf volume because of the low viscoelastic strength of triticale protein. Loaf volume from 100% triticale flour doughs was improved by adding dough conditioners and wheat-flour supplements (Tsen *et al.* 1973). Finely ground flours from three triticale varieties gave air-classification patterns similar to those of various wheats (Stringfellow *et al.* 1976). Air-classification data for these varieties, a rye, and two wheats are summarized in Table 9.5. Fas Gro 204 gave a protein content range among the eight classified fractions of 4.4– 34.3%. The greater ease of fractionation is indicated by the protein shift of 73% of 204 flour as compared to 56– 58% from other triticale flours, 41% for rye flour, and 29– 50% for the two hard wheats. Fas Gro 204 is less vitreous, which results in its greater response to fine grinding and air classification. Fas Gro 204 gave a 25% yield of the three high-protein fractions which is slightly less than that of soft wheat flour (29%). It is equal to rye flour (26%) and greater than hard wheat flours (19– 21%). Protein content of the three finest fractions (29%) from 204 was higher than either the rye fractions (20%) or the hard and soft wheat fractions (19– 24%).

Currently, whole ground triticale is being marketed in various parts of the U.S. as flour for pancake mixes, cereal flakes, and bread. Studies on making high-protein snacks by direct extrusion-cooking are reported by Anderson *et al.* (1974). Protein isolates of triticale with protein contents between 82– 87% (Wu *et al.* 1976) have good emulsifying activity and emulsion stability.

Barley and Rye Proteins

Barley is used for human food in the form of parched grain, pearled grain for soups, flour for flat bread, and ground grain for porridge. Barley flour is milled generally by conventional roller-milling (Pomeranz *et al.* 1971). Air classification can be used to separate barley flours into high-protein and low-protein fractions.

Conventional roller-milling of barley gives four major products: flour, tailings flour, shorts, and bran. The flour contains primarily

TABLE 9.5

COMPARISON OF AIR CLASSIFIED FRACTIONS OF REGROUND FLOURS
FROM TRITICALE, RYE, AND WHEAT
(EXPRESSED AS PERCENT ON 14% MOISTURE BASIS)

| | Triticale Variety Fas Gro | | | Rye, Com-mercial | Wheat | | |
| | | | | | Hard Red Spring | Hard Red Winter | Soft Red Winter |
Sample	204	131B	385	Mix	Selkirk	Wichita	Vermillion
Straight-grade flour, protein	12.7	10.2	10.4	10.9	12.3	10.9	9.4
Range of protein	34.3–4.4	27.6–4.7	30.2–5.1	21.5–6.2	23.7–7.6	29.4–5.5	26.7–2.3
Combined high-protein fractions 1–3							
Yield	25.1	21.7	18.5	25.8	18.8	21.0	29.4
Protein	29.0	22.1	24.5	19.8	19.4	24.4	21.4
Combined starchy fractions 4–7							
Yield	65.3	64.6	64.2	56.8	49.5	52.9	64.5
Protein	5.5	5.6	5.9	7.4	8.7	6.5	3.3
Coarse residue fraction 8							
Yield	9.6	13.7	17.3	17.4	31.7	26.1	6.1
Protein	13.3	12.8	12.4	9.6	14.2	9.8	6.2
Protein shifted, total	73	58	56	41	29	50	82

Source: Anderson *et al.* 1974.

the starchy endosperm; the shorts and tailings flour, a mixture of aleurone and pericarp with some germ and endosperm; and the bran, hulls and pericarp. A 65% extraction of barley flour contains 9.8% protein (Robbins and Pomeranz 1972).

A screening program of the World Barley Collection for genetic varieties having high lysine and high protein was successful, and the most promising variety, CI 3947 (Hagberg and Karlsson 1969) was later called Hiproly. The opportunities offered by this improved barley variety and its properties have been reviewed by Munck (1972).

Protein bodies from barley have been isolated by Ory and Henningsen (1969) and Adams *et al.* (1976). The bodies had little evidence of any internal structure.

In the United States about 5.6 million bu of rye are processed in a year for food products. The principal products are rye flours and meals for bread, crackers, and snack foods. Rye bread is favored by the U.S. population with a recent European background. Cracker and biscuit manufacturers frequently use about 10% white rye flour in their flour blend. Meat processors may use rye flours as fillers in ground meat products.

Basically, two grades of rye flour are made—white rye flour or patent and dark rye flour or clear. About 80% of the milled flour is white rye flour. Many grades of rye flour are possible by combining the white and dark rye flour streams in varying percentages (Shaw 1970). Ash specifications on rye flours are one criterion for selling the flours. Color and ash contents are fairly closely related. White flours range from 0.6–0.7% ash, whereas the dark rye flours will range from 2.2–3.0% ash. Little attention is given to the protein content of rye flours.

Although protein composition of rye endosperm is sparse, a recent study on the subunit structure and amino acid composition of rye gliadins fractionated by gel filtration was reported by Preston and Woodbury (1975). Considerably more research appears to be needed on rye proteins.

BIBLIOGRAPHY

ADAMS, C. A., NOVELLIE, L., and LIEBENBERG, N.v.d.W. 1976. Biochemical properties and ultrastructure of protein bodies isolated from selected cereals. Cereal Chem. *53*, No. 1, 1–12.

ANDERSON, R. A., PFEIFER, V. F., and LANCASTER, E. B. 1958. Continuous batter process for separating gluten from wheat flour. Cereal Chem. *35*, 449–457.

ANDERSON, R. A., PFEIFER, V. F., LANCASTER, E. B., VOJNOVICH, C., and GRIFFIN, E. L., JR. 1960. Pilot-plant studies on the continuous batter process to recover gluten from wheat flour. Cereal Chem. *37*, 180–188.

ANDERSON, R. A., PEPLINSKI, A. J., and PFEIFER, V. F. 1965. Gluten and starch from long-extraction wheat flours. Cereal Sci. Today *10*, 106–108, 110, 198.

ANDERSON, R. A., STRINGFELLOW, A. C., WALL, J. S., and GRIFFIN, E. L., JR. 1974. Milling characteristics of triticale. Food Technol. *28*, No. 11, 66–76.

BARBER, S., FLORS, A., TORTOSA, E., CAMACHO, J. M., and CERNI, R. 1974. High protein flours from rice bran by wet fractionation process. Int. Conf. on Rice By-products Utilization, Valencia, Spain.

BIETZ, J. A., HUEBNER, F. R., and WALL, J. S. 1973. Glutenin—the strength protein of wheat flour. Bakers' Dig. *47*, No. 1, 26–35.

BIETZ, J. A., and WALL, J. S. 1972. Wheat gluten subunits: Molecular weights determined by sodium dodecyl sulfate-polyacrylamide gel electrophoresis. Cereal Chem. *49*, 416–430.

BIETZ, J. A., and WALL, J. S. 1973. Isolation and characterization of gliadin-like subunits from glutenin. Cereal Chem. *50*, 537–547.

BLESSIN, C. W., GARCIA, W. J., DEATHERAGE, W. L., CAVINS, J. F., and INGLETT, G. E. 1973. Composition of three food products containing defatted corn germ flour. J. Food Sci. *38*, 602–606.

BLESSIN, C. W., GARCIA, W. J., DEATHERAGE, W. L., and INGLETT, G. E. 1974. An edible defatted germ flour from a commercial dry-milled corn fraction. Cereal Sci. Today *19*, 224–226.

BLESSIN, C. W., INGLETT, G. E., GARCIA, W. J., and DEATHERAGE, W. L. 1972. Defatted germ flour—food ingredient from corn. Food Prod. Dev. *6*, No. 4, 34–35.

BUTTROSE, M. S. 1963. Ultrastructure of the developing wheat endosperm. Aust. J. Biol. Sci. *16*, 305–317.

CAGAMPANG, G. B., CRUZ, L. J., ESPIRITU, S. G., SANTIAGO, R. G., and JULIANO, B. O. 1966. Studies on the extraction and composition of rice proteins. Cereal Chem. *43*, 145–155.

CHRISTIANSON, D. D., NIELSEN, H. C., KHOO, U., WOLF, M. J., and WALL, J. S. 1969. Isolation and chemical composition of protein bodies and matrix proteins in corn endosperm. Cereal Chem. *46*, 372–381.

CLARK, W. L., and POTTER, G. C. 1972. The compositional and nutritional properties of protein in selected oat varieties. *In* Symposium: Seed Proteins, G. E. Inglett (Editor). Avi Publishing Co., Westport, Conn.

CLUSKEY, J. E., WU, Y. V., WALL, J. S., and INGLETT, G. E. 1973. Oat protein concentrates from a wet-milling process: Preparation. Cereal Chem. *50*, 475–481.

DUVICK, D. N. 1961. Protein granules of maize endosperm cells. Cereal Chem. *38*, 374–385.

EYNON, L., and LANE, J. H. 1928. Starch: Its Chemistry, Technology, and Uses. W. Heffner and Sons, Ltd., Cambridge, England.

FELLERS, D. A., JOHNSTON, P. H., SMITH, S., MOSSMAN, A. P., and SHEPHERD, A. D. 1969. Process for protein-starch separation in wheat flour. Food Technol. *23*, No. 4, 162–166.

HAGBERG, A., and KARLSSON, K. E. 1969. Breeding for high protein content and quality in barley. Symposium: New Approaches to Breeding for Improved Plant Protein. Int. Atomic Energy Agency, Vienna (1968).

HILBERT, G. E., DIMLER, R. J., and RIST, C. E. 1944. Wheat flour: A potential raw material for the expanded production of starch and sirups. Am. Miller Process. *72*, 32–37, 78–80.

HISCHKE, H. H., JR., POTTER, G. C., and GRAHAM, W. R. 1968. The nutritive value of oat protein. 1. Varietal differences as measured by amino acid analysis and rat growth response. Cereal Chem. *45*, 374–378.

HOUSTON, D. F., and MOHAMMAD, A. 1966. Air-classification and sieving of rice bran and polish. Rice J. *69*, 20.

INGLETT, G. E. 1970. Corn: Culture, Processing, Products, G. E. Inglett (Editor). AVI Publishing Co., Westport, Conn.

INGLETT, G. E. 1972. Seed proteins in perspective. *In* Symposium: Seed Proteins, G. E. Inglett (Editor). AVI Publishing Co., Westport, Conn.

INGLETT, G. E. 1974. Wheat in perspective. *In* Wheat: Production and Utilization, G. E. Inglett (Editor). AVI Publishing Co., Westport, Conn.

INGLETT, G. E. 1975. Effects of refining operations on cereals. *In* Nutritional Evaluation of Food Processing, R. S. Harris and E. Karmas (Editors). AVI Publishing Co., Westport, Conn.

JENNINGS, A. C., MORTON, R. K., and PALK, B. A. 1963. Cytological studies of protein bodies of developing wheat endosperm. Aust. J. Biol. Sci. *16*, 366.

JOHNSON, V. A., MATTERN, P. J., and SCHMIDT, J. W. 1972. Genetic studies of wheat proteins. *In* Symposium: Seed Proteins, G. E. Inglett (Editor). Avi Publishing Co., Westport, Conn.

JOHNSTON, P. H., and FELLERS, D. A. 1971A. Process for protein-starch separation in wheat flour. 2. Experiments with a continuous decanter-type centrifuge. J. Food Sci. *36*, 649–652.

JOHNSTON, P. H., and FELLERS, D. A. 1971B. Process for separating starch and protein from wheat flour wherein the flour is agitated with water and NH_4OH and centrifugation is applied to the resulting slurry. U.S. Pat. 3,576,180. April 6.

JONES, R. W., and BECKWITH, A. C. 1970. Proximate composition and proteins of three grain sorghum hybrids and their dry-mill fractions. J. Agric. Food Chem. *18*, 33–36.

JONES, R. W., BECKWITH, A. C., KHOO, U., and INGLETT, G. E. 1970. Protein composition of proso millet. J. Agric. Food Chem. *18*, 37–39.

JULIANO, B. O. 1972. Studies on protein quality and quantity of rice. *In* Symposium: Seed Proteins, G. E. Inglett (Editor). AVI Publishing Co., Westport, Conn.

KHOO, U., CRISTIANSON, D. D., and INGLETT, G. E. 1975. Scanning and transmission microscopy of dough and bread. Baker's Dig. *49*, No. 4, 24–26.

KHOO, U., and WOLF, M. J. 1970. Origin and development of protein granules in maize endosperm. Am. J. Bot. *57*, No. 9, 1042–1050.

LEW, E. J-L., HOUSTON, D. F., and FELLERS, D. A. 1975. A note on protein concentrate from full-fat rice bran. Cereal Chem. *52*, 748–750.

MERTZ, E. T., BATES, L. S., and NELSON, O. E. 1964. Mutant gene that changes protein composition and increases lysine content of maize endosperm. Science *145*, 279–280.

MUNCK, L. 1972. Barley seed proteins. *In* Symposium: Seed Proteins, G. E. Inglett (Editor). Avi Publishing Co., Westport, Conn.

NIELSEN, H. C., INGLETT, G. E., WALL, J. S., and DONALDSON, G. L. 1973. Corn germ protein isolate—preliminary studies on preparation and properties. Cereal Chem. *50*, 435–443.

ORY, R. L., and HENNINGSEN, K. W. 1969. Enzymes associated with protein bodies isolated from ungerminated barley seeds. Plant Physiol. *44*, 1488–1498.

PENCE, J. W., MOHAMMAD, A., and MECHAM, D. K. 1953. Heat denaturation of gluten. Cereal Chem. *30*, 115–126.

POMERANZ, Y., KE, H., and WARD, A. B. 1971. Composition and utilization of milled barley products. 1. Gross composition of roller-milled and air-separated fractions. Cereal Chem. *48*, 47–58.

PRESTON, K. R., and WOODBURY, W. 1975. Amino acid composition and subunit structure of rye gliadin proteins fractionated by gel filtration. Cereal Chem. *52*, 719–726.

REINERS, P. A., WALL, J. S., and INGLETT, G. E. 1973. Corn proteins: Potential for their industrial use. Symposium: Industrial Uses of Cereals. Am. Assoc. Cereal Chem., St. Paul, Minn.

ROBBINS, G. S., and POMERANZ, Y. 1972. Composition and utilization of milled barley products. 3. Amino acid composition. Cereal Chem. *49*, 240–246.

SANDSTEDT, R. M., SCHAUMBERG, L., and FLEMING, J. 1954. The microscopic structure of bread and dough. Cereal Chem. *31*, 43–49.

SAUNDERS, R. M., CONNOR, M. A., EDWARDS, R. A., and KOHLER, G. O. 1975A. Preparation of protein concentrates from wheat shorts and wheat millrun by a wet alkaline process. Cereal Chem. *52*, 93–101.

SAUNDERS, R. M., BETSCHART, A. A., and KOHLER, G. O. 1975B. By-products utilization. Preparation of cereal protein concentrates. Bakers' Dig. *49*, No. 1, 49–52.

SECKINGER, H. L., and WOLF, M. J. 1973. Sorghum protein ultrastructure as it relates to composition. Cereal Chem. *50*, 455–465.

SHAW, M. 1970. Rye milling in the United States. Assoc. Operative Millers Bull., 3203–3207.

SHOUP, F. K., DEYOE, C. W., CAMPBELL, J., and PARRISH, D. B. 1969. Amino acid composition and nutritional value of milled sorghum. Cereal Chem. *46*, 164–171.

STRINGFELLOW, A. C., WALL, J. S., DONALDSON, G. L., and ANDERSON, R. A. 1976. Protein and amino acid compositions of dry-milled and air-classified fractions of triticale grain. Cereal Chem. *53*, No. 1, 51–60.

TECSON, E. M. S., ESMAMA, B. V., LONTOK, L. P., and JULIANO, B. O. 1971. Studies on the extraction and composition of rice endosperm glutelin and prolamin. Cereal Chem. *48*, 168–181.

TSEN, C. C., MOJIBIAN, C. N., and INGLETT, G. E. 1974. Defatted corn germ flour as a nutrient fortifier for bread. Cereal Chem. *51*, 262–271.

TSEN, C. C., HOOVER, W. J., FARRELL, E. P. 1973. Baking quality of triticale flours. Cereal Chem. *50*, 16–26.

USDA. 1975. Agricultural Statistics. U.S. Government Printing Office, Washington, D.C.

VIRUPAKSHA, T. K., and SASTRY, L. V. S. 1968. Studies on the protein content and amino acid composition of some varieties of grain sorghum. J. Agric. Food Chem. *16*, 199–203.

WOERMAN, J. H., and SATTERLEE, L. D. 1974. Extraction and nutritive quality of wheat protein concentrate. Food Technol. *28*, 50–52.

WRENSHALL, C. L. 1974. Research on rice bran utilization in Thailand. Int. Conf. on Rice By-products Utilization, Valencia, Spain.

WU, Y. V., CLUSKEY, J. E., WALL, J. S., and INGLETT, G. E. 1973. Oat protein concentrates from a wet-milling process: Composition and properties. Cereal Chem. *50*, 481–488.

WU, Y. V., and INGLETT, G. E. 1974. Denaturation of plant proteins related to functionality and food applications. J. Food Sci. *39*, 218–225.

WU, Y. V., SEXSON, K. R., CAVINS, J. F., and INGLETT, G. E. 1972. Oats and their dry-milled fractions: Protein isolation and properties of four varieties. J. Agric. Food Chem. *20*, 757–761.

WU, Y. V., and SEXSON, K. R., and CLUSKEY, J. E. 1974. Wet milling of wheat and oats by alkaline extraction. Proceedings of the 8th National Conference on Wheat Utilization Research, Denver. Colo., Oct. 10–12, 1973.

WU, Y. V., SEXSON, K. R., and WALL, J. S. 1977. Triticale protein concentrate: Preparation, composition, and properties. J. Agric. Food Chem. (in press).

WU, Y. V., and STRINGFELLOW, A. C. 1973. Protein concentrates from oat flours by air classification of normal and high-protein varieties. Cereal Chem. 50, 489–496.

10

Legumes: Seed Composition and Structure, Processing into Protein Products and Protein Properties

W. J. Wolf

The family Leguminosae includes more than 12,000 species in about 500 genera that are widely distributed from the tropics to the arctic regions. Although abundant in the temperate zones, legumes are especially numerous in the tropics. Man has cultivated some members of this family since remote antiquity and few families of dicotyledons are of greater importance than the legumes. Nonetheless, only a few legumes are of economic significance either as feeds or foods. Clovers, vetches, and alfalfa are among our most prominent forage crops, but of even more importance are peas, beans, lentils, peanuts, and soybeans that are grown as food plants.

Peas, beans, and lentils are consumed directly as foods in many parts of the world and are important sources of dietary protein in countries where animal proteins are scarce and expensive or are not consumed for religious or cultural reasons. Few of these legumes are processed industrially to concentrate the protein fraction. In Thailand, the mung bean (*Phaseolus aureus*) is processed for its starch for conversion into a spaghetti-like "bean thread" which is transparent rather than opaque when cooked. The protein-rich residue has been used to prepare a textured product. In England, the broad bean (*Vicia faba*) has been fractionated on a pilot scale into protein isolates which in turn were spun into fibrous meat analogs. Field peas and broad beans are being developed as new sources of protein for food and feed used in Western Canada. Studies on flours, protein concentrates, and isolates from these two legumes have been reported (Fleming *et al.* 1975).

Large quantities of peanuts are consumed directly as well as processed into oil and meal. In the United States, for example, about one-half of the crop is used domestically as salted peanuts, peanut candy, and peanut butter; the latter accounts for about 50% of the edible uses. The remainder of the crop is either crushed for oil or exported. The high-protein meal obtained as a by-product in oil crushing goes largely into animal feeds. Edible grade defatted peanut flours and grits became available in the United States in the early 1970's.

Soybeans are rarely eaten in the United States, but have been processed into comparatively simple food products for many centuries in the Orient. Processing includes water extraction (soy milk) with coagulation by a calcium salt (tofu), roasting (kinako), and fermentation (miso, tempeh, and soy sauce). The bulk of the world soybean crop is processed into edible oil and soybean meal. As with peanut meal, the bulk of defatted soybean meal goes into animal feeds, but increasing amounts are being converted into edible flours and grits, protein concentrates, and protein isolates. Of all the legumes, soybean proteins have reached the highest degree of refinement and extent of development and are added to a wide variety of processed foods (Wolf and Cowan 1975).

Because of their growing importance as food ingredients, proteins from several legumes are receiving increased attention from biochemists and food scientists who are gradually unraveling their complex properties and trying to relate them to problems encountered in food applications. Only selected topics are reviewed here. Other sources should be consulted for more comprehensive information on production, composition, and processing (Deschamps 1958; Smith and Circle 1972; Wolf 1975); food uses, and nutritional properties (Bressani and Elias 1974); and biochemical, physical, and chemical properties of the proteins (Derbyshire *et al.* 1976; Millerd 1975).

SEED COMPOSITION

Proximate analyses for a selection of edible legumes are given in Table 10.1. A distinguishing feature of legume seeds is their high protein content; most legumes fall into the range of 20–30% protein. Another common feature of many legumes is a low fat (2–5%) and high carbohydrate (55–60%) content. Starch is often a major

TABLE 10.1
COMPOSITION OF SEVERAL LEGUMES[1]

Legume	Protein[2] (%)	Fat (%)	Ash (%)	Fiber (%)	Carbohydrate[3] (%)	Reference
Chick pea (*Cicer arietinum*)	20.6	5.4	2.8	10.3	61	Verma *et al.* (1964)
Lentil (*Lens esculenta*)	29.6	3.1	2.4	3.2	62	Singh *et al.* (1968)
Pea (*Pisum sativum*)	27.9	3.2	2.8	5.9	60	Singh *et al.* (1968)
Broad bean (*Vicia faba*)	31.8	0.9	3.6	8.5	55	Marquardt *et al.* (1975)
Peanut (*Arachis hypogaea*)	30.0	50.0	3.1	3.0	14	Hoffpauir (1953)
Soybean (*Glycine max*)	43.9	21.0	4.9	—[4]	30[4]	Kawamura (1967)

[1] Moisture-free basis.
[2] Kjeldahl N $\times$ 6.25.
[3] Measured by difference.
[4] Fiber is included in carbohydrate value.

constituent of the carbohydrate fraction. Broad beans, for example, may contain as high as 40% starch (Bhatty 1974).

When compared to other legumes, soybeans and peanuts are anomalous in their compositions. Soybeans have high protein and oil content while peanuts are very high in oil. Because of their high oil contents, soybeans and peanuts are classified as oilseeds for purposes of commerce. Their high oil contents are counterbalanced by low carbohydrate contents. Soybeans, for example, contain little or no starch; the carbohydrates are primarily cell wall polysaccharides plus the oligosaccharides sucrose, raffinose, and stachyose.

In addition to having a high protein content, most legumes have a reasonably good balance of essential amino acids (Table 10.2). They are generally a good source of lysine. Although peanuts are low in this amino acid, a more common deficiency is methionine which is the first limiting amino acid in many legumes, including soybeans.

SEED STRUCTURE

Legume seeds have a characteristic structure consisting of three major parts: seed coat (hull), cotyledons, and hypocotyl (including plumule). Table 10.3 shows the distribution of these parts for several selected legumes. Compositions of the three seed fractions for soybeans are given in Table 10.4. Similar studies on the distribution of nutrients in anatomical parts of other legumes are available (Lal *et al.* 1963; Singh *et al.* 1968). In many of the legumes, the cotyledons constitute about 90% of the seed. The seed coat is the next largest fraction, but because it is high in fiber, it contains an insignificant part of the total food value of the whole seed. In the preparation of

TABLE 10.2

ESSENTIAL AMINO ACID CONTENTS OF SEVERAL LEGUMES

	Chick Pea[1] (*Cicer arietinum*)	Lentil[2] (*Lens esculenta*)	Broad Bean[3] (*Vicia faba*)	Pea[4] (*Pisum sativum*)	Soybean[5] (*Glycine max*)	Peanut[6] (*Arachis hypogaea*)
Protein content, %	17.5	24.1	31.8	31.6	61.4	56.9
Amino acids			g/16 gN			
Arginine	7.98	8.45	10.6	9.2	8.42	—[7]
Histidine	2.57	3.81	2.8	2.5	2.55	—[7]
Isoleucine	4.53	6.30	4.5	4.4	5.10	3.2
Leucine	7.63	10.9	7.7	7.4	7.72	5.9
Lysine	7.72	7.96	7.0	7.7	6.86	3.1
Methionine	1.16	0.70	0.6	1.3	1.56	0.9
Phenylalanine	6.46	6.25	4.3	4.9	5.01	3.8
Threonine	3.86	4.47	3.7	3.8	4.31	2.3
Tryptophan	1.78	1.22	—[7]	1.3	1.28	—[7]
Valine	4.63	5.42	5.2	4.9	5.38	4.1

[1] Shehata and Fryer (1970).
[2] Khan and Baker (1957).
[3] Marquardt *et al.* (1975).
[4] Data for dehulled pea flour (Anon. 1974).
[5] Data for defatted soy flour (Rackis *et al.* 1961).
[6] Data for defatted peanut (Virginia variety) flour on as-is moisture basis (Conkerton and Ory 1976).
[7] Not reported.

TABLE 10.3
DISTRIBUTION OF SEED PARTS IN LEGUMES

Legume	Proportion of Whole Seed		
	Seed Coat (%)	Cotyledons (%)	Hypocotyl (%)
Pea (*Pisum sativum*)[1]	10.0	89.3	1.3
Lentil (*Lens esculenta*)[1]	8.1	90.0	2.0
French bean (*Phaseolus vulgaris*)[1]	8.6	90.4	1.0
Mung bean (*Phaseolus aureus*)[1]	12.1	85.6	2.3
Soybean (*Glycine max*)[2]	7.3	90.3	2.4

[1] Singh *et al.* (1968).
[2] Kawamura (1967).

TABLE 10.4
PROXIMATE COMPOSITIONS OF SOYBEANS AND SEED PARTS[1]

Fraction	Protein (N × 5.71) (%)	Fat (%)	Carbohydrate (%)	Ash (%)
Whole bean (100%)	40	21	34	5
Seed coat (8%)	9	1	86	4
Cotyledon (90%)	43	23	29	5
Hypocotyl (2%)	41	11	44	4

Source: Kawamura (1967).
[1] Moisture-free basis.

edible soy flours and grits, the hulls (seed coat) are removed and standards for defatted soy flour permit a maximum of 3.5% fiber. Undehulled, defatted soybean meal contains about 6% fiber.

Legume seeds are highly organized structures when examined at the subcellular level. The major constituents, proteins, lipids, and starch (when present) are neatly packaged as exemplified for the soybean (Fig. 10.1). Protein storage sites are called protein bodies and in soybeans vary from 2–20 μm in diameter, although many are about 5–8 μm. The lipid deposits are called spherosomes and in soybeans are about 0.2–0.5 μm in diameter. Peanuts contain protein bodies 2–10 μm in diameter along with occasional starch grains ranging from 5–10 μm in size (Bagley *et al.* 1963). Spherosomes in peanuts are about five times (1–2 μm) as large as those in soybeans (Jacks *et al.* 1967).

In the nonoil-bearing legumes, the prominent subcellar structures are protein bodies and starch granules. For example, in peas the protein bodies average 2 μm in diameter and the starch granules are egg shaped with lengths up to 15 μm (Varner and Schidlovsky 1963).

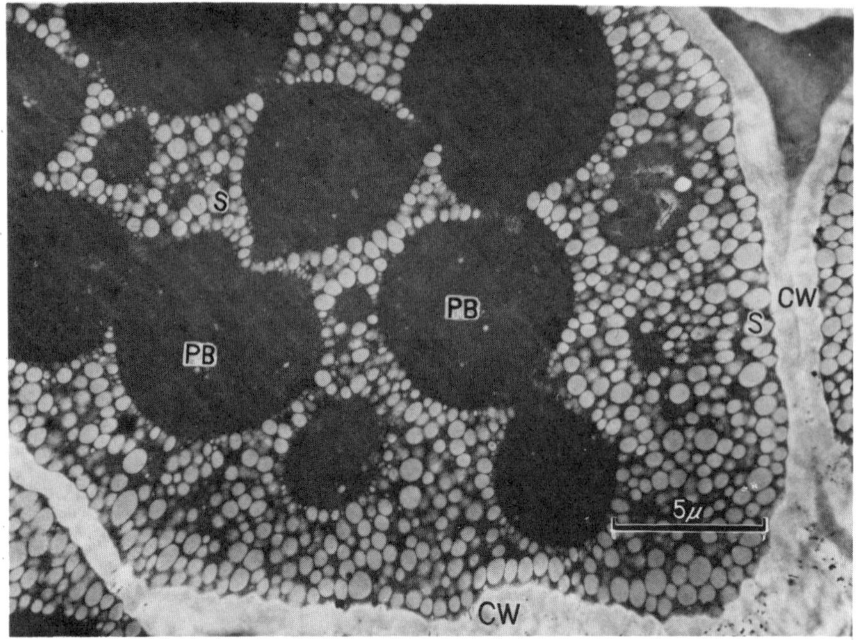

From Saio and Watanabe (1968)

FIG. 10.1. ELECTRON MICROGRAPH OF A SECTION OF A MATURE SOYBEAN COTYLEDON

Seed was soaked in water overnight, fixed with osmium tetroxide, and stained with uranyl acetate and lead citrate. Structures identified are protein bodies (PB), spherosomes (S), and cell wall (CW).

Starch granules are very prominent in broad beans (*Vicia faba*) and range from 10–40 μm in length to 10–25 μm in width; protein bodies in this legume are about 1–5 μm in diameter (McEwen *et al.* 1974). Cotyledon cells of lima beans (*Phaseolus lunatus*) likewise are very high in starch, and granules averaging about 25 μm in diameter are closely packed into the cells (Rockland and Jones 1974).

PROCESSING INTO PROTEIN PRODUCTS

Conversion of legumes into protein products suitable for incorporation into processed foods involves disruption of the highly ordered seed structure (Fig. 10.1) and fractionation by physical methods. These methods may be as simple as screening and aspiration to remove seedcoats or as complex as extraction of defatted flakes with aqueous solutions to dissolve proteins. In simple processing, some of original structures of the seed are detectable in the final products (Wolf and Baker 1975). Processing of soybeans has reached the highest degree of development, but the same basic principles have been applied to other legumes on laboratory and pilot-plant scales.

TABLE 10.5
PROXIMATE ANALYSES AND BIOCHEMICAL PARAMETERS OF SOY
FLOURS, PROTEIN CONCENTRATES, AND PROTEIN ISOLATES

	Full-Fat Flour[1]	Toasted Defatted Flour[2]	Protein Concentrate[2,3]	Protein Isolate[2]
Moisture, %	3.4	6.5	8.0	4.8
Protein (N × 6.25), %	41.0	53.0	65.3	92.0
Crude fat, %	22.5	1.0	0.3	—
Crude fiber, %	1.7	3.0	2.9	0.25
Ash, %	5.1	6.0	4.7	4.0
PER[4]	2.15	2.3	2.3	1.1–1.6
Inactivation of trypsin inhibitors, %	89	—	—	—
Urease activity, pH change	0.1	—	—	—
Nitrogen solubility index	16	15–25	5	75

[1]Experimental sample (Mustakas *et al.* 1970).
[2]Tech. Serv. Manual, Central Soya Co., Chicago.
[3]Prepared by extraction with aqueous alcohol.
[4]Protein efficiency ratio corrected to casein = 2.50.

Full-Fat Soy Flours

These products are made on an industrial scale in the United States and England. Beans are cleaned, steamed, dried, cracked between corrugated rolls, dehulled by aspiration and screening, ground, and sieved (Pringle 1974). Alternatively, the beans may be cracked and dehulled prior to steaming. The steam treatment inactivates antinutritional factors plus enzymes such as lipoxygenases that can catalyze oxidation of lipids and thereby develop undesirable flavors (Mustakas *et al.* 1969).

An alternate process for preparing full-fat soy flours involves extrusion cooking. Beans are cracked, dehulled, heated, raised to 18% moisture, extruded (121–143°C), cooled, and ground (Mustakas *et al.* 1970).

Full-fat soy flours prepared by either method have a composition (Table 10.5) closely resembling that of whole soybeans because only the hulls (seedcoats) are removed during processing. Heat treatment during processing decreases solubility of the proteins as measured by the nitrogen solubility index and inactivates the bulk of the trypsin inhibitors. The protein efficiency ratio is about 90% of the value for the casein standard.

Full-Fat Peanut Flakes

Pilot-plant studies on preparation of full-fat peanut flakes were announced recently (Anon. 1976A). Developed at Clemson Uni-

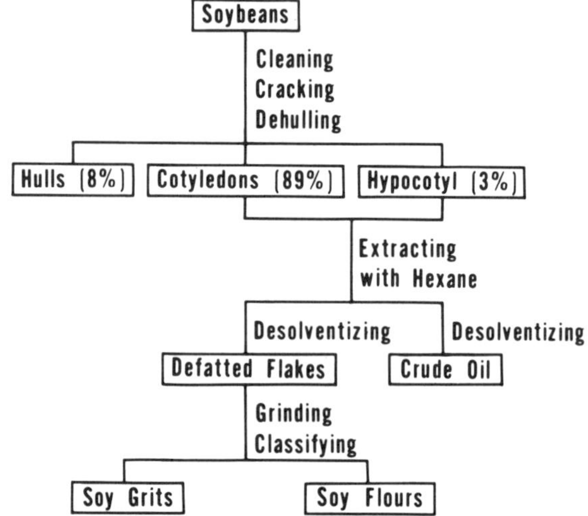

FIG. 10.2. OUTLINE OF COMMERCIAL PROCESSING OF SOYBEANS INTO OIL AND DEFATTED FLOURS AND GRITS

versity, the process has been licensed for commercialization. The peanuts are steamed, dried to 2–3% moisture, ground, precooked with water, and finally drum-dried to a bland, light-colored flake (Mitchell 1974).

Defatted Soy Flours and Grits

These products are the major form of legume proteins used today as food ingredients. Commercial processing is outlined in Fig. 10.2. Soybeans are cleaned, cracked, and dehulled. The dehulled fraction is then conditioned by mild steaming, flaked, and extracted with hexane to remove the oil. Hexane is removed by desolventizing under conditions where the amount of moist heat treatment given to the flakes is controlled according to the desired end use (Becker 1971). The final step is grinding and classifying into grits (particles>100 mesh) and flours (particles≤100 mesh).

By definition defatted soy flours and grits have a minimum protein content of 50%, but often they will analyze higher (Table 10.5).

Defatted Peanut Flours and Grits

Edible peanut flours and grits are manufactured by one U.S. company by a combined prepress-solvent extraction method (Ayres *et al.* 1974). In the process (Fig. 10.3), peanuts are hammer milled, adjusted to 10% moisture, and cooked. The cooked peanuts are then fed to an expeller to press out the bulk of the oil and obtain a cake containing 8–12% oil. The press cake is ground, conditioned to 10% moisture, flaked, and extracted with hexane. After desolventizing,

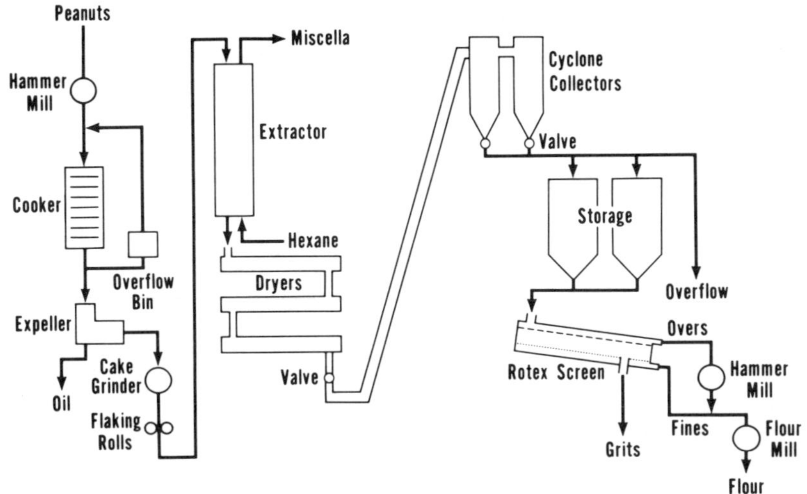

From Ayres et al. (1974)

FIG. 10.3. OUTLINE OF COMMERCIAL PROCESSING OF PEANUTS INTO EDIBLE FLOURS AND GRITS

the flakes are screened to remove grits (16–60 mesh) and the fines are milled into a flour. The resulting flour has a crude protein content of 57%.

A pilot study has demonstrated that the prepressing step in conventional handling of peanuts (Fig. 10.3) can be eliminated by cutting the peanuts into thin slivers (Anon. 1976B). The thin slivers permit direct extraction of the oil as currently carried out on soybeans. Because the cooking and expelling step is avoided, defatted flakes or flours with high protein solubilities can be prepared.

A third alternative for making defatted peanut products consists of coarsely grinding skinned peanuts, mixing with water, homogenizing, and then drum-drying to form flakes. Extraction with a fat solvent yields defatted flakes (Mitchell 1976).

Pea Flours

Pilot studies on conversion of field peas into flours have been conducted by the Prairie Regional Laboratory in Saskatoon, Saskatchewan. These studies led to building of a pilot plant by Newfield Seeds Ltd. in Nipawin, Saskatchewan, and a commercial-scale plant is scheduled for completion in Saskatoon in the fall of 1976 (Anon. 1975). Whole or dehulled peas are pin-milled to yield a pale golden flour containing almost 32% protein on a moisture-free basis when dehulled peas are used. Product characteristics, composition, and functional and nutritional properties have been summarized (Anon. 1974; Youngs 1975). The flour has potential use as an ingredient for

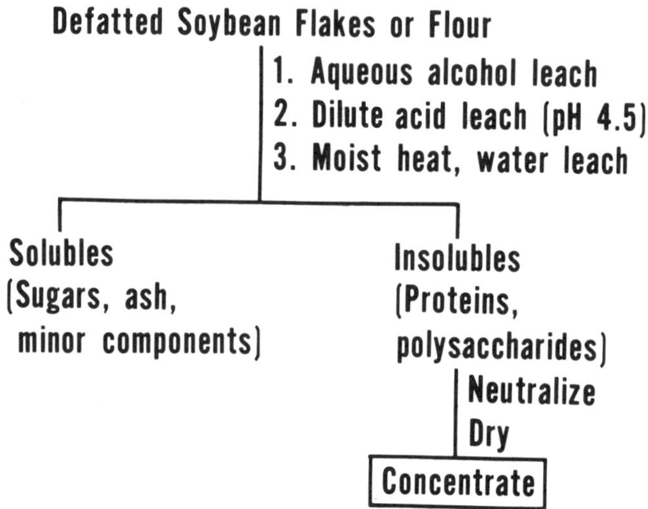

FIG. 10.4. OUTLINE OF PROCESSES FOR MANUFACTURING SOYBEAN PROTEIN CONCENTRATES

Initial extraction is made with one of the three solvent systems listed as described in the text.

snack foods, soup mixes and baby foods, but it will be primarily converted into protein concentrates as described below.

Soy Protein Concentrates

Removal of the soluble sugars (sucrose, raffinose, and stachyose) and other low-molecular-weight components from defatted soy flakes or flours raises the protein content to 70% or higher. The resulting products are called protein concentrates. Protein concentrates are made by three processes as outlined in Fig. 10.4. The first process employs aqueous alcohol to dissolve the sugars and nonprotein constituents, leaving the insoluble proteins and polysaccharides that constitute the protein concentrate after desolventizing (Mustakas *et al.* 1962). The second process uses a dilute acid leach at pH 4.5 to remove the soluble sugars. At this pH the globulins that make up the bulk of the proteins are at their isoelectric point and insoluble. After neutralizing and drying, the second type of concentrate is obtained (Sair 1959). In the third process, defatted flakes or flours are steamed to heat-denature and insolubilize the proteins. Washing with water then removes the sugars, and the denatured protein-polysaccharide residue is dried to yield the third type of protein concentrate (McAnelly 1964). A fourth process developed recently is a variation of the first. When the hexane-wet flakes leave the extractor in the usual hexane extraction (Fig. 10.2), they are mixed with alcohol. The hexane-alcohol mixture is then separated to remove residual lipids and flavor components not removed by hexane alone. Residual hexane is then removed by selective desolventizing, and a final wash with aqueous alcohol yields a protein concentrate (Hayes and Simms 1973).

Concentrates made by the three processes have similar chemical compositions. Proximate analyses for a concentrate made by the aqueous alcohol extraction method are given in Table 10.5. The protein efficiency ratio of this concentrate is the same as that for flours, and protein solubility is low because of alcohol and heat denaturation resulting from processing.

Pea Protein Concentrates

The Prairie Regional Laboratory has extended its pilot studies on pea flours to the preparation of protein concentrates by two methods (Anon. 1974). In the first method, pin-milled pea flour (100 lb) is air classified to yield a starch fraction (65 lb) containing 2.5% protein and concentrate (35 lb) consisting of 56% protein. This method is being commercialized at present (Anon. 1975). In the second process, pea flour is slurried in lime water to give a pH of 9 and centrifuged. The resulting supernatant is spray or drum dried to yield one-third of the starting flour as a concentrate (60% protein). The starch fraction from the centrifuge is reslurried to wash it and again centrifuged. The starch solids (two-thirds of starting flour) containing 1.8% protein are then dried. Compositions, functional properties, nutritional properties, and product applications were evaluated (Anon. 1974; Bell and Youngs 1970; Youngs 1975).

Pea protein concentrate made by the aqueous process described above differs from soy protein concentrates in the nature of the carbohydrates present. In the pea concentrate, the insoluble polysaccharides (starch plus cell wall material) are removed and the soluble sugars retained, whereas in soy concentrates the insoluble polysaccharides are retained and the soluble sugars removed.

Peanut Protein Concentrates

An aqueous processing method developed on a pilot scale at Texas A&M University starts with blanched (skinned) peanuts which are ground and then dispersed in dilute acid to give a pH of 4.0. The acid-precipitated proteins plus insoluble polysaccharides are removed by screening and on drying yield a protein concentrate. The acid-soluble sugars (whey) and oil are separated by centrifuging (Rhee *et al.* 1973).

Soy Protein Isolates

These protein forms are made (Fig. 10.5) by extracting defatted flakes or flour having a high protein solubility [desolventized with a

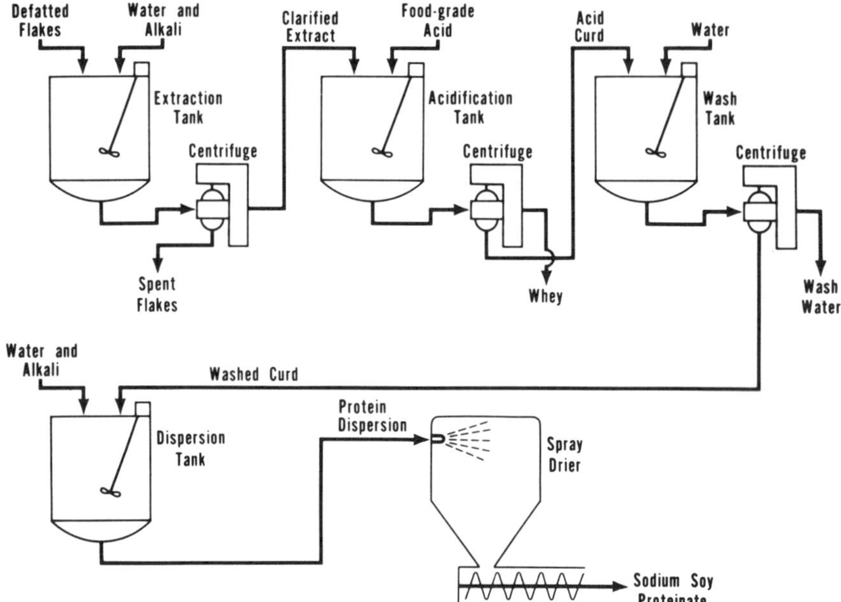

FIG. 10.5. OUTLINE OF PROCESS FOR COMMERCIAL PRODUCTION OF SOYBEAN PROTEIN ISOLATES

minimum of moist heat treatment as described by Becker (1971)] with dilute alkali (pH 7–9). The spent flakes (water-insoluble polysaccharides plus residual protein) are centrifuged off and resulting extract containing the soluble proteins and sugars is then acidified to pH 4.5. This is the isoelectric point for the bulk of the proteins and the pH of minimum solubility, hence the proteins precipitate. Centrifuging separates the whey (soluble sugars, some proteins, peptides, salts, and minor constituents) from the protein curd which is then reslurried with water to wash it. The washed curd can be concentrated and dried to yield the isoelectric (water insoluble) form of isolate. Usually the washed curd is resolubilized by neutralizing and then spray dried. This results in the proteinate form of isolate that is often preferred because of its water dispersibility. Proximate analyses for a commercial soy protein isolate are given in Table 10.5. Protein efficiency ratios for isolates are lower than for flours because of loss of essential amino acids in the proteins removed in the whey, residual antinutritional factors, and possibly low digestibility (Meyer 1971).

Broad Bean Protein Isolates

Pilot-scale preparations of isolates from broad beans (*Vicia faba*) are described in the literature (Flink and Christiansen 1973; Guinat

1969). Flink and Christiansen extracted bean flour at pH 8–10, centrifuged to remove starch and other insoluble materials, and then recovered the protein by precipitation at pH 3.5. Guinat extracted a flour with $NaHSO_3$ solution at pH 7.4 and screened off the insoluble residue. The resulting protein extract and suspended starch were then separated in a hydroclone. The clarified extract was acidified to pH 4.5 and centrifuged to isolate the protein.

Several years ago, isolates from *Vicia faba* were made in a commercial pilot plant by Rank, Hovis, McDougal, Ltd., in England. A secondary processor spun the isolates into a protein fiber that was converted into a meat analog. The process was technically feasible, but soy protein isolates are now used instead for the protein fibers because of uncertainty of an economical supply of broad bean (Smith 1976).

Peanut Protein Isolates

The aqueous method for processing peanuts developed at Texas A&M University can be conducted to yield isolates instead of protein concentrates as described earlier. Ground peanuts are dispersed in dilute alkali to give a final pH of 8.0. Screening removes the insoluble polysaccharide residue from the aqueous extract which is then acidified to precipitate the peanut globulins. Screening separates the protein curd, and centrifuging the aqueous phase separates the oil and whey fractions. Pilot-scale studies are available (Rhee *et al.* 1972).

PHYSICAL AND CHEMICAL PROPERTIES OF THE PROTEINS

Solubility

The bulk of the proteins in legume seeds are globulins which are characterized by being insoluble in water at their isoelectric points but are soluble in the presence of salts. The proteins may, however, be highly soluble in water if the pH is sufficiently removed from the isoelectric pH. This behavior is best illustrated by the classical studies of Smith and Circle (1938) on extractability of soybean meal proteins in water as a function of pH (Fig. 10.6). In water, the meal slurry gives a pH of 6.5–6.8 and over 80% of the proteins dissolve. If the pH is lowered progressively, less protein is dissolved and extractability reaches a minimum in the range pH 4–5, which is the isoelectric region for the bulk of the proteins. This is the basic principle employed in the preparation of soy protein isolates: extractions are made at pH 7–9 and the proteins are then recovered by acidifying the extracts to pH 4.5.

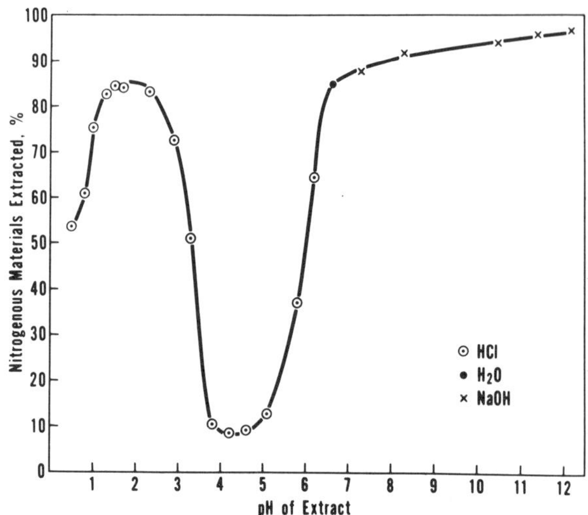

From Smith and Circle (1938)

FIG. 10.6. EXTRACTABILITY OF DEFATTED SOYBEAN
MEAL PROTEINS AS A FUNCTION OF pH

Solubility-pH relationships similar to Fig. 10.6 have been reported for a large number of legumes (Ayres *et al.* 1974; Bhatty 1974; Fan and Sosulski 1974; Hang *et al.* 1970; Pusztai 1965).

The insolubility of soybean proteins at pH 4.5 can be overcome, at least partially, by adding sodium or calcium chloride to the extraction solvent (Fig. 10.7). Maximum solubility occurs with 0.3 N calcium chloride or 0.7 N sodium chloride. Higher concentrations of salts have no further effect and the remaining proteins apparently are denatured by the pH 4.5 treatment (Anderson *et al.* 1973).

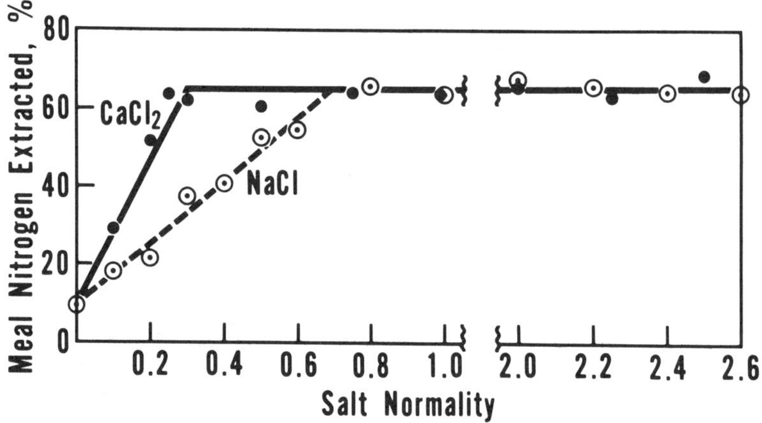

From Anderson et al. (1973)

FIG. 10.7. EXTRACTABILITY OF DEFATTED SOYBEAN MEAL PROTEINS
AT pH 4.5 AS A FUNCTION OF SALT CONCENTRATION

Molecular Size and Distribution of Proteins

Nearly 30 years ago, Danielsson (1949) examined the globulins of peas in the ultracentrifuge and found two major fractions that he designated vicilin and legumin. Vicilin sedimented as a 7S entity and legumin had a sedimentation coefficient of about 12S. He was able to separate the two by isoelectric precipitation at pH 4.5 in the presence of 0.2 *M* sodium chloride. Under these conditions, vicilin dissolved whereas legumin precipitated. Hence, it was simple to separate them on the basis of their differences in solubility at the isoelectric point. Molecular weights of the two proteins were estimated to be 186,000 (vicilin) and 331,000 (legumin).

In surveying the globulin fraction from 34 species of Leguminosae including peanuts (*Arachis hypogea*), soybeans (*Glycine max*), beans (*Phaseolus vulgaris*), and broad beans (*Vicia faba*), Danielsson found a remarkable similarity in the composition of the proteins. Practically all of the species contained counterparts of vicilin and legumin. These two fractions represent a large proportion of the total protein in a legume seed. For example, in *Vicia faba*, vicilin and legumin account for 90% of the protein found in the protein bodies (Bailey *et al.* 1970). A detailed review of the physical, chemical, and immunological properties of vicilin and legumin in various legumes appeared recently (Derbyshire *et al.* 1976). Further discussion of these protein fractions occurs below.

Protein Body Composition

Because of their high content in seeds and lack of demonstrated biological activity, vicilin and legumin are generally considered to be storage proteins and are probably packaged together in the protein bodies. In soybeans, the protein bodies consist almost exclusively of 7S and 11S fractions (Wolf 1970; Koshiyama 1972). Fig. 10.8A shows the ultracentrifuge pattern for total proteins extractable from defatted flour; 7S and 11S fractions make up about 70% of the total protein. The remainder of the proteins exists as 2S and 15S fractions. When the defatted flour is suspended in 20% sucrose at pH 5 and centrifuged in a sucrose density gradient (Tombs 1967), a homogenate fraction (proteins soluble at pH 5 and probably very similar to soybean whey proteins) and the protein bodies are obtained. The homogenate fraction (Fig. 10.8B) clearly consists of portions of the 2S and 7S fractions found in the total protein mixture. The protein bodies in turn contain a small amount of 2S protein, a high proportion of 7S protein, and all of the 11S protein found in soybean flour (Fig. 10.8C). Although the 7S protein found in unfractionated soybean protein is distributed between the

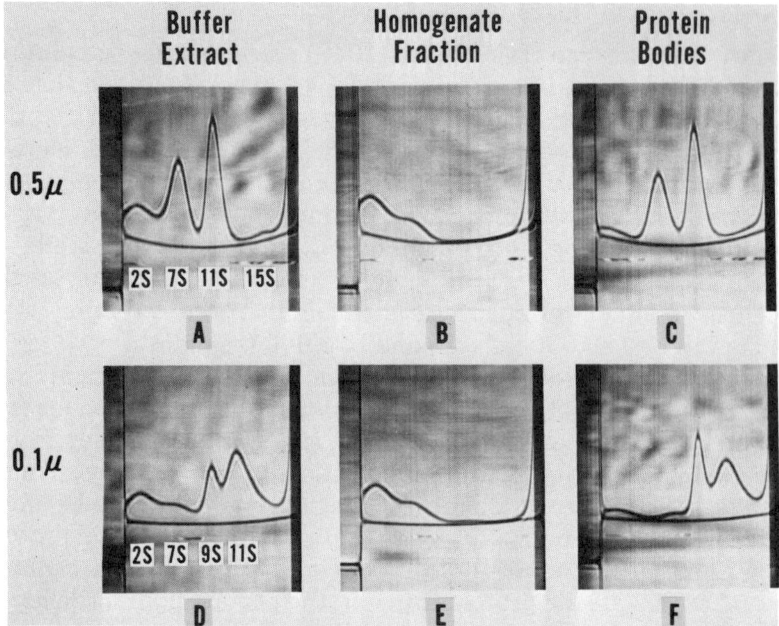

FIG. 10.8. ULTRACENTRIFUGE PATTERNS FOR UNFRACTIONATED SOY-
BEAN PROTEINS, HOMOGENATE FRACTION AND PROTEIN BODY
PROTEINS

The unfractionated proteins were obtained by extracting undenatured, defatted flour with pH 7.6, 0.5 ionic strength buffer containing 0.01 M mercaptoethanol and dialyzing against the same solvent. Homogenate fraction and protein bodies were prepared by sucrose density gradient centrifugation (Tombs 1967). Analyses were conducted at pH 7.6, 0.5 ionic strength (A-C), and 0.1 ionic strength (D-F).

homogenate fraction and the protein bodies, a significant separation occurred. Koshiyama (1968A) isolated a 7S globulin that exists as a monomer (mol wt 180,000–210,000) at pH 7.6, 0.5 ionic strength, and forms a dimer ($S_{20,w} \sim$ 9S, mol wt 370,000) at 0.1 ionic strength. This property is readily demonstrated even in the un-fractionated mixture by analyzing the sample at 0.1 ionic strength (Fig. 10.8D); about two-thirds of the total 7S fraction dimerizes. Under the same conditions, the homogenate fraction shows the formation of only a barely detectable amount of dimer (Fig. 10.8E). In contrast, almost all of the 7S material in the protein bodies at 0.5 ionic strength (Fig. 10.8C) was converted into the dimer form at 0.1 ionic strength (Fig. 10.8F). Obviously the protein bodies contain predominantly the 7S fraction that is capable of forming a dimer at low ionic strength and the 11S protein.

7S Globulins

Although vicilin-like proteins occur in a large number of legumes (Derbyshire *et al.* 1976), comparatively few have been purified and

TABLE 10.6
7S GLOBULINS FROM LEGUMES

Legumes	Sedimentation Coefficient	Molecular Weight	References
Peanuts (*Arachis hypogaea*)	8.7 (S_{20})	190,000	Johnson and Naismith (1953)
Soybean (*Glycine max*)	7.9 ($S_{20,w}^{o}$)	193,000	Koshiyama (1968B)
Beans (*Phaseolus vulgaris*)	7.6 ($S_{20,w}^{o}$)	140,000	Pusztai and Watt (1970)
Broad bean (*Vicia faba*)	7.1 ($S_{20,w}$)	150,000	Wright and Boulter (1973)

characterized in detail. Table 10.6 lists four 7S globulins for which molecular weights are reported. Molecular weights for these four 7S globulins range from 140,000–190,000 and many 7S globulins fall into this range. The preparations from soybeans and beans appear to be glycoproteins. These proteins have subunit structures which can be disrupted by various means. For example, the 7S globulin from soybeans has nine N-terminal residues and in 8 *M* urea or 4 *M* guanidine hydrochloride dissociates into subunits with a molecular weight of 22,500–24,000 in accord with the presence of nine subunits (Koshiyama 1971). Not all of the 7S globulins possess the ability to associate at low ionic strength as observed for this protein from soybeans (Fig. 10.8D and 10.8F).

Recent work indicates the 7S globulin fraction in soybeans is much more complex than previously believed. Five components were isolated by chromatography on DEAE-Sephadex A-50 (Thanh *et al.* 1975). There are probably only subtle differences between them because all five components appear to be glycoproteins and dimerize at pH 7.6, 0.1 ionic strength. Likely, other legumes will exhibit similar heterogeneity of the 7S fraction when they are examined in sufficient detail.

11S Globulins

The legumin-like fraction of legumes appears less heterogeneous than the vicilin-like proteins and more representatives of this fraction have been isolated and characterized. Many of them have molecular weights of 300,000–400,000 (Table 10.7), and it has long

TABLE 10.7
11S GLOBULINS FROM LEGUMES

Legume	Sedimentation Coefficient	Molecular Weight	Reference
Peanuts (*Arachis hypogaea*)	13.2 ($S_{20,w}^{o}$)	340,000	Brand *et al.* (1955)
Soybeans (*Glycine max*)	12.3 ($S_{20,w}^{o}$)	320,000	Badley *et al.* (1975)
Beans (*Phaseolus vulgaris*)	11.6 ($S_{20,w}$)	340,000	Derbyshire *et al.* (1976)
Peas (*Pisum sativum*)	13.1 (S_{20}^{o})	398,000	Johnson and Richards (1962)
Broad beans (*Vicia faba*)	11.4 ($S_{20,w}^{o}$)	328,000	Derbyshire *et al.* (1976)

been recognized that these complex molecules are made up of subunits. A surprising feature observed in several 11S proteins to date is the presence of acidic and basic subunits. This structural detail has been found in the 11S or legumin fractions of *Vicia faba* (Wright and Boulter 1974), *Vicia sativa* (Vaintraub and Tuen 1971), and *Glycine max* (Catsimpoolas *et al.* 1971).

Glycinin, the 11S protein of soybeans, contains three acidic and three basic subunits with isoelectric points as follows (Catsimpoolas 1969):

Subunits	**pI**
Acidic	4.75, 5.15, 5.40
Basic	8.00, 8.25, 8.50

Each of these subunits apparently occurs twice in the 11S molecule because Badley *et al.* (1975) found 12 N-terminal residues per 320,000 g consisting of: 6 glycines, 2 leucines, 2 isoleucines, and 2 phenylalanines. The basic subunits terminate with glycine whereas the other N-terminal residues are found in the acidic subunits.

Electron microscopy using negative staining revealed that the 11S molecule consists of two hexagonal rings stacked on top of each other with six subunits in each ring (Fig. 10.9). When viewed from the top, there appears to be a hole in the center. Side views show six (view from D) or eight subunits (view from C) depending on the viewing position. Catsimpoolas (1969) and Badley *et al.* (1975) proposed that the acidic (A) and basic (B) subunits alternate in the

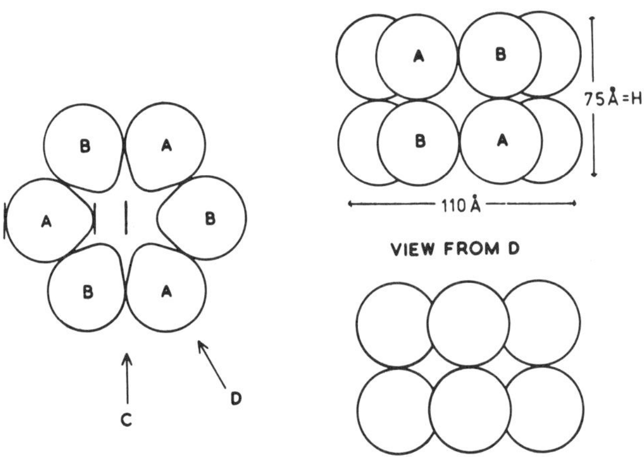

From Badley et al. (1975)

FIG. 10.9. MODEL OF SUBUNIT ARRANGEMENT IN SOYBEAN 11S PROTEIN

Left is top view and right shows side views from positions C and D as indicated by arrows on left.

hexagonal ring arrangement as well as in the two-layered structure (Fig. 10.9).

Small angle X-ray scattering measurements combined with electron microscopy indicate that the hollow oblate cylinder model for the glycinin molecule measures 110 × 110 × 75 Å. Pea legumin likewise appears to be a hollow cylinder; it is 85 Å in diameter and 85 Å high (Valentine 1959).

Although the quaternary structures of the 11S globulins of legumes are comparatively stable, it is possible to disrupt them stepwise:

$$11S\ (A_6 B_6) \quad \rightarrow \sim 7S(2A_3 B_3) \quad \rightarrow \sim 3S(6AB) \quad \rightarrow \sim 2S(6A + 6B)$$

Designations within parentheses indicate the extent of subunit interaction in terms of six acidic (A) and six basic (B) subunits as typified by glycinin. The conditions under which the various reactions occur vary widely depending on the particular protein. All of the interactions are noncovalent with the possible exception of the $\sim$3S form where disulfide bonds have been implicated (Derbyshire *et al.* 1976; Badley *et al.* 1975).

Primary, Secondary and Tertiary Structures

Broad outlines of the quaternary structures of the major globulins of several legumes are emerging as just described. However, corresponding information about the primary, secondary, and tertiary structures of most legume proteins is very meager. In soybeans, for example, the primary structures are known for just two proteins: Bowman-Birk trypsin inhibitor (Odani and Ikenaka 1973) and Kunitz trypsin inhibitor (Koide and Ikenaka 1973). Although of physiological importance as antinutritional factors in their native state, both of these proteins are readily inactivated by moist heat and are present in small amounts as compared to the major globulins.

Significant details of the secondary and tertiary structures of soybean proteins are known only for the Kunitz trypsin inhibitor as a result of X-ray crystallographic studies of the crystalline complex between porcine trypsin and the inhibitor (Sweet *et al.* 1974). The X-ray studies confirmed previous optical rotatory dispersion and circular dichroism studies (Ikeda *et al.* 1968) that indicated the absence of α-helical structures. The molecule approximates a sphere of about 35 Å in diameter made up of crisscrossing loops wrapped around a core of hydrophobic side chains (Fig. 10.10). The polypeptide chain is folded in approximate β-sheet structures with little regular sheet formation.

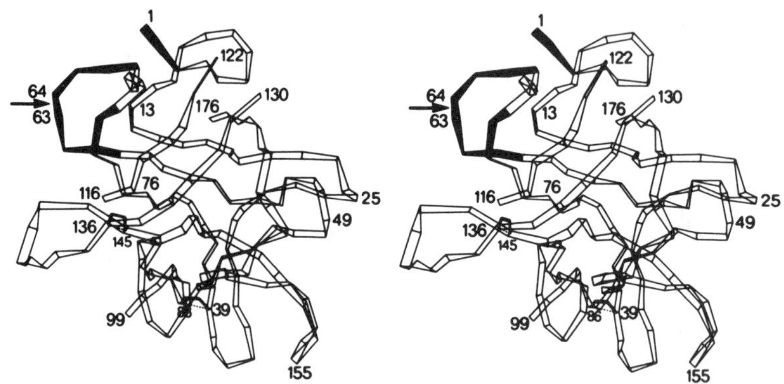

From Sweet et al. (1974)

FIG. 10.10. STEREO PAIR OF MODEL OF POLYPEPTIDE CHAIN FOLDING
IN KUNITZ SOYBEAN TRYPSIN INHIBITOR

A more complex protein, whose structure may be indicative of the major legume globulins having subunit structures, is concanavalin A, the agglutinin found in jack beans (*Canavalia ensiformis*). Below pH 6 concanavalin A exists as a dimer (2 identical subunits of 25,500 mol wt) and above pH 7 as a tetramer. X-ray crystallography and amino acid sequence determinations have yielded its structure at 2-Å resolution. The predominant structural elements are β-pleated sheets and random coils; no α-helices are present. The sites of interactions between subunits to form the dimers consist of β-pleated sheets, whereas side chain interactions contribute to the formation of the tetramers from two dimers (Edelman *et al.* 1972; Hardman and Ainsworth 1972).

Optical rotatory dispersion studies of soybean 11S globulin indicate an α-helical content of 5% and β-structure of 35% (Koshiyama 1972). Circular dichroism measurements suggest 9% α-helix, 33% β-pleated sheet, and 58% unordered structure for the 11S globulin; similar values were obtained for α-conarachin from peanuts (Jacks *et al.* 1973). Absence of appreciable amounts of α-helical structure in soybean and peanut globulins suggests that the gross structural features of Kunitz trypsin inhibitor and concanavalin A may be representative of the legume globulins.

CONCLUSIONS

Much has been learned about the properties of legume seed proteins, but clearly we have only scratched the surface. A few of the major globulins have been purified to a high degree and methods are being developed to separate the individual subunits. The next logical step is the determination of the amino acid sequences. Ultimately, the proteins must also be crystallized to permit an unequivocable determination of their three-dimensional structures by X-ray crystallography. These are obviously formidable tasks. Nonetheless, such information should be invaluable to the food scientist in relating structure to functional properties and to the plant scientist confronted with the task of understanding the biosynthesis of specific seed proteins and their control mechanisms in order to improve the protein quality of legumes (Millerd 1975).

BIBLIOGRAPHY

ANDERSON, R. L., WOLF, W. J., and GLOVER, D. 1973. Extraction of soybean meal proteins with salt solutions at pH 4.5. J. Agric. Food Chem. *21*, 251–254.

ANON. 1974. Pea flour and pea protein concentrate. PFPS Bull. No. 1, Prairie Regional Laboratory, National Research Council, and College of Home Economics, University of Saskatchewan, Saskatoon, Saskatchewan.

ANON. 1975. Field peas yield quality protein. Can. Chem. Process. *59*, No. 12, 16–18.

ANON. 1976A. Peanut flakes duplicate texture/taste of egg, meat and dairy products. Food Process. *37*, No. 1, 42–43.

ANON. 1976B. High protein peanut flour is white, highly functional bland flavored. Food Process. *37*, No. 1, 39–42.

AYRES, J. L., BRANSCOMB, L. L., and ROGERS, G. M. 1974. Processing of edible peanut flour and grits. J. Am. Oil Chem. Soc. *51*, 133–136.

BADLEY, R. A., ATKINSON, D., HAUSER, H., OLDANI, D., GREEN, J. P., and STUBBS, J. M. 1975. The structure, physical and chemical properties of the soybean protein glycinin. Biochim. Biophys. Acta *412*, 214–228.

BAGLEY, B. W., CHERRY, J. H., ROLLINS, M. L., and ALTSCHUL, A. M. 1963. A study of protein bodies during germination of peanut (*Arachis hypogaea*) seed. Am. J. Bot. *50*, 523–532.

BAILEY, C. J., COBB, A., and BOULTER, D. 1970. A cotyledon slice system for the electron autoradiographic study of the synthesis and intracellular transport of the seed storage protein of *Vicia faba*. Planta *95*, 103–118.

BECKER, K. W. 1971. Processing of oilseeds to meal and protein flakes. J. Am. Oil Chem. Soc. *48*, 299–304.

BELL, J. M., and YOUNGS, C. G. 1970. Studies with mice on the nutritional value of pea protein concentrate. Can. J. Anim. Sci. *50*, 219–226.

BHATTY, R. S. 1974. Chemical composition of some faba bean cultivars. Can. J. Plant Sci. *54*, 413–421.

BRAND, B. P., GORING, D. A. I., and JOHNSON, P. 1955. The attempted preparation of monodisperse seed globulins. Trans. Faraday Soc. *51*, 872–876.

BRESSANI, R., and ELIAS, L. G. 1974. Legume foods. *In* New Protein Foods, Vol. 1A, Technology, A. M. Altschul (Editor). Academic Press, New York.

CATSIMPOOLAS, N. 1969. Isolation of glycinin subunits by isoelectric focusing in urea-mercaptoethanol. FEBS Lett. *4*, 259–261.

CATSIMPOOLAS, N., KENNY, J. A., MEYER, E. W., and SZUHAJ, B. F. 1971. Molecular weight and amino acid composition of glycinin subunits. J. Sci. Food Agric. *22*, 448–450.

CONKERTON, E. J., and ORY, R. L. 1976. Peanut proteins as food supplements: A compositional study of selected Virginia and Spanish peanuts. J. Am. Oil Chem. Soc. *53*, 754–756.

DANIELSSON, C. E. 1949. Seed globulins of the Gramineae and Leguminosae. Biochem. J. *44*, 387–400.

DERBYSHIRE, E., WRIGHT, D. J., and BOULTER, D. 1976. Legumin and vicilin, storage proteins of legume seeds. Phytochemistry *15*, 3–24.

DESCHAMPS, I. 1958. Peas and beans. *In* Processed Plant Protein Foodstuffs, A. M. Altschul (Editor). Academic Press, New York.

EDLEMAN, G. M., CUNNINGHAM, B. A., REEKE, G. N., JR., BECKER, J. W., WAXDAL, M. J., and WANG, J. L. 1972. The covalent and three-dimensional structure of concanavalin A. Proc. Natl. Acad. Sci. U.S.A. *69*, 2580–2584.

FAN, T. Y., and SOSULSKI, F. W. 1974. Dispersibility and isolation of proteins from legume flours. Can. Inst. Food Sci. Technol. J. *7*, 256–259.

FLEMING, S. E., SOSULSKI, F. W., and HAMON, N. W. 1975. Gelation and thickening phenomena of vegetable protein products. J. Food Sci. *40*, 805–807.

FLINK, J., and CHRISTIANSEN, I. 1973. The production of a protein isolate from *Vicia faba*. Lebensm.-Wiss. Technol. *6*, 102–106.

GUINAT, E. 1969. Separation of starch and proteins from pulverized plant materials. Ger. Offen. 1, 911, 107; C.A. *72*, 45263 (1970).

HANG, Y. D., STEINKRAUS, K. H., and HACKLER, L. R. 1970. Comparative studies on the nitrogen solubility of mung beans, pea beans and red kidney beans. J. Food Sci. *35*, 318–320.

HARDMAN, K. D., and AINSWORTH, C. F. 1972. Structure of concanavalin A at 2.4-Å resolution. Biochemistry *11*, 4910–4919.

HAYES, L. P., and SIMMS, R. P. 1973. Defatted soybean fractionation by solvent extraction. U.S. Pat. 3,734,901. May 22.

HOFFPAUIR, C. L. 1953. Peanut composition: Relation to processing and utilization. J. Agric. Food Chem. *1*, 668–671.

IKEDA, K., HAMAGUCHI, K., YAMAMOTO, M., and IKENAKA, T. 1968. Circular dichroism and optical rotatory dispersion of trypsin inhibitors. J. Biochem. (Tokyo) *63*, 521–531.

JACKS, T. J., BARKER, R. H., and WEIGANG, O. E., JR. 1973. Conformations of oilseed storage proteins (globulins) determined by circular dichroism. Int. J. Pept. Protein Res. *5*, 289–291.

JACKS, T. J., YATSU, L. Y., and ALTSCHUL, A. M. 1967. Isolation and characterization of peanut spherosomes. Plant Physiol. *42*, 585–597.

JOHNSON, P., and NAISMITH, W. E. F. 1953. 3. High molecular weight systems. The physicochemical examination of the conarachin fraction of the groundnut globulins (*Arachis hypogaea*). Discuss. Faraday Soc. *13*, 98–109.

JOHNSON, P., and RICHARDS, E. G. 1962. The study of legumin by depolarization of fluorescence and other physicochemical methods. Arch. Biochem. Biophys. *97*, 260–276.

KHAN, N. A., and BAKER, B. E. 1957. The amino-acid composition of some Pakistani pulses. J. Sci. Food Agric. *8*, 301–305.

KAWAMURA, S. 1967. Quantitative paper chromatography of sugars of the cotyledon, hull, and hypocotyl of soybeans of selected varieties. Kagawa Univ. Fac. Tech. Bull. *18*, 117–131.

KOIDE, T., and IKENAKA, T. 1973. Studies on soybean trypsin inhibitors. 3. Amino acid sequence of the carboxyl-terminal region and the complete amino-acid sequence of soybean trypsin inhibitor (Kunitz). Eur. J. Biochem. *32*, 417–431.

KOSHIYAMA, I. 1968A. Factors influencing conformation changes in a 7S protein of soybean globulins by ultracentrifugal investigations. Agric. Biol. Chem. (Tokyo) *32*, 879–887.

KOSHIYAMA, I. 1968B. Chemical and physical properties of a 7S protein in soybean globulins. Cereal Chem. *45*, 394–404.

KOSHIYAMA, I. 1971. Some aspects of subunit structure of a 7S protein in soybean globulins. Agric. Biol. Chem. (Tokyo) *35*, 385–392.

KOSHIYAMA, I. 1972. A comparison of soybean globulins and the protein bodies in the protein composition. Agric. Biol. Chem. (Tokyo) *36*, 62–67.

LAL, B. M., PRAKASH, V., and VERMA, S. C. 1963. The distribution of nutrients in the seed parts of Bengal gram. Experientia *19*, 154–155.

MC ANELLY, J. K. 1964. Method for producing a soybean protein product and the resulting product. U.S. Pat. 3,142,571. July 28.

MC EWEN, T. J., DRONZEK, B. L., and BUSHUK, W. 1974. A scanning electron microscope study of faba bean seed. Cereal Chem. *51*, 750–757.

MARQUARDT, R. R., MC KIRDY, J. A., WARD, T., and CAMPBELL, L. D. 1975. Amino acid, hemagglutinin, and trypsin inhibitor levels, and proximate analyses of faba beans (*Vicia faba*) and faba bean fractions. Can. J. Anim. Sci. *55*, 421–429.

MEYER, E. W. 1971. Oilseed protein concentrates and isolates. J. Am. Oil Chem. Soc. *48*, 484–488.

MILLERD, A. 1975. Biochemistry of legume seed proteins. Annu. Rev. Plant Physiol. *26*, 53–72.

MITCHELL, J. H., JR. 1974. Process for making peanut flakes. U.S. Patent 3,800,056. March 26.

MITCHELL, J. H., JR. 1976. Process for making flavorless food extenders derived from peanuts, and a method of recovering peanut oil. U.S. Patent 3,947,599. March 30.

MUSTAKAS, G. C., ALBRECHT, W. J., BOOKWALTER, G. N., McGHEE, J. E., KWOLEK, W. F., and GRIFFIN, E. L., JR. 1970. Extruder-processing to improve nutritional quality, flavor, and keeping quality of full-fat soy flour. Food Technol. *24*, 1290–1296.

MUSTAKAS, G. C., ALBRECHT, W. J., McGHEE, J. E., BLACK, L. T., BOOK-WALTER, G. N., and GRIFFIN, E. L., JR. 1969. Lipoxidase deactivation to improve stability, odor, and flavor of full-fat soy flours. J. Am. Oil Chem. Soc. *46*, 623–626.

MUSTAKAS, G. C., KIRK, L. D., and GRIFFIN, E. L., JR. 1962. Flash desolventizing defatted soybean meals washed with aqueous alcohols to yield a high-protein product. J. Am. Oil Chem. Soc. *39*, 222–226.

ODANI, S., and IKENAKA, T. 1973. Studies on soybean trypsin inhibitors. 8. Disulfide bridges in soybean Bowman-Birk proteinase inhibitors. J. Biochem. (Tokyo) *74*, 697–715.

PRINGLE, W. 1974. Full-fat soy flour. J. Am. Oil Chem. Soc. *51*, 74A–76A.

PUSZTAI, A. 1965. Studies on the extraction of nitrogenous and phosphorus-containing materials from the seeds of kidney beans (*Phaseolus vulgaris*). Biochem. J. *94*, 611–616.

PUSZTAI, A., and WATT, W. B. 1970. Glycoprotein. 2. The isolation and characterization of a major antigenic and nonhemagglutinating glycoprotein from *Phaseolus vulgaris*. Biochim. Biophys. Acta *207*, 413–431.

RACKIS, J. J., ANDERSON, R. L., SASAME, H. A., SMITH, A. K., and VanETTEN, C. H. 1961. Amino acids in soybean hulls and oil meal fractions. J. Agric. Food Chem. *9*, 409–412.

RHEE, K. C., CATER, C. M., and MATTIL, K. F. 1972. Simultaneous recovery of protein and oil from raw peanuts in an aqueous system. J. Food Sci. *37*, 90–93.

RHEE, K. C., CATER, C. M., and MATTIL, K. F. 1973. Aqueous process for pilot plant scale production of peanut protein concentrate. J. Food Sci. *38*, 126–128.

ROCKLAND, L. B., and JONES, F. T. 1974. Scanning electron microscope studies on dry beans. Effects of cooking on the cellular structure of cotyledons in rehydrated large lima beans. J. Food Sci. *39*, 342–346.

SAIO, K., and WATANABE, T. 1968. Observation of soybean foods under electron microscope. J. Food Sci. Technol. Jpn. *15*, 290–296.

SAIR, L. 1959. Proteinaceous soy composition and method of preparing. U.S. Pat. 2,881,076. April 7.

SHEHATA, N. A., and FRYER, B. A. 1970. Effect on protein quality of supplementing wheat flour with chickpea flour. Cereal Chem. *47*, 663–670.

SINGH, S., SINGH, H. D., and SIKKA, K. C. 1968. Distribution of nutrients in the anatomical parts of common Indian pulses. Cereal Chem. *45*, 13–18.

SMITH, P. R. 1976. Personal communication. The British Food Manufacturing Industries Research Association, Leatherhead, Surrey, England.

SMITH, A. K., and CIRCLE, S. J. 1938. Peptization of soybean proteins. Extraction of nitrogeneous constituents from oil-free meal by acids and bases with and without added salts. Ind. Eng. Chem. *30*, 1414–1418.

SMITH, A. K., and CIRCLE, S. J. 1972. Soybeans: Chemistry and Technology, Vol. 1, Proteins. Avi Publishing Co., Westport, Conn.

SWEET, R. M., WRIGHT, H. T., JANIN, J., CHOTHIA, C. H., and BLOW, D. M. 1974. Crystal structure of the complex of porcine trypsin with soybean trypsin inhibitor (Kunitz) at 2.6-Å resolution. Biochemistry *13*, 4212–4228.

THANH, V. H., OKUBO, K., and SHIBASAKI, K. 1975. Isolation and characterization of the multiple 7S globulins of soybean proteins. Plant Physiol. *56*, 19–22.

TOMBS, M. P. 1967. Protein bodies of the soybean. Plant Physiol. *42*, 797–813.

VAINTRAUB, I. A., and TUEN, N. T. 1971. Quarternary structure of vetch seed legumin. Mol. Biol. (Engl. Transl.) *5*, 46–54.

VALENTINE, R. C. 1959. The shape of protein molecules suggested by electron microscopy. Nature *184*, 1838–1841.

VARNER, J. E., and SCHIDLOVSKY, G. 1963. Intracellular distribution of proteins in pea cotyledons. Plant Physiol. *38*, 139–144.

VERMA, S. C., LAL, B. M., and PRAKASH, V. 1964. Changes in the chemical composition of the seed parts during ripening of Bengal gram (*Cicer arietinum* L.) seed. J. Sci. Food Agric. *15*, 25–31.

WOLF, W. J. 1970. Scanning electron microscopy of soybean protein bodies. J. Am. Oil Chem. Soc. *47*, 107–108.

WOLF, W. J. 1975. Effects of refining operations on the composition of foods. 2. Effects of refining operations on legumes. *In* Nutritional Evaluation of Food Processing, 2nd Edition, R. S. Harris and E. Karmas (Editors). Avi Publishing Co., Westport, Conn.

WOLF, W. J., and BAKER, F. L. 1975. Scanning electron microscopy of soybeans, soy flours, protein concentrates, and protein isolates. Cereal Chem. *52*, 387–396.

WOLF, W. J., and COWAN, J. C. 1975. Soybeans as a Food Source. Revised Edition, CRC Press, Inc., Cleveland, Ohio.

WRIGHT, D. J., and BOULTER, D. 1973. A comparison of acid extracted globulin fractions and vicillin and legumin of *Vicia faba*. Phytochemistry *12*, 79–84.

WRIGHT, D. J., and BOULTER, D. 1974. Purification and subunit structure of legumin of *Vicia faba* L. (broad bean). Biochem. J. *141*, 413–418.

YOUNGS, C. G. 1975. Primary processing of pulse. *In* Oil Seed and Pulse Crops in Western Canada, Western Cooperative Fertilizers Ltd., Calgary, Alberta.

11

Single-Cell Protein

Steven R. Tannenbaum

GENERAL ASPECTS OF SINGLE-CELL PROTEIN

Single-cell protein (SCP) is a generic term for crude or refined protein whose origin is unicellular or multicellular organisms, *i.e.*, bacteria, yeast, molds, and algae. There have been a number of recent reviews published on this subject but the most comprehensive treatments of organisms, processes, and nutritional and food technological aspects of utilization are the books based upon two conferences held at the Massachusetts Institute of Technology (Mateles and Tannenbaum 1968; Tannenbaum and Wang 1975).

The need for SCP as a supplement to the world protein source is well-established and research and development on SCP production have been intense for over a decade. As a consequence of this work there are today a number of large SCP plants in operation and several more under construction. Therefore, we are no longer talking about a protein source of the future, but rather a protein source in use today. The problems being addressed today are the problems of the future, second and third generation, processes for SCP production for both animal feed and human food.

SCP as Food

There are a number of reasons why it is reasonable to expect that some form of SCP will ultimately fulfill a portion of the major food needs for mankind:

(1) The population of the earth may reach a point where agriculture becomes insufficient or uneconomic as a sole source of food. We can only guess when this will occur. It has been calculated that a 10% supplement to the world's food protein supply could be provided by a fermentor of an area equivalent to 1/2 square mile of the earth's surface (Humphrey 1966).

(2) SCP produced from nonagricultural raw materials is the only potential food which has no dependency on agricultural inputs, although cellulosic wastes may become an important substrate. It can be a truly synthetic yet complete source of food whose composition can be controlled. It is important to stress that SCP contains numerous nutrients in addition to protein.

(3) SCP processes cause markedly fewer and simpler waste disposal problems than other food processes because almost all of the product can be consumed as food and the waste is mostly in the form of heat when the spent growth medium can be recycled after harvesting the cells.

(4) Some forms of SCP have been used as human food for millenia. Any fermented foods will contain significant quantities of cellular mass as diverse as bacteria, yeast and fungi. Thus, the use of such organisms as a basic protein food is a logical extension of our previous experience.

SCP for Animal Feeds

There is good scientific evidence that various types of SCP can be useful as additional sources of protein and vitamins in the feeding of animals. During the last few years scientists have gathered extensive data about nutritive value and safety of different kinds of yeasts and bacteria, including those grown on *n*-alkanes and methanol. There have also been developed technological schemes for industrial production of these SCPs and good estimates are available of the economic aspects and safety of their utilization for feeding swine, broilers, and calves. Their amino acid patterns, content of nucleic acid, lipids, as well as data on possible toxic substances, have been studied in detail (Duthie 1975; Pokrovsky 1975; Shacklady 1975).

These data indicate that in many countries which need additional protein for animal feeds, it will be feasible to begin SCP production for this purpose. At the same time it is recognized that the USA is relatively fortunate in having an abundance of traditional animal protein feed resources, so that SCP will play a lesser role here in animal feeding for the near to middle term.

CHARACTERISTICS OF SCP PROCESSES

A complete description of individual processes is outside the scope of this paper. It would be useful, however, to indicate the general components of any SCP process with the schematic in Fig. 11.1. This indicates not only the key pieces of equipment in any process, but also important economic factors such as the necessity of oxygen transfer and heat removal.

There are many factors involved in the choice of a process. Choosing a process, including the organism and the substrate, is extremely complex, and it should not be surprising to find that there are many potential processes. In fact, flexibility in process design is an important attribute of SCP.

Raw Materials

The raw materials chosen for SCP production depend, to a very large extent, on locale. Thus, natural gas or methane would be of great interest in areas where it is available in abundance, or where it may even be discarded (flare gas). With regard to petroleum fractions, local economics of the petroleum industry will determine to some extent whether crude, semi-purified (*e.g.*, gas-oil) or refined fractions (*e.g.*, *n*-paraffins) are the more desirable substrates. For example, in areas which use large quantities of diesel fuel, it is

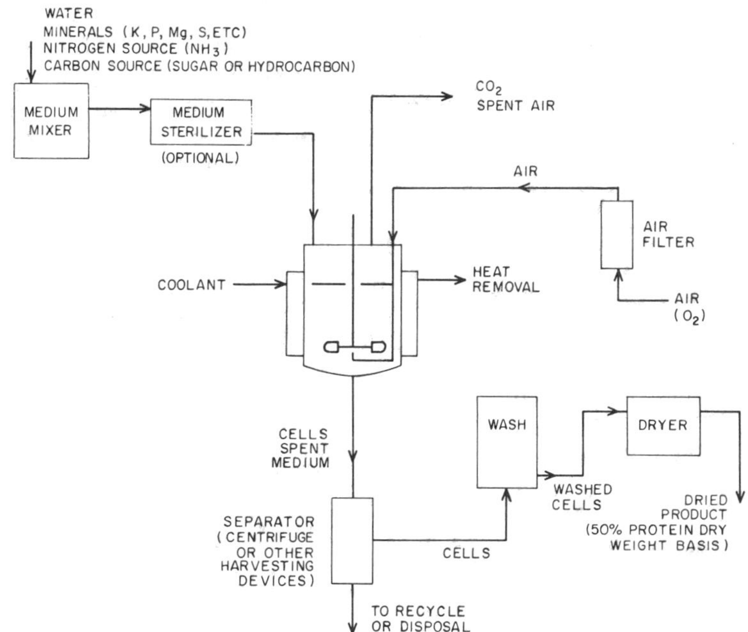

FIG. 11.1. SIMPLIFIED FLOWSHEET OF PRODUCTION OF SINGLE-CELL PROTEIN

important to remove the paraffinic fraction to lower the pour point of the oil. In areas which use large quantities of gasoline, it is unnecessary to remove the paraffins prior to the cracking operation, and a credit would not be allowed for removal of that fraction. Since gas-oil contains only approximately 10–40% paraffins (the only fraction readily used as a carbon source), a large unmetabolized portion passes through the fermentor creating problems in cell recovery and in removal of residual unmetabolized hydrocarbons from the cells. The removal of this residual material is a significant problem in the use of petroleum fractions as carbon sources and requires that the final biomass be washed extensively with detergent solution, and preferably with organic solvents such as hexane and lower boiling alcohols (Lainé and du Chaffaut 1975). Because of potential problems of toxicity, all of the products produced on gas-oil seem destined for animal feeds.

A number of research groups have concentrated on processes which use a relatively pure paraffin distillate for production of either food or feed (Cooper *et al.* 1975; Kanazawa 1975). Since all of the substrate is theoretically metabolizable in this case, it is possible to run the fermentation in such a manner that no unmetabolized hydrocarbon remains in the effluent. In practice, however, some hydrocarbon may be adsorbed to the cells and will have to be removed later in the process.

Methane.—This substrate is available at very low costs in some regions of the world and also permits high cell yields. Its direct fermentation appears limited to bacteria, however, and no large-scale process using pure microculture has apparently yet been developed which provides an economically attractive combination of growth rate and concentration in continuous operation. The probability of economical mixed cultures is more likely (Hamer *et al.* 1975). Since methane is a gaseous substrate, it would have the advantage over paraffinic hydrocarbons of leaving no residue in the final product.

Alcohol.—Methanol and ethanol, in particular, are derived from gaseous hydrocarbons via catalytic hydration. In the future, this conversion may become so efficient with inorganic catalysts that the corresponding biological step could not compete. Therefore, methanol and not methane would be the logical starting material for fermentation.

Both methanol and ethanol can be utilized by a wide variety of microorganisms. They are prepared in very high states of purity, are totally water soluble, and they leave no residues in the cell mass leaving the fermentor. They are also intermediate in cell yield and oxygen yield between hydrocarbons and carbohydrates. Since the lower alcohols maintain some of the advantages of both hydrocarbon

and carbohydrate fermentations they have significant potential for
SCP processes, and are particularly attractive for consideration as a
source of food protein for both the long and short terms (Cooney
and Levine 1975; Dostalek and Molin 1975; Gow *et al.* 1975).

Carbohydrates.—There are also a variety of carbohydrate sources.
Those, such as sulfite waste liquor (Romantschuk 1975) and cheese
whey (Pace and Goldstein 1975), are already operative in very special
locales and, along with starch and molasses, may be used to produce
cells by conventional fermentation technology.

Cellulose.—As the most abundant raw material, available from
waste paper, bagasse, wood pulp, etc., cellulose is of particular
importance as a renewable resource. In the USSR, hydrolyzed wood
pulp supports a feed yeast industry estimated to be approximately
one million tons per year. From an economic point of view, the most
crucial step in utilization of cellulose is conversion to metabolizable
sugars. Although proposals for possible economic processes exist,
they have not been proven on a suitable scale. In the case of
agricultural by-product materials such as solid cannery waste and
citrus waste, problems exist because of seasonal processing schedules,
conversion to usable form, and the sometimes dilute nature of the
fermentable material. Biological utilization is frequently rate-limited
by the hydrolysis of cellulose, and extensive pretreatment may be
necessary. However, the possibility of simultaneously addressing the
problems of waste material disposal and protein production makes
these materials attractive (Bellamy 1975; Dunlap 1975; Rolz 1975).

Starch.—Starch is much more readily hydrolyzed than cellulose,
and the potential exists for direct fermentation of starch by
amylolytic fungi in a continuous process (Anderson *et al.* 1975). An
alternative approach is to use a mixed culture for waste starch.
Starch may be of particular interest in tropical areas which can
produce high yields of starchy root crops, such as cassava, provided
that such subsistence crops can be accumulated economically for
industrial processing.

Sugars.—Sugars, such as molasses, are similarly useful substrates,
and waste crops such as carob may be an important source of sugars
for fermentation (Imrie and Vlitos 1975).

The common difficulty with substrates derived from waste plant
material is the limited market potential available in small tropical or
semitropical countries. This suggests the need for small fermentation
plants, which may not be able to operate economically. In such
circumstances, the utility of simple, uncontrolled, nonaseptic
processes to produce animal feeds should be examined.

Organisms

Bacteria, yeasts and fungi are all in active consideration as the organisms to be used for pilot or commercial scale processes. Each has advantages and disadvantages relative to the others.

Yeasts.—The most popular organisms for food applications appear to be yeasts. These organisms seem somehow more familiar to human experience. They have been used in foods as vitamin additives and flavoring agents for a very long time, have reasonable protein concentrations, and are easier to recover from fermentation media than bacteria. They also have the considerable advantage of current availability in the open market, being produced from a variety of carbohydrate substrates which are recognized as sources of food themselves. The genera of interest are mainly *Saccharomyces* and *Candida*.

Bacteria.—Bacteria have some advantages over other organisms, particularly for animal feed. They have higher growth rates, higher protein contents, and higher content of sulfur amino acids. They also have the disadvantages of being susceptible to phage, having high nucleic acid content and being less well-known in human nutritional experience. A major technical and economic problem is the high cost of cell recovery due to the smaller cell size. However, they are a good long range prospect because of the potential economic advantage they offer.

Fungi.—The higher fungi have received very little attention in industrial-scale projects until recently. The possibility exists of successful growth in continuous culture on very inexpensive waste carbohydrate sources, but contamination is more of a problem because of relatively slow growth rates. Fungi might be of considerable interest for their potent enzymatic capabilities, ease of harvesting from fermentation media, and mycelial nature which provides natural texture.

ECONOMIC CONSIDERATIONS

A detailed treatment is beyond the scope of this paper; therefore, I will present only a relatively brief consideration of some of the more important aspects.

Of the several ways to examine process economics, one of the more interesting approaches is through a breakdown of the costs of the process. Giacobbe *et al.* (1975) have examined several processes in this manner, and I have selected their analysis of a process producing bacteria from methanol for illustration.

The process in question was developed by Imperial Chemical Industries, Agricultural Division, and utilizes a unique pressure cycle air-lift fermentor (Gow *et al.* 1975). For estimating the operating variables of this process the following assumptions were made:

(a) The fermentation is conducted at 40°C;

(b) The oxygen transfer rate is 11 kg/hr-m^3;

(c) Fermentor oxygen utilization is 50%; and

(d) Substrate yield is 0.46 kg of cells/kg of methanol.

The overall cost of the process may be viewed as consisting of the cost of the manufacturing facility and the cost of operating this facility. The breakdown of the operating cost is shown in Table 11.1. It consists of the cost of raw materials, utilities, labor, maintenance, other overhead, and depreciation. Note that almost three-quarters of the operating cost for this process is in raw materials, and that most of this cost is for the source of carbon.

TABLE 11.1

OPERATING COST INDEX FOR BACTERIA GROWN IN METHANOL[1]

Raw Materials:		
methanol	47.4	
phosphate	11.8	
ammonia	12.0	
salts	2.6	
	73.8	73.8
Utilities		14.2
Labor, maintenance, overhead		6.2
Depreciation		5.8
		100.0

[1] Giacobbe *et al.* (1975).

The basis of this cost is strictly related to the composition of the organism and the yield of cells per unit of substrate. Since all cells have grossly similar composition in terms of C, H, N, S, P, etc., the key factor is cell yield, and this in turn is determined by the elemental composition of the substrate and its metabolic route to cellular constituents. A more detailed consideration of these principles for the case of methanol can be found in the paper by Cooney and Levine (1975).

The cost of utilities is in fact the energy cost of the process. A breakdown of this cost is shown in Table 11.2, which clearly indicates that the bulk of this cost is in the biosynthesis of the cells and in the dewatering and drying of the cells. The term biosynthesis in this regard refers to the fermentation process, and the energy is consumed in mixing, oxygen transfer, and in refrigeration for temperature control. The extent of these costs are a function of the substrate, the organism, the fermentor design, and even the location of the facility. For example, if the plant is constructed in an area where cooling water is available around the year, mechanical refrigeration would be unnecessary.

TABLE 11.2

COST INDICES OF BACTERIA GROWN ON METHANOL[1]

Operation	Investment	Utilities
Cell biosynthesis	43.4	61.7
Cell drying	23.1	28.8
SCP storage, bagging, shipping	15.0	0.6
Cell recovery	11.0	3.8
Raw material handling and storage	4.9	0.2
Culture medium preparation	1.5	2.4
Inoculum preparation	1.1	2.5
Total	100.0	100.0

[1] Giacobbe *et al.* (1975).

The cost of drying has generally been overlooked as a significant part of the overall cost of the process. This is an area in which research could play a significant dividend in the form of cost reduction.

NUTRITIONAL VALUE

The nutritional properties of a variety of types of SCP have been extensively treated elsewhere. Studies in the USSR have been summarized by Pokrovsky (1975). Yeasts grown on gas-oil and normal alkanes have been investigated in over a dozen species of animals (Shacklady 1975). A microfungal protein has been studied in several species, including the baboon (Duthie 1975).

The nutritional and safety requirements for testing SCPs have been the subject of extensive discussion within the Protein Advisory Group of the United Nations. Out of these discussions has emerged an international consensus of the appropriate steps to be taken for a new material (Oser 1975).

The amino acid content of a variety of SCPs tends to resemble that of plant proteins. An example of an alkane grown yeast is shown in Table 11.3 compared to soya meal and to fish meal. Yeast in particular tends to be limiting in the sulfur amino acids, but quite a good source of lysine. The effect of methionine supplementation is dramatic as shown in Table 11.4, in which the NPU and Biological Value are increased to approximately the level for whole egg.

The supplemental value of yeast as a source of protein is shown in Fig. 11.2. Yeast is an excellent supplement for corn and sesame, both of which are deficient in lysine. It is of no supplemental value to soy protein, since both proteins are deficient in the sulfur amino acids.

TABLE 11.3
AMINO ACID CONTENT OF AN ALKANE-GROWN YEAST
COMPARED TO FISH MEAL AND SOYA BEAN MEAL[1]

Protein Source, Amino Acid in g/16 g N

Amino Acid	Yeast	Fish Meal	Soya Bean Meal
Isoleucine	5.1	4.6	5.4
Leucine	7.4	7.3	7.7
Phenylalanine	4.3	4.0	5.1
Tyrosine	3.6	2.9	2.7
Threonine	4.9	4.2	4.0
Tryptophan	1.4	1.2	1.5
Valine	5.9	5.2	5.0
Arginine	5.1	5.0	7.7
Histidine	2.1	2.3	2.4
Lysine	7.4	7.0	6.5
Cystine	1.1	1.0	1.4
Methionine	1.8	2.6	1.4
Total S-amino acids	2.9	3.6	2.8

[1] Shacklady (1975).

TABLE 11.4
NUTRITIONAL VALUE OF AN ALKANE-GROWN YEAST WITH AND WITHOUT
METHIONINE SUPPLEMENTATION COMPARED TO DRIED WHOLE EGG[1]

Material	Net Protein Utilization	Digestibility	Biological Value
Yeast	59	96	61
Yeast + 0.3% methionine	88	96	91
Dried whole egg	90	100	90

[1] Shacklady (1975).

The nutritive values of some SCPs for man have also been studied, and are summarized by Young and Scrimshaw (1975). Some of these properties are shown in Table 11.5. As discussed by Young and Scrimshaw (1975) the Biological Value is a function of nitrogen intake for all proteins. Both algal protein and yeast protein are of relatively good value to man, whose sulfur amino acid requirements are somewhat less than that of the rodent.

NUCLEIC ACIDS

Compared to conventional food sources, the nucleic acid content of SCP is relatively high. For example, the nucleic acid (NA) content

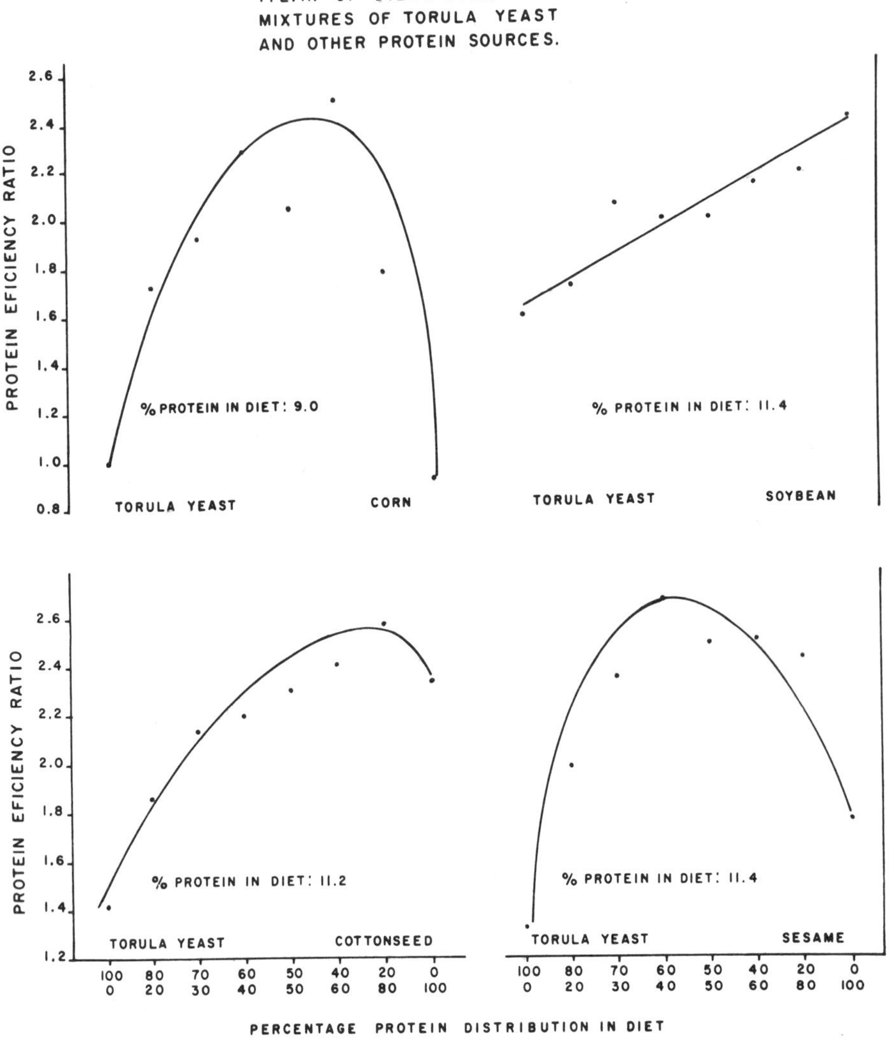

From Bressani (1968)

FIG. 11.2.

PROTEIN EFFICIENCY RATIO OF DIETS MADE FROM MIXTURES OF TORULA
YEAST AND OTHER PROTEIN SOURCES

of SCP can vary from 8 to 25 g/100 g of cells with most of the
nucleic acid being RNA. In animal tissues, highest levels are found in
liver (4 g NA/100 g protein), sardines (2.2 g NA/100 g protein) and
in fish roe (5.7 g NA/100 g protein).

As discussed by Kilhberg (1972), components of dietary nucleic acid, guanine and adenine, are metabolized to uric acid. Man has lost the enzyme uricase which oxidizes uric acid to the soluble and excretable metabolite allantoin. Consumption of a protein source high in purines results in an elevated level of uric acid in plasma and in urine (Table 11.6). Because of the low solubility of uric acid, it is suspected that an increased plasma level of uric acid may result in urate precipitating in tissues and joints, a condition similar to that which occurs in gout.

Consequently, investigators have recommended that if SCP is to be used as a primary protein source for human populations, the nucleic acid content should be reduced to a level which would allow a maximum intake in the range of 2 g of nucleic acid per day (Edozien *et al.* 1970).

The advantages and disadvantages of proposed methods for the removal of RNA in SCP are summarized in Table 11.7 (Sinskey and Tannenbaum 1975).

TABLE 11.5

BIOLOGICAL VALUE OF ALGAE AND YEAST COMPARED
WITH THAT OF CASEIN IN YOUNG MEN[1]

Material	N-Intake (g/day)	True Digestibility (%)	Biological Value
Casein + RNA	4.25	95 ± 8	66 ± 4
	7.84	99 ± 1	52 ± 6
Chlorella sorokiniana	4.83	89 ± 4	79 ± 12
	7.81	82 ± 6	60 ± 6
Candida utilis	4.51	83 ± 7	70 ± 5
	8.20	87 ± 3	58 ± 6

[1] Waslein *et al.* (1970).

TABLE 11.6

EFFECT OF YEAST NUCLEIC ACID INTAKE ON SERUM
AND URINARY URIC ACID IN YOUNG MEN[1]

Yeast Intake (g/day)	Uric Acid Content	
	Serum (mg/100 ml)	Urine (mg/24 hr)
0 (−)[2]	4.5 ± 0.2[3]	510 ± 81
45 (2.9)	7.2 ± 0.3	1192 ± 109
90 (5.8)	8.9 ± 0.4	1853 ± 121
135 (8.7)	9.4 ± 0.7	1871 ± 460

[1] Edozien *et al.* (1970); Young and Scrimshaw (1975)
[2] Values in parenthesis refer to g nucleic acid intake per day
[3] Mean ± sd for four subjects

TABLE 11.7

SUMMARY OF METHODS AVAILABLE FOR RNA REDUCTION IN SCP[1]

Methods		Advantages	Disadvantages
(1)	Growth and cell physiology growth rate substrate limitation	Only proper fermentation design	Limited reduction economics
(2)	Base-catalyzed hydrolysis	Simple and rapid	Loss of weight and N, salt addition, deleterious effects of high pH
(3)	Chemical extraction	Simple, rapid, remove polymerized RNA	Chemical residue, loss of weight and N
(4)	Cell disruption	Only if protein isolate desired	Economics, others specific to process
(5)	Exogenous RNase	Rapid, simple, choice of enzyme	Cost and availability of enzyme, loss of dry matter
(6)	Endogenous RNase heat shock anions, etc.	Simple, cells direct from fer., no added chemicals	Weight loss, slow, only certain cells, added chemical

[1]Sinsky and Tannenbaum (1975).

UTILIZATION OF CELLS IN FOOD

There are numerous applications for SCP in food as flavoring agents, emulsifiers, nutrient supplements, etc. There is an extensive use of a variety of yeast products in both United States, European and Eastern markets.

If an extension of SCP properties is desired, there are a number of routes to produce soluble or insoluble protein isolates, SCP-derived protein fibers, and SCP-based extruded products (Table 11.8).

The production of protein isolates has been extensively investigated by several groups (referenced in Table 11.8). The basic technology and operations are summarized in Fig. 11.3. The disruption of cells and the recovery of their proteins has been recently summarized by Dunnill and Lilly (1975). An example of the type of distribution of protein which can be achieved by this type of approach is shown in Tables 11.9 and 11.10 for *B. megaterium*. The

TABLE 11.8

SOME PROCESSES FOR EXTENSION OF SCP FUNCTIONALITY

Process	Reference
Protein isolates	Tannenbaum *et al.* (1966); Hedenskog and Ebbinghaus (1972); Vananuvat and Kinsella (1975); Newell (1975)
Protein fibers	Huang and Rha (1971); Hayakawa *et al.* (1975)
Extruded SCP	Tannenbaum (1975)

main technical problem is the loss in overall yield which occurs as a result of the distribution of the protein between several fractions. The amino acid composition of the extracted protein resembles that of the whole cell and there is little nutritional gain in cell disruption or in production of protein isolates (Tannenbaum and Miller 1967).

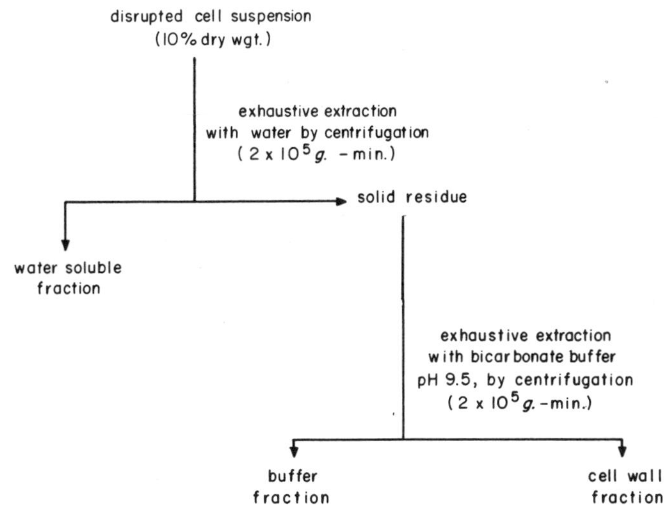

From Tannenbaum et al. (1966)
FIG. 11.3. PREPARATION OF PROTEIN ISOLATES FROM SCP

TABLE 11.9
AMINO ACID CONTENT OF CELL FRACTIONS OF *B. megaterium*[1]

	% N[2]	% Protein by Biuret	% Amino Acids[3]	% of Total Cell Amino Acids[4]
Whole cells	6.3	—	38	—
Water soluble fraction				
TCA ppt		68	68	15
HCl ppt		77	—	—
Buffer soluble fraction				
TCA ppt		100	52	37
HCl ppt		99	—	—
Cell wall fraction	5.3	—	33	48

[1] Tannenbaum *et al.* (1966).
[2] N Content of that fraction.
[3] % by weight which is amino acids. The biuret technique gives false high values.
[4] Distribution of amino acids in the various fractions.

TABLE 11.10
AMINO ACID COMPOSITION OF WHOLE CELLS AND
WATER SOLUBLE FRACTION OF *B. megaterium*[1]

Amino Acid	Whole Cells	Water Soluble Fraction (g/16 g Nitrogen)	Cow's Milk
Arginine	6.7	5.4	3.7
Histidine	2.3	2.3	2.7
Lysine	6.3	8.6	7.9
Leucine	11.0	9.3	10.0
Isoleucine	4.4	6.3	6.5
Methionine	3.1	2.6	2.5
Cystine	*	1.0	0.9
Phenylalanine	5.4	4.9	4.9
Tyrosine	5.0	3.8	5.1
Threonine	4.8	4.3	4.7
Valine	6.5	6.6	7.0
Aspartic Acid	9.1	9.2	*
Glutamic Acid	18.9	13.7	*
Serine	3.4	3.9	*
Proline	*	3.1	*
Glycine	4.0	3.8	*
Alanine	10.1	5.7	*

[1] Tannenbaum *et al.* (1966).
*Not determined.

As a source of food protein, SCP is in its infancy. Its situation can be compared to that of soy 25 years ago. As new sources of SCP become economically feasible and of proven safety and nutritive value, uses and technology will expand to fill the available market.

BIBLIOGRAPHY

ANDERSON, C., LONGTON, J., MADDIX, C., SCAMMELL, G. W., and SOLOMONS, G. L. 1975. The growth of microfungi on carbohydrates. *In* Single-Cell Protein II, S. R. Tannenbaum and D. I. C. Wang (Editors). M.I.T. Press, Cambridge, Mass.

BELLAMY, W. D. 1975. Conversion of insoluble agricultural wastes to SCP by thermophilic microorganisms. *In* Single-Cell Protein II, S. R. Tannenbaum and D. I. C. Wang (Editors). M.I.T. Press, Cambridge, Mass.

BRESSANI, R. 1968. The use of yeast in human foods. *In* Single-Cell Protein II, S. R. Tannenbaum and D. I. C. Wang (Editors). M.I.T. Press, Cambridge, Mass.

COONEY, C. L., and LEVINE, D. W. 1975. SCP production from methanol by yeast. *In* Single-Cell Protein II, S. R. Tannenbaum and D. I. C. Wang (Editors). M.I.T. Press, Cambridge, Mass.

COOPER, P. G., SILVER, R. S., and BOYLE, J. P. 1975. Semi-commercial studies of a petroprotein process based on n-paraffins. *In* Single-Cell Protein II, S. R. Tannenbaum and D. I. C. Wang (Editors). M.I.T. Press, Cambridge, Mass.

DOSTALEK, M., and MOLIN, N. 1975. Studies of biomass production of methanol oxidizing bacteria. *In* Single-Cell Protein II, S. R. Tannenbaum and D. I. C. Wang (Editors). M.I.T. Press, Cambridge, Mass.

DUNLAP, C. E. 1975. Production of single-cell protein from insoluble agricultural wastes by mesophiles. *In* Single-Cell Protein II, S. R. Tannenbaum and D. I. C. Wang (Editors). M.I.T. Press, Cambridge, Mass.

DUNNILL, P., and LILLY, M. D. 1975. Protein extraction and recovery from microbial cells. *In* Single-Cell Protein II, S. R. Tannenbaum and D. I. C. Wang (Editors). M.I.T. Press, Cambridge, Mass.

DUTHIE, I. F. 1975. Animal feeding trials with a microfungal protein. *In* Single-Cell Protein II, S. R. Tannenbaum and D. I. C. Wang (Editors). M.I.T. Press, Cambridge, Mass.

EDOZIEN, J. C., UDO, U. U., YOUNG, V. R., and SCRIMSHAW, N. S. 1970. Effect of high levels of yeast on uric acid metabolism in young men. Nature *228*, 180.

GIACOBBE, F., PUGLISI, P., and LONGOBARDI, G. 1975. Economic evaluation of new trends in SCP manufacture. Paper presented at a Symposium on Single-Cell Protein, American Chemical Society National Meeting, Philadelphia, April 9.

GOW, J. S., LITTLEHAILES, J. D., SMITH, S. R. L., and WALTER, R. B. 1975. SCP production from methanol: Bacteria. *In* Single-Cell Protein II, S. R. Tannenbaum and D. I. C. Wang (Editors). M.I.T. Press, Cambridge, Mass.

HAMER, G., HARRISON, D. E. F., HARWOOD, J. H., and TOPIWALA, H. H. 1975. SCP production from methane. *In* Single-Cell Protein II, S. R. Tannenbaum and D. I. C. Wang (Editors). M.I.T. Press, Cambridge, Mass.

HAYAKAWA, I., KAWASAKI, S., and NOMURA, D. 1975. Spinnability of yeast protein and its viscoelastic properties. Agr. Biol. Chem. *49*, 641–646.

HEDENSKOG, G., and EBBINGHAUS, L. 1972. Reduction of nucleic acid content of single-cell protein concentrates. Biotechnol. Bioengr. *14*, 447–457.

HUANG, F., and RHA, C. K. 1971. Rheological properties of SCP protein concentrate: Dope formation and its flow behavior. J. Food Sci. *36*, 1131–1134.

HUMPHREY, A. E. 1966. Starvation: Chemical engineering can help fight it. Chem. Eng. 149.

IMRIE, F. K. E., and VLITOS, A. J. 1975. Production of fungal protein from carob (*Ceratonia siligua* L.). *In* Single-Cell Protein II, S. R. Tannenbaum and D. I. C. Wang (Editors). M.I.T. Press, Cambridge, Mass.

KANAZAWA, M. 1975. The production of yeast from n-paraffins. *In* Single-Cell Protein II, S. R. Tannenbaum and D. I. C. Wang (Editors). M.I.T. Press, Cambridge, Mass.

KILHBERG, R. 1972. The microbe as a source of food. Annu. Rev. Microbiol. *26*, 427–466.

LAINE, B. M., and DU CHAFFAUT, J. 1975. Gas-oil as a substrate for single-cell protein production. *In* Single-Cell Protein, II, S. R. Tannenbaum and D. I. C. Wang (Editors). M.I.T. Press, Cambridge, Mass.

MATELES, R. I., and TANNENBAUM, S. R. 1968. Single-Cell Protein. M.I.T. Press, Cambridge, Mass.

NEWELL, J. A. 1975. Manufacture of yeast protein isolate having a reduced nucleic acid content by an alkali process. U.S. Pat. 3,867,555. Feb. 18.

OSER, B. L. 1975. Guidelines for the evaluation of SCP for human consumption. *In* Single-Cell Protein II, S. R. Tannenbaum and D. I. C. Wang (Editors). M.I.T. Press, Cambridge, Mass.

PACE, G. W., and GOLDSTEIN, D. J. 1975. Economic analysis of ultrafiltration—fermentation plants producing whey protein and SCP from cheese whey. *In* Single-Cell Protein II, S. R. Tannenbaum and D. I. C. Wang (Editors). M.I.T. Press, Cambridge, Mass.

POKROVSKY, A. 1975. Some results of SCP medico-biological investigations. *In* Single-Cell Protein II, S. R. Tannenbaum and D. I. C. Wang (Editors). M.I.T. Press, Cambridge, Mass.

ROLZ, C. 1975. Utilization of cane and coffee processing by-products as microbial protein substrates. *In* Single-Cell Protein II, S. R. Tannenbaum and D. I. C. Wang (Editors). M.I.T. Press, Cambridge, Mass.

ROMANTSCHUK, H. 1975. The Pekilo process: Protein from spent sulfite liquor. *In* Single-Cell Protein II, S. R. Tannenbaum and D. I. C. Wang (Editors). M.I.T. Press, Cambridge, Mass.

SHACKLADY, C. A. 1975. Value of SCP for animals. *In* Single-Cell Protein II, S. R. Tannenbaum and D. I. C. Wang (Editors). M.I.T. Press, Cambridge, Mass.

SINSKEY, A. J., and TANNENBAUM, S. R. 1975. Removal of nucleic acids in SCP. *In* Single-Cell Protein II, S. R. Tannenbaum and D. I. C. Wang (Editors). M.I.T. Press, Cambridge, Mass.

TANNENBAUM, S. R. 1975. Texturizing process for single-cell protein containing protein mixtures. U.S. Pat. 3,925,562. Dec. 9.

TANNENBAUM, S. R., MATELES, R. I., and CAPCO, G. R. 1966. Processing of bacteria for production of protein concentrates. Adv. in Chem. Series, No. 57.

TANNENBAUM, S. R., and MILLER, S. A. 1967. Effect of cell fragmentation on nutritive value of *Bacillus megaterium* protein. Nature *214*, No. 5094, 1261–1262.

TANNENBAUM, S. R., and WANG, D. I. C. 1975. Single-Cell Protein II. M.I.T. Press, Cambridge, Mass.

VANANUVAT, P., and KINSELLA, J. E. 1975. Some functional properties of protein isolates from yeast, *Saccharomyces fragilis*. J. Agr. Food Chem. *23*, 613–616.

WASLIEN, C. I., CALLOWAY, D. H., MARGEN, S., and COSTA, F. 1970. Uric acid levels in men fed algae and yeast as protein sources. J. Food Sci. *35*, 294–298.

YOUNG, V. R., and SCRIMSHAW, N. S. 1975. Clinical studies on the nutritional value of single-cell proteins. *In* Single-Cell Protein II, S. R. Tannenbaum and D. I. C. Wang (Editors). M.I.T. Press, Cambridge, Mass.

12

Some Observations on the Roles of Basic Research in Development of High Protein Food Products

J. M. McIntire

It is most fitting in the bicentennial year that this important protein symposium should be held and that the results should become available to all through this book. The need for education and scientific investigation in agriculture was recognized by some of the founding fathers of this country. Jefferson was one of the foremost supporters and quite a food scientist himself. A little over a hundred years ago (in 1862) the Land Grant College Bill was enacted which provided for science and education in agriculture and for the establishment of the Department of Agriculture. In the United States, much of the scientific knowledge regarding foods and specifically the important protein field has come from institutions that derive support and even their beginning from the Land Grant Act. It is interesting that several of the authors of chapters in this book probably derive some or all of their support from the original Land Grant source. Many other outstanding protein chemists are located in private and state universities.

Process art has often preceded understanding of the basic scientific principles. Some of the information reported in this book adds to our understanding of processing art that was developed long ago. Salt preservation was practiced in prehistoric times. Appert began his experiments on canning nearly two hundred years ago. Pasteur explained the microbiological basis for preservation over a hundred years ago. The sudden turn in events, where the U.S. food supply has proven to be a most important asset in international power and politics, places much importance on the advancement of our knowledge in food research. Undoubtedly, our greatest challenge in the years to come will be the world food supply. We will be called upon to fully utilize our food processing and food production capacity. Food preservation and food utilization will play a most important role in our future.

In order to fully utilize our protein resources it will be necessary to better understand protein structure and how to modify it to optimize the functional and nutritional properties. Higher yielding varieties of grains and legumes are being developed through genetic engineering. Generally, it has been found that higher yields of protein result in decreased nutritional quality. It may be necessary to learn how to fractionate the proteins, to modify the fractions to increase nutritional and functional properties and then to recombine these fractions, or fractions from diverse sources, in order to tailor-make the high protein foods best suited for proper diets of children, adults and the aged, those with different genetically derived metabolic diseases, specific health problems, etc. These possibilities, which appear impractical now, could become the foundations of formulation of high protein foods of the future. But first we must understand the nature of individual proteins in various common food sources and how these proteins react to changes in their environments—changes involving pH, temperature, chemical composition, solvent interaction and compositional properties.

I should like to comment on a few of the highlights as I see them of the chapters presented to this point. Understanding the mechanism of protein folding into three-dimensional structures is important from several points of view. The three-dimensional protein structure is predetermined by the amino acid sequence and micro-environment. Genetic manipulation leading to amino acid sequence changes would be expected to result in changes in the three-dimensional structure and thus other properties. The three-dimensional structure may not be determined by the global energy minimum. The conformation of globular proteins is probably attained by a two-stage process which involves rapid local formation of secondary structures followed by the folding together of these (LINCS). Prediction of structure based on local amino acid sequences has been successful which provides support for the LINCS concept.

Monomer units of globular proteins will associate in aqueous solutions if they contain more than 28% hydrophobic amino acids (Van Holde, Chapter 1). Dr. Van Holde described a schizophrenic structure characterized by the C-terminal end having more hydrophobic amino acids and few basic ones; thus with more than 28%

hydrophobic amino acids it has a tendency to associate with other molecules whereas the N-terminal end, having a high content of basic amino acid residues, has a tendency to remain unfolded and not to associate into aggregates. Van Holde suggested that the folding of a protein molecule into its natural conformation is a much simpler process than previously believed. The assembly of polypeptide chains into larger multi-subunit proteins may parallel on a large scale the mechanism of folding of local regions in individual chains.

Understanding of protein structure is important in providing answers to many of the questions related to food processing and biological processes (Whitaker, Chapter 2). Basic protein structural information has answered many important questions concerning the properties and functionality of foods. Some examples of importance involve enzyme inactivation and reactivation, emulsification, whipping and clotting, etc.

Many food processes involve changes of the protein structure. Some of these are the processes involved in the production of such common foods as bread, cheese, evaporated milk and preserved fruits and vegetables, inactivation of enzymes that affect color, texture, flavor and other sensory properties of many foods, the isolation of proteins and reprocessing of these into new foods such as meat and dairy product replacers, and the compatible combination of proteins with other food products to provide desired functional properties such as whey and vegetable proteins as milk replacers, whipping mixtures, etc. The involvement of proteins in adaptation or alteration of food systems is almost as common as the presence of proteins in food systems. Understanding of protein structure and how it is affected by environmental changes requires knowledge of the primary, secondary, tertiary and quaternary structure of proteins as well as knowledge of the nature of bonds and the energy relationships involved in stabilizing structure (Whitaker, Chapter 2). Solvent involvement is an important contributor to the structure of protein relative to its native and denatured states and its hydrophobic and hydrophilic properties.

The renaturation or refolding of proteins is of great significance with respect to enzyme inactivation by different processing methods. Several enzymes are known to undergo reactivation in certain environments; the rates of reactivation are temperature, pH and substrate dependent.

Proteins without water are not proteins at all since native structure and functional properties do not exist in absence of water (Fennema, Chapter 3). Water behaves abnormally compared to other substances of similar molecular weight. Hydrogen bonding capabilities explain many of its unusual properties. There are several states of water in relation to protein (Fennema, Chapter 3). The importance of water in food processing and food reactions is difficult to overstate. It is interrelated with many properties such as stability, color, flavor, microbial and enzymatic activity. Protein interaction with many other components cannot take place without it. Indeed, as Dr. Fennema indicated, without water, food is not food at all. Undoubtedly, research in this area will be more intensive and more productive in providing answers regarding the function of proteins and especially in relation to the preservation of food.

Proteins are not nutritionally important and useful unless they are consumed and converted by specific hydrolytic enzymes to the amino acid stage. Hydrolysis of proteins occurs at two levels in the human. In the gastrointestinal tract food proteins are converted to amino acids which are then transported across the small intestinal wall into the blood stream. In the blood they are transported to sites of biosynthesis in the cell. The cell is also capable of hydrolyzing tissue proteins to amino acids through the action of the proteolytic enzymes found in the lysosomal organelles of the cell.

The coordinated effort of the intracellular proteolytic enzymes of the lysosome in breaking down proteins is a most interesting one (Tappel, Chapter 4). Relating the intracellular proteolytic enzymes to the better understood digestive-tract enzymes provides a basis for their understanding by the food scientist and an appreciation for these little understood enzymes. Although little interest has been shown in these enzymes in the past, these enzymes cannot be disregarded in future food research. They are of much more importance in food protein degradation than has been appreciated. Deficiencies and improper balance in amounts of these enzymes may be important in several nutritional diseases that are only recently being investigated.

Understanding of protein function and properties depends on development of adequate methodology for protein isolation and analysis (Catsimpoolas, Chapter 5). As one who was involved in the microbiological analyses of amino acids in the early forties, I can state with authority that we have come a long, long way in amino acid and protein analyses. The tools that are available for separation based on various applications of electric charge, separations based on molecular size including membrane systems, gel filtration, combinations of size and electric charge, chromatographic adsorption, partition coefficients, and the latest methods of two-dimensional separation, are certainly an array of sophisticated techniques. Proper utilization of these elegant tools imposes a demanding test of talent of the user.

I will now turn to some summary remarks on those chapters dealing with several high protein raw food materials. In all cases, it has been the aim of the researcher to understand these systems at the molecular level and how the molecules are organized into the system. Great progress has been made in this direction.

Structure of muscle and organization of the proteins into the system has reached a high level of sophistication (Goll, Chapter 6). This information is of considerable value to the scientists concerned with meat processing. With the increased cost of feed, factors relating to meat production and meat quality assume paramount importance. The discovery by Dr. Goll and his collaborators of the calcium activation of a protease which attacks the Z disks of muscle fiber is most exciting and suggests new fundamental approaches to solving meat quality problems, especially changes that occur during the post rigor aging of meat.

The present remarkable understanding of the proteins of milk and their interaction is an example of the importance of fundamental research on an important high protein food (Brunner, Chapter 7). The basic protein structural explanation for inactivation and re-activation of lipases and phosphatase is rewarding to those concerned with the processing art. Studies concerning the structures of the caseins and the role of κ-casein and calcium in stabilizing the casein micelle have been most important to developments in milk tech-nology. For example, these have been useful in solving the age-related thickening problem in sterilized milk products.

Knowledge of specific peptide bond cleavage by rennet has provided exact information regarding an old art and permits a high degree of selectivity in choosing other proteolytic enzymes for this treatment as the supply of rennet has diminished. The interaction of β-lactoglobulin and κ-casein on heat treatment is another example of an important reaction in milk processing.

Undoubtedly, the very detailed review of the properties of egg proteins will be of value to those food scientists interested in this field (Osuga and Feeney, Chapter 8). The tremendous amount of data accumulated is indicative of this interest. One might wonder why so many proteins considered to have adverse effects on man such as avidin, inhibitors of trypsin and other enzymes, and lysozyme occur in such an important food. Perhaps there are functional purposes for these substances yet to be discovered.

Much information is available on the proteins of cereals (Inglett, Chapter 9). The role of the different proteins in providing specific functional properties to formulated food products points the direction that will be so important in other formulated foods as more combinations of plant and animal proteins are used in order to provide better high protein diets with limited animal protein resources. Most revealing is the great opportunity for further work in this field that will benefit the food scientist.

While not as important volume-wise as the cereal proteins, legume proteins are an essential component of the diet of a substantial part of the world's population. Techniques for concentration and preparation of the proteins from various legume sources are in general well developed (Wolf, Chapter 10). However, there is a need for a greater amount of research on several of these products.

Two most important problems in utilization of soy protein are flavor and functionality in certain food systems. Basic information regarding soy protein structure may reveal the source of the beany flavor component and supply the answer to this perplexing problem. Although soy proteins are noted for their emulsification properties, they are not suitable replacements for casein in some dry formulated lipid-protein systems. Blocking of the free amino groups in protein isolates, by acylation for example, appears to improve functionality. Antinutritional factors, *i.e.*, trypsin inhibitors, may be more effectively inactivated with better understanding of the protein structure of these inhibitors. It is also possible that the flatulence factor in soy protein may be associated with protein and could be removed if the structures were understood.

The potentialities of single-cell protein for meeting the world's need for high protein foods are exciting (Tannenbaum, Chapter 11). Substrate economics for protein production and regulatory hurtles may prove to be serious future problems. Other problems yet to be adequately solved are release of protein from the cell and the reduction of levels of nucleic acids. Solutions to these problems may come through genetic engineering to maximize autolytic destruction of the cell wall following death of the cells and reduction of nucleic acid content through use of nucleases and processing techniques.

In the foregoing chapters, the authors have demonstrated repeatedly the usefulness of basic protein research in identifying and solving problems encountered in practical food processing. It is becoming more and more evident that industry must get involved and support this type of basic research to meet future challenges of providing adequate high quality protein foods, to remain competitive in the market place and to provide food for the expanding world population.

13

Nutrition and Technology: The Role of Sensory Properties of Proteins

C. F. Erik von Sydow
Ingmar H. Qvist

The sensory properties of the proteins and the resulting products are related on one hand to the chemistry and physics of the proteins, and on the other hand to consumer preference and acceptance. The development and the commercialization of unconventional proteins have been slower than anticipated. Perhaps we can find at least part of the explanation for this fact by analyzing the nature and importance of the sensory properties, such as flavor, in the protein raw materials and in the finished products.

Sensory properties, such as flavor, texture and appearance, are important for several obvious reasons. Eating is an important part of social and psychological well-being, implying the sensory properties are important contributors to the spectrum of satisfaction/dissatisfaction. More important, however, are the direct and indirect links between sensory properties and nutrition.

It can be accepted without much argument that flavor is important for food choice and food consumption, making flavor important from the nutritional point of view: a nutritious food with bad flavor properties will never or rarely be consumed, and even if it is consumed, digestion will be negatively affected.

In processed foods the constituents, such as the proteins, act as precursors for compounds responsible for different sensory properties of the finished product, such as color, taste and aroma. If for economical reasons, for example, one or more of the ingredients are replaced, it is apt to cause changes in the sensory properties. It is not enough to produce a new ingredient with the same sensory properties as the one to be replaced, as the processing and storage of the finished product may result in new, undesirable sensory properties. Thus, a flavorless protein ingredient is not necessarily properly replaced by another flavorless protein preparation.

Amino acid composition, amino acid sequence and perhaps peptide chain conformation of a protein are factors of importance when considering the role played by the protein as aroma precursor. Other factors are degree of hydrolysis of a protein and the resulting composition of the peptide and amino acid mixture. A protein preparation for food use always contains other components such as lipids and carbohydrates, and these components will also play their role as precursors of compounds affecting the sensory properties of the finished product.

Replacement of the protein component has so far been done mostly by trial and error and the result has therefore not always been satisfactory, as reflected by consumer rejections and protests. If we are to make the replacement procedure more rational we must get more objective data about the chemical and sensory properties of the ingredients and the resultant products and learn how these data change during processing and storage.

Strangely enough the literature is not very informative about aroma properties of protein concentrates or isolates from various sources or about their role as aroma precursors in processed foods (Qvist and von Sydow 1974). A few years ago we felt that in order to improve the situation with regard to rational replacement of one protein ingredient by another without undesirable or with desirable changes in sensory properties, we must acquire more knowledge about what is going on chemically in the food systems and how the human being—the consumer—perceives the changes resulting from substitution.

The unconventional proteins investigated were: a soy protein isolate, sodium caseinate, a fish protein concentrate, and a rapeseed protein concentrate. They were analyzed in mixture with water and salt and the samples were either unheated or heated. Other models contained also fat and starch. Finally, two of the unconventional proteins were used in a varying amount together with beef protein in model samples containing water and salt.

The analytical techniques employed were gas chromatography of the headspace volatiles and mass spectrometry for identification (Qvist and von Sydow 1974, 1976A). Roughly 100 compounds of high, medium and low volatility were identified in each sample and those compounds important from the sensory point of view were also quantified.

Sensorically, an odor quality assessment technique was used (Qvist and von Sydow 1976B). It was originally suggested by Harper *et al.* (1968A, B) but further developed and applied by us (von Sydow *et al.* 1970; von Sydow and Karlsson 1971; Persson *et al.* 1973A).

To interpret the sensory data and their changes it has been essential to analyze the material instrumentally for chemical properties and to develop models for correlating the sensory and instrumental data in a meaningful way (Qvist *et al.* 1976). The practical purpose of determining such correlations may be to supplement or complement the panel service with an instrumental technique in quality control work. Other applications can be found in product and process development work to be discussed later.

One may distinguish between two types of approaches:

1) Quantification of sensory qualities enabling the application of parametric statistical techniques, *e.g.*, like analysis of variance and regression analysis, to identify predictors. This approach has been reviewed and used by Persson *et al.* (1973A,B).

2) Categorization of sensory qualities applying cluster analysis, discriminant analysis and related techniques to identify true instrumental predictors.

Discriminant analysis of gas chromatographic data has been successfully used by several food scientists. Some examples are classification of roasted coffee and potato chips (Powers and Keith 1968); tea aroma (Gianturco *et al.* 1974); olive oils (Olias Jiménez *et al.* 1974); grape jelly (Quinlan *et al.* 1974); and blueberry-whey beverage, brands of bourbon and brands of canned peaches (Powers and Quinlan 1974). Dravnieks *et al.* (1973) successfully applied stepwise discriminant analysis to classify corn odor by the gas chromatographic technique. This work is of special value since the generality of the relations, calculated from a set of reference samples, was tested on an independent set of "unknown" samples.

With these methods the volatile compounds in unheated and heated model systems containing soy, fish, rapeseed protein or casein were studied (Qvist and von Sydow 1974, 1976A). The sensory changes caused by heat treatment in models containing soy or rapeseed protein were also investigated (Qvist and von Sydow 1976B). The volatile compounds in a canned meat product containing soy or rapeseed protein were analyzed using the instrumental and sensory methods. The possibility of "correlating" the sensory and instrumental data with the use of stepwise discriminant analysis has been tested (Qvist *et al.* 1976).

Some results of a more general interest are presented here.

COMPARISON OF BEEF, SOY, RAPESEED, CASEIN AND FISH

The concentrations of some sensorically important volatile compounds present in the headspace of heated samples containing beef, soy protein isolate, rapeseed protein concentrate, sodium caseinate and fish protein concentrate, respectively, are given in Table 13.1. The data have been extracted from publications by Persson and von Sydow (1973) and by Qvist and von Sydow (1974, 1976A). The table also includes odor threshold data obtained from the literature.

When comparing the chemical data from the different protein samples with each other and with the odor threshold values, it is obvious that the different proteins can be expected to play very different roles as aroma precursors. Compared with beef as a "standard," hydrogen sulfide is high in concentration in fish protein concentrate and low in casein. The branched chain aliphatic aldehydes, known to have "burnt, smoky" odor attributes, are too high in soy and fish, the thiols and sulfides are too low in soy and casein and the furans, having "burnt," "bread" and "cardboard" odor notes, are too high in soy and rapeseed protein. Normal

TABLE 13.1

VOLATILE COMPOUNDS IN BEEF AND IN NEW PROTEINS[a]

		Concentration in the Headspace Gas ppb (v/v)				
	Beef	Soy Protein	Sodium Caseinate	Fish Protein	Rapeseed Protein	Odor Threshold
Ethanal	890	4225	4500	8250	1280	210
n-Butanal	2	125	4	15	92	9[b]
n-Pentanal	4	780	4	3	290	12[b]
n-Hexanal	6	1710	15	1	1100	5[b]
n-Heptanal	1	41	5		53	3[b]
2-Methyl propanal	47	230	75	125	60	1[b]
2-Methyl butanal	43	41	10	84	14	
3-Methyl butanal	56	36	15	70	20	0.2[b]
Hydrogen sulfide	1300	1690	310	15,500	1550	5
Methane thiol	2400	47	89	895	240	2
Dimethyl sulfide	610	43	19	40	830	1
Dimethyl disulfide	4	2	72	14	4	8
Furan	760	370	1800	120	330	4500[b]
2-Methyl furan	82	310	165	52	440	3500[b]
2-Ethyl furan		4000	40	710	1370	8000[c]
2-Pentyl furan	9	2700	15	2	340	6[b]

[a]Samples were heated at 121°C to $F_c = 28$.
[b]Parts per billion in water.
[c]Parts per billion in cottonseed oil.

aliphatic aldehydes, contributing to some plant-oriented odor notes, are too high in concentration in soy and rapeseed protein, compared with beef protein samples. It may also be mentioned that some samples contain compounds specific for that type of protein. An example is the presence of isothiocyanates in the rapeseed protein samples. These compounds decrease in concentration on heating, while the concentrations of some nitriles increase as a result of heating. These compounds contribute to the "mustard-like" odor.

Soy and Rapeseed Protein in Meat Product

The chemical changes in, and differences between, the various samples investigated are, of course, also noticeable in changes and differences in the aroma properties as perceived by the consumer during food consumption or by the panel member in panel sessions. Figure 13.1 gives the mean panel intensities of preference, odor strength, and a few typical odor attributes for samples of canned beef, where part of the beef protein has been replaced by a soy protein isolate and a rapeseed protein concentrate, respectively (Qvist *et al.* 1976). Some similarities and differences in sensory properties and their changes are easily seen.

The odor quality, which is specific for the rapeseed protein samples, is "mustard-like." Those typical for the soy protein samples are "wort-like" and "glue-like."

Most of the odor qualities increased or decreased with increasing amount of added protein material. However, some qualities change in an irregular way, *e.g.*, "odor strength," "sulfurous," "sharp, pungent" and "sour acid," possibly because these odor qualities are generated by several chemical compounds.

The odor qualities decreasing in intensity with increasing amount of added protein material are those associated with beef, such as "cooked meat."

The odor qualities increasing in intensity with increasing amount of added protein material are the ones associated with the protein material included, such as "bread, cereal-like," "cardboard-like," "wort-like," "glue-like" and "mustard-like." The preference value for both sample groups decreased with increasing amount of added protein material.

The two types of samples differed from each other in most of the intensities of the common odor qualities. Thus, the intensities of "sulfurous," "dry, powdery," "wet wool" and "hay-like" were stronger, and "cooked meat" and "retort off-flavor" weaker in the group containing rapeseed protein than in the one containing soy protein. Odor qualities with higher intensities in the soy group than in the rapeseed group were "cardboard-like," "musty," "bread, cereal-like" and "sour, acid."

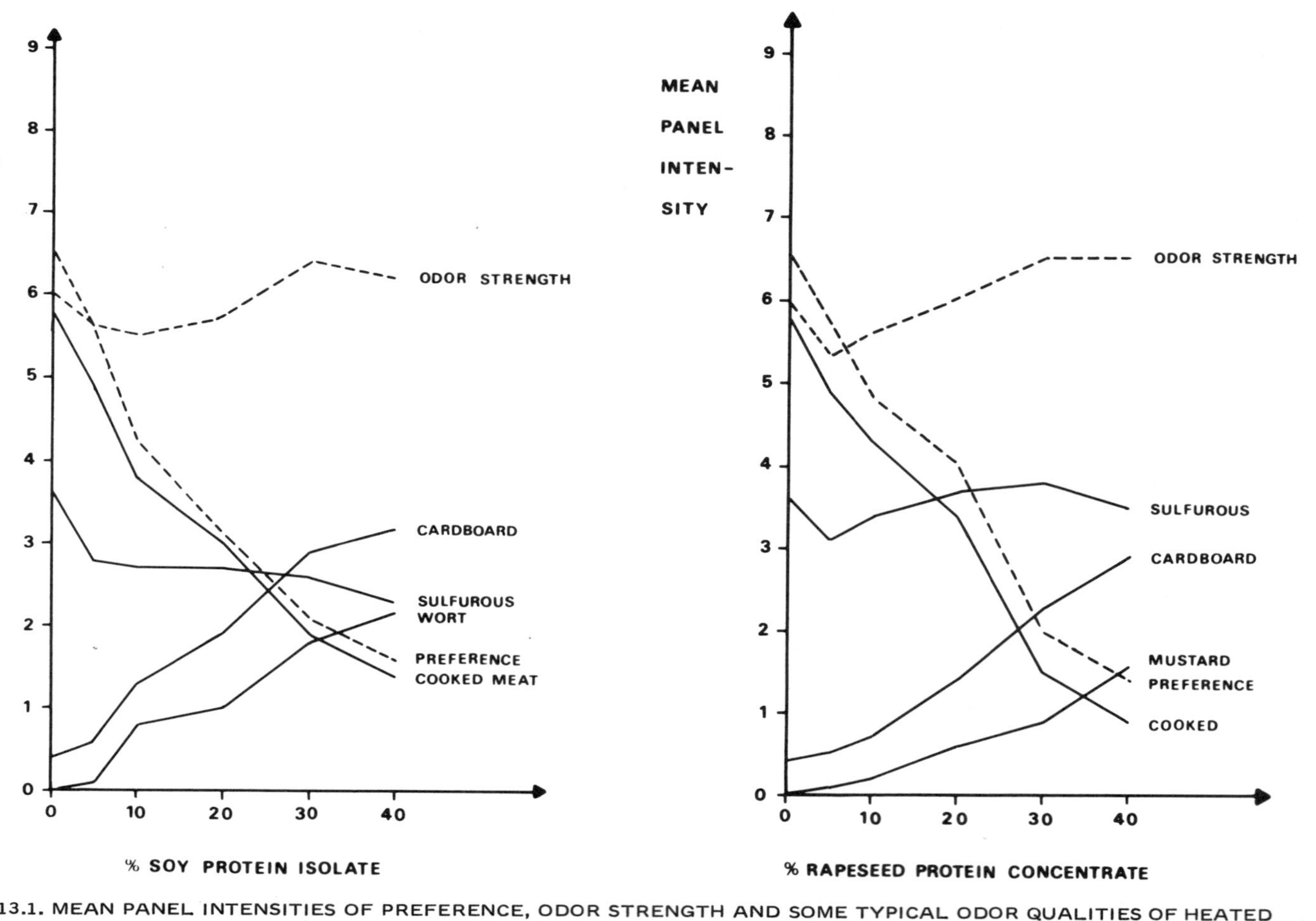

FIG. 13.1. MEAN PANEL INTENSITIES OF PREFERENCE, ODOR STRENGTH AND SOME TYPICAL ODOR QUALITIES OF HEATED BEEF SAMPLES CONTAINING VARYING AMOUNTS OF SOY AND RAPESEED PROTEIN

Correlations

One way to study the interrelation between the various sensory parameters is to calculate a correlation matrix for all the odor qualities. Usually, one finds that some odor qualities correlate very well, which means that the number of attributes used when studying one type of sample can be reduced without loss of descriptive power.

A correlation matrix was calculated for the odor qualities for each group of protein samples. It was found that only a few odor qualities correlated well, suggesting that the number of qualities used, could be reduced only to a minor extent without loss of descriptive power. For instance, the preference value correlated well with "cooked meat" for both sample groups. "Dry, powdery" correlated well with "bread, cereal-like" and "cardboard-like" for the rapeseed group but not for the soy group.

Psychophysical relations between sensory and instrumental data may be classified into three main categories (Persson *et al.* 1973B), namely accidental (*ad hoc*), predictive or causative ones. While the classification of models as predictive or nonpredictive is based on statistical criteria, the identification of causative models is based on sensory characteristics of the compounds under consideration.

The possibility of using stepwise discriminant analysis when estimating relations between odor qualities and the corresponding gas chromatographic data is based on the assumption that the samples could be grouped in a number of distinct classes according to odor intensity or quality. Moreover, it must be possible to describe the classified samples chemically, *e.g.*, with gas chromatographic data. The aim was to find those combinations of concentrations of compounds, represented by these data, which give the best "correlation" with the samples grouped according to intensity/quality, and to approximate these relations in terms of discriminant functions.

The meat samples were divided into two groups, one consisting of those containing soy protein and the other including those containing rapeseed protein. Each group was treated separately.

Especially large numbers of correct classifications in the soy protein group were obtained for the odor qualities "bread, cereal-like," "wort-like," "wet wool," "hay-like," "musty," "cooked meat" and "retort off-flavor" and in the rapeseed protein group for "cardboard-like," "wet wool," "musty" and "mustard-like" (Table 13.2) (Qvist *et al.* 1976).

The classification of "sulfurous" (Table 13.2), "cooked vegetables" and "sour, acid" was not so successful. The relative failure with these odor qualities was probably due to the fact that they have been difficult to define unambiguously, the intensity therefore having a large standard deviation and small mean value differences between the samples when assessed by the panel.

The odor quality "musty" has a high number of correct classifications. This was, however, obtained by the use of a relatively large number of classifiers (peaks) from several chemical classes, indicating that the odor quality is very complex.

Generally, misclassifications depend mainly on the relatively high frequency of overlapping between samples of the sensory assessments and to some extent to the variability of the classifiers (peaks).

Since the relevant sensory data are more or less incomplete, the identification of causative compounds can only be tentative. The classification of models as accidental (*ad hoc*) or predictive ones, on the other hand, is determined solely by the statistical decision rules applied and is a completely computerized procedure.

In this study discriminant analysis was preferred to regression analysis for the following reasons. In regression analysis the dependent variable (sensory variable) has to satisfy the condition of being continuous (normally distributed) and measurable at interval or ratio scale level (*i.e.*, having additivity properties). At least the last criterion was probably not met by the kinds of complex odor qualities under consideration. However, in discriminant analysis none of these criteria is necessary.

The results suggest that the use of discriminant analysis is suitable for instrumental separation of sensory variables on the basis of gas chromatographic data. For practical purposes of quality control work, ranking or classification is sufficient in most applications, for which purpose discriminant analysis is adequate.

TABLE 13.2

EXAMPLES OF THE OUTCOMES OF THE DISCRIMINANT ANALYSIS OF GC-DATA FOR PREDICTING INTENSITY CLASSES OF THE ODOR QUALITY[1] (CANNED BEEF/RAPESEED PROTEIN SAMPLES)

Odor Quality	No. of Intensity Classes	No. of Samples	No. of Correct Classifications	No. of Misclassifications	Gas Chromatographic Variables
Mustard-like	6	36	34	2	*n*-Hexanal 4-Pentenonitril Methyl pentyl sulfide Butenyl furan Methylethyl sulfide
Cardboard-like	6	36	31	5	2-Pentyl furan Thiophene *n*-Hexanal 2-Methyl furan
Sulfurous	6	36	24	12	*n*-Hexanal Methane thiol Dimethyl disulfide

[1] For details see Qvist *et al.* (1976).

CONCLUSIONS

The results and methods developed so far can be used in essentially two ways. One is in quality control work, the other in product and process development.

In quality control a limited number of components can be picked out and used as predictors of certain desirable or undesirable sensory properties. In this way tedious panel work can be supplemented with instrumental techniques.

A more important application is probably in product and process development work, where the results can be used in optimizing sensory and other properties. Possible parameters are:

—composition of protein mixture;
—degree of hydrolysis of the different proteins;
—other components of a nonprotein nature; and
—time-temperature-pressure relationships during processing.

Using optimization procedures a desirable product composition can be arrived at by starting with a number of components acceptable for food use. Alternatively, the results can be used to suggest changes in the production of the components, such as the unconventional protein preparations, so that they serve better as precursors of sensory and other properties in the finished products.

BIBLIOGRAPHY

DRAVNIEKS, A., REILICH, H. G., and WHITFIELD, J. 1973. Classification of corn odor by statistical analysis of gas chromatographic patterns of headspace volatiles. J. Food Sci. *38*, 34–39.

GIANTURCO, M. A., BIGGERS, R. E., and RIDLING, B. H. 1974. Seasonal variations in the composition of the volatile constituents of black tea. A numerical approach to the correlation between composition and quality of tea aroma. J. Agr. Food Chem. *22*, 758–764.

HARPER, R., BATE-SMITH, E. C., LAND, D. G., and GRIFFITHS, N. M. 1968A. A glossary of odour stimuli and their qualities. Perfum. Essent. Oil Rec. *59*, 22–37.

HARPER, R., LAND, D. G., GRIFFITHS, N. M., and BATE-SMITH, E. C. 1968B. Odour qualities: A glossary of usage. Brit. J. Psychol. *59*, 231–252.

OLÍAS JIMÉNEZ, J. M., CARBRERA MARTIN, J., and GUTIERREZ GONZÁLEZ-QUIJANO, R. 1974. Relación entre C.G.L.-percepción sensorial de los aromas de aceitas de oliva envasados. Grasas y Aceitas *25*, 34–41.

PERSSON, T., and VON SYDOW, E. 1973. The aroma of canned beef: Gas chromatographic and mass spectrometric analysis of the volatiles. J. Food Sci. *38*, 377–385.

PERSSON, T., VON SYDOW, E., and ÅKESSON, C. Å 1973A. The aroma of canned beef: Sensory properties. J. Food Sci. *38*, 386–392.

PERSSON, T., VON SYDOW, E., and ÅKESSON, C. Å 1973B. The aroma of canned beef. Models for correlation of instrumental and sensory data. J. Food Sci. *38*, 682–689.

POWERS, J. J., and KEITH, E. S. 1968. Stepwise discriminant analysis of gas chromatographic data as an aid in classifying the flavor quality of foods. J. Food Sci. *33*, 207–213.

POWERS, J. J., and QUINLAN, M. C. 1974. Refining of methods for subjective-objective evaluation of flavor. J. Agr. Food Chem. *22*, 744–749.

QUINLAN, M. C., BARGMANN, R. E., EL-GALALLI, Y. M., and POWERS, J. J. 1974. Correlation between subjective and objective measurements applied to grape jelly. J. Food Sci. *39*, 794–799.

QVIST, I. H., and VON SYDOW, E. C. F. 1974. Unconventional proteins as aroma precursors. Chemical analysis of the volatile compounds in heated soy, casein, and fish protein model systems. Agric. & Food Chem. *22*, 1077–1084.

QVIST, I. H., and VON SYDOW, E. C. F. 1976A. Unconventional proteins as aroma precursors. Chemical analysis of the volatile compounds in a heated rapeseed protein system. Agric. & Food Chem. *24*, 437–442.

QVIST, I. H., and VON SYDOW, E. C. F. 1976B. Unconventional protein as aroma precursors. Sensory analysis of heat treated soy and rapeseed protein model systems. Lebensm.-Wiss. Technol. *9*, 299–393.

QVIST, I. H., VON SYDOW, E. C. F., and ÅKESSON, C. Å. 1976. Unconventional proteins as aroma precursors: Instrumental and sensory analysis of the volatile compounds in a canned meat product containing soy or rapeseed protein. Lebensm.-Wiss. Technol. *9*, 311–320.

VON SYDOW, E., ANDERSSON, J., ANJOU, K., KARLSSON, G., LAND, D., and GRIFFITHS, N. 1970. The aroma of bilberries (*Vaccinium myrtillus* L.). 1. Identification of the press juice by sensory methods and by gas chromatography and mass spectrometry. Lebensm.-Wiss. Technol. *3*, 11–17.

VON SYDOW, E., and KARLSSON, G. 1971. The aroma of black currants. 5. The influence of heat measured by odor quality assessment techniques, Lebensm.-Wiss. Technol. *4*, 152–157.

14

Protein Quality and Its Determination

D. M. Hegsted

It has long been clear that proteins differ in nutritional quality and that this is largely, if not entirely, explained by differences in their content of essential amino acids. It is particularly important to emphasize that the amount of protein needed for growth or other functions rises as the nutritive quality of the protein falls. Thus, quantity and quality are involved in defining protein needs and the problem in defining protein quality is to find a method which allows us to deal with these two variables quantitatively, such that,

$$f_1 \times \text{g of protein A} = f_2 \times \text{g of protein B} = f_3 \times \text{g of protein C}$$

where f is the measure of protein quality.

The need for such a method is obvious since protein quality is a continuum which ranges from the highest quality, say 100%, to zero. We ought to be able to define "available protein" as the nutritive value multiplied by the protein content. An unsatisfactory compromise due to the unsatisfactory methodology available is represented by current labeling requirements for the protein content of foods. This defines the U.S. RDA as 45 g if the protein has a Protein Efficiency Ratio (PER) equivalent to casein or higher, whereas the U.S. RDA is 65 g if the PER is less than casein. Clearly, the protein need does not jump 45% as soon as quality drops fractionally below that of casein. It would be much more satisfactory and understandable if the label simply indicated the amount of "available protein" as defined above.

I propose to discuss briefly the various methods that have been used to evaluate protein quality and comment on their weaknesses. I shall close by recommending a modified slope-ratio assay and present some of our data on the utility of this assay.

Estimation of Biological Value

Up until very recently the method presumed to be the most adequate of all the methods for evaluating protein quality was the estimation of Biological Value (BV) as defined by Thomas (1909) and Mitchell (1924) and, indeed, this method has been the only method available for direct estimation of protein quality in man. The method, in essence, consists of the measure of "retention of nitrogen" in the body which is obtained by the difference in the urinary nitrogen excretion at intakes of zero protein and under the test situation divided by the "absorbed nitrogen," the absorbed nitrogen being obtained by the difference in the fecal nitrogen excretion under zero and test protein intakes. A hypothetical example is shown in Table 14.1.

It is apparent that this is a "two-point assay." It can be readily shown that BV is the slope of the line relating nitrogen balance to the absorbed nitrogen (Allison and Anderson 1945) (Fig. 14.1). Thus, if the change in nitrogen balance were proportional to the change in absorbed nitrogen, all of the nitrogen is "retained" (replaces endogenous losses); if the urinary nitrogen rises in proportion to absorbed nitrogen, the nitrogen balance does not change and the BV is zero.

Obviously, the assumption upon which this assay is based, as is true of all two-point assays, is that there is a linear relationship between nitrogen balance and absorbed nitrogen. The errors in this

TABLE 14.1

CALCULATION OF BIOLOGICAL VALUE (BV)

	Intake (g/day)	
	0	4
Fecal nitrogen	0.5	1.0
Urinary nitrogen	3.0	4.0
Nitrogen balance	−3.5	−1.0

Absorbed nitrogen = 4.0 − (1.0−0.5) = 3.5
Retained nitrogen = 3.5 − (4.0−3.0) = 2.5
Biological Value = 2.5 ÷ 3.5 × 100 = 71.4%

assumption have become obvious in recent years. Figure 14.2 is taken from data produced by Said in our laboratory (Said and Hegsted 1969) by feeding lactalbumin and wheat gluten at various levels to adult rats and measuring body water. Body protein and

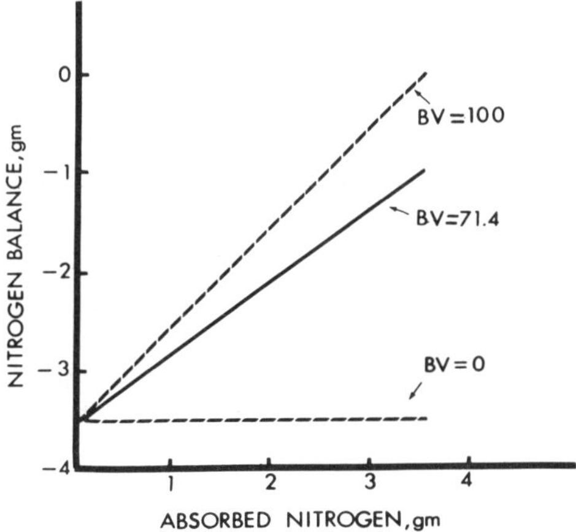

FIG. 14.1. THE CALCULATION OF BV ASSUMES LINEAR DOSE-RESPONSE LINES AS SHOWN IN THIS FIGURE
The data for the calculation of a BV of 71.4 are from Table 14.1 and compared to the dose-response curves for BVs of 100 and 0.

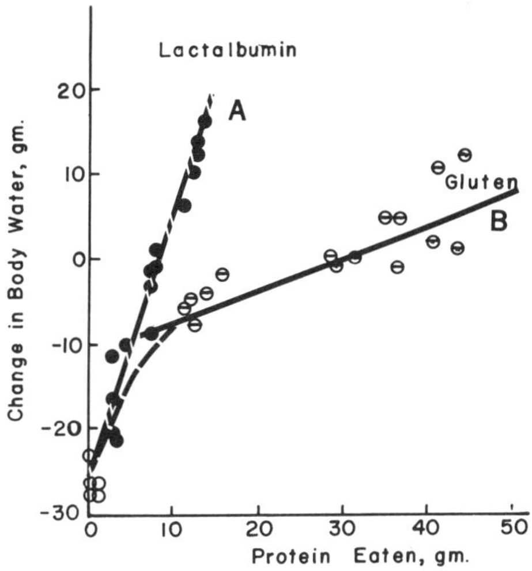

Data from Said and Hegsted (1969)
FIG. 14.2. CHANGES IN BODY WATER IN ADULT RATS CONSUMING DIFFERENT AMOUNTS OF LACTALBUMIN AND WHEAT GLUTEN

body water are highly correlated with each other. Lactalbumin approximates the supposition upon which BV is based, *i.e.*, that there is a linear relationship between intake and body protein from zero intake up to the requirement level. The situation is quite different with wheat gluten. It is obvious that if one computes BV, the slope of the line between zero intake and any selected intake, the BV of lactalbumin will be essentially constant. The BV of gluten will be entirely dependent upon the level which happens to be selected for test. At very low levels, the BV of wheat gluten will approximate that of lactalbumin and it will fall progressively as higher levels are fed. These kinds of data have been reproduced exactly in adult human subjects by Inoue and his co-workers (1974).

Limiting Amino Acids

Wheat gluten is a protein in which lysine is the first limiting amino acid. We (Chu and Hegsted 1976) have been able to show recently that the enzyme which is presumed to be the first step in lysine degradation—lysine ketoglutarate reductase—falls to very low levels in the liver of animals fed wheat gluten. This adapative response presumably results in the conservation of body lysine. Furthermore, on diets containing gluten, the level of the enzyme is much lower than in animals fed a protein-free diet. We interpret this as meaning that on a protein-free diet, the breakdown of body protein is relatively large and endogenously produced levels of lysine are relatively high, so that maximal adaptation of the enzyme level is not called for. Thus as one provides lower and lower levels of wheat gluten, one reaches a point where lysine is no longer the limiting amino acid. Total protein or some other amino acid becomes limiting.

When one feeds diets in which each essential amino acid is varied independently (Said and Hegsted 1970; Said *et al.* 1974), the data indicate that whereas the adaptation to lysine is very marked and the most obvious of all the essential amino acids, some adaptation is apparent for most amino acids. It appears, however, that the body has no capacity to adapt to low levels of threonine.

It should be noted that these kinds of effects are relatively more difficult to distinguish in young rats. This is not because they do not occur but simply because the proteins are usually tested at higher levels, well above the maintenance requirement, and modest differences in weight gain at very low levels of intake are often obscured by the variability of weights of animals within groups. Nevertheless, there is abundant evidence that such adaptation does occur. This is demonstrated by the fact that the dose-response line does not extrapolate to the value obtained by feeding a protein-free diet (Hegsted and Worcester 1966; Hegsted *et al.* 1968).

It should be noted that when Mitchell and Block (1946) developed the idea of an amino acid score the inherent assumption was that equivalent deficiencies of any essential amino acid would have the same effect upon BV. That is to say, if one knew the amino acid requirements exactly, reducing the intake of any essential amino acid to 50% of the requirement would reduce the BV to 50%. This was rationalized by the reasonable assumption that since all amino acids must be present for protein synthesis to occur, similar reductions in any amino acid should have the same effect upon protein synthesis. Conversely, a reduction in BV to 50% would result in a doubling of the protein requirement, or a reduction to 25% would quadruple the requirement regardless of which amino acid was low.

This would presumably be true if no adaption occurred, but it is clear that it is no longer tenable. Indeed, evidence of adaption was apparent in the Mitchell and Block paper. They found, for example, that gelatin gave a BV of 20- 25%. This would mean if the theory and data were correct, that adequate protein would be provided if one raised the level of gelatin in the diet four to five times that required for high quality protein. Gelatin will, of course, not support life at any level.

It should be apparent that the ability of the animal to adapt to low levels of an amino acid like lysine does not mean that lysine is any less essential than an amino acid for which adaptation is minimal. Rather, it only means that it takes different periods of time for the development of an equivalent degree of deficiency. Animals fed diets completely lacking in lysine, methionine, threonine and a protein-free diet are compared in Fig. 14.3 (Said *et al.* 1974).

The lack of adaptation to the threonine-free diet is apparent. These animals act the way Mitchell and Block assumed they would, *i.e.*, when an essential amino acid is lacking the nutritive value would be equivalent to that of a protein-free diet. Adaptation to a lysine-free diet is rapid and effective; adaptation to the methionine-free diet is less rapid and less effective. Nevertheless, deficiency of any of these will eventually kill the animal.

The problem here is somewhat similar to that which occurred in the early days of vitamin research before the multiplicity of the factors involved was known. Suppose one attempted to develop an assay for estimating the overall "vitamin potency" of a food material by feeding animals an otherwise complete diet but completely lacking in all vitamins. Then we would observe the growth over a limited period when differing quantities of the product to be assayed were added to the diet. Obviously, the response would be quite different if the limiting vitamin were vitamin B_6 where deficiency occurs very early, than if it were vitamin A where body stores may allow growth for a considerable period even though none is supplied in the diet. In an absolute sense one cannot assay for one nutrient in terms of another. Ideally, therefore, we should assay for each amino acid independently rather than try to deal with protein quality. The situation with proteins providing varying mixtures of amino acids is, however, not as complex as if we were dealing with varying mixtures of vitamins. At least in the growth process the primary function of each amino acid is to form new tissue and the proportions of amino acids deposited in protein remain rather constant in the body tissue.

It should be apparent, however, that the ability to adapt to limiting amounts of essential amino acids to varying degrees does pose problems in defining requirements of proteins and amino acids, especially in adult animals where the requirement is only for the maintenance of the body tissue proteins and, possibly, in slowly growing animals like man. The only way we know to investigate lysine requirements, for example, in adults is to feed a low lysine diet and observe some kinds of response at increasing intakes. Thus we, of necessity, work with animals adapted to some degree to a low intake. Is there any inherent health disadvantage associated with minimal levels of the catabolizing enzymes? It is not obvious why there should be but we are dealing with a complex system and we do not understand much about it. The ability to adapt to low lysine intakes almost certainly explains why human populations perform better on high cereal diets than would be expected from studies in animals that grow rapidly.

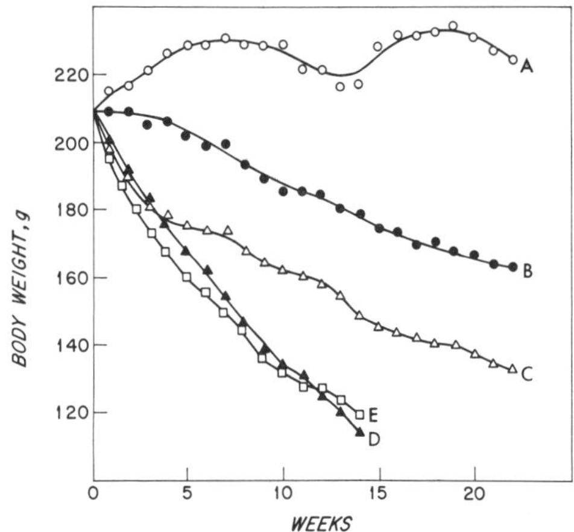

Data from Said et al. (1974)
FIG. 14.3. AVERAGE BODY WEIGHTS OF GROUPS OF ADULT RATS FED DIFFERENT DIETS
A, a diet containing approximately a maintenance amount of lactalbumin (1.77% lactalbumin); B, a lysine-free diet; C, a methionine-free diet; D, a threonine-free diet; E, a protein-free diet.

Admitting that there may not be an ideal solution to the problem it seems readily apparent that we do need estimates of protein quality. I believe that the most advantageous procedure is to utilize the slope of the line relating response to intake *over the range of intakes where response is essentially linearly related to intake* (Fig. 14.4). If the response measured is body protein, the slope of the line is the increment in body protein per unit protein consumed. One may use other estimates of body protein rather than measuring protein itself if these are demonstrated to provide satisfactory estimates of body protein. Since the slopes of the dose-response lines are dependent in part upon experimental conditions, they must always be compared to some standard protein as in any satisfactory biological assay.

Now I wish to comment on various other assays in use.

Net Protein Utilization (NPU)

NPU has all of the characteristics of BV except that it is based upon young rats and the change in body protein is estimated by carcass analysis rather than nitrogen balance, and the denominator has been defined as the intake rather than the absorbed nitrogen. Two groups of animals are used: one fed a protein-free diet and one the test protein. Food intake is measured to obtain protein eaten. The difference in body protein in the two groups divided by the protein eaten is NPU. The methodology actually described (Food and Nutrition Board 1963) is poor from the statistical point of view

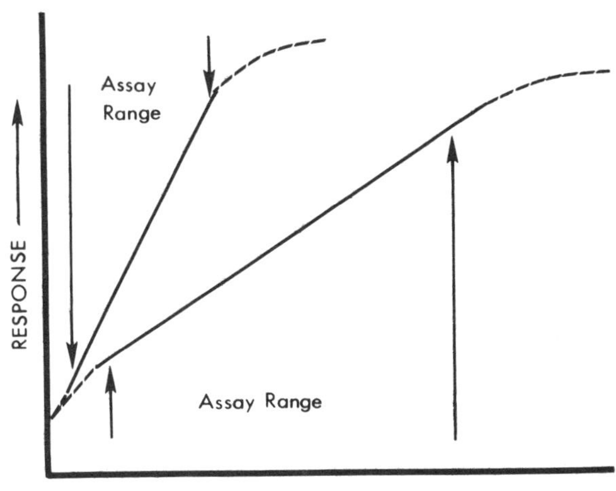

FIG. 14.4. THE APPROXIMATE NATURE OF THE DOSE-RESPONSE CURVES OBTAINED WITH DIFFERENT PROTEINS

Appropriate comparisons can only be made within the dose range where the response is approximately linear.

since it deals with groups of animals rather than individuals. The major point, however, is that this method has all of the faults described for the measurement of BV.

Net Protein Ratio (NPR)

NPR is exactly the same as NPU except that body weight is utilized rather than body nitrogen. This is probably no disadvantage in most situations although it is in some (Samonds and Hegsted, unpublished data). The theoretical objections, however, are the same as for BV and NPU.

Protein Efficiency Ratio (PER)

The AOAC method in essence consists of feeding weanling rats a diet containing 10% of the test protein for 28 days and dividing the gain in weight by the protein eaten. A group of rats is also fed a diet containing casein as a control and the PERs "corrected" to an assumed PER of casein of 2.5.

This method originated with Osborne *et al.* (1919) who were confused by the chicken and egg question of the relationship between food intake and growth. In this context this is an unproductive argument. Since animals do grow at different rates on different proteins, the method can differentiate good from bad proteins. It does not provide a quantitative measure of differences in protein quality, *i.e.*, PERs are not proportional to nutritive quality.

Some of the specific criticisms are:

(a) Jane Worcester and I (Hegsted and Worcester 1947) demonstrated almost 30 yr ago that PER is a function of weight gain and that introducing protein eaten into the calculation provides no new information. Animals fed exactly the same diet demonstrate PERs proportional to their rate of growth.

(b) It has long been recognized that since the calculation is based upon gain in weight only, it does not consider the value of the protein required for maintenance. A PER of zero does not indicate zero nutritive value. Thus, the method penalizes poor quality proteins.

(c) Because of the attempt to choose a single arbitrary level of protein, the level of protein chosen is too high for good quality proteins. Thus, the PER also underestimates the quality of the best quality proteins.

Because of both (b) and (c) above, PER is not proportional to nutritive value and PERs cannot be used to estimate "available protein" or to calculate protein requirements.

(d) Finally, the use of casein as a standard and the attempt to "correct" values to a standard PER to account for variability in rats, environmental conditions, etc., generally does not work. The original publication in the *Journal of the Association of Analytical Chemists* (Derse 1962) demonstrated, in fact, that the correction did not decrease the interlaboratory variance.

Slope-Ratio Assay

I would also note that the methods in use, with the exception of the attempt to utilize casein as a standard for the correction of PER, are absolute rather than comparative assays. This is inherently dangerous. It recalls the original attempts to quantitate vitamins, units being defined as the amount which would allow a certain growth rate. Practically all analytical methods require the use of standards whether they be chemical, physical or biologic. In biological assays, particularly those involving animals, we are always dealing with inherent variability from various sources which is large and which can only be controlled to a limited degree. Satisfactory assays attempt to deal with this by the appropriate use of standards in each assay so that comparisons are only made within assay runs at the same time with the similar animals and under similar conditions. Since we cannot clearly identify the causes of the variability, it is not possible to control or eliminate them.

We (Hegsted and Chang 1965A, B; Hegsted and Worcester 1966; Hegsted *et al.* 1968) originally proposed a slope-ratio assay for the assessment of protein quality. This involved feeding three or more levels of protein, measuring the protein consumed and the body protein, and analyzing the data according to Finney (1964). This method of statistical analysis involved the assumption that all dose-response curves have a common origin and intercept defined at zero dose. This seemed a reasonable assumption—everyone else had made this assumption up to that time. We noted, however, that the analysis of variance consistently indicated that the assays were to varying degrees unsatisfactory with the lines deviating statistically from a common intercept. Furthermore, with many proteins the deviations from an ideal assay were of little importance. Since we did not understand the cause we minimized its importance and, indeed, for many proteins it did not appreciably affect the conclusion. It is now clear, however, as I have explained that this is a serious limitation particularly if one is working with adult animals whether they be rats or man. Since it is now clear that the data do not fit an ideal assay, and we are beginning to understand why they do not, the statistical analysis cannot assume that they are ideal assays. The only alternative analysis appears to be a comparison of each protein with the standard individually rather than combining the data into an analysis of variance.

Thus as shown in the example of Fig. 14.5 the slope of each regression line is compared with that of the standard, lactalbumin. The characteristics of the regression lines in Fig. 14.5 are shown in Table 14.2. Note that the intercept of several of the proteins, particularly proteins 6 and 7 which are known to be limiting in lysine, differ substantially from the expected intercept of −19.1 which is defined by feeding the protein-free diet. The relatively small

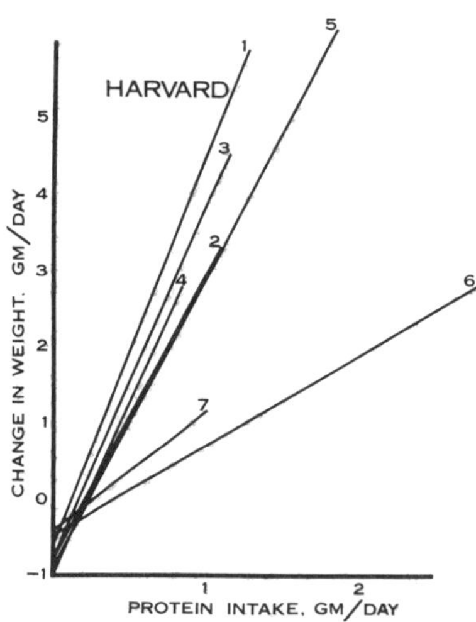

Data from
Samonds and Hegsted (1977)
FIG. 14.5. CALCULATED DOSE-RESPONSE CURVES FOR THE HARVARD DATA FROM THE COLLABORATIVE STUDY
The various proteins and the characteristics of the regression lines are given in Table 14.2.

TABLE 14.2
EXAMPLE OF ASSAY FOR RELATIVE PROTEIN VALUE[1]

Protein	Intercept	Slope	Relative Slope	Av. Relative Slope
Column	1	2	3	4
1. Lactalbumin	− 11.2	5.07	100	100
2. Soy flour	− 14.2	3.64	71.8 ± 2.5	70.0
3. Defatted meat	− 16.3	4.60	90.7 ± 3.1	76.7
4. ANRC casein	− 20.7	4.56	89.9 ± 3.6	82.6
5. Soy concentrate	− 18.8	3.79	74.8 ± 3.1	70.0
6. Wheat gluten	− 10.6	1.20	23.7 ± 1.0	22.6
7. White flour	− 9.3	1.60	31.6 ± 1.6	30.9
Zero protein	− 19.1			

[1] Columns 1, 2 and 3 are data from one laboratory in a collaborative assay. Average data from the seven collaborating laboratories are shown in column 4.

value of -11.2 obtained for lactalbumin in this assay is unusual in our experience. The calculated slopes and the slope relative to the standard lactalbumin are shown in the next two columns.

Figure 14.5 is only an example of a particular assay. It is, however, data from our laboratory as part of a collaborative assay in which the seven proteins were compared in several different laboratories. The average relative slopes obtained from the collaborative assay are shown in the final column of Table 14.2. This report is presumably to be published soon (Samonds and Hegsted, 1977). I can only summarize the major conclusions of this study which do, however, agree with our general experience.

(1) Substantial differences in growth rates and food consumption were observed in different laboratories using the same diet. These result in substantial differences in PER, NPR, and absolute slopes of the regression lines.

(2) Analysis of variance indicates substantial interlaboratory variance. The calculation of the relative slope is more nearly successful in eliminating interlaboratory differences than any other method.

(3) Maximal accuracy appears to be achieved with a feeding period between 2 and 3 weeks. Little is gained from feeding longer than 2 weeks but a 1-week period is definitely insufficient.

We have previously reported that little is gained in accuracy by using more than four animals per group (Hegsted *et al.* 1968). Thus near maximal efficiency would appear to be obtained with 12 animals (3 groups $\times$ 4 animals) fed for a 2-week period.

In terms of the utility of an assay a number of questions arise. Clearly, one would expect that the more uniform everything is, the more comparable the results obtained in one assay versus another or in one laboratory versus another would be. I think we can accept that for any assay whether it be chemical, physical or biological. It is not a very useful statement, however. The important question is what needs to be controlled.

An animal assay that required animals of the same strain, for example, would have limited utility. I simply note that in the collaborative assay above which seemed quite successful, the laboratories utilized whatever rats they were accustomed to using. Although there were rather large differences in growth rates—which presumably reflect strain or environmental conditions—these effects were largely removed by the internal comparison with the standard protein. This is, of course, what one would expect a standard to do if the assay is appropriate.

Questions might, of course, be asked about the crude fiber or fat content of the diets since these may affect food intake and thus protein intake. One would not expect this to cause variation in the response if the animals are, in fact, responding to protein. Limited data indicate that this is true. Data from diets which contained three levels of lactalbumin and from diets containing the same levels of lactalbumin with the addition of 10% celluflour are indicated in Table 14.3. As would be expected, the addition of celluflour resulted in some increase in food intake, especially at the higher level of protein. However, as shown in Fig. 14.6 when these data are plotted it is apparent that the change in food intake has not modified the dose-response curve. Figure 14.7 shows the data from another assay in which a soy flour preparation was tested and the same preparation in diets to which celluflour was added. Clearly, this had minimal, if any, effect upon the dose-response relationship.

TABLE 14.3
EFFECT OF DIETARY CELLULOSE ON RESPONSE TO LACTALBUMIN

	Food Eaten	Protein Eaten	Body Nitrogen
Lactalbumin (3.08%)	131.6 ± 19.5	4.05	1.35
Lactalbumin + 10% celluflour	143.7 ± 15.1	4.42	1.49
Lactalbumin (5.39%)	176.9 ± 16.3	9.53	2.09
Lactalbumin + 10% celluflour	186.6 ± 22.9	10.05	2.04
Lactalbumin (7.7%)	211.4 ± 10.7	16.26	2.84
Lactalbumin + 10% celluflour	268.3 ± 13.9	20.65	3.29

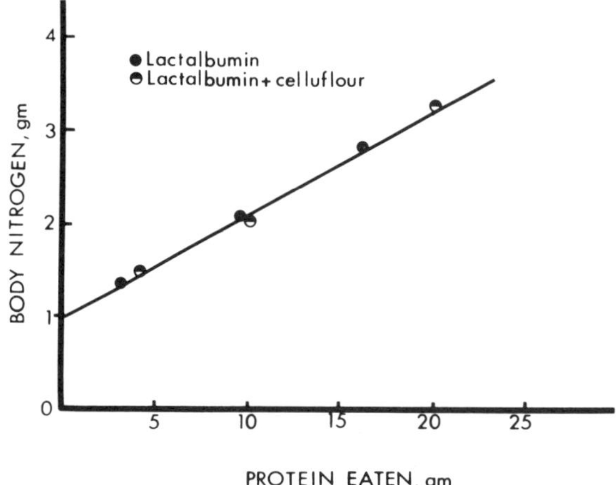

FIG. 14.6. DATA DEMONSTRATING THAT THE ADDITION OF CELLULOSE TO A DIET MAY MODIFY THE AMOUNT OF PROTEIN EATEN BUT DOES NOT APPRECIABLY MODIFY THE DOSE-RESPONSE LINE

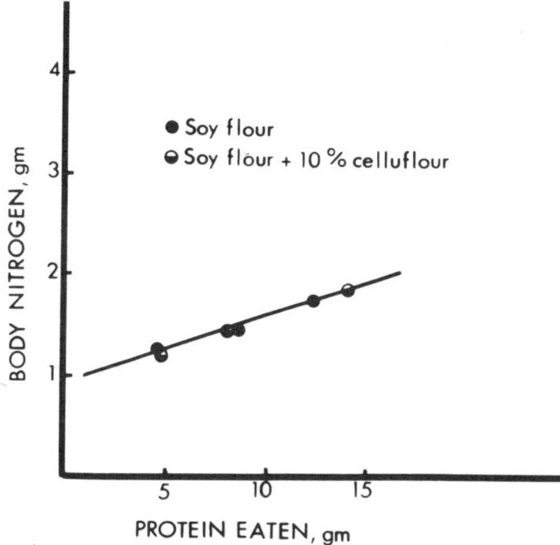

FIG. 14.7. DATA OBTAINED WITH ANOTHER PROTEIN DEMONSTRATING THAT THE ADDITION OF CELLULOSE DOES NOT MODIFY THE DOSE-RESPONSE LINE

Data from assays in which the fat content of the diet was doubled from the usual level to 10% are shown in Fig. 14.8. Even in those instances where the addition of fat appears to have diminished the food intake (and consequently the protein intake), the dose-response relationship appears unaffected.

Finally, a question might be asked as to the prior treatment of the animals. In one assay we divided the animals into two groups. One group was started on the assay the day after the animals were received in the laboratory as is our usual practice. The other group was fed a diet containing wheat gluten which just allowed weight maintenance for a two-week period. Then they were given the same test diets as in the previous groups.

The results are shown in Fig. 14.9. Although feeding the prior low protein diet did in many instances decrease the total food intake during the subsequent test periods and, thus, the protein intake, the dose-response lines are again not appreciably influenced. One must consider a diet which does not allow weight gain for a two-week period a rather severe dietary treatment and thus it does not appear that the prior diet the animals received is likely to have much effect.

With regard to animal strains we have recently completed a study in which we compared the Charles River strain with the Zucker strains, both obese and lean. A preliminary analysis of the data does not indicate that there are marked differences in the efficiency with which these widely different strains utilize protein (Seronde, Samonds and Hegsted, unpublished data).

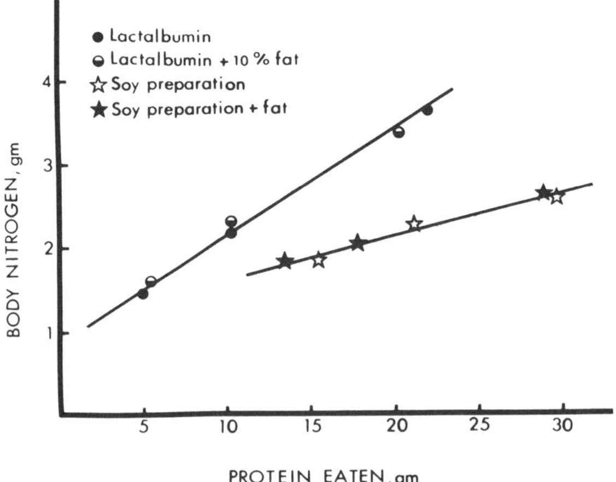

FIG. 14.8. THE ADDITION OF FAT TO A DIET MAY MODIFY FOOD AND PROTEIN CONSUMPTION BUT DOES NOT APPEAR TO MODIFY THE DOSE-RESPONSE LINE

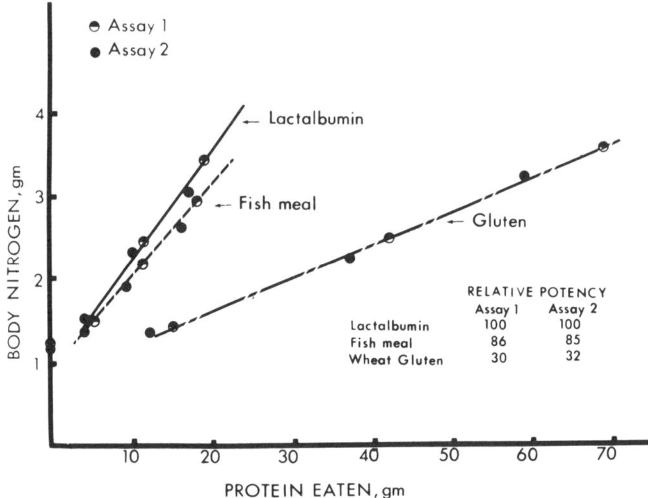

FIG. 14.9. ASSAY 1 WAS PERFORMED WITH WEANLING RATS IN THE USUAL FASHION
Additional animals were fed a diet containing 10% wheat gluten for two weeks before the assay was started. The prior dietary treatment appears to have little or no effect upon the results of the assay.

Although the data I have presented are, of course, preliminary and need confirmation, the assay appears exceptionally rugged and unaffected by a variety of factors. This would, of course, be expected if the assay is as we expect it to be—almost entirely dependent upon the protein eaten and not a variety of extraneous factors.

We now have a fair idea of the difficulties with the various procedures which have been used to estimate protein quality. All of the methods which use as a base line the feeding of a protein-free diet are inherently incorrect. This includes BV, NPU, and NPR.

All two-point assays are inherently dangerous. This includes the above as well as PER (PER utilizes the original weight as the first point on the assay). A satisfactory assay includes a range of doses so that the linearity of the dose-response curve can be tested, not assumed.

The measurement of PER is perhaps the least satisfactory of all assays available. Although it can distinguish between proteins of differing nutritional quality, it does not do so effectively and the results are not proportional to nutritive value.

The modified slope-ratio assay as described here and elsewhere appears to be the most useful and sensible approach available at this time.

ACKNOWLEDGMENTS

This work has been supported in part by grants-in-aid from the National Institutes of Health (AM-09520 and 5-K6-AM18455) and the Fund for Research and Teaching, Department of Nutrition, Harvard School of Public Health.

BIBLIOGRAPHY

ALLISON, J. B., and ANDERSON, J. A. 1945. The relationship between absorbed nitrogen, nitrogen balance and biological value of proteins in adult dogs. J. Nutr. *29*, 413–420.

CHU, S. W., and HEGSTED, D. M. 1976. Adaptive response of lysine and threonine degrading enzymes in adult rats. J. Nutr. *106*, 1089–1096.

DERSE, P. H. 1962. Evaluation of protein quality (biological method). J. Assoc. Off. Anal. Chem. *45*, 418–422.

FINNEY, D. J. 1964. Statistical Method in Biological Assay, 2nd Edition. Hafner Publishing Co., New York.

FOOD AND NUTRITION BOARD. 1963. Evaluation of Protein Quality. NAS-NRC Publ. 1100, Washington, D.C.

HEGSTED, D. M., and CHANG, Y. 1965A. Protein utilization in growing rats. 1. Relative growth index as a bioassay procedure. J. Nutr. *85*, 159–168.

HEGSTED, D. M., and CHANG, Y. 1965B. Protein utilization in growing rats at different levels of intake. J. Nutr. *87*, 19–25.

HEGSTED, D. M., NEFF, R. and WORCESTER, J. 1968. Determination of the relative nutritive value of proteins. Factors affecting precision and validity. J. Agri. Food Chem. *16*, 190–195.

HEGSTED, D. M., and WORCESTER, J. 1966. Asseessment of protein quality with young rats. Proc. 7th Int. Cong. Nutr. *4*, 1–8.

HEGSTED, D. M., and WORCESTER, J. 1947. A study of the relation between protein efficiency and gain in weight on diets of constant protein content. J. Nutr. *33*, 685–702.

INOUE, G., FUJITA, Y., KISHI, K., YAMAMOTO, S., and NIIYAMA, N. 1974. Nutritive values of egg protein and wheat gluten in young men. Nutr. Rep. Int. *10*, 201–207.

MITCHELL, H. H. 1924. A method of determining the biological value of protein. J. Biol. Chem. *58*, 873-903.

MITCHELL, H. H., and BLOCK, R. J. 1946. Some relationships between the amino acid content of proteins and their nutritive values for the rat. J. Biol. Chem. *163*, 599-620.

OSBORNE, T. B., MENDEL, L. B., and FERRY, E. L. 1919. A method of expressing numerically the growth-promoting value of proteins. J. Biol. Chem. *37*, 223-229.

SAID, A. K., and HEGSTED, D. M. 1969. Evaluation of dietary protein quality in adult rats. J. Nutr. *99*, 474-480.

SAID, A. K., and HEGSTED, D. M. 1970. Response of adult rats to low dietary levels of essential amino acids. J. Nutr. *100*, 1363-1376.

SAID, A. K., HEGSTED, D. M., and HAYES, K. C. 1974. Response of adult rats to deficiencies of different essential amino acids. Brit. J. Nutr. *31*, 47-57.

SAMONDS, K. W., and HEGSTED, D. M. 1977. A collaborative study to evaluate four methods of estimating protein quality: Protein efficiency ratio, net protein ratio, protein value and relative protein value. *In* Evaluation of Protein Quality (rev. Publ. 1100). NAS-NRC, Washington, D.C. (In Press).

THOMAS, K. 1909. Ueber die biologische Wertigkeit der stickstoff-substanzen in verschiedenen Nahrungsmitteln. Arch. Physiol. 219-302.

15

Human Amino Acid and Nitrogen Requirements as the Basis for Evaluation of Nutritional Quality of Proteins

A. E. Harper

Although knowledge—be it of science, art, music or literature—has its own intrinsic value, a major objective of the pursuit of scientific knowledge is the betterment of human welfare. Knowledge of human nutrient requirements is of basic interest, but the major stimulus for investigation of nutritional needs is the potential for using the information obtained to solve practical problems relating to food and health.

The protein and amino acid requirements of man have been investigated in considerable detail. The information obtained has been used to evaluate the nutritional status of human populations, in the development of proposals for the alleviation of malnutrition and as the basis for establishing standards for the nutritional quality of foods. Despite the extensive accumulation of information on the subject and its application in a variety of practical programs, differences of opinion continue to exist about the appropriate use of knowledge of protein nutrition in programs to improve protein nutriture (Waterlow and Payne 1975; Harper and Hegsted 1974), and even about its use in assessing protein needs (Waterlow and Payne 1975; Scrimshaw 1976; Harper *et al.* 1973). It is important that the basis for controversies about protein nutrition be recognized and that differences of opinion be resolved through both research and discussion in order to ensure that programs for solving practical problems will be as effective as is possible. In the meantime, until fully adequate answers to unresolved questions are obtained, it is important to define the limits of accuracy of current knowledge so that what is known will be used appropriately within those limits.

The requirement for protein is a dual requirement. Food proteins must provide nitrogen to serve as a precursor for the many nitrogenous compounds that are synthesized in the body. They must also provide the nutritionally essential or indispensable amino acids that cannot be synthesized by the body. Besides, amino acids can be oxidized for energy (calories). Because the proteins of the diet provide nine amino acids that are essential for tissue protein formation as well as nitrogen for many purposes and because amino acids can be diverted from their primary use as building materials for tissue proteins to serve as energy sources, unique difficulties are encountered both in determining human protein requirements and in assessing differences in nutritional value among food proteins.

Many methods for evaluating the nutritional quality of proteins have been devised (Hegsted 1974; Campbell 1963; NAS/NRC 1963); these can be used to rank proteins with considerable confidence. However, as Hegsted (1974) has pointed out, few of these methods provide satisfactory quantitative comparisons among proteins or predict accurately how much of a particular food protein will be required to meet both the nitrogen and amino acid requirements of man. Also, despite an immense amount of work on the estimation of protein quality, there is no simple way of predicting the nutritional quality of mixtures of food proteins, such as occur in ordinary diets, from measurements made on individual foods. The need for solutions to these problems in order to obtain accurate information for nutritional labeling of foods and for the development of nutritional standards for food products has been an impetus for renewed interest in this subject (NAS/NRC 1974A; FAO/WHO 1973). Another important stimulus for reassessment of knowledge of protein nutrition generally is the widespread occurrence of malnutrition throughout the poor countries of the world, malnutrition that is associated with deficits in intake of both energy and protein (Gopalan 1969), and the need to alleviate this condition.

My objectives in this paper are to present information about the nitrogen and indispensable amino acid requirements of man and to discuss the potential for using information about these requirements to predict the effectiveness of food proteins in meeting human protein needs.

Protein (Nitrogen) Requirements and Allowances

Attention has been given regularly over the years to assessment and reassessment of protein requirements. In 1923, Lusk reviewed experimental results of Voit, Siven, Chittenden and others and concluded that adult man could be maintained in good health and nitrogen equilibrium with an intake of somewhat above 4 and somewhat below 6 g of nitrogen (equivalent to 25–40 g of protein) daily. Human protein requirements have been reassessed frequently since then and most recently the available literature has been reviewed by a Food and Agriculture Organization/World Health Organization (FAO/WHO) Expert Committee (1973) and by a US National Academy of Sciences/National Research Council (NAS/NRC) Committee (1974A; Williams *et al.* 1974). These committees concluded that a reasonable average value for the amount of nitrogen from high quality proteins needed to maintain adults in nitrogen equilibrium was 75 mg/kg of body wt/day or about 33 g of protein/day for a person weighing 70 kg (Table 15.1), close to the midpoint of the range proposed by Lusk (1923). Since diets throughout the world rarely provide less than 10–12% of the energy as protein (Fig. 15.1) (Perisse *et al.* 1969), the average protein intake of a 70-kg person who is performing only light activity and consuming the Recommended Dietary Allowance (NAS/NRC, 1974B) for energy of 2700 kcal is 67–81 g of protein/day. Assuming that this protein is used with an efficiency of only 75% of that of high quality proteins, this would be the equivalent of 50–60 g of high quality protein. It is, therefore, not surprising that protein deficiency is rare in adults unless they do not have enough food to meet their energy requirements because of poverty, famine or some other catastrophe. This bears out the adage that if the energy (calorie) requirement of the adult is met, the protein requirement will take care of itself.

TABLE 15.1

AVERAGE NITROGEN INTAKES THAT MAINTAIN NITROGEN
EQUILIBRIUM IN ADULTS[1]

Source	Amount (mg/kg/day)	Efficiency of Utilization (%)
Milk, egg, soya flour	68–74	60–70
Vegetable-meat, cereal-milk and some vegetable mixtures	75–85	59–72
Wheat, corn, oats, rice	80–90	52–58
White flour	106	40

Average of lower values for requirement = 75 mg N/kg body wt = 33 g protein/70-kg person/day

[1] Based on FAO/WHO (1973) and NAS/NRC (1974A).

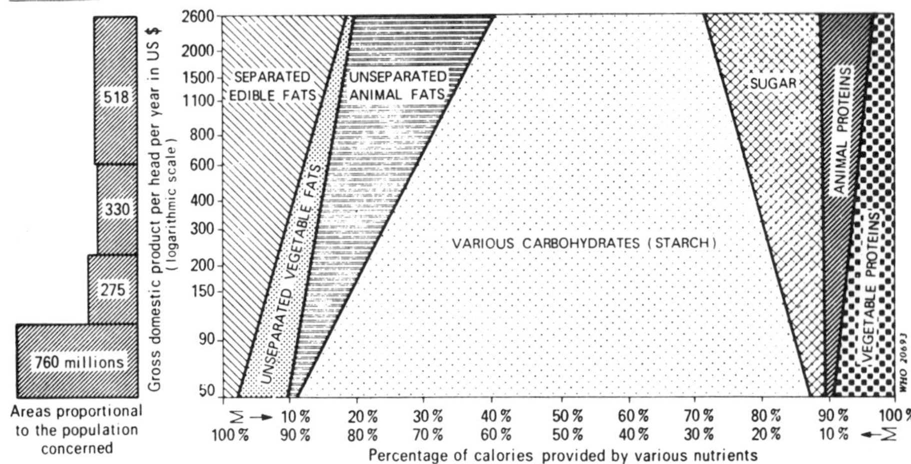

* Correlation based on 85 countries.

From Périssé et al. (1969)

FIG. 15.1. LIMITED EFFECT OF INCOME ON PROPORTION OF CALORIES FROM
PROTEIN

Note, however, that as income rises vegetable proteins tend to be displaced by animal proteins.

Since adults who are maintaining constant body weight should be
in nitrogen equilibrium, *i.e.*, the amount of nitrogen lost from the
body should equal the amount consumed, a previous FAO/WHO
Expert Committee (FAO/WHO 1965) proposed that the amount of
nitrogen lost from the body when none is being consumed should
provide an estimate of the minimum nitrogen requirement for
maintenance. Numerous estimates of the amounts of nitrogen lost
through urine, feces, skin and various body secretions and excretions
have been made (FAO/WHO 1973; Williams *et al.* 1974). A summary
of these (Table 15.2) indicates that to replace nitrogen losses would
require only 23 g of protein/day (70-kg person), well below the
estimated average protein requirement of 33 g/day obtained from
nitrogen balance studies. The difference is evidently due to inefficient utilization of even the highest quality proteins when the
amount of nitrogen consumed approaches the requirement value
(Calloway and Margen 1971).

Protein requirements during the first year of life can be estimated
by calculating the amount of protein consumed by infants who have
been fed enough of a formula diet which is relatively low in protein
but which enables them to grow at a satisfactory rate (FAO/WHO
1973; Williams *et al.* 1974). Table 15.3 provides a summary of values
from studies by Foman and associates (FAO/WHO 1973; Williams *et*

al. 1974) for protein requirements of infants from one to six months of age. Values for older children are estimated by interpolation between infant and adult values from information about body composition and rates of growth. Estimates for these, made independently by Hegsted (1959) and by Waterlow (1970), are in close agreement. The requirement falls quite rapidly during the first year of life; then more slowly as maturity is approached. It is worth noting (Fig. 15.2) that by two years of age, 90% of the protein requirement is for maintenance and only 10% for growth (Foman 1967).

Average nutritional requirements are of basic interest but from a practical viewpoint we must be concerned about meeting the needs of those whose requirements exceed the average. Since the requirement of an individual can be determined only through elaborate metabolic studies, it is necessary to devise a procedure for estimating "safe intakes" for an entire population from the information available about the requirements of a limited number of individuals.

TABLE 15.2

AVERAGE OBLIGATORY NITROGEN LOSSES OF
ADULT (70 KG) MAN[1]

Source	Nitrogen Loss (g/day)
Urine	2.35
Feces	0.85
Integuments	0.30
Other	0.15
Total	3.65

[1] Based on FAO/WHO (1973) and NAS/NRC (1974A).
$(3.65 \times 6.25) = 22.8$ g of protein.

TABLE 15.3

CHANGE IN PROTEIN REQUIREMENTS WITH AGE[1]

Age (months)	Requirement for Protein (g/kg body wt/day)	Age (yr)	Requirement for Protein (g/kg body wt/day)
1	2.2	2	0.9
2	2.0	6	0.8
4	1.7	12	0.7
6	1.5	15	0.6
9	1.25	adult	0.5
15	1.0		

[1] Based on information in Williams *et al.* (1974).

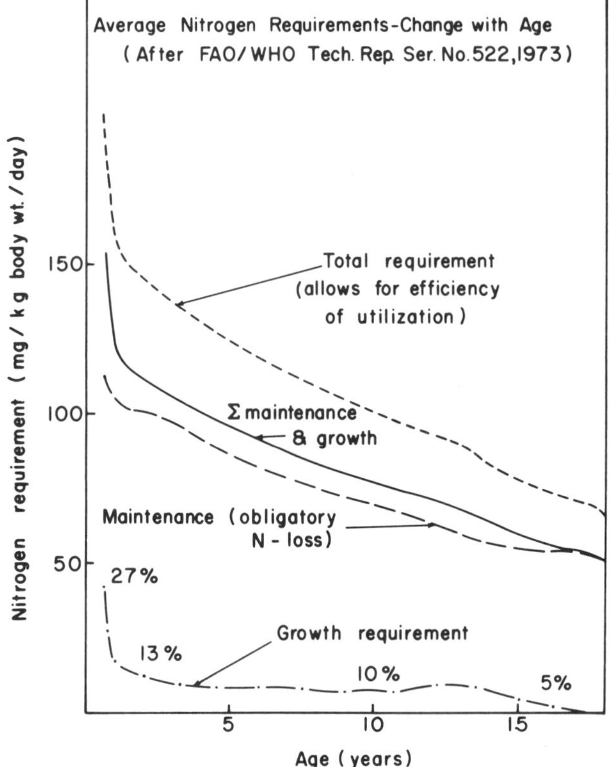

FIG. 15.2. CHANGE WITH AGE IN PROPORTION OF DIETARY NITROGEN REQUIRED FOR GROWTH AND MAINTENANCE
Based on information in FAO/WHO (1973).

This type of estimate is needed as a standard for evaluating the adequacy of nutritional intakes of populations, for developing nutritional programs for groups who are considered to be at risk, and for planning food supplies. The Canadians call such values "dietary standards" (Minister of National Health and Welfare 1975); FAO/ WHO call them "safe intakes" (FAO/WHO 1973); they are our "recommended dietary allowances" (NAS/NRC 1974B). In essence, they are the amounts of nutrients each person in a healthy population must consume in order to ensure that the nutritional needs of those with the highest requirements will be met. This is a public health concept. The underlying basis for it is satistical but, as with the establishment of practical standards generally, decisions about safe intakes involve an element of judgment.

Measurements of most biological variables on a large number of individuals are found to follow a Gaussian distribution, *i.e.*, they fit the normal distribution curve. With this type of distribution, 95% of the population will fall within the range of the mean plus or minus twice the coefficient of variation (standard deviation as a percent of the mean) and almost all of the population within plus or minus three times the coefficient of variation. With enough individual measurements, a reliable estimate of the standard deviation and coefficient of variation can be obtained. Coefficients of variation for a wide variety of biological variables are between 10 and 20%. If the coefficient of variation exceeds 15%, it is usually indicative of the population not being homogeneous, of the method of measurement being inadequate or of some other variable that influences the results not being controlled effectively. For measurements of protein requirements, coefficients of variation between 10 and 20% have been reported, with the preponderance of values falling within 15% (Waterlow and Payne 1975; FAO/WHO 1973).

In developing safe intakes (FAO/WHO 1973) and recommended allowances (NAS/NRC 1974B) for protein, a coefficient of variation of 15% has been accepted. Waterlow and Payne (1975) have concluded that the value of 1.27 g/kg of body wt/day of high quality protein (increasing to 1.7 g/kg of body wt/day when dietary protein quality is low) obtained in this way for the one year old child is very well founded. The safe intake (FAO/WHO 1973) or recommended dietary allowance (NAS/NRC 1974B) for adults of 0.8 g of protein/kg of body wt/day was derived similarly but assumes that the mixture of proteins in usual diets are utilized only 70–75% as well as the high quality proteins of the experimental diets used for estimating requirements. This amount should cover the requirements of 97.5% of the adult population. In view of the capacity of organisms to adapt to nutrient intakes somewhat below those commonly recommended (Mitchell 1944; Waterlow 1968; Harper 1971; Waterlow and Garlick 1975) and in view of the generous allowance for protein quality corrections (see Table 15.1 for milk and egg vs mixed diets), the FAO/WHO safe intake and the NAS/NRC recommended allowance would be expected to meet the needs of all individuals in a healthy population.

Both requirements and allowances for protein increase in several other conditions besides growth (FAO/WHO 1973; NAS/NRC 1974B). Requirements increase during pregnancy and lactation because of the need for additional protein for the synthesis of fetal and maternal tissues or milk. During recovery from protein depletion after periods of protein deprivation or as a result of body losses due to illness, trauma or stress, or in any other condition in which new tissue must be synthesized, requirements are elevated just as they are during periods of growth. It has been proposed that requirement figures should be increased to provide for effects of minor infections and stresses (Scrimshaw 1976). However, since energy requirements also increase when the body has been depleted owing to illness or stress, the increase in food consumption that occurs as the result of such depletion (Harper and Boyle 1976) ensures that extra protein will be consumed during the repletion period. There would, therefore, seem to be no logical basis for increasing recommended protein allowances generally to compensate for nitrogen losses due to stress or illness. The rational approach is the direct one, *i.e.*, to prevent or cure illness and relieve stress and, until this has been accomplished, to deal with them as special situations if they are severe enough to elevate nutritional needs appreciably.

There have been many suggestions that work increases protein requirements, but evidence obtained from experiments in which energy requirements have been met adequately from nonprotein sources do not support this view (NAS/NRC 1974B). Nevertheless, since energy needs have primacy over others and since protein can serve as an energy source, both dietary and body protein will be used as a source of energy if energy needs are not met. Nitrogen retention falls when energy intake is inadequate and if this occurs during experiments in which nitrogen (protein) requirements are being determined, the estimated requirement will be low. Conversely, if energy intake is excessive in such experiments, nitrogen retention tends to be high and the requirement for protein may be under-estimated (Calloway and Spector 1954).

Recently, three groups (Scrimshaw 1976; Inoue *et al.* 1973; Calloway 1975) have questioned the adequacy of current estimates of protein requirements and allowances for adults because many of the values on which these were based were taken from studies in which subjects consumed amounts of energy that were considered to exceed their needs. The degree of underestimation suggested by these observations is not clearly quantified owing to the limited number of appropriate comparisons, but one study (Inoue *et al.* 1973) on which some weight is placed (Scrimshaw 1976) would lead to the conclusion that the average adult male requirement for high quality protein should be 39 rather than 33 g/day.

How significant is this difference from a practical viewpoint? Quantification of nitrogen requirements is not as accurate as is desirable (Williams *et al.* 1974), especially since establishing the point at which an adult achieves nitrogen equilibrium depends upon the accuracy with which the significance of a small difference between two large values for excretion and intake of nitrogen can be established. If the coefficient of variation of estimates of nitrogen requirements is only 10% rather than 15% (Waterlow and Payne

1975), the recommended safe intakes would be unchanged. Also, quantification of energy needs is notoriously difficult (FAO/WHO 1973; NAS/NRC 1974B) as energy expenditure can vary considerably from day to day. Accurate estimates of nitrogen requirement depend upon accurate estimates of energy needs.

Recent observations on the relationship between energy intake and nitrogen balance (Inoue *et al.* 1973; Calloway 1975) have resulted in the surfacing (Scrimshaw 1976) of an old controversy (Lusk 1923) and, since many of the questions raised do not have adequate answers, nutrition scientists must try to deal with these questions rationally until the answers are obtained. Is nitrogen retention improved by a high energy intake if the energy is expended rather than being stored? Do fluctuations in food intake from day to day owing to fluctuations in energy expenditure result in periods of low nitrogen retention alternated with periods of high nitrogen retention so that, over long periods of time, overall nitrogen equilibrium is achieved despite periods of negative nitrogen balance? Is there a valid way of relating protein requirements quantitatively to the highly fluctuating energy requirement? These and other related questions require answers that can be obtained only through meticulous research. In the meantime, the practical question that must be faced before those answers are obtained is whether recommendations made on the basis of currently available information are likely to be seriously in error.

Recommended dietary allowances and other dietary standards, it should be emphasized, are not recommendations for the formulation of practical diets. Recommended dietary allowances are designed as guides for assessing the probability of nutrient intakes of populations being inadequate (NAS/NRC 1974B). Therefore, it is important that safe intakes (FAO/WHO), dietary standards (Canadian Dietary Standard) and recommended dietary allowances (NAS/NRC) not be overestimates of protein needs, as will be the case if protein allowances are increased arbitrarily.

The NAS/NRC Recommended Dietary Allowance (RDA) for protein of 56 g/day for an adult male is identical with the FAO/WHO safe intake for protein of moderate quality. Assuming that the average requirement for high quality protein is 39 g/day as has been suggested (Scrimshaw 1976) and that some people may have requirements as much as 30% above this (FAO/WHO 1973; NAS/NRC 1974B), the recommended allowance for high quality protein would then become 51 g/day. Published reports indicate that various mixtures of proteins, including mixtures that contain a large proportion of vegetable proteins, are used 85–90% as efficiently as egg or milk proteins for meeting adult protein needs (FAO/WHO

1973; Williams *et al.* 1974). Thus the RDA of 56 g or 0.8 g/kg of body weight for adults should be a satisfactory guide to the adequacy of protein intake. If the diet is composed largely of staples with protein quality less than that of mixed diets, then, as the FAO/WHO report (1973) emphasizes, a further upward correction in the recommended safe intake is required. Even assuming that average requirements of adults for high quality protein have been somewhat underestimated, recommendations for safe intakes of high quality protein are of theoretical interest only. There is no evidence that the practical allowances and safe intakes from mixed diets are inadequate.

Values for protein requirements and allowances for young children, as was mentioned, are considered highly reliable. The young child's needs are met by diets that provide between 7 and 8% of the dietary energy (calories) from high quality protein. It is, therefore, difficult to accept the conclusion that diets for adults should have to provide more than that proportion of protein (Scrimshaw 1976), even if the quality were somewhat below the best, unless energy intake were inadequate. Diets for adults based on present recommended allowances or safe intakes from a mixture of foods of moderate protein quality would provide from 7.2–8.3% of the energy from protein. Most diets provide 10% or more of the energy from protein. Thus, the probability that protein intakes of adults will be inadequate if they are consuming enough food to meet their energy needs is highly unlikely. This is particularly so if they are performing more than light activity as is the case for almost all but the economically well-to-do whose diets almost invariably contain more and higher quality protein than those of the economically deprived.

Finally, malnutrition in the presence of an adequate national food supply is a disease of the pre-school child. Raising the safe intake of protein for adults will do nothing to relieve it. If national nutrition policies that are directed toward alleviating malnutrition emphasize the need for improving total food and energy intakes rather than just protein intake because the major deficit among young children who are susceptible to malnutrition is considered to be a deficit of energy (Waterlow and Payne 1975; Harper and Hegsted 1974; Gopalan 1969), protein supplies will be improved at the same time. To propose, as has been done (Scrimshaw 1976), that average adult requirements for high quality protein have been underestimated to an extent that may result in nutrition policies that will be a hazard to public health and that this has been done capriciously by FAO/WHO and NAS/NRC committees is clearly unwarranted, even though there is a need for a better understanding of the relationship between nitrogen retention and energy intake and for continuing reconsideration of recommendations for "safe" intakes of protein.

Requirements for Amino Acids

Information about amino acid requirements has also been reviewed by the NAS/NRC (Williams *et al.* 1974) and FAO/WHO (1973) committees that reviewed protein requirements. The two major studies of amino acid requirements of adults are those of Rose and his associates on young men (1957) and those of Leverton and her associates on young women (1959). As well, there have been several studies by others of requirements for one or more of the amino acids (Williams *et al.* 1974). Hegsted (1963) has estimated requirements for each of the indispensable amino acids by regression analysis of the relationship between amino acid intake and nitrogen retention, using all of the individual values for women. The estimated amino acid requirements of adults are listed in Table 15.4. The values for men tend to be high, as Rose and associates accepted as the requirement the highest of the values for the individuals they studied. Although within each study there was no clear relationship between requirement and body weight, when the values for men and those for women are expressed per kg of body weight, differences between the sexes are small (FAO/WHO 1973; Williams *et al.* 1974). This suggests that if body weight differences are large, a relationship between requirement and body weight can be demonstrated. In view of the variability observed and the different criteria used for estimation of requirements, differences between sexes are of doubtful significance. FAO/WHO (1973) proposed a combined average value (Table 15.4, last column) as representative of requirements. Williams *et al.* (1974) accepted the Hegsted (1963) values (Table 15.4, column 6) as the most reliable estimates.

TABLE 15.4
AMINO ACID REQUIREMENTS OF HUMAN ADULTS

	Male[1]		Female[2]			Proposed Requirement[3]
	mg/day	mg/kg body wt	mg/day	mg/kg body wt		mg/kg/day
Ile	700	10.1	450	7.8	9.5	10
Leu	1100	15.7	620	10.7	12.5	14
Lys	800	11.4	500	8.6	9.4	12
Met + Cys	1010	14.4	700	12.1	12.1	13
Phe + Tyr	1100	15.7	700	12.1	12.1	14
Thr	500	7.1	305	5.3	6.5	7
Trp	250	3.6	160	2.8	2.9	3.5
Val	800	11.4	650	11.2	10.7	10

[1] Rose (1957).
[2] Leverton (1959); Hegsted (1963).
[3] FAO/WHO (1973); the values of Hegsted (1963), preceding column, were accepted as the most reliable estimates by Williams *et al.* (1974).

Hegsted (1963) has emphasized that variations among the amino acid requirements of individuals are large. This raises a question about the reliability of the estimated requirement values. However, if the expected upper limits for the observed requirements of men are calculated on the assumption that requirements are distributed normally and that the coefficient of variation is 15%, the predicted upper limit (midpoint + 45%) is seen to be higher than that observed (Table 15.5). In a similar analysis of the values for women, the highest values for leucine and threonine requirements exceeded the predicted range but the discrepancies were small. Thus, the range of requirements observed, although quite wide, is not unusually so.

Amino acid requirements, like protein requirements, fall with age as shown in Table 15.6. Amino acid requirements of infants have been estimated by calculating the amounts of amino acids provided by quantities of milk or formula diets that support satisfactory growth rates (Williams *et al.* 1974). Also, infant requirements for most of the amino acids have been determined directly by Holt and his associates (1960) using highly purified diets containing crystalline amino acids. Nakagawa and his associates (references in Williams *et al.* 1974) have estimated amino acid requirements of children but the increments between the intakes they tested were large, so the estimates are less accurate than those for the other age groups. It is evident from Table 15.6 that, although the protein (nitrogen) requirement falls by maturity to one-quarter of that in early infancy, amino acid requirements fall to one-eighth to one-tenth. The requirement for the sulfur-containing amino acids is an exception. The adult requirement for these is comparatively high and may have been overestimated (Zezulka and Calloway 1976).

TABLE 15.5
RANGE OF AMINO ACID REQUIREMENTS OF MEN

	Requirement Range (mg/day)			Expected Upper Limit Assuming Requirements are Normally Distributed and C.V. = 15% of Midpoint Value (Midpoint + 45%)
	Min.	Midpoint	Max.	
Ile	650	675	700	979
Leu	500	800	1100	1160
Lys	400	600	800	870
Met	800	950	1100	1378
Met + Cys	910	960	1010	1392
Phe + Tyr	800	950	1100	1378
Thr	300	400	500	580
Trp	150	200	250	290
Val	400	600	800	870

TABLE 15.6
CHANGE IN AMINO ACID REQUIREMENTS WITH AGE[1]

	Infant	Child	Adult
		(mg/kg of body wt/day)	
Protein	1500	750	470
His	33	?	?
Ile	83	28	9.5
Leu	135	42	12.5
Lys	99	44	9.4
Met + Cys	49	22	12.1
Phe + Tyr	141	22	12.1
Thr	68	28	6.5
Trp	21	3.3	2.9
Val	92	25	10.7
Total	721	214	76
Total as % of protein	48	30	16

[1] After Williams *et al.* (1974).

If the requirements for each age group are divided by a factor based on the average fall in requirement for that age group, the pattern of requirements, with the exception of the value for the sulfur-containing amino acids again, does not change greatly with age (Williams *et al.* 1974). The agreement is particularly good for infants and adults and, as would be expected from the comment above, less so for the child. An important point to note here is that, if the amino acid requirement values are realistic, the six-month-old infant requires between 40 and 50% of the amino acids in protein as essential when protein intake is just adequate to meet the nitrogen requirement, whereas the adult requires less than 20% as essential. Protein quality is thus much more critical for the infant than for the adult. This can account for the fact that differences in efficiency of utilization of high quality proteins and the proteins of mixed diets by adults are small (FAO/WHO 1973; Williams *et al.* 1974; and Table 15.1). It also indicates that programs for the improvement of protein quality cannot be expected to benefit adults and that if such programs are undertaken, they should focus on the needs of the youngest children (Waterlow and Payne 1975; Harper and Hegsted 1974).

Nutritional Value of Food Proteins

Since food proteins must meet the dual requirement for nitrogen and essential amino acids, proteins that meet the requirement for each essential amino acid, when they are consumed in an amount that just meets the nitrogen requirement, will be of high nutritional quality. If the amount of food protein that will just meet the nitrogen requirement does not carry with it enough of one or more of the indispensable amino acids, a larger amount of such a protein must be consumed in order to meet the requirements for amino acids that are limiting. This will provide other amino acids in excess and the extra amino acids will then be used as a source of energy. Efficiency of utilization of such proteins will be less than for high quality proteins, especially when total protein intake is low. As nitrogen requirements are determined using diets containing proteins of high nutritional quality, whereas diets ordinarily consumed contain proteins of lower quality, it is of considerable practical importance to be able to determine how much the nutritional value of a particular food protein deviates from that of the highest quality proteins.

Besides amino acid composition, other factors influence the efficiency with which food proteins are used. Obviously efficiency of nitrogen utilization falls when the amount of protein consumed is above the nitrogen requirement value, regardless of the quality of the protein. Figure 15.3, taken from a paper by Bressani *et al.* (1973), illustrates the fall in efficiency of utilization of proteins by children with increasing protein intake. Biological value, or any other measure of protein utilization, will decrease as protein intake increases.

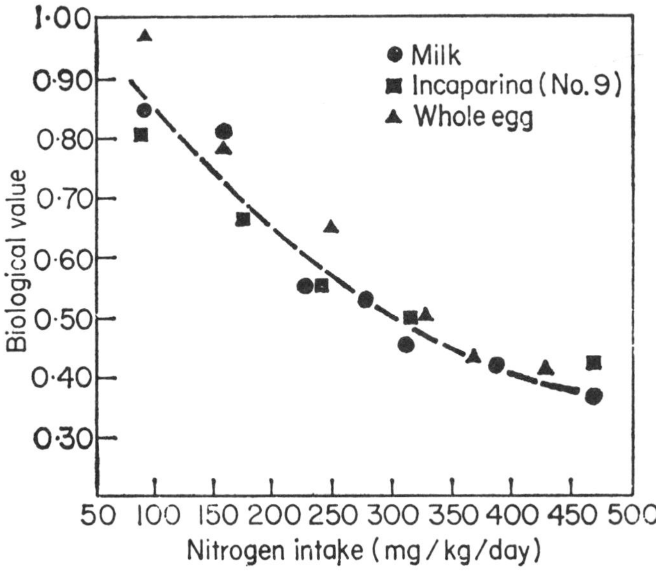

From Bressani et al. (1973)
FIG. 15.3. EFFECT OF N INTAKE ON BIOLOGICAL VALUE OF PROTEINS FOR CHILDREN

Also, and perhaps more importantly, the highest values for measures of efficiency of protein utilization are obtained when protein intake is below the requirement. As the requirement value is approached, efficiency of utilization decreases. Even for milk protein (and the same is true for egg) at intakes that meet human adult requirements, efficiency of nitrogen utilization is only 65–70% (Table 15.1). As was mentioned earlier many diets made up of mixtures of proteins are used by adults as efficiently as the highest quality proteins when both are consumed in quantities that meet the nitrogen requirement.

The observations that efficiency of utilization of high quality protein falls below the maximum as the requirement is approached is not surprising in view of the tendency for most, if not all, biological variables to show diminishing increments in response per unit of limiting factor applied as the maximum response is approached (Brody 1945). The growth response per increment of amino acid consumed by rats falls as maximum growth rate is approached (Harper 1959). Similar response curves have been obtained with human subjects (Young and Scrimshaw 1974). Ordinarily, when proteins are compared for nutritional quality the measurements are made at levels of intake that are inadequate, *i.e.*, levels of intake at which efficiency of utilization is maximum or close to it and at which values for high quality proteins approach 100%.

However, intrinsic differences in nutritional quality among food proteins are attributable mainly to differences in amino acid composition. White flour proteins, as shown in Table 15.1, are utilized by man with an efficiency of only 40% compared to 70% for those of milk. This is largely due to the deficit of lysine in wheat proteins. Bricker *et al.* (1945) demonstrated many years ago that efficiency of utilization of flour proteins by women was improved if the flour was supplemented with lysine. There have been many such demonstrations since (Harper and Hegsted 1974; NAS/NRC 1959).

The first effort to evaluate proteins directly on the basis of their ability to meet amino acid requirements was the Chemical Score procedure developed by Block and Mitchell (1946–47). The basis of the procedure is illustrated in Table 15.7. They used whole egg proteins as their standard since it is utilized with an efficiency of close to 100% in rat assays of nutritional quality and is considered to be one of the most effective food sources for meeting protein requirements. The essence of the procedure was to calculate the per cent deficit of each amino acid in the test protein with respect to the amount in the proteins of whole egg. The amino acid that was in greatest deficit could then be identified and the amount, expressed as a percentage of the amount in whole egg, gave the value for the Chemical Score. This was 40 for whole wheat.

TABLE 15.7
PERCENT DEVIATIONS OF THE ESSENTIAL AMINO ACID
CONTENT OF WHOLE WHEAT PROTEINS FROM
AMINO ACID CONTENT OF WHOLE EGG PROTEINS
AND CALCULATION OF CHEMICAL SCORE[1]

| | Amino Acid Content (g/16 g N) | | % Deviation From Value |
	Egg	Wheat	For Egg
Phe + Tyr	10.1	8.1	−20
His	2.4	1.9	−21
Trp	1.2	1.6	−25
Leu	8.8	6.3	−28
Met + Cys	5.5	3.5	−36
Ile	6.6	4.0	−40
Val	7.4	4.3	−42
Thr	5.0	2.7	−46
Lys	6.4	2.6	−60

[1] Limiting amino acid is lysine. Chemical Score $= \dfrac{2.6}{6.4} \times 100 = 40.$

The main problem with the method is that egg is a particularly rich source of indispensable amino acids. When egg is fed in an amount that meets the nitrogen requirement, it provides surpluses of most of the indispensable amino acids, particularly for the human adult. Thus, if the standard is high, the nutritional value of a protein that is compared with such a standard may be seriously under-estimated. This is especially true when whole egg is used as the standard for assessing proteins that are low in total sulfur-containing amino acids or tryptophan. Nevertheless, there is a high correlation between Chemical Scores determined in this way and values obtained by biological assays for protein quality (Hegsted 1974).

A major step toward substituting human amino acid requirements for whole egg proteins as the standard for evaluating food proteins was taken by the Food and Agriculture Organization in 1957. An expert committee of FAO proposed that an amino acid scoring pattern, in which the amounts of amino acids per gram of protein were about double the estimated adult amino acid requirements as known at that time, should serve as an appropriate standard for evaluating food proteins. Proteins having this composition, it was reasoned, should provide the required amounts of all of the essential amino acids when they were consumed in an amount that would meet the nitrogen requirement. Amino acid requirements of infants had not been established before the Committee made its proposal and there was little information about amino acid requirements other than the original observations of Rose on adult men. It is not surprising, therefore, that the proposed scoring pattern has required revision. In 1965, however, a joint FAO/WHO expert committee proposed that the amino acid patterns of whole egg or human milk be substituted for the proposed scoring pattern. This proposal had all of the disadvantages of the original chemical score procedure. The rationale for the proposal is difficult to understand.

Between 1965 and 1970 a NAS/NRC committee considered the possibility of developing a more satisfactory amino acid scoring pattern. In 1971 another FAO/WHO expert committee initiated a similar effort. As there was overlap in the membership of the two committees the exchange of information that occurred resulted in great similarity in the proposed amino acid scoring patterns published in (FAO/WHO) 1973 and (NAS/NRC) 1974, respectively (Table 15.8). The one significant difference between the two patterns is the lower value for the sulfur-containing amino acids in the NAS/NRC committee's pattern, as that committee placed more weight on the requirements of infants determined using cow's milk formulas than did the FAO/WHO committee. The FAO/WHO committee was influenced more by the rather high estimated sulfur-containing amino acid requirement of school children. Both patterns were designed so that 2 g of protein having an amino acid composition resembling that of the scoring pattern should meet the estimated amino acid requirements of all age groups beyond that of infancy. Interestingly, the amino acid in greatest surplus for infants consuming protein with the proposed FAO/WHO pattern would be methionine plus cystine, the amino acid thought to be too high in the 1957 pattern.

Both committees emphasized the tentative nature of the proposed scoring patterns as neither had been tested directly. Nevertheless, from the information in Table 15.8, both would be expected to predict quite well whether a food protein would meet amino acid requirements provided the protein was highly digestible. However, since amino acid requirements fall with age more rapidly than does the nitrogen requirement and the fall is quite sharp even during the first year of life (Holt and Snyderman 1967), proteins resembling the scoring patterns in composition should provide essential amino acids in excess when they are consumed by older age groups in amounts that meet the nitrogen requirement. The scoring pattern would, thus, be expected to underestimate the nutritional quality of food proteins for adults, as is evident from Table 15.8, last three columns, and probably for all age groups other than the young child. This should not be considered a serious shortcoming as protein quality is of concern primarily in meeting the needs of young children.

After the scoring patterns had been developed, Arroyave (1974) examined the results of some feeding trials that had been done with human subjects to assess retrospectively how well the results

TABLE 15.8

ADEQUACY OF PROTEINS RESEMBLING IN COMPOSITION NAS/NRC OR FAO/WHO AMINO ACID
SCORING PATTERNS FOR MEETING ESTIMATED AMINO ACID REQUIREMENTS
OF INFANTS AND ADULTS

	Scoring Pattern		Infants			Adults		
			Amino Acids Provided by 2 g of Protein		Estimated Amino Acid Requirements[1]	Amino Acids Provided by 0.5 g of Protein		Estimated Amino Acid Requirements[1]
	NAS/NRC	FAO/WHO	NAS/NRC Pattern	FAO/WHO Pattern		NAS/NRC Pattern	FAO/WHO Pattern	
	(mg/g of protein)		(mg)	(mg)	(mg/kg/day)	(mg)	(mg)	(mg/kg/day)
Histidine	17	—	34	—	33 (28)	8	—	? (?)
Isoleucine	42	40	84	80	83 (70)	21	28	9.5 (10)
Leucine	70	70	140	140	135 (161)	35	35	12.5 (14)
Lysine	51	55	102	110	99 (103)	25	27	9.4 (12)
Met + Cys	26	35	52	70	49 (58)	13	17	12.1 (13)
Phe + Tyr	73	60	146	120	141 (125)	36	30	12.1 (14)
Threonine	35	40	70	80	68 (87)	17	20	6.5 (7)
Tryptophan	11	10	22	20	21 (17)	5.5	5	2.9 (3.5)
Valine	48	50	96	100	92 (93)	24	25	10.7 (10)

[1]NRC estimates of amino acid requirements with FAO/WHO estimates in parentheses.

obtained would have been predicted if the FAO/WHO amino acid scoring pattern had been used to evaluate the diets. The estimated average requirement of the two-year-old child for high quality protein is 0.9 g/kg body wt/day (FAO/WHO 1973). This amount of protein from milk provides all of the indispensable amino acids in amounts equal to or in excess of the amounts that would be provided by 0.9 g of protein with an amino acid composition identical with that of the FAO/WHO scoring pattern (Table 15.9). It should, therefore, meet the amino acid requirements of children of this age. Experimental observations on children consuming milk as the entire source of protein indicated that average nitrogen retention was in excess of 20.7 mg/kg body wt/day (considered adequate for that age) when the average intake of milk proteins was 0.9 g/kg/day or more but not if intake was appreciably below this (Table 15.10).

In a similar trial in which a corn-bean mixture was the source of protein, the amounts of lysine and tryptophan it provided were much less than would be provided by a protein with an amino acid composition identical with that of the scoring pattern. Calculations indicated, as shown in Table 15.9, that 1.33 g of the corn-bean mixture would provide enough lysine and tryptophan. In an experimental study, 1.33 g or more of the proteins in the corn-bean mixture supported satisfactory growth and nitrogen retention; in fact, many children showed satisfactory nitrogen retention with between 1.0 and 1.33 g of protein (Table 15.10). The agreement between the experimental results and the predicted amounts of milk and corn-bean proteins needed is remarkable.

In another study in which adult men were fed a diet in which 80% of the protein was from corn, the average amount of protein required to maintain the subjects in nitrogen equilibrium was 0.57 g/kg body wt/day. This is very close to the FAO/WHO "safe" intake of high quality proteins proposed for adults. Calculation of the amounts of amino acids that were provided by this amount of protein and comparison with adult amino acid requirements rather than with the proposed scoring patterns indicated that no amino acid was limiting in the diet. The amount of tryptophan, the limiting amino acid, was just 10% in excess of the estimated requirement of 3.7 mg/kg/day for adults. The amount of protein required to maintain the subjects in N-equilibrium was much less than would have been predicted from estimates based on classical Biological Value or Chemical Score values. The amount of protein that would be predicted as necessary from such a diet if the FAO/WHO scoring pattern had been used to evaluate it would have been 0.74 g/kg/day, well above the amount actually required, showing experimentally that use of the scoring pattern tends to underestimate efficiency of utilization of proteins by adults.

TABLE 15.9
COMPARISON OF AMINO ACID COMPOSITION OF PROTEINS
OF COW'S MILK OR CORN-BEAN MIXTURE WITH THAT OF
PROTEIN BASED ON FAO/WHO AMINO ACID SCORING PATTERN [1]

	Amounts of Amino Acids Provided By:		
	FAO/WHO Scoring Pattern = 0.9 g Protein	Cow's Milk = 0.9 g Protein	Corn-bean Mixture = 1.33 g Protein
	mg	mg	mg
Ile	36	48	51
Leu	63	89	151
Lys	50	66	51
Met + Cys	32	33	41
Phe + Tyr	54	88	113
Thr	36	38	49
Trp	9	13	9
Val	45	58	64

[1] Based on Arroyave (1974).

TABLE 15.10
AVERAGE NITROGEN RETENTION OF CHILDREN FED COW'S MILK
OR CORN AND BEANS AS PROTEIN SOURCE [1]

	Range of Protein Intake (mg/kg body wt/day)	Nitrogen Retention (mg/kg body wt/day)
Cow's milk	0.5–0.7	−28
	0.7–0.9	11
	0.9–1.2	33
Corn-bean mixture	0.8–1.0	−18
	1.0–1.33	20
	1.33–1.55	52

[1] Based on Arroyave (1974).

In experiments on adults in which white flour was the source of protein (Bricker *et al.* 1945), the amount of nitrogen (6.2 g/day required for maintenance of nitrogen equilibrium was much higher than that required from milk (3.6 g/day) (Williams *et al.* 1974). From comparison of the amino acid composition of flour with the estimated amino acid requirements of adults it would be predicted that flour protein intake would have to exceed considerably milk protein intake in order to meet the lysine requirement and that lysine supplementation of flour would reduce the need for nitrogen from flour as has been observed (Bricker *et al.* 1945). However, if the amino acid scoring patterns (Table 15.8) were used instead of the lysine requirement of the adult to predict the amount of nitrogen from flour that would be required to replace high quality protein, a figure of about 12 g of nitrogen per day from flour would be predicted rather than the 6.2 observed.

As the likelihood of either protein or amino acid deficiencies occurring in adults is negligible unless they have an inadequate intake of food owing to disease, a natural or an economic catastrophe, it would be inappropriate, as both committees concluded, to base scoring patterns for assessing the quality of food proteins on the amino acid requirements of adults. It is the young child who suffers from malnutrition. Also, the amino acid needs of debilitated persons are more likely to resemble those of the child than of the healthy adult. The scoring patterns based on the amino acid needs of the young child have given encouraging results in predicting the adequacy of dietary proteins from this age group. So far only a limited number of such comparisons have been made but the method deserves to be explored fully to assess its validity and limitations in predicting the adequacy of food proteins for meeting human protein needs.

This method has potential for assessing quantitatively the nutritional value of products proposed for use in feeding programs, for predicting the nutritional value of diets from the amino acid composition of the individual components and for predicting the nutritional value of protein supplements that represent only one component of a mixture in a diet. It also has the potential for providing a meaningful standard for nutritional labeling.

It has one obvious shortcoming that is common to all methods of protein evaluation based on chemical analysis. It depends upon the amounts of amino acids measured chemically representing the amounts that are biologically available. Low digestibility of protein or reduced availability of specific amino acids due to processing or other damage are not taken into account. Digestibility must be considered, either directly or indirectly, in all measurements of protein quality and appropriate corrections to allow for less than complete digestibility can be made with the scoring procedure. The problem of low availability of one or more specific amino acids is a more serious problem. This subject has recently been reviewed (Friedman 1977). There are chemical methods for assessing biologically available lysine and according to Friedman there is promise of new methods for other amino acids. These could eventually be substituted for direct chemical analysis. Since only four amino acids, lysine, methionine, tryptophan and threonine, are likely to be low in human diets, methods for these four would be sufficient for nutritional quality assessment.

In reviewing this subject I have been led to the conclusion that it is time to move vigorously toward an amino acid scoring procedure for assessing the nutritional quality of proteins. Only with such a procedure will it become possible to express the nutritional potential of food proteins in terms of amino acids, just as the vitamin and mineral contributions of foods are expressed in terms of the specific individual essential nutrients. Although limited ability to measure the biological availability of amino acids is a shortcoming of such a procedure at the present time, the ability to measure availability of vitamins and minerals still leaves much to be desired.

BIBLIOGRAPHY

ARROYAVE, G. 1974. Amino acid requirements by age and sex. *In* Nutrients in Processed Foods—Proteins, P. L. White and D. C. Fletcher (Editors). Publishing Sciences Group, Inc., Acton, Mass.

BLOCK, R. J. and MITCHELL, H. H. 1946–47. The correlation of the amino acid composition of proteins with their nutritive value. Nutr. Abstr. Revs. *16*, 249–278.

BRESSANI, R., VITERI, F., and ELIAS, L. G. 1973. *In* Proteins in Human Nutrition, J. W. G. Porter and B. A. Rolls (Editors). Academic Press, London.

BRICKER, M., MITCHELL, H. H., and KINSMAN, G. M. 1945. The protein requirement of adult human subjects in terms of the protein contained in individual foods and food combinations. J. Nutr. *30*, 269–283.

BRODY, S. 1945. Bioenergetics and Growth. Reinhold Publishing Corp., New York.

CALLOWAY, D. H. 1975. Nitrogen balance of men with marginal intakes of protein and energy. J. Nutr. *105*, 914–923.

CALLOWAY, D. H., and MARGEN, S. 1971. Variation in endogenous nitrogen excretion and dietary nitrogen utilization as determinants of human protein requirement. J. Nutr. *101*, 205–216.

CALLOWAY, D. H., and SPECTOR, H. 1954. Nitrogen balance as related to calorie and protein intake in active young men. Am. J. Clin. Nutr. *2*, 405–412.

CAMPBELL, J. A. 1963. Methodology of protein evaluation. A critical appraisal of methods for evaluation of protein in foods. Publication *21*, Faculty of Agricultural Sciences, American University of Beirut, Lebanon.

FOOD AND AGRICULTURE ORGANIZATION. 1957. Protein requirements. Nutrition Studies *16*. FAO, Rome, Italy.

FOOD AND AGRICULTURE ORGANIZATION/WORLD HEALTH ORGANIZATION. 1973. Energy and protein requirements. WHO Technical Report Series No. 522. FAO/WHO, Geneva, Switzerland.

FOOD AND AGRICULTURE ORGANIZATION/WORLD HEALTH ORGANIZATION. 1965. Protein requirements. WHO Technical Report Series No. 301 or FAO Nutrition Meeting Report Series *37*. FAO/WHO, Geneva, Switzerland.

FOMAN, S. J. 1967. Infant Nutrition. W. B. Saunders Co., Philadelphia.

GOPALAN, C. 1969. Observations on some epidemiological factors and biochemical features of protein-calorie malnutrition. *In* Protein Calorie Malnutrition, A. von Muralt (Editor). Springer-Verlag, Berlin.

HARPER, A. E. 1971. Adaptability and amino acid requirements. *In* Metabolic Adaptation and Nutrition. Scientific Publication No. 222, Pan American Health Organization, Washington, D.C.

HARPER, A. E. 1959. Advances in our knowledge of protein and amino acid requirements. Fed. Proc. *18*, Suppl. 3, 104–113.

HARPER, A. E., and BOYLE, P. C. 1976. Nutrients and food intake. *In* Dahlem Workshop on Appetite and Food Intake, T. Silverstone (Editor). Dahlem Konferenzen, Berlin.

HARPER, A. E., and HEGSTED, D. M. 1974. Improvement of protein nutriture. *In* Improvement of Protein Nutriture. NAS/NRC, Washington, D.C.

HARPER, A. E., PAYNE, P. R., and WATERLOW, J. C. 1973. Assessment of human protein needs. Am. J. Clin. Nutr. *26*, 1168–1169.

HEGSTED, D. M. 1974. Assessment of protein quality. *In* Improvement of Protein Nutriture. NAS/NRC, Washington, D.C.

HEGSTED, D. M. 1959. Protein requirement in man. Fed. Proc. *18*, 1130–1136.

HEGSTED, D. M. 1963. Variation in requirements of nutrients—amino acids. Fed. Proc. *22*, 1424–1430.

HOLT, L. E., JR., GYORGY, P., PRATT, E. L., SNYDERMAN, S. E., and WALLACE, W. M. 1960. Protein and Amino Acid Requirements in Early Life. New York University Press, New York.

HOLT, L. E., JR., and SNYDERMAN, S. E. 1967. The amino acid requirements of children. *In* Amino Acid Metabolism and Genetic Variation, W. L. Nyhan (Editor). McGraw-Hill Book Co., New York.

INOUE, G., FUGITA, Y., and NIIYAMA, Y. 1973. Studies on protein requirements of young men fed egg protein and rice protein with excess and maintenance energy intakes. J. Nutr. *103*, 1673–1687.

LEVERTON, R. M. 1959. Amino acid requirements of young adults. *In* Protein and Amino Acid Nutrition, A. A. Albanese (Editor). Academic Press, New York.

LUSK, G. 1923. The Science of Nutrition, 3rd Edition. W. B. Saunders Co., Philadelphia.

MINISTER OF NATIONAL HEALTH AND WELFARE OF CANADA. 1975. Dietary Standard for Canada. Bureau of Nutritional Sciences, Food Directorate, Health Protection Branch, Dept. of National Health and Welfare, Ottawa, Canada.

MITCHELL, H. H. 1944. Adaptation to undernutrition. J. Am. Dietet. Assoc. *20*, 511–515.

NATIONAL ACADEMY OF SCIENCES/NATIONAL RESEARCH COUNCIL. 1959. Evaluation of protein nutrition. NAS/NRC Publication 711, Washington, D.C.

NATIONAL ACADEMY OF SCIENCES/NATIONAL RESEARCH COUNCIL. 1963. Evaluation of protein quality. NAS/NRC Publication 1100, Washington, D.C.

NATIONAL ACADEMY OF SCIENCES/NATIONAL RESEARCH COUNCIL. 1974A. Improvement of protein nutriture. Committee on Amino Acids, Food and Nutrition Board, NAS/NRC, Washington, D.C.

NATIONAL ACADEMY OF SCIENCES/NATIONAL RESEARCH COUNCIL. 1974B. Recommended dietary allowances: Revised, 8th Edition. NAS/NRC, Washington, D.C.

PÉRISSÉ, J., SIZARET, F., and FRANCOIS, P. 1969. The effect of income on the structure of the diet. FAO Nutr. Newsletter 7, No. 3, 1–9.

ROSE, W. C. 1957. The amino acid requirement of adult man. Nutr. Abstr. Revs. *27*, 631–647.

SCRIMSHAW, N. S. 1976. An analysis of past and present recommended dietary allowances for protein in health and disease. New England J. Med. *294*, 136–142, 198–203.

WATERLOW, J. C. 1970. Human protein requirements and malnutrition. *In* Evaluation of Novel Protein Products, A. E. Bender, R. Kihlberg, B. Löfqvist and L. Munck (Editors). Proc. IBP Werner-Gren Symp., Stockholm, 1968. Pergamon Press, New York.

WATERLOW, J. C. 1968. Observations on the mechanism of adaptation to low protein intakes. Lancet *2*, 1091–1097.

WATERLOW, J. C., and GARLICK, P. J. 1975. Metabolic adaptations to protein deficiency. *In* Alcohol and Abnormal Protein Biosynthesis, M. A. Rothschild, M. Oratz and S. S. Schreiber (Editors). Pergamon Press, New York.

WATERLOW, J. C., and PAYNE, P. R. 1975. The protein gap. Nature *258*, 113–117.

WILLIAMS, H. H., HARPER, A. E., HEGSTED, D. M., ARROYAVE, G., and HOLT, L. E., JR. 1974. Nitrogen and amino acid requirements. *In* Improvement of Protein Nutriture, A. E. Harper and D. M. Hegsted (Editors). NAS/NRC, Washington, D.C.

YOUNG, V. R. and SCRIMSHAW, N. S. 1974. Relation of animal to human assays of protein quality. *In* Nutrients in Processed Foods—Proteins, P. L. White and D. C. Fletcher (Editors). Publishing Sciences Group, Inc., Acton, Mass.

ZEZULKA, A. Y., and CALLOWAY, D. H. 1976. Nitrogen retention in men fed varying levels of amino acids from soy proteins with or without added *L*-methionine. J. Nutr. *106*, 212–221.

16

Functional Properties of Proteins and Their Measurement

Herbert M. Schoen

This discussion is not intended to be a comprehensive review of the large body of literature relating to measuring the functional properties of proteins. Rather, it will address the scope and nature of functionality research highlighting the problems and difficulties to be considered. Some suggestions are proposed, hopefully aimed at advancing the state of the art.

FACTORS OF IMPORTANCE TO PROTEINS

Much of the present food research interest in proteins stems from the concept of providing a less expensive source of proteins to replace more expensive ones. One of the more specific goals in protein research is to use vegetable protein to extend or replace meat protein.

Extension of meat is technologically less demanding than replacement. At low levels the extender can be *relatively* inert; that is, the functional properties as well as the organoleptic properties need not mimic the meat being extended. As one increases the extension level, for example, TVP in hamburger, the functional and organoleptic properties necessarily must be closer to meat. In the case of 100% analogs, the vegetable proteins must approach their meat counterpart closely if consumers are to accept them well.

Analogs based on nonanimal proteins should, in the long term, offer significant economic advantages over meat. The overall energy conversion efficiency, say, in comparing a TVP to hamburger, favors the analog by a significant factor.

The analog offers some other potential advantages to meat in terms of consumer products, namely the potential of a better overall nutritional profile. Meat analogs could be developed which have a better polyunsaturated to saturated fat ratio than the natural meat counterpart, and are lower in cholesterol, or higher in certain minerals. Depending on the nature of the analogs developed, they could offer advantages in terms of storage conditions required as well as extension of usable shelf life.

Collectively these factors paint a very attractive picture of the importance of vegetable proteins and their growth potential. Although reference has been made to meat analogs, similar possibilities exist for other protein-based analogs such as milk and cheese.

One of the keys, therefore, in developing successful analogs lies in the functional properties of the proteins that comprise the analog. However, analogs will be produced not by single vegetable proteins, but by mixtures; each contributing some attribute to the total—sometimes an organoleptic one, other times to enhance the PER. Analogs are generally not pure protein, in the case of meat analogs. Fats are present to produce the desired product; hence, the functional interactions of fats with proteins and other components must also be considered.

OBJECTIVES OF FUNCTIONALITY TESTING RESEARCH

How proteins interact and function in the finished product is clearly of prime importance; hence, the interest in measuring and modifying functionality.

The objectives of research on functionality may be stated to be one or more of the following:

—To determine how the protein has been affected by the processing conditions.

—To screen proteins and extrapolate functionality test results to performance in the finished product.

—To understand why specific proteins function as they do.

—To understand how proteins interact with each other and with other components in mixtures or systems.

PROBLEMS TO BE CONSIDERED

With a set of possible research objectives stated, what are some of the problems facing scientists in this area of study?

—Processing parameters used to obtain the protein from its source can markedly affect functional properties, hence performance in the end application.

—Lack of standarization of functionality tests hinders the use of published data.

—Methods of reporting results can lead to erroneous conclusions about functional performance in end application.

—Functional properties are "functions" rather than "points"; hence, reporting individual values has limited utility.

—The effects of mixing different proteins are not simply additive; hence, evaluation of individual proteins is at best only a partial predictive tool.

—Protein functionality depends strongly on the total system. Interactions between salts, gums, fats, carbohydrates, etc., markedly affect the final results.

—Storage and aging of proteins can markedly affect functional properties.

With the objectives of this research stated, and some of the problems connected with functionality research delineated, let us now examine the state of the art. Since the body of literature is large, this will be done by choosing representative papers rather than attempting to be comprehensive.

INTERACTIONS OF PROTEINS WITH OTHER COMPONENTS

Circle (1964) studied the rheological properties of concentrated aqueous dispersions of soy isolate. When the higher concentration dispersons were heated, they formed gels, whose viscosity was also measured. When various sodium salts were added at a concentration of 0.05 and 0.1%, there was little effect on the ungelled and gelled viscosity. At the higher concentrations of sodium salts (0.5 and 1%) the viscosity of the ungelled was lowered and the gelled increased.

When various lipids and polysaccharides were added, some significant increases were obtained in ungelled viscosity values and some of the gelled viscosities. For example, when 1% carrageenan or carboxymethylcellulose is added to a 10% soy dispersion, the ungelled viscosity rises from 27 poise to 2,600 and 2,130, respectively.

EFFECT OF SALTS ON SOLUBILITY

Mattil (1971) discussed the solubility of a number of proteins as a function of pH with and without the presence of various salts. He discussed how these solubility profiles could be useful in predicting how the proteins would function in real food systems. Figure 16.1 shows the effect of Na_2SO_3 and Na_2HPO_4 on the solubility of coconut meal and sunflower seed meal. The salts chosen were ones which could be used in protein food systems. At increasing concentrations both salts enhanced the solubility of the coconut meal and decreased the solubility of the sunflower seed meal.

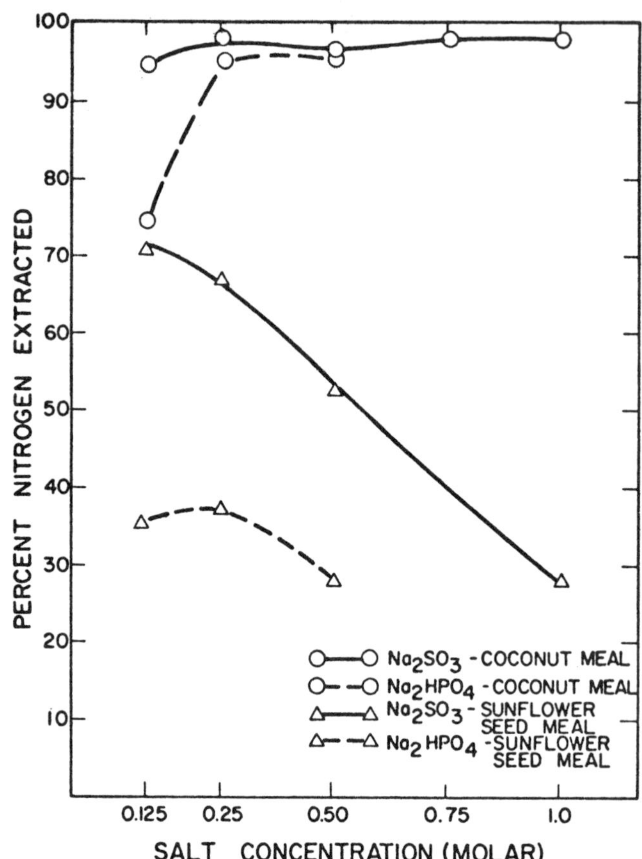

FIG. 16.1. EFFECT OF VARIOUS CONCENTRATIONS OF Na_2SO_3 AND Na_2HPO_4 ON THE PROTEIN SOLUBILITY PROFILES OF COCONUT MEAL AND SUNFLOWER SEED MEAL

The point being made is that misleading solubility values could be obtained if determined in pure water, whereas the true solubility—in which one is interested—is the one obtained in the presence of ions which could occur in the final food system.

Although it is well known that ionic strength is important in influencing solubility, the differences brought about by the nature of specific ions, *e.g.*, SO_3^{-2} and HPO_4^{-2}, would not readily be predictable.

EFFECT OF CHEMICAL MODIFICATION

The functional properties of proteins are strongly dependent on the isolation methods used. For example, the pH, ionic strength, type of salt, and temperature of extraction would selectively extract certain proteins and not others. Hence, the resulting extracts would have different functional properties. In addition, one can modify the functional properties by chemical means. Groninger (1973) succinylated fish myofibrillar protein and found a relationship between degree of succinylation and emulsion capacity (EC). A control isolate had an EC of 0.35 g oil/mg protein; when 30% amino groups were succinylated, the EC rose to 1.3, and at 77% reaction to 2.6.

STORAGE EFFECTS

To further complicate the task of measuring functionality of proteins is the fact that the proteins may undergo significant changes upon storage. This is described by Koury and Spinelli (1975) with regard to emulsification capacity of fish protein isolates. They conclude that maximum stability occurs when the isolates are stored with moisture contents of 2.5%. Above and below these moisture levels there is a rapid decline of emulsification capacity. Whether the isolates were stored under vacuum, air or nitrogen, the loss of emulsion capacity was similar, depending essentially on the moisture content rather than the atmosphere.

Additionally, they investigated the effect of co-drying the isolates with various carbohydrates and found that there were varying degrees of improved retention of the emulsification capacity on storage depending on the carbohydrate used.

NONSTANDARDIZED METHODOLOGY

Huffman *et al.* (1975) investigated the emulsification capacity, water retention and whippability of sunflower meal, developing optimum conditions for maximizing each of these functional properties.

For example, emulsion capacity was optimum at pH 7, a low speed of mixing and a fast rate of oil addition. Optimum foam production was obtained at pH 9, meal concentration of 8%, a temperature of 15°C and a 12-min whipping time. Hence, the values they report are unique for their conditions and cannot quantitatively be compared to other workers. This is the general problem we encounter, since test methods are not standardized. They then examined these functional properties in the presence of various additives and in certain instances obtained major increases or decreases in the values (Fig. 16.2).

METHOD OF REPORTING VALUES

Swift *et al.* (1961) developed an emulsification capacity (EC) method for application to sausage and meat systems. It consists of blending an aliquot of oil with a meat slurry to form an oil-in-water emulsion. Additional oil is added until the emulsion breaks. Many variations of this method have been reported by others. In addition to methodology variations, there are several different ways of reporting the final emulsification values, *e.g.*, ml oil/g meat, ml oil/100 mg soluble protein. Depending on how the data are reported, the relative ranking of EC values in the test system as compared to EC performance in real meat systems can reverse.

DIFFICULTY OF PREDICTING FUNCTIONALITY

Care must be taken in using a particular functionality test to predict its functional properties in the finished product. This was pointed out by Thomas *et al.* (1974). For example, Promine D (Soy Isolate, Central Soya Co.) has been reported as a poor emulsifier in the pH range of sausage meat and hence, should not function well. Promine D does in fact function well in sausage systems, not by virtue of its emulsification capacity, but perhaps because of its water binding capacity.

Promine D binds fats tightly in analog systems such that the product is dry and has poor "fat release." This may not have much relationship with EC, since the test method concentrations and the concentrations in an analog are significantly different. Hence, distinctions should be made between "fat building" and "fat releasing" properties and EC.

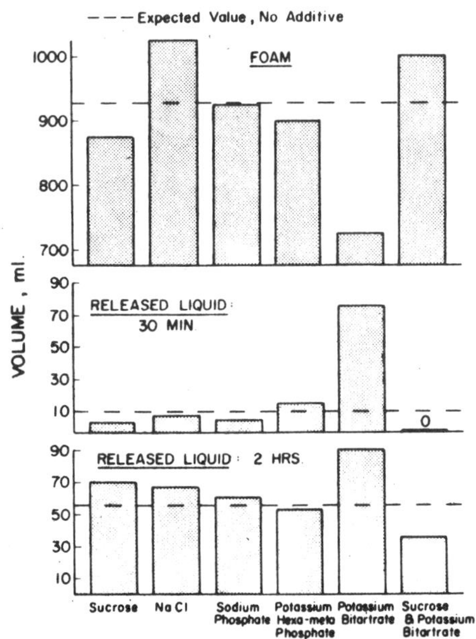

FIG. 16.2. VOLUME AND STABILITY OF SUNFLOWER MEAL FOAMS AS AFFECTED BY ADDITIVES

In a like manner water binding (A_w) should also be distinguished from hydration or water-holding properties, the latter being "free water."

FUNCTIONALITY CAN BE VARIED SIGNIFICANTLY BY TREATMENTS

One of the most significant means of modifying the functional properties of a protein is by enzyme treatment. Many investigators have studied this. In one study, Puski (1975) enzymatically treated soy with *Aspergillis oryzae* neutral protease enzyme at various levels, and measured a number of functional properties. He found in general that enzyme treatment increases solubility, prevents heat coagulation, and increases foam volume. The degree of enzyme treatment must be controlled to optimize the particular property desired—one can overtreat and thereby go past the optimum.

ISOLATION METHODS ARE IMPORTANT

The processing method by which a protein is obtained from its source can markedly affect the subsequent functional properties. Many researchers have reported on this. One study by Lawton and Cater (1971) examines functionality obtained by various processes on glandless cottonseed. Significant changes in whippability, heat gelation, solubility and foaming were found. pH solubility profiles for example were significantly altered by changing the pH at which the proteins were originally precipitated. This suggests that selective extraction of protein might be taking place as pH of extraction is varied.

COMMENTS ON THE VALUE OF FUNCTIONALITY TESTS

The representative examples of the state of the art described above clearly point out the complexity of the subject, and thereby raise some questions. Below are some quotes from publications in this field.

Wolf (1970) gives an excellent review of the functional properties of soybean proteins giving some 65 references. Although he summarizes a considerable body of knowledge he concludes the following:

> This brief summary of some of the physical and chemical properties of soybean proteins emphasizes the complexity of these materials and suggests that they may undergo a wide range of reactions in food systems. Although progress is being made in determining the chemistry of soybean proteins, we are still unable to explain their functional properties on a physical or chemical basis. Further studies are needed on the individual proteins and on their reactions, not only with one another, but with constituents like water, lipids, and starch, before we can attempt to explain such properties as water and fat absorption and gelation of foods.

In a paper by Mattil (1974), he states his viewpoint:

> We know of no analytical tests that will characterize the concentrates and isolates adequately so that a potential user can determine which would be best for his purposes. Evidence is lacking that these tests correlate positively with performance in specific food systems. As a matter of fact, in at least one food system, meat emulsions, where such correlations have been tested, diametrically opposed results and conclusions have been reported. Therefore, we have chosen to make only a few of these tests, principally to make the point that there are substantial differences among the concentrates and the isolates.

A recent study (1975) compiled under the auspices of NSF addresses the subject of functionality. It states:

> New approaches are required in the area of testing procedures for the evaluation of the properties of protein materials, especially with respect to tests for the so-called functional properties to which this report refers as performance properties. For the most part, tests that are currently in use have been used for a number of years. The validity and rationale for the effectiveness of these tests are very doubtful, but they appear continuously in the literature.
>
> Other considerations of major importance are the lack of standardization of the basic tests and the proliferation of unqualified, modified procedures. While trends in the behaviour of proteins can be discerned from these non-standarized tests, there should be more systematic generalization of procedures in order to develop more quantitative comparative information.
>
> New protein uses and new protein sources will limit the effectiveness of using procedures of times past. New situations will require the consideration and adoption of new approaches.

In a discussion by Hermansson (1973) she states:

> The problem of functional characterization can be approached in different ways. The traditional method is to test the new products in a wide range of different types of foods. This trial-and-error method is both time-consuming and expensive, and gives very limited information about the protein.

This author concurs with the concerns expressed above, and clearly takes the position that the traditional functionality research approaches offer little hope of advancing the state of the art. New approaches must be sought if we expect any meaningful progress.

NEW APPROACHES TO FUNCTIONALITY RESEARCH

With the difficulties and limitations faced in the traditional research approaches, what new directions might be taken? The approaches of Yasumatsu *et al.* (1972A, B, C, D, E), Hermansson (1972, 1975), Hermansson and Åkesson (1975A, B) and Hermansson *et al.* (1971, 1972) appear to offer the most promise in shedding light on this complex subject of protein functionality. Both research groups apply statistics in their data analyses, which appears a necessity when the variables, interactions and end results are so large in number. Both research groups relate the individual functional properties to the end property of the product. It is only by an integration of individual functionality studies that we can expect to understand the subject.

EXAMPLES OF SOME NEWER APPROACHES

Yasumatsu *et al.* (1972A, B, C, D, E) conducted an extensive study primarily of soybean proteins. They studied a large number of commercially available soy products, applying principal component analysis to group proteins with regard to their functional properties. In the first paper (Yasumatsu *et al.* 1972D), they measured the chemical composition (crude ash, crude fat, crude protein, N-free extract, sucrose, K, Ca, Mg, and Na) and attempted to relate these to a number of properties (water dispersible N, amount of native protein, percent dispersibility in several solutions and in water at several pHs). As would be expected they obtained a high correlation between the chemical composition of the material and how it was processed. Water dispersible N and amount of native protein had an unexpectedly low correlation; therefore, they are both important in determining functional properties.

In the second paper (Yasumatsu *et al.* 1972C) they attempted to correlate chemical composition with flavor. The undersirable "defatted soy flavor" correlated with the amount of crude fiber. "Raw soy flavor correlated with the amount of native protein and enzyme activity."

In the third paper (Yasumatsu *et al.* 1972E) heat coagulated gels were studied. Hard gels were formed by those materials which have a crude protein content in excess of 75%; soft gels were formed with those materials having a fiber content below 1%; and paste-like gels when crude fiber content was greater than 1%.

Whipping and emulsification properties were treated in the fourth study. Emulsification properties (Yasumatsu *et al.* 1972A) correlated positively with protein and negatively with fiber content. Whipping correlated positively with water dispersible nitrogen, and foams were stable when the dissolved proteins were native.

In the study on doughs (Yasumatsu *et al.* 1972B), soy protein was added at a 5% level to wheat flour. Dough stability correlated positively with the amount of native protein. Water absorption of the dough correlated positively with the protein dispersibility and protein content and negatively with the amount of native protein.

Hermansson and her co-workers conducted a series of functionality studies. The earlier work began with the more traditional approaches. In one study (Hermansson *et al.* 1971) they attempted to relate solubility, foaming and swelling of fish protein concentrate as a function of pH, and ionic strength. This was done with the isolate before and after enzyme and alkaline hydrolysis. Although FPC had a low solubility, it had good foaming properties. The presence of insolubles is known to stabilize foams. Hence solubility per se is not necessarily a meaningful index to predict foaming.

Swelling of several proteins was studied by Hermansson (1972) as a function of pH, ionic strength and temperature. Seven methods of hydrolyzing rapeseed protein concentrate were studied by Hermansson *et al.* (1972): alkaline, acid and enzymatic hydrolysis increased solubility, emulsion stability, swelling, foamability and foam stability. All of the hydrolyzed products improved in their solubility, emulsification and foaming properties, and three of the seven improved in their swelling properties. In three more recent studies by Hermansson, functional properties of added proteins to meat were studied. The effects of adding Promine D, caseinate and wheat protein concentrate to a model meat system were studied (Hermansson and Åkesson 1975A). Solubility was found to be positively correlated, whereas swelling and viscosity were negatively correlated with moisture loss. Statistical regression models were used.

Building on this model system (Hermansson and Åkesson 1975B), salt was added to more closely resemble a finished product and additional functional properties were measured. Due to the more complex system, hierarchical clustering analysis was used to interpret the data. Good correlation was obtained between functional properties and moisture loss even though the salt influenced the functional properties differently for the three proteins studied, namely Promine D, whey protein concentrate, and caseinate.

In a final paper (Hermansson 1975), these three proteins were studied in a complete meatball recipe. Statistical models were used to successfully predict textural changes in a finished meat product, with measured functional properties as the independent variables. The textural properties were measured both objectively and subjectively (Table 16.1). In this series of papers, it was shown that well controlled measurements in model systems are useful in understanding relationships between variables, and further, that model systems can be useful in predicting changes in complex finished products. The usefulness of a statistical approach to this very complex subject was clearly demonstrated.

RECOMMENDATIONS

Based on the foregoing discussion let us suggest a possible framework for functionality research in the future.

TABLE 16.1
TEXTURAL PROPERTIES OF SOY PROTEIN ISOLATE

Meat Systems	Swelling of Added Proteins in Pure Water (μl/mg)	Texture Measurements on Meatballs with 4% Added Protein			
		Instrumental Measurements		Sensory Measurements on Firmness	
		Extrusion Force (Kp)	Compression Work (g/cm) (2nd cycle)	Scoring	Ranking
Reference					
Without added proteins		34.9	305.0	4.8	3.3
With Promine-D[1]	9.6	38.1	236.6	3.7	2.4
With Promine-D[1] Heat-treated at 65°C	14.6	38.5	282.6	5.2	3.7
With Promine-D[1] Heat-treated at 80°C	20.0	48.5	305.0	5.3	3.9
With Promine-D[1] Heat-treated at 100°C	15.8	39.2	228.0	3.2	1.7
Least significant difference (5%)	<1.0	3.6	28.4	1.2	1.2

[1] Soy Isolate, Central Soya Co.

Simplify and Standardize Methods and Reporting of Data

Since existing screening tests in water, *e.g.*, emulsification capacity, gelation, etc., may correlate poorly with performance in finished products, it is suggested that these methods be simplified and used only as a *very gross indication* of possible performance. Reporting of data would then simply be in terms of "very stable foam," "stable foam," "unstable foam," and "no foam" under a standard set of test conditions of pH, concentration, etc.

Measure More Basic Properties and Develop Models

Attempts could be made at measuring basic properties and then attempt to develop mathematical models interrelating the measured variables as a predictive tool. One could hypothesize, for example, that performance in whipped toppings could be predicted by a function composed of such factors as molecular weight distribution, solubility, surface tension and viscosity.

TABLE 16.2
PROTEIN PROPERTIES

1. Molecular Properties
   a. Mol wt determination
   b. Mol wt distribution
   c. Ionization properties
   d. Reactive side chains
   e. Primary, secondary, tertiary, quaternary structure morphology

2. Physical Properties
   a. Hydration properties
      (1) Solubility
      (2) Suspendability
      (3) Swelling
      (4) Gelling

   b. Rheological properties
      (1) Viscosity or apparent viscosity
      (2) Yield stress
      (3) Time dependency
      (4) Shear dependency
      (5) Creep and relaxation
   c. Thermal properties
      (1) Gelation, coagulation
      (2) Thermal conductivity, heat capacity
   d. Surface properties
      (1) Hydrophilic and lipophilic properties
      (2) Surface absorption and adsorption
      (3) Foaming

One cannot ignore the importance of amino acid composition, primary and secondary structure of different proteins as they impact on functional properties.

An NSF study (1975) also suggested that more properties be measured, with the hopes of developing improved prediction of performance in products. Their suggested list is presented in Table 16.2.

More Systematic and Comprehensive Studies

Rather than performing isolated studies, consider an experimental protocol that starts with measurement of fundamental properties in the ideal state (normally pure water), then carries forward through increasingly complex model systems and finally into the finished product. Then mathematical and statistical techniques can be applied to interrelate data along the spectrum from the "ideal" to the "real." Yasumatsu and Hermansson have taken the first steps in this direction.

BIBLIOGRAPHY

CIRCLE, S. J., MEYER, E. W., and WHITNEY, R. W. 1964. Rheology of soy protein dispersions. Effect of heat and other factors on gelation. Cereal Chem. *41*, 157–172.

GRONINGER, H. 1973. Preparation and properties of succinylated fish myofibrillar protein. J. Agr. Food Chem. *21*, No. 6, 979–981.

HERMANSSON, A.-M. 1972. Functional properties of proteins for foods—Swelling. Lebensm.-Wiss. u. Technol. *5*, No. 1, 24–29.

HERMANSSON, A.-M. 1973. Proteins in Human Nutrition, J. W. G. Porter and B. A. Rolls (Editors). Academic Press, London, 407–420.

HERMANSSON, A.-M. 1975. Functional properties of added proteins correlated with properties of meat systems. Effect and texture of a meat product. J. Food Sci. *40*, 611–614.

HERMANSSON, A.-M. and ÅKESSON, C. 1975. Functional properties of added proteins correlated with properties of meat systems. Effect of concentration and temperature on water-binding properties of model meat systems. J. Food Sci. *40*, 595–602.

HERMANSSON, A.-M., and ÅKESSON, C. 1975. Functional properties of added proteins correlated with properties of meat systems. Effect of salt on water-binding properties of model meat systems. J. Food Sci. *40*, 603–610.

HERMANSSON, A.-M., OLSSON, D., and HOLMBERG,B. 1972. Functional properties of proteins for foods—Modification studies on rapeseed protein concentrate. Lebensm.-Wiss. u. Technol. 7, No. 3, 175–183.

HERMANSSON, A.-M., SIVIK, B., and SKJOLDEBRAND, C. 1971. Functional properties of proteins for foods—Factors affecting solubility, foaming and swelling of fish protein concentrate. Lebensm.-Wiss. u. Technol. *4*, No. 6, 201–204.

HUFFMAN, V. L., LEE, C. K., and BURNS, E. E. 1975. Selected functional properties of sunflower meal (*Helienthus annuus*). J. Food Sci. *40*, 70–74.

KOURY, B., and SPINELLI, J. 1975. Effect of moisture, carbohydrate and atmosphere on the functional stability of fish protein isolates. J. Food Sci. *40*, 58–61.

LAWHON, J. T., and CATER, C. M. 1971. Effect of processing method and pH of precipitation on the yields and functional properties of protein isolates from glandless cottonseed. J. Food Sci. *36*, 372–377.

MATTIL, K. F. 1971. The functional requirements of proteins for foods. J. Am. Oil Chem. Soc. *48*, No. 9, 447–480.

MATTIL, K. F. 1974. Composition, nutritional, and functional properties, and quality criteria of soy protein concentrate and soy protein isolate. J. Am. Oil Chem. Soc. *51*, No. 1, 81A–84A.

NSF. 1977. Basic food science and technology problems affecting the properties and processing of protein resources. *In* Protein Resources and Technology: Status and Research Needs. National Science Foundation Study (In press).

PUSKI, G. 1975. Modification of functional properties of soy proteins by proteolytic enzyme treatment. Cereal Chem. *52*, No. 5, 655–664.

SWIFT, C. E., LOCKETT, C., and FRYAR, A. J. 1961. Comminuted meat emulsion—The capacity of meats for emulsifying fat. Food Technol. *15*, 468–473.

THOMAS, M. A., BAUMGARTNER, P. A., and HYDE, K. A. 1974. A study of some of the functional properties of calcium co-precipitates in a model system. Australian J. Dairy Technol. June, 59–64.

WOLF, W. J. 1970. Soybean proteins: Their functional, chemical and physical properties. J. Agr. Food Chem. *18*, No. 6, 969–976.

YASUMATSU, K., SAWADA, K., MORITAKA, S., MISAKI, M., TODA, J., WADA, T., and ISHII, K. 1972. Whipping and emulsifying properties of soybean products. Agr. Biol. Chem. *36*, No. 5, 719–727.

YASUMATSU, K., SAWADA, K., MORITAKA, S., TODA, J., WADA, T., and ISHII, K. 1972. Effects of addition of soybean products on dough properties. Agr. Biol. Chem. *36*, No. 5, 729–735.

YASUMATSU, K., TODA, J., AOKI, H., WADA, T., and ISHII, K. 1972. Studies on the functional properties of food-grade soybean products. 2. Flavor profile. Agr. Biol. Chem. *36*, No. 4, 532–536.

YASUMATSU, K., TODA, J., KAJIKAWA, M., OKAMOTO, N., MORI, H., KUWAYAMA, M., and ISHII, K. 1972. Studies on the functional properties of food-grade soybean products. 1. Classification of soybean products by their chemical constituents and protein properties. Agr. Biol. Chem. *36*, No. 4, 523–531.

YASUMATSU, K., TODA, J., TAKEO, W., MISAKI, M., and ISHII, K. 1972. Studies on the functional properties of food-grade soybean products. 3. Properties of heat-coagulated gels from soybean products. Agr. Biol. Chem. *36*, No. 4, 537–543.

17

Chemical and Nutritional Modifications of Food Proteins Due to Processing and Storage

J. Claude Cheftel

The effects of processing on the nutritive value of food proteins have been studied very extensively and a few general reviews on the subject have been published (Bender 1970, 1972; Mauron 1972; Carpenter 1973).

Some technological treatments are beneficial to protein quality; thus the heating of many vegetable protein foods enhances the digestibility, as a result of protein "denaturation." Most food processing, of course, is not expected to significantly affect the protein value. A number of detrimental effects have, however, been demonstrated as a result of either severe heat or pH treatments, or of close contact with reactive substances present in the food or used as additives. Such detrimental nutritional effects are always related to chemical modifications of the primary structure of food proteins. These modifications have been tentatively classified here into three groups, according to the type of chemical reaction, rather than according to food processing operation or to food class.

Chemical modifications of food proteins may reduce the content of intact essential amino acids and slow down or decrease the release of these amino acids during digestion; in some cases, they may even lead to the formation of toxic substances. However, from the standpoint of the nutritive value, these modifications may have little practical importance if they affect (1) a food protein which does not contribute significantly to the diet; (2) a small number of amino acid residues; or (3) an essential amino acid which is not the limiting nutritional factor in the diet.

SIMPLE MODIFICATIONS OF AMINO ACID RESIDUES

Oxidation

The effects of strong oxidizing agents on the amino acid residues of proteins are well known: sulfur amino acids are oxidized into methionine sulfone and cysteic acid, tryptophan to N-formyl-kynurenine; tyrosine, serine, and threonine can also be partly oxidized, but only under more severe conditions (Means and Feeney 1971).

Although protein foods are not usually submitted to strong oxidizing agents, hydrogen peroxide is presently used for a number of processes: sterilization of milk (for cheese and for bread manufacture), of whey, and of containers for milk and other foods; bleaching and detoxifying of various protein concentrates. Benzoyl peroxide is used for the bleaching and maturation of wheat flour. Oxidative compounds such as peroxides, superoxide anion radical, or quinones may occur naturally in foods as a result of enzymatic or nonenzymatic reactions, such as lipid oxidation. Physical processes such as γ or light irradiation in the presence of air, hot-air drying, or even lengthy storage, may also oxidize the amino acid residues of proteins, together with other food constituents.

Chemical Effects of Various Oxidizing Agents or Processes.—Mild hydrogen peroxide treatments (up to 0.2 M, at 50°C for 30 min), similar to those suggested for milk sterilization, progressively oxidize most of the methionine residues of casein into residues of methionine sulfoxide (Cuq *et al.* 1973). No methionine sulfone appears. Whey proteins appear to be more sensitive: methionine sulfone and cysteic acid were formed, and the content in tryptophan was lowered as a result of treatment with 0.1 M hydrogen peroxide (Munyua 1975). It would be useful to investigate whether hydrogen peroxide treatments of different strength and duration oxidize extensively tryptophan, methionine and cysteine/cystine residues in milk.

In the presence of a sensitizing dye such as riboflavin, photo-oxidation of residues of sulfur amino acids, tryptophan, histidine and tyrosine may occur (Means and Feeney 1971). N-formyl-kynurenine and/or kynurenine have been identified as photooxidation products of tryptophan-containing proteins (Gomyo and Fujimaki 1970).

γ Irradiation of foods in the presence of oxygen gives rise to hydrogen peroxide, through radiolysis of water. γ Irradiation of food proteins is known to induce some radiolysis of sulfur and aromatic amino acids. Volatile sulfur compounds formed may be responsible for off-flavors in irradiated milk, meats, and vegetables. These reactions are reduced when irradiation is performed in the absence of oxygen or in the frozen state. γ Irradiation of fish and other protein foods at up to 3 Mrads does not appear to lower the protein nutritive value (Kennedy and Ley 1971; Ley 1970).

Contact with oxidizing lipids (such as methyl linoleate) has been shown to oxidize methionine residues into methionine sulfoxide (Tannenbaum *et al.* 1969; Cuq 1973; Cuq *et al.* 1974). This is probably due to the presence of lipid peroxides.

Tryptophan destruction occurring during the acid hydrolysis of proteins appears to be of an oxidative nature; it is enhanced by cystine. The main oxidation product appears to be β-3-oxindolylalanine (Nakai and Ohta 1976).

Up to 20% of methionine residues have been found to be oxidized to sulfoxide during the production of single cell protein (yeast fermentation) (Mauron 1975).

Analytical Methods.—Methionine sulfoxides and methionine are unstable during acid hydrolysis. Methionine sulfone, however, is stable. The most accurate method for the determination of the sum of methionine, methionine sulfoxide and methionine sulfone residues relies on performic acid oxidation to methionine sulfone. Performic acid also oxidizes cystine and cysteine to cysteic acid, which is stable during acid hydrolysis. When carboxymethylation of nonoxidized methionine residues is carried out before performic acid treatment, the sum of methionine sulfoxides and methionine sulfone can be determined after acid hydrolysis (Neumann 1967; Cuq *et al.* 1973). Methionine, methionine sulfoxides and methionine sulfone initially present in a protein can best be determined after alkaline hydrolysis (Neumann 1967; Cuq *et al.* 1973).

Some investigators have performed surface analysis by X-ray photoelectron spectroscopy under vacuum of unhydrolyzed samples of skim milk and hydrogen peroxide-treated skim milk (Walker *et al.* 1975); it was possible to determine methionine + cysteine + cystine, methionine sulfoxides, and methionine sulfone + cysteic acid (Table 17.1).

More correlations of X-ray photoelectron spectroscopy data with data from other analytical methods are needed.

Many of the derivatives of cystine residues which are thought to form upon oxidation at neutral pH are unstable, and cannot be determined as of yet (Fig. 17.1) (Savige and MacLaren 1966).

Nutritional Implications of the Oxidation of Amino Acid Residues.—Free cysteic acid, methionine sulfone and formylkynurenine cannot replace cysteine, methionine and tryptophan, respectively, in the diet of rats. Free L-cystine mono- and disulfoxides and cysteine sulfenic acid (but not cysteine sulfinic acid) prove capable of partly replacing L-cystine for the growth of rats; free L-methionine sulfoxides (two stereoisomers) can also partly replace methionine (Njaa 1962; Miller and Samuel 1970). The nutritional value of methionine sulfoxide appears to depend on the age of the animal (Miller *et al.* 1970). An explanation for this might be the induction, with age, of enzymes able to convert larger amounts of methionine sulfoxide into methionine *in vivo*. This should be studied further in various animal species. The possible toxicity of excess methionine sulfoxide should also be tested.

In this laboratory, casein with 98% of methionine residues oxidized into methionine sulfoxide, but devoid of methionine sulfone, was fed to young rats as the sole source of protein in the diet, and at various levels of protein intake. Control diets included

TABLE 17.1
SULFUR AMINO ACID COMPOSITION OF
OXIDIZED SKIM MILK SAMPLES
(MOLE % OF TOTAL SULFUR AMINO ACIDS)

	MET + CYS	MET O	MET O_2 + CYS. ACID
Dry skim milk	78	8.5	13
Dry skim milk (heated)	84	8	7
Dry skim milk (0.015M H_2O_2-treated)	61	29	10
Dry skim milk (0.15M H_2O_2-treated)	42	37	21

According to Walker *et al* (1975).

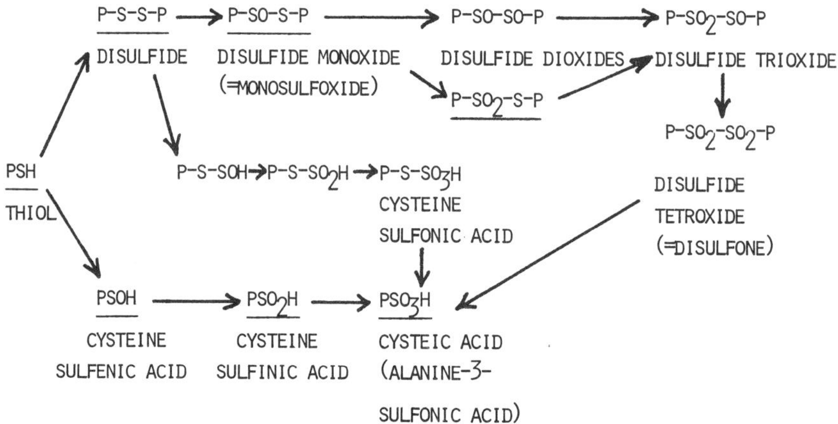

FIG. 17.1. OXIDATION DERIVATIVES FROM CYSTEINE AND CYSTINE RESIDUES
(STABLE DERIVATIVES ARE UNDERLINED)

oxidized casein supplemented with free methionine. Only a small (10%) decrease in PER and NPU of oxidized casein was detected (Cuq 1973: Cuq *et al.* 1974). It thus appears that, *in vivo* methionine sulfoxide is released during digestion, absorbed, and to a large extent reduced to methionine and utilized for protein synthesis. Plasmatic and muscular free amino acids were also determined. Free methionine sulfoxide accumulated at a high level in animals fed oxidized casein (Cuq 1973; Cuq *et al.* 1974). These findings indicate that the nutritional role of methionine is slightly decreased when the source of methionine consists of methionine sulfoxide residues.

Slump and Schreuder (1973) treated casein and fish meal with hydrogen peroxide in the presence of perchloric acid, and found that methionine residues had been oxidized to methionine sulfoxide and methionine sulfone; a reduced net protein utilization was determined in rats and attributed solely to the presence of methionine sulfone and cysteic acid.

Anderson *et al.* (1975) found that 0.9 and 2 *M* hydrogen peroxide treatment of rapeseed flour for 1 hr at room temperature was effective in destroying most of the thyroid-toxic glucosinolates. However, it also caused extensive formation of methionine sulfoxide (ca 50% of initial methionine residues), methionine sulfone (ca 50%), and cysteic acid (ca 70% of initial half-cystine). The tryptophan content also decreased by up to 20%. Rats were fed a diet containing 10% of casein or rapeseed flour proteins; the PER of casein, of nonthyroid-toxic (water-washed) rapeseed flour, and of a similar flour treated with 0.9 *M* hydrogen peroxide were 3.5, 4.0 and 1.5, respectively. The decrease in nutritional value of the treated flour could not be completely reversed by methionine supplementation. This was perhaps due to the presence of methionine sulfone, since very small additions of free methionine sulfone to diets were found to depress food intake.

In spite of some uncertainties in the reported results concerning the nutritional availability of methionine sulfoxide residues for the rat, it can probably be concluded that mild oxidative treatments do not significantly reduce the nutritional value of food proteins, as long as no methionine sulfone is formed. The absence of formyl-kynurenine or other oxidation products of tryptophan should also be ascertained. It is felt that it would be useful to further study the nutritional value of methionine residues in other animal species than the rat.

Desulfuration

Severe heating of moist foods with a low carbohydrate content has been repeatedly found to cause a marked destruction of cystine-cysteine together with a decrease in the availability of many amino acids.

Some investigators reported a 44% loss in cystine in canned meat, after severe sterilization. Similar losses in cystine were found on heating pork meat for 24 hr at 110°C (Donoso *et al.* 1962). Heating of cod muscle for 27 hr at 116°C also caused cystine destruction (Miller *et al.* 1965B). Overheating of soybean protein also causes cystine destruction (Iriarte and Barnes 1966).

The formation of H_2S and other volatile sulfur-containing compounds has been demonstrated in heated milk and meat, where they contribute to flavor. Similar compounds form when dry bovine plasma albumin is heated (Bjarnason and Carpenter 1970).

The above mentioned loss of cystine probably occurs through desulfuration reactions, like that due to alkaline treatments.

Isomerization

Alkaline treatments of proteins may cause the isomerization of various amino acid residues (see Provansal *et al.* 1975).

Amino acid isomerization is initiated by the dissociation, in an alkaline medium, of the hydrogen on the asymmetric carbon in the α position.

Methionine isomerization has been studied by microbiological analysis in NaOH-digested fish protein concentrate (Tannenbaum *et al.* 1970). Complete racemization occurred for strong alkali treatment.

D-Lysine residues have been determined in NaOH-treated sunflower protein isolate by enzymatic and microbiological techniques after acid hydrolysis (Provansal *et al.* 1975). A marked degree of isomerization was found for treatments in 0.2 *M* NaOH for 1 hr at 80°C or above, but no isomerization occurred for milder treatments (0.05 and 0.1 *M* NaOH, 1 hr, 55°C) comparable to those used for vegetable protein purification.

Isomerization of amino acid residues has also been observed on protein treatment with strong acids at high temperatures. However, it occurs much less readily in acid than in alkaline solution.

Roasting of protein has also been shown to cause amino acid isomerization. Casein, lysozyme and poly L-amino acids were dry-heated at 180–300°C for 20 min under air or nitrogen; decomposition and isomerization of amino acid residues were investigated by gas chromatography. Aspartic acid, glutamic acid, alanine, and lysine residues were found to be extensively isomerized, and the other amino acid residues except proline were also isomerized to a considerable extent at higher temperatures. Free amino acids and oligopeptides formed from roasted casein were found to be largely or completely racemized (Hayase *et al.* 1973, 1975).

Isomerization of amino acid residues may partly inhibit the proteolytic digestion of the protein. Studies by Berg have shown that free D-lysine and many D-amino acids have almost no nutritional value. Some free D-amino acids, namely methionine, tryptophan and arginine, can partly replace the corresponding L-amino acids for rats, probably because of the presence of D-amino acid oxidases in the organism; growth is however generally more or less retarded. DL-Methionine is frequently used in animal feeding. D-Methionine appears to have little or no nutritional value for man (Kies *et al.* 1975).

Naturally Occurring Amino Acid Substitutions

Individual amino acids of some unprocessed food proteins, especially of plant origin, are known to be only partly available nutritionally: methionine and cystine in beans (Seidl *et al.* 1969; Evans *et al.* 1974) and in leaf protein concentrates, lysine in wheat, and isoleucine in corn. The reasons for this unavailability are not known, but it is likely that resistance of the proteins to digestion plays a major role.

By contrast, the existence of some naturally occurring amino acid substitutions is well documented, but their nutritional implications have not been studied to a great extent.

Hydroxylysine and hydroxyproline residues are found in collagen. Hydroxylysine has no nutritional value for animals. Phosphoserine and phosphothreonine residues exist in caseins. The carbohydrate units of glycoproteins and mucoproteins are covalently attached to residues of threonine, serine, asparagine and hydroxylysine. Methyl-lysine, histidine and arginine have been identified in some proteins. Small amounts of ϵ-N-mono-methyllysine, ϵ-N-trimethyllysine and 3-methylhistidine have been obtained from acid hydrolyzates of chick actin and myosin.

The nutritive value of selectively acylated casein or lysine is reported in a later section.

PROTEIN-PROTEIN INTERACTIONS:
FORMATION OF CROSS-LINKS BETWEEN AMINO ACID RESIDUES

Naturally Occurring Protein Cross-links

Many unprocessed protein foods are known to be only partially digestible. The main factors involved are probably the following: close contact with indigestible polysaccharide cell walls, fibers, starches; presence of protease inhibitors (generally heat-labile); protease-resistant nature of protein or peptide fractions (even after heat denaturation) (Seidl *et al.* 1969; Evans *et al.* 1974). The latter phenomenon may be due to cross-links occurring naturally within or between polypeptide chains.

The best known examples of such cross-links are those of keratin, elastin and collagen.

Keratin.—Numerous disulfide bonds in hair and feather keratin make this protein totally insoluble in water and resistant to proteolytic attack. Treatment with reducing and dispersing agents, alkaline, acid, or enzymatic hydrolysis, autoclaving, or even ball milling increases the digestibility to the point that such feather meals can be used as protein feeds.

ϵ-N(γ-Glutamyl)-lysine cross-links have been identified in keratin. They are formed by transglutaminases. ϵ-N-(γ-Glutamyl)-lysine was not split after almost complete *in vitro* proteolysis, but was released from adjoining amino acid residues (Pisano *et al.* 1971). ϵ-N-(γ-Glutamyl)-lysine has nevertheless been found to be a nutritionally available source of lysine.

Elastin.—Elastin is unusual by its elasticity, its insolubility, its content of lysine-derived cross-links, and its general resistance to proteolysis by many mammalian proteases.

Elastin contains a number of desmosine and isodesmosine cross-links; these could cross-link a maximum of four peptide chains in the elastin structure:

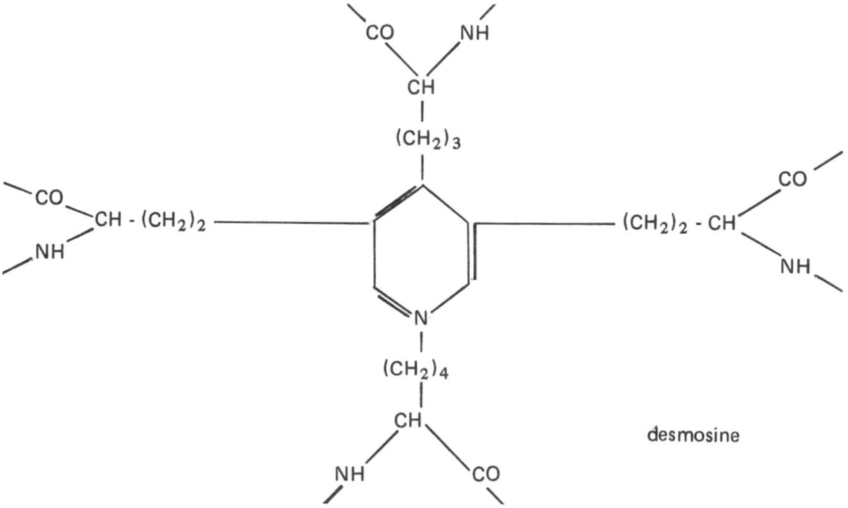

CO   NH

CH

(CH₂)₃

CO
CH - (CH₂)₂ ———————— ———— (CH₂)₂ - CH
NH   NH
CO

N

(CH₂)₄

CH

NH   CO

desmosine

A diamino-dicarboxylic amino acid, "lysinonorleucine," was also isolated from proteolytic digests of elastin.

Biosynthetic studies have shown the following: lysyl residues are oxidatively deaminated into the corresponding aldehyde ("allysine") by lysyloxidase. Lysine-derived aldehydes then condense spontaneously by aldol and carbonylamine reactions to form the various cross-links.

Substituted lysine residues are probably nutritionally unavailable. Desmosines are stable during acid hydrolysis.

Collagen.—Various types of cross-links occur between the polypeptide chains of the tropocollagen molecules (Feeney *et al.* 1975). Apart from hydrogen bonds, which can easily be split by moist heat, making meat collagen partly digestible, one can mention: (1) ester cross-links between ω-carboxylic groups of aspartic or glutamic acid and hydroxy groups of hydroxyamino acid residues (especially serine); (2) carbonyl-amine cross-links (lysinonorleucine, desmosine, etc., as in keratin); (3) ϵ-$N(\gamma$-glutamyl)-lysine cross-links; (4) phosphodiester cross-links between two hydroxyaminoacid residues; and (5) carbohydrate cross-links, involving galactose or glucose and residues of lysine or hydroxylysine.

Beneficial Effect of the "Denaturation" of Vegetable Proteins by Moderate Heat

It has been reported that the protein digestibility and the availability of various amino acids are often low in many plant foods (Bressani *et al.* 1972). In several of these, especially in legumes, heat brings about an improvement in protein quality, as well as the destruction of growth inhibitors and other toxic factors (Table 17.2) (Liener 1975A, B; 1976).

The enhancement of the nutritive value of soybean proteins by moderate heat was generally ascribed to the destruction of antitryptic factors and phytohemagglutinins, both of protein nature. From studies on rats and chicks with raw soybean proteins freed from, or enriched in trypsin inhibitors or phytohemagglutins, the following conclusions have been drawn (Liener 1975A, B; 1976): both trypsin inhibitors and undigestible soybean globulins bind trypsin in the intestine. The resulting hypersecretion of trypsin and other pancreatic enzymes adds to the fecal loss of amino acids from the undigested soybean proteins, and brings about a severe nutritional deficiency in sulfur amino acids. The poor digestibility of the globulins appears to be the most important factor in growth retardation and fecal loss of nitrogen. Moist heat treatment of raw soybean causes a denaturation of the proteins, inactivating the trypsin inhibitors and making the globulins more digestible (Table 17.3).

TABLE 17.2
EFFECT OF HEAT ON THE PROTEIN DIGESTIBILITY
AND THE "PER" OF LEGUMES POSSESSING
TRYPSIN INHIBITOR ACTIVITY[1]

Legume	Scientific Name	Trypsin Inhibitor Activity $\times 10^{-4}$ Units /g	Digestibility (rats) %		PER (rats)	
			Raw	Heated	Raw	Heated
Kidney beans	*Phaseolus vulgaris*	4.25	56	79	loss in weight	0.8
Soybeans	*Glycine max*	4.15	70	85	1.3	2.4
Pigeon Peas	*Cajanus cajan*	2.77	59	60	0.7	1.6
Cow Peas	*Vigna sinensis*	1.91	79	83	1.4	2.2
Lentils	*Lens esculenta*	1.78	88	93	0.4	1.2

[1] After Liener 1975A.

TABLE 17.3
EFFECT OF AUTOCLAVING[1] AND OF METHIONINE SUPPLEMENTATION
ON THE "PER" OF SOYBEANS

Protein Source	PER (rats)
Raw soybeans	1.40
Autoclaved soybeans	2.63
Raw soybeans + 0.6% met	2.42
Autoclaved soybeans + 0.6% met	2.99

After Liener 1975A.
[1] 20 min. 115°C.

Industrial heat treatment of soybean can be done in a number of ways, at various stages of the preparation of full-fat meals or of defatted flours, grits, protein concentrates or isolates. Experiments have shown that conditions of soybean autoclaving giving the best nutritional value are over 1 hr at 100°C or 30 min at 121°C. This steam processing appears to be sufficient for the proper heat denaturation of proteins, in spite of the low initial moisture content of the bean. Extrusion techniques allow accurate control of moisture content and precise timing of the heat application. There is however a large variability in the protein quality of commercial soy products.

Mustakas *et al.* (1970) found the PER of full-fat soy flours to reach a maximum after an extrusion time of 2 min at 135°C (with a moisture content of the flour of 20%). These conditions were found to give an 89% trypsin inhibitor destruction. The contents of available lysine and of thiamin were practically unchanged. The high temperature-short time extrusion cooking of high protein-containing mixtures has also been reported as less detrimental than cooking on a drum dryer. In this laboratory, a protein-enriched snack food prepared by extrusion of a mixture containing 50% corn starch, 20% soy protein concentrate, 20% casein, and 10% water was analyzed for its content of available lysine both by the indirect chemical FDNB method and by bioassay. Extrusion was performed in a double screw extruder. With the highest temperature of ca 200°C in the food, chemically unavailable and nutritionally unavailable lysine were found to represent 5 and 10% of total lysine, respectively.

The nutritional value of various soy protein fibers commercially prepared by spinning alkaline protein isolates into an acid bath was evaluated in animals and in men, by the nitrogen balance method (Bressani *et al.* 1967; Kies and Fox 1971; Korslund *et al.* 1973; Turk *et al.* 1973). Spun soy protein fibers and a textured food made from them showed similar digestibility and only slightly lower biological values than dehydrated beef, casein and whole milk proteins, especially when the soy proteins were supplemented with methionine and lysine, or fed above the critical level of intake (Bressani *et al.* 1967). Digestibility values of 92% for the textured food are worth noting, since the spinning process has been said to induce numerous disulfide bonds between tightly packed parallel protein fibrils (Kelley and Pressey 1966). Kies and Fox (1971) reported PER values in growing rats on the basis of casein (2.5) to be 2.12, 2.82 and 2.37 for textured vegetable protein, methionine-supplemented textured vegetable protein and beef, respectively. Spun soy protein containing 1/3 egg albumin had a high nutritive value for man (Turk *et al.* 1973).

Trypsin inhibitor activity was found to persist partially in a soy protein isolate, and to a lesser extent in some spun fibers. This may, however, be of little nutritional significance, since soybean trypsin inhibitors have but a very limited effect on human trypsin (Liener 1975A, B; 1976).

Overheating of defatted soybean flours, as may occur during desolventizing, partly destroys cystine, lysine, arginine, tryptophan, and serine, making cystine the limiting amino acid. The content of available lysine may also be reduced.

Heating of peanut seeds was found to increase the PER of the defatted meal, and to reduce the content of contaminating aflatoxin. However both wet and dry treatments at 121°C for 4 hr reduced PER and the content of chemically available lysine (Neucere *et al.* 1972). The latter appears to be due to the reaction of ϵ-amino groups of lysine with reducing sugars formed by the hydrolysis of sucrose present in the peanut (Anantharaman and Carpenter 1971).

The beneficial, then detrimental, effects of increasing the heat treatment of cotton seed protein meals are due to the presence of gossypol and of some oligosaccharides such as raffinose, and the reaction of these compounds with lysine residues.

A nonheat-treated rapeseed meal contains the enzyme system myrosinase, which hydrolyzes glucosinolates, producing toxic compounds. Heat treatment of the seed or meal is necessary to produce a meal with a high nutritional value. The optimum temperature and duration of seed treatment for proper inactivation of the myrosinase and of other deleterious factors, and in order to obtain a maximum nutritional value (PER), appears to be 100–110°C and 15–60 min respectively, at a moisture content of 8% (Josefsson 1975). Higher temperatures or moisture contents bring about a decrease in available lysine.

Heating of wheat flours in a heat exchanger for 2–10 min was found to have marked effects only at or above 150°C; the content in lysine, arginine and cystine fell by 50%, both protein breakdown and aggregation took place, and the *in vitro* protein digestibility by proteases was reduced (Hansen *et al.* 1975). Similar results were obtained with overheated bread crust, puffed wheat, wheat granules and flakes. No effects were noticed with normal bread crumb and wheat shreds.

It can be concluded that a moderate heat treatment of many plant protein foods enhances the nutritional value of the proteins. This effect is mainly the result of protein denaturation: (1) toxic protein substances are inactivated; and (2) major protein constituents are made more digestible. Overheating of these relatively moist vegetable proteins, however, may cause a decrease in the content of the most thermolabile amino acid, cystine, and a partial unavailability of the most reactive amino acid, lysine, mainly through reactions with reducing carbohydrates present in the seed, meal or flour.

ε-N-(γ-Glutamyl)-Lysine and Other Cross-links in Heated Animal Food Proteins

Severe heating in the moist or dry state of pure model proteins or of protein foods with a low carbohydrate content (such as fish and meat) brings about a marked destruction of cystine. The content of other amino acids is usually unchanged, as seen after acid hydrolysis, except for small losses of lysine, and sometimes of methionine (Donoso *et al.* 1962; Miller *et al.* 1965A, C). However, the nitrogen digestibility is often severely diminished, together with the availability (40–60% decrease) of a number of amino acids, and the overall nutritional value. Thus when bovine plasma albumin or presscake herring meal (free of lipid proxides) was heated at 130°C for 27 hr, a sharp decrease in the availability of methionine, tryptophan, arginine, lysine and even leucine was noticed (Carpenter *et al.* 1962). The decrease was most marked at 5–14% moisture content. Many comparable results were obtained with cod muscle, meat, and various pure proteins (Ford and Salter 1966; Ford and Shorrock 1971; Carpenter 1973).

Why does the severe heat treatment of protein foods with a very low carbohydrate content lower the overall digestibility of the protein, and the availability of several amino acids simultaneously? The simplest explanation is that such a treatment gives rise to a number of new protease-resistant cross-links within and between polypeptide chains. Parts of these chains then become biologically unavailable due to the masking of the site of protease attachment. While this picture is true to a large extent, as reported later, it is appropriate to first consider some of the complex nutritional effects of severe heat treatments.

Miller *et al.* (1965A) indicated that the decrease in protein and amino acid digestibility, as measured in rats, cannot account for the much larger decrease in nutritional value (PER, NPU in rats) and in amino acid, especially methionine, availability (chick and *Streptococcus zymogenes* tests).

Other investigators (Ford and Salter 1966; Ford 1973) studied the *in vitro* release of amino acids by proteases, and the availability of essential amino acids in dry-heated cod meal. They demonstrated differences in the rate of release of individual amino acids and postulated that such differences may prevent nutritional supplementation and explain the low availability of methionine and lysine. Further studying the digestion and the metabolism of heat damaged fish proteins in rats, they found abnormally high levels of undigested peptides, and also of free amino acids and of small peptides in the gut; they suggested that the inefficient digestion resulted from the resistance of heated proteins to proteases, from an inhibition of peptide absorption and from a fecal loss of digestive enzymes simultaneously (Buraczewski *et al.* 1967; Ford and Shorrock 1971). Finally they found small amounts of lysine, aspartic acid and glutamic acid-containing peptides in the urine; these peptides are probably nutritionally unavailable.

Such peptides may correspond to the ϵ-N-(γ-glutamyl)-lysyl amide cross-links which Bjarnason and Carpenter (1970) demonstrated to form, from lysyl and glutamine residues, with release of ammonia, when bovine plasma albumin was heated for 27 hr at 110–145°C in the presence of 14% water:

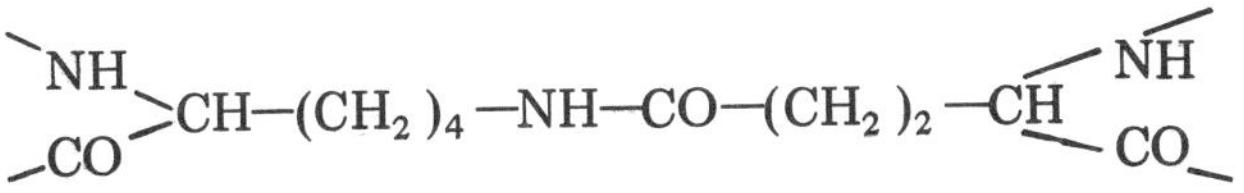

$$\begin{array}{c}{}^{\diagdown}NH \\ {}_{\diagup}CO\end{array}{>}CH-(CH_2)_4-NH-CO-(CH_2)_2-CH{<}\begin{array}{c}{}^{\diagup}NH \\ {}_{\diagdown}CO\end{array}$$

It was first thought that the ϵ-amino substituted lysine of such cross-links was biologically unavailable, although hydrolysis with hydrochloric acid was able to regenerate lysine. However, Mauron (1975) and then Waibel and Carpenter (1972) found that ϵ-N-(γ-glutamyl)-L-lysine can be fully used as a source of lysine by the rat and the chick. This compound is mainly split in the kidney (Raczynski *et al.* 1975).

These findings prompted detailed studies of the nutritional effects of lysine and protein acylation. The results of these studies partly explain the effects of severe heat processing.

It is interesting to note here that acetylation and succinylation of various food proteins has been suggested as a means of improving solubility and/or functional properties (Spinelli *et al.* 1975). In addition to the ϵ-amino groups, SH groups, and OH groups of tyrosine, serine and threonine may also be acylated.

Bjarnason and Carpenter (1969), and Mauron (1972) studied the availability for the chick and/or for the rat of several acylated lysine derivatives; the results are summarized in Table 17.4.

TABLE 17.4

NUTRITIONAL AVAILABILITY OF ACYLATED LYSINE DERIVATIVES

Derivative	% Utilization as a Source of Lysine	Animal
ϵ-N-formyl-L-lysine	$\sim$ 50	rat
ϵ-N-acetyl-L-lysine	$\sim$ 50	rat, chick
ϵ-N-(γ-glutamyl)-L-lysine	$\sim$ 100	rat, chick
ϵ-N-(α-glutamyl)-L-lysine	$\sim$ 100	rat
ϵ-N-glycyl-L-lysine	$\sim$ 80	rat
α-N-glycyl-L-lysine	$\sim$ 100	rat
ϵ-N-(N-acetylglycyl)-L-lysine	0	rat
ϵ-N-propionyl-L-lysine	0	rat
ϵ-N-propionyl-L-lysine	$\sim$ 70	chick

Lactalbumin (or bovine plasma albumin) in which over 3/4 of the ϵ-amino groups had been formylated, acetylated or propionylated (without any cross-link formation) is partly deacylated *in vivo*, after ingestion by rats (Bjarnason and Carpenter 1969). Deacylation could occur both in the intestine and in the kidney (if acylated peptides were absorbed). A high amount of bound lysine was found in the urine.

In a further study, Varnish and Carpenter (1975A, B) compared four protein samples: fully or nonpropionylated lactalbumin, and chicken muscle autoclaved or not at 116°C for 27 hr. Propionylation decreased the FDNB-reactive lysine content by 100% (the decrease upon autoclaving was only 40%). The amino acid content was unchanged except for a 20% loss in cystine in both cases. Propionylation decreased the nutritional availability (chick growth) of lysine only, by 50%. Autoclaving decreased the availability of most amino acids including that of chemically inert leucine; a 50% decrease was noted for lysine and tryptophan, 30% for methionine. The digestibility of crude protein and of most individual amino acids was determined by the ileal technique: they were all 10% reduced by propionylation and 40% by autoclaving. It is of interest to note that the acylation of ϵ-amino groups, known to prevent proteolytic cleavage by trypsin at the carboxyl end of lysine, does not significantly reduce the protein or the lysine digestibility *in vivo*; there is however a large amount of nonavailable lysine in the absorbed peptides.

It is therefore clearly apparent that the effects of severe autoclaving cannot be explained simply by an acylation of ϵ-amino groups of lysyl residues. They may, however, be explained by the formation of intra- and intermolecular cross-links of ϵ-N-(γ-glutamyl)-lysine (or of ϵ-N-(δ-aspartyl)-lysine) since these cross-links do not appear to be digested in the gastrointestinal tract (Mauron 1975). The presence of many of these cross-links may therefore prevent easy access of proteolytic enzymes by steric hindrance and slow down or even reduce proteolytic attack. Such a mechanism may be partly responsible for the concomitant decrease in availability of several amino acids in severely heated proteins. A time delay in the release of some amino acids, and the absorption of unavailable peptides could explain why, in some cases, the protein digestibility remains higher than the availability of essential amino acids. As suggested by Carpenter (1973), it would be interesting to study the digestibility of proteins where ϵ-N-(γ-glutamyl)-lysine cross-links have been introduced without the use of heat.

Other cross-links may be formed as a result of the heat treatment of protein foods containing little carbohydrate, and may accentuate the steric hindrance effect (Bjarnason and Carpenter 1970). Mauron (1972) has hypothesized that severe heating could induce the formation of imide, ester and thioester cross-links; such links would not resist acid hydrolysis.

A recent finding of interest is the formation of lysinoalanine cross-links (known to occur in alkali-treated proteins) during the heat processing of a variety of foods that have never been submitted to a pH above 7 (Sternberg *et al.* 1975A). Similar results were found by Aymard (unpublished 1976) in this laboratory (Table 17.5). It would be useful to investigate if lanthionine and other cross-links found in alkali-treated proteins occur as a result of heat treatment.

TABLE 17.5
LYSINOALANINE[1] CONTENT OF HEAT-TREATED PROTEINS

Protein Food or Protein	Conditions of Heat Treatment in the Laboratory	Lysinoalanine (μg/g Protein)
Strasbourg sausage	no treatment	0
Strasbourg sausage	boiling water, 10 min	50
Evaporated milk	no treatment	700
Canned corned beef	no treatment	<25
Dried skim milk	no treatment	<25
Whipping agents (from whey proteins spray-dried at high pH)		
Brand A	no treatment	30,000 (20,000)
Brand B	no treatment	5,000
Bovine serum albumin[2]	120°C, 6 hr, pH 6	8,000
Casein[2]	NaOH 0.2 M, 75°C, 1 hr	1,000–1,500 (1,300)
Sodium caseinate		
Brand A[2]	120°C, 6 hr, pH 6	600
Brand B[2]	120°C, 6 hr, pH 6	200

[1] Determined by thin-layer chromatography (adapted from Sternberg *et al.* 1975B). Numbers within brackets refer to determination by ion exchange chromatography.

[2] Pure proteins were heat treated as a 0.5 or 1% aqueous solution; casein as a 7% solution.

Another question of interest concerns the possibility of a direct alkylation of the methyl-thio group of methionine during the heat processing of proteins (Pieniazek *et al.* 1975A, B). The nutritional availability of protein-bound or free methionine sulfonium derivatives is not known. Lipton and Bodwell (1975, 1976) showed that methionine sulfonium derivatives can be desulfurized quantitatively into homoserine lactone and a methyl sulfide. This reaction could be used for the determination (after S-alkylation) of methionine residues not initially substituted on the methyl-thio group (available methionine?).

It is now appropriate to question whether or not cross-linking and nutritional damage demonstrated in model systems also take place during domestic or industrial heat processing of protein foods. Meat canning was found to cause very little damage (see Mauron 1972). There is however a lack of recent data with improved nutritional techniques. Milder heat treatments corresponding to normal domestic cooking did not induce any appreciable loss in nutritive value in meat or fish (De Groot 1963).

Fish, meat and bone meals can, however, be seriously damaged when the heat processing is not properly controlled. A precise knowledge of the availability of essential amino acids in such meals is necessary for the exact calculation of low-cost rations for poultry and pigs.

Commercial fish meals and flours frequently display a marked variability in protein quality (Mauron 1972; Carpenter 1973). Firstly, the content in lysine and methionine may vary depending upon the proportion of collagen from skin and bones; secondly, the oil content may be high or low. Different drying equipment and techniques are also used, such as flame, steam tube or rotary vacuum dryers, at various temperatures. As a result, NPU is found to vary from 90 to 50. Lipid oxidation of the oil, when it occurs, decreases the nutritional value. Both overheating during drying and long storage of defatted meals are known to reduce the overall digestibility of the protein and the availability of lysine, methionine, and other amino acids. Chemical tests for lysine availability were found to correlate with nutritional value only when severe damage had occurred (Carpenter 1973; Carpenter and Opstvedt 1976).

Lysinoalanine and Other Cross-links Caused by Alkaline Treatments

Alkaline treatments of food proteins are being increasingly used for protein solubilization and purification in order to prepare vegetable and other protein concentrates and isolates. Viscous alkaline solutions of soy and other proteins are used for making spun fibers. Partial alkaline hydrolysis has been used as a means of solubilizing fish protein concentrate, feathers, hooves and hair. Chemical peeling of grain with sodium hydroxide has been advocated. Treatment of contaminated peanut or cottonseed meals with gaseous ammonia has been suggested for aflatoxin destruction. Preservation of fish with ammonia has also been suggested. Alkali cooking of maize is a traditional procedure to increase digestibility. The pH of egg white may reach 9 during refrigerated storage (see Provansal *et al.* 1975; Sternberg *et al.* 1975A).

Formation of New Amino Acids and Cross-links.—It is well known that severe alkaline processing of proteins provokes a progressive destruction of various amino acids, cystine, arginine, threonine and serine being the most sensitive.

Some modifications of amino acid residues take place with the simultaneous formation of new amino acids such as lysinoalanine (Patchornik and Sokolovsky 1964; Bohak 1964), lanthionine (Horn *et al.* 1941) and ornithinoalanine (Ziegler *et al.* 1967), responsible for intra- or intermolecular cross-links.

Lysinoalanine has been found in acid hydrolyzates of numerous

$$\begin{array}{ccc} COOH & & COOH \\ | & & | \\ CH-(CH_2)_4-NH-CH_2-CH \\ | & & | \\ NH_2 & & NH_2 \end{array}$$

proteins previously submitted to an alkaline treatment (De Groot and Slump 1969; Provansal *et al.* 1975). Lysinoalanine forms through condensation of ϵ-amino groups of lysyl residues with dehydroalanyl residues.

$$\begin{array}{c} | \\ CO \\ | \\ C = CH_2 \qquad \text{dehydroalanyl residue} \\ | \\ NH \\ | \end{array}$$

The latter probably come from the alkaline degradation of cystine residues through β-elimination and desulfuration with H_2S formation, or from the decomposition of serine or phosphoserine residues. Lysinoalanine has also been found in heated foods that have not been submitted to any alkaline treatment (Sternberg *et al.* 1975A).

Lanthionine forms through condensation between dehydroalanyl

$$\begin{array}{ccc} COOH & & COOH \\ | & & | \\ CH - CH_2 - S - CH_2 - CH \\ | & & | \\ NH_2 & & NH_2 \end{array}$$

and cysteinyl residues.

Ornithinoalanine forms through condensation of dehydroalanyl

$$\begin{array}{ccc} COOH & & COOH \\ | & & | \\ CH -(CH_2)_3 - NH - CH_2 - CH \\ | & & | \\ NH_2 & & NH_2 \end{array}$$

and ornithine residues. The latter may come from alkaline decomposition of arginyl residues.

Lysinoalanine, lanthionine, ornithine (and possibly ornithinoalanine) and alloisoleucine (from isoleucine) have been identified in a 0.05–1 *M* sodium hydroxide-treated sunflower protein isolate (55–80°C, 1–16 hr) (Provansal *et al.* 1975). The formation of ornithine and the destruction of arginine are exactly proportional and appear to be the best indicators for the severity of the alkaline treatment.

The chromatographic behavior of most of the inhabitual amino acids in a given ion-exchange chromatography system is given in Fig. 17.2 (Mauron 1975). Damage to proteins from alkaline processing is relatively easily assessed by such chromatography after acid hydrolysis.

Lysinoalanine and lanthionine can also be determined by thin-layer chromatography (Sternberg *et al.* 1975B).

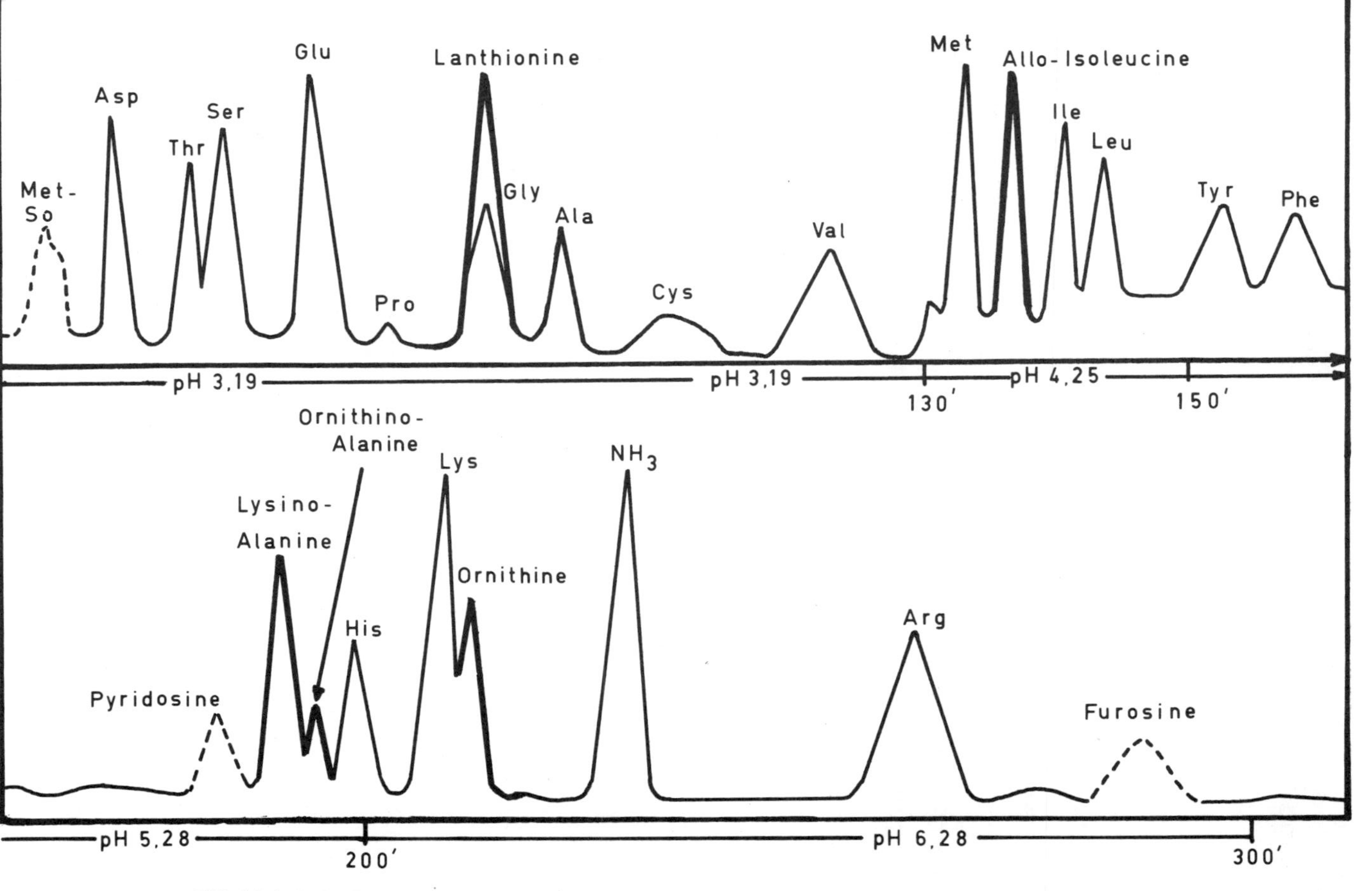

FIG. 17.2. SEPARATION OF UNUSUAL AMINO ACIDS BY ION EXCHANGE CHROMATOGRAPHY

Nutritional Effects.—In most experiments, alkali-treated proteins were found to have a decreased nutritional value. The loss or substitution of cystine, lysine or other amino acids may account for this decrease, depending on the limiting amino acid in the diet. Amino acid isomerization may also cause a decrease in availability of some amino acids.

It is likely that inter- or intramolecular cross-links due to lysinoalanine, lanthionine and ornithinoalanine reduce the accessibility of the protein to proteolytic enzymes. The nitrogen digestibility of severely NaOH-treated casein (0.2 and 0.5 M, 80°C, 1 hr) was determined on rats and found to be 71 and 47%, respectively, as compared to 90% for intact casein (Cheftel *et al.* 1976). Lysinoalanine (probably in protein-bound form) could be detected in the kidneys and in the liver, but not in the blood of the rats. Most lysinoalanine, however, was present in the feces. *In vivo* protein digestibility of mildly NaOH-treated soybean proteins was much less markedly affected (De Groot and Slump 1969).

Severely alkali-treated herring meals do not support adequate growth in chicks, and induce toxic effects at high levels of feeding. Growth retardation has also been reported in lambs fed NaOH-dispersed soybean protein concentrates. A decrease in the NPU of NaOH-treated soybean proteins was observed in experiments with rats (De Groot and Slump 1969).

The possible toxicity of the latter proteins has been studied by feeding rats with a diet containing 20% of such proteins (De Groot and Slump 1969; Van Beek *et al.* 1974). No histological or clinical particularities were noted (when enough calcium was present in the diet). Other investigators, however, reported cytoplasmic and nuclear enlargement of renal tubular cells (Woodard and Short 1973). Similar cytomegalic renal lesions were noted in rats fed a diet containing up to 0.3% of synthetic lysinoalanine (Woodard *et al.* 1975). Absorbed lysinoalanine is the nephrotoxic agent. The alkali-treated soybean proteins of both groups of investigators contained ca 0.7% of lysinoalanine residues. Recently De Groot *et al.* (1975) also found marked nephrocytomegalia when rats were fed acid-hydrolyzed alkali-treated soy proteins or synthetic lysinoalanine. This appeared to be specific to the rat and was not observed with mice, hamsters, quails, dogs or monkeys.

At this point, it can be tentatively concluded that mild alkaline treatment of protein is acceptable, but that the processing should remain carefully controlled.

BINDING OF NONPROTEIN MOLECULES TO AMINO ACID RESIDUES

Reaction with Carbohydrates

Maillard Reactions and Nonenzymatic Browning.—The main features of the reactions between proteins and carbonyl compounds, particularly reducing sugars, is that they result in the binding of ϵ-amino groups of lysine residues, and that they may occur even during storage at ordinary temperature.

The scheme of Fig. 17.3 summarizes the initial interactions between reducing sugars and ϵ-amino groups, up to the formation of "Amadori products," with no or little browning. Later reactions, which may occur without delay, induce the formation of reactive polycarbonyl compounds, which polymerize and cross-link protein chains, with concomitant intense browning (Mauron 1972; Carpenter 1973; Finot 1973; Feeney *et al.* 1975).

Maillard reactions have a relatively high energy of activation; as a result, when protein foods containing reducing sugars are moderately heated (as in the drying of milk), Maillard reactions occur in preference to any other type of protein damage. Protein foods containing small amounts of reducing sugar and submitted to a more severe heating (toasted breakfast cereals, bread crust, some biscuits, oilseed meals, glucose-protein model systems at high temperature) may undergo both late Maillard-type and protein-protein damage. In such foods, there is not only a loss of available lysine, but also an accompanying decrease in the overall nitrogen digestibility of the protein and therefore in the availability of most amino acids (Miller *et al.* 1965A).

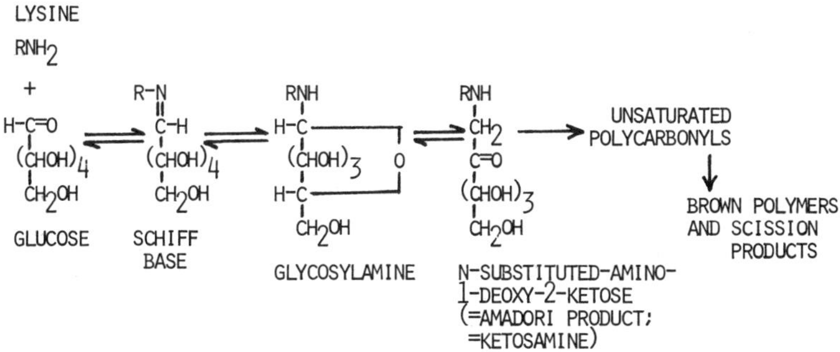

FIG. 17.3. EARLY MAILLARD REACTIONS

The rate of Maillard reactions is maximal at a water activity of 0.6–0.8, and decreases at higher concentrations of water (as in liquid milk); it is still appreciable at an a_w as low as 0.2. This type of deterioration is therefore enhanced during the concentration and the dehydration of various protein foods (milk concentration and drum drying, dehydration of egg white or whole eggs, drying of oilseed meals). Maillard reactions may also occur during the storage at ambient temperature of some dehydrated or intermediate moisture protein foods. The most efficient browning inhibitor, SO_2, which blocks carbonyl groups, can be added to some foods; for others it is necessary to carefully control temperature, duration and relative humidity of both processing and storage.

Pentoses are usually more active than hexoses in promoting Maillard reactions. The nonreducing disaccharide, sucrose, will induce browning only after it is inverted by heat or by spontaneous hydrolysis in acid medium. Low pH, however, partly inhibits browning since the first step of Maillard reactions rests in the condensation of carbonyl groups with nonionized ϵ-amino groups. Alkaline pH generally enhances Maillard reactions.

From the nutritional standpoint, initial and later Maillard reactions do not induce the same type of damage. However, both types of reactions may occur almost simultaneously in the same food.

Initial Maillard reactions lead to the substitution of the ϵ-amino groups of lysine residues, and to a marked decrease in the availability of this amino acid. Amadori products are the main substituted lysine derivatives in food or model proteins undergoing initial Maillard reactions; ϵ-N-(1-deoxy-lactulosyl)-lysine residues represent 70–75% of Maillard compounds in samples of overheated milk (Finot 1973).

Synthetic ϵ-N-(1-deoxy-D-fructosyl)-L-lysine, corresponding to the reaction of glucose with lysine, is totally unavailable as a source of lysine for the rat; it is extensively absorbed from the intestine, and then excreted in the urine (Finot 1973). ϵ-N-(Deoxy fructosyl)-lysine residues from glucose-protein model systems appear to be partly released during digestion *in vitro*, absorbed and then excreted in the urine under the form of unavailable peptides (Valle-Riestra and Barnes 1970; Ford and Shorrock 1971; Pronczuk *et al.* 1973). This could partly explain the frequent high protein digestibility of Maillard-damaged proteins with a very low content of available lysine. Deoxy-lactulosyl-lysine residues from milk proteins reacted with lactose are, however, found in the feces rather than in the urine (Finot 1973). ϵ-N-(Deoxy-ketosyl)-lysyl peptides appear to be partly metabolized by the intestinal flora (Finot 1973; Erbersdobler 1973).

In many studies of protein-glucose model systems kept at $37°C$ for various periods of time, it was found that the nutritive value of the protein was severely decreased although its digestibility was much less affected (see Carpenter 1973). The addition of lysine to the diet fully restored the nutritional value. Schematically, in such cases of initial Maillard reactions, the protein damage can be exclusively ascribed to the unavailability of lysine residues with attached carbohydrates (Hurrel and Carpenter 1974). Some other amino acid residues, primarily cysteine, may react with reducing sugars, but to a much smaller extent than lysine.

The classic food example for such initial Maillard reactions is drum-dried milk powder. It has been well established by the Nestlé scientists that drum-drying can be markedly detrimental to the nutritive value of milk powder (10–40% drop in available lysine depending on processing equipment and conditions), while properly conducted spray drying has no effect (Mauron 1972). Other examples include biscuits enriched with milk powder, and sterilized fluid and evaporated milks. In sterilized but not in pasteurized milks, the availability of lysine decreases by 10–20%; a partial destruction of cystine also occurs.

In the more advanced stages of nonenzymatic browning, reactive unsaturated polycarbonyl compounds are formed from the reducing sugars; these reactive compounds polymerize and bind simultaneously to the α terminal, ϵ and other amino groups of different polypeptide chains, bringing about the formation of colored, high molecular weight, highly cross-linked protein-carbohydrate polymers with low solubility, digestibility and nutritional value (Clark and Tannenbaum 1974). "Limit peptides" undigestible *in vivo* may contain various essential amino acids, including lysine residues with free ϵ-amino groups. This would explain that in proteins in an advanced stage of browning reactions, the overall protein digestibility and nutritional value may be reduced. Several amino acids, and not only lysine, may become unavailable (Boctor and Harper 1968; Valle-Riestra and Barnes 1970).

In a study of the nutritional value of protein autoclaved in the presence of various reducing sugars, it was found that egg proteins are much more sensitive to damage than soy proteins or casein (Knipfel *et al.* 1975).

Acid hydrolysis of severely damaged proteins regenerates much less lysine than during early stages of Maillard reactions.

Late nonenzymatic browning reactions occur extensively in heated or long stored model systems of proteins with reducing sugars, and also in a number of heated protein-containing foods, especially in baked goods (various types of breads, biscuits, cookies, cakes, and toasted or puffed breakfast cereals). They are largely responsible for the desirable color and flavor of these foods, and therefore their effect on the nutritional value (quite detrimental in the case of several breakfast cereals) is not considered to be a problem. It is however recommended to control baking temperature and sugar content in order to avoid large decreases in the availability of lysine. Added synthetic lysine is similarly lost. In foods such as milk powders, browning, which may occur during improper storage, renders the product totally unacceptable for consumption.

From a toxicological standpoint, Adrian and co-workers (Adrian *et al.* 1966; Adrian and Susbielle 1975) have shown that some soluble brown melanoidins prepared by autoclaving mixtures of free amino acids with reducing sugars are biologically active substances able to stimulate or inhibit rat growth and disturb reproduction when given in the diet. These findings are probably relevant only in the case of foods containing appreciable amounts of free amino acids, such as protein hydrolyzates, flavor precursors, amino acid supplemented cereals or some types of single cell proteins (Mauron 1975). Compounds from Maillard reactions in heated proteins are only partly digestible and absorbed (see above), and do not influence animal or infant growth, as long as enough available lysine is present or added to the diet (see Mauron 1975).

Chichester and co-workers (Tanaka *et al.* 1974; Lee *et al.* 1974) have reported a preliminary study of the toxic effects of heated mixtures of amino acids and glucose, of pure browning model compounds such as α-N-(deoxyfructosyl)-leucine or tryptophan, of apricot browning pigments, and of browned egg albumin, when added to the diet of rats.

Dillard and Tappel (1976) studied the possible occurrence of carbonylamine reactions *in vivo*. It appears that no fluorescent brown products are formed in blood from glucose and plasma proteins. It would be interesting to investigate the possible presence of browning inhibitors in the blood.

Chemical Methods for the Determination of Lysine Availability.— It has been reported in preceding sections that either through severe heating or through reactions with carbonyl compounds, lysine residues of food proteins can become almost completely unavailable to animals (rat or chick assays). However, the total lysine content, as determined after acid hydrolysis, is never reduced to such a large extent.

ϵ-N-(1-Deoxy-D-fructosyl)-L-lysine, a pure synthetic model com-

$$\begin{array}{c} \text{COOH} \\ \diagdown \\ \diagup \quad \text{CH} - (\text{CH}_2)_4 - \text{NH} - \overset{\displaystyle O}{\overset{\|}{C}} - (\text{CHOH})_3 - \text{CH}_2\text{OH} \\ \text{NH}_2 \qquad\qquad\qquad \epsilon \\ \alpha \quad \text{lysine} \qquad\qquad\qquad \text{deoxy-fructose} \end{array}$$

pound of initial Maillard reactions which has no nutritional value as a source of lysine, can regenerate lysine with a yield of approximately 50% on hydrochloric acid hydrolysis (Finot and Mauron 1972); it is likely that many ϵ-amino substituted lysine residues in heated or otherwise treated proteins similarly regenerate lysine during protein hydrolysis, although they are not, or little available nutritionally *in vivo*.

Various investigators have therefore attempted to evaluate the nutritional availability of lysine by the chemical determination of either the ϵ-*N*-substituted or, more frequently, the nonsubstituted lysyl residues. Only the principal methods of determination will be briefly reviewed (Finot and Mauron 1972; Carpenter 1973; Hurrel and Carpenter 1974).

The Direct Fluorodinitrobenzene Method (Carpenter's method) *(Carpenter 1960).*—The reaction of free ϵ-amino groups of lysine residues with fluorodinitrobenzene (FDNB) is followed by the colorimetric determination of the ϵ-*N*-dinitrophenyl lysine (DNP-lysine) released from the protein by acid hydrolysis, and extracted. Various interferences, mainly from carbohydrates, and partial degradation of DNP-lysine during acid hydrolysis can be overcome (see Booth 1971). Some free ϵ-amino groups may however be sterically inaccessible to FDNB.

In addition, Mauron and co-workers (Finot and Mauron 1972; Mauron 1972), in contrast to Hurrell and Carpenter (1974) find that synthetic α-*N*-formyl-(ϵ-*N*-1-deoxy-D-fructosyl)-L-lysine (FFL) and α-*N*-formyl-(ϵ-*N*-1-deoxy-D-lactulosyl)-L-lysine (FLL) partly react with FDNB. These models of initial Maillard reactions between lysine residues and glucose or lactose possess a blocked α amino group, and probably react with FDNB at the ϵ nitrogen and/or at the keto group. When FFL and FLL are treated with FDNB and then submitted to hydrochloric acid hydrolysis, 5–10% of their constitutive lysine is regenerated as free lysine, and 5–30% appears as DNP-lysine (see Fig. 17.4).

In practice, however, the direct method has been used with success by Mauron and co-workers for the evaluation of lysine availability in scorched roller-dried milk powder; results were in close agreement with those of *in vivo* and *in vitro* enzymatic evaluations (Table 17.6) (Mottu and Mauron 1967). Good results were also obtained with this method in a collaborative study for the evaluation of dried fish and meat meals (Boyne *et al.* 1961).

The Indirect Fluorodinitrobenzene Method.—Free ϵ-amino groups of the protein are reacted with FDNB. The difference in free lysine content between the protein acid hydrolyzed before and after reaction with FDNB supposedly corresponds to FDNB-reactive lysine (and therefore hopefully to nutritionally available lysine) (Milner and Westgarth 1973; Hurrell and Carpenter 1974). This method gives an accurate result only in initially substituted lysine residues regenerate the same percentage of lysine upon acid hydrolysis, whether or not the protein has been reacted with FDNB. This is the case with heat-treated animal protein foods of low carbohydrate content, or with either unheated or severely heat-processed vegetable proteins.

TABLE 17.6

LYSINE[1] CONTENT AND AVAILABILITY IN CONCENTRATED OR DRIED COW'S MILK[2]

Method of Preparation	Total Lysine (Acid Hydrolysis)	FDNB-Reactive	Available from *in vitro* Proteolysis	Available from Rat Growth Assay
Freeze-dried	8.3	8.4	8.3	8.4
Spray-dried	8.0	8.2	8.3	8.1
Evaporated	7.6	6.4	6.2	6.1
Roller-dried (mild)	7.1	4.6	5.4	5.9
Roller-dried (severe)	6.1	1.9	2.3	2.0

[1] Lysine values in g/16 g N.

[2] After Mottu and Mauron 1967.

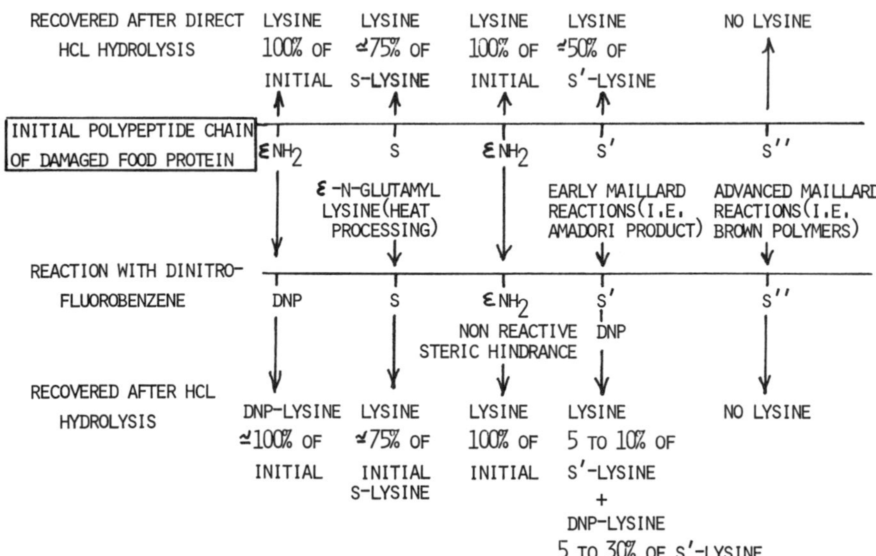

FIG. 17.4. REGENERATION OF LYSINE FROM BLOCKED LYSINE RESIDUES DURING HCl HYDROLYSIS BEFORE OR AFTER REACTION WITH FLUORODINITROBENZENE (S, S', S'' = ε-N-SUBSTITUTED LYSINE)

However, it is not the case for mildly heat-treated reducing sugar-protein mixtures such as milk powder (as demonstrated with FFL and FLL) (Fig. 17.4); this leads to an overestimation of FDNB-reactive and nutritionally available lysine.

The Methylisourea Method.—The protein is reacted with *o*-methylisourea; lysine residues with free ϵ-amino groups are quantitatively guanidinated into residues of homoarginine; these are released from the protein by acid hydrolysis, and determined by ion-exchange chromatography (Finot and Mauron 1972; Hurrell and Carpenter 1974; Mauron 1975). This method is lengthy but gives accurate results with all types of lysine substitution, including derivatives from initial Maillard reactions.

The Determination of Furosine.—Furosine can be determined by ion-exchange chromatography of proteins initially heated in the presence of carbohydrates (Fig. 17.2). It occurs as one of the breakdown products of FFL and other derivatives from Maillard reactions (see Finot 1973). Furosine and another basic amino acid, pyridosine, have been used as an index of Maillard-type unavailability of lysine in scorched drum-dried milk powder (Finot 1973).

$$\text{furosine:} \quad \underset{O}{\overset{\displaystyle \bigcirc}{\bigvee}}\!\!-CO-CH_2-NH-(CH_2)_4-CH \underset{\diagdown COOH}{\overset{\diagup NH_2}{}}$$

$$\text{pyridosine:} \quad \underset{CH_3}{\overset{OH}{\underset{\displaystyle O=\!\!\!\bigcirc\!\!-N}{}}}-(CH_2)_4-CH \underset{\diagdown COOH}{\overset{\diagup NH_2}{}}$$

Other methods have been suggested for the chemical determination of reactive lysine (Hurrell and Carpenter 1974). The binding of various types of dyes on proteins has been used for the rapid and inexpensive determination of protein quantity and quality (evaluation of the lysine content of cereals and evaluation of available lysine) (Hurrell and Carpenter 1975).

Reaction with Lipids

Lipoproteins consisting of noncovalent complexes of proteins and lipids or phospholipids occur widely in living tissues (as cellular and subcellular membranes, etc.) and considerably influence the physical and textural properties of foods (Karel 1973). The lipid constituents can, in most cases, be fully extracted with solvents and do not affect the nutritional value of the protein constituents.

There are cases, however, when covalent bonds are known to occur between lipids and proteins. Such interactions take place mainly as a result of the oxidation of unsaturated lipids. This oxidation is a free radical reaction, the course of which is shown approximately in Fig. 17.5 (Schultz *et al.* 1962; Labuza 1971). The major initial products consist largely of lipid peroxides. As oxidation progresses, breakdown products of peroxides accumulate.

Lipid oxidation and protein-lipid covalent interaction are known to take place *in vivo* in various aging tissues where they can damage the integrity of protein polymers including collagen, enzymes and membranes. They can also take place in some foods and feeds, such as frozen or dehydrated fish, fish meal, oilseeds, etc. While it is likely that foods having undergone extensive lipid oxidation are organoleptically unacceptable, it also appears that protein may act as a trap for lipid peroxides and their breakdown products. Some degree of protein damage may therefore occur before the food is rendered unacceptable.

The mechanisms of protein damage have been studied in various model systems where a protein is reacted with an oxidizing lipid (generally an ester of linoleic acid) under given conditions of temperature, relative humidity, aeration, presence of antioxidants, etc. Noncovalently-bound lipid constituents are extracted from the

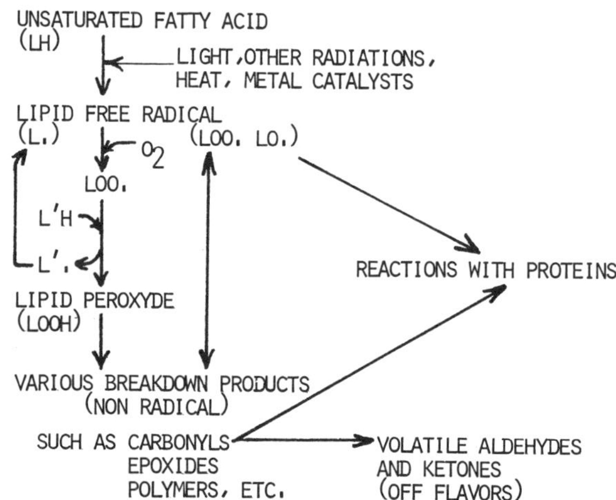

FIG. 17.5. SCHEMATIC REPRESENTATION OF LIPID OXIDATION REACTIONS

mixture with solvents, and various properties of the protein are determined.

Mechanisms of Reaction.—Two types of mechanisms appear to be at work in the covalent binding of peroxidizing lipids on proteins and the lipid-induced polymerization of protein: free radical reactions on the one hand, carbonylamine reactions on the other.

Free Radical Reaction.—Lipid free radicals [such as L., or LO. (alkoxy) or LOO. (peroxy)] react with protein (P_H), resulting in addition onto a protein (1):

$$LOO. + P_H \rightarrow .LOOP$$

followed by polymerization through cross-linking of protein chains by multifunctional lipid peroxy free radicals:

$$.LOOP + O_2 \rightarrow .OOLOOP$$

$$.OOLOOP + P_H \rightarrow \rightarrow POOLOOP \text{ (dimer) etc.}$$

$$(17.1)$$

This implies that the ratio of lipid incorporated per weight of protein is high.

Reaction of lipid free radical with protein can also result in the formation of a protein free radical (2):

$$LO. + P_H \rightarrow LOH + P.$$

$$LOO. + P_H \rightarrow LOOH + P.$$

Protein free radicals have been found by electron spin resonance to be located mainly on the α carbon atoms, but also on the sulfur atoms of cysteine residues (Karel *et al.* 1975). The concentration of protein free radicals was found to decrease when the water activity of the model system increased (Schaich and Karel 1975).

Relatively stable free radicals can also be formed from free amino acids-linoleate peroxides mixtures (Karel *et al.* 1975).

The formation of protein free radicals is followed by direct polymerization of protein chains (without lipid incorporation):

$$P. + P \rightarrow P-P. \quad \text{(dimer)} \quad (17.2)$$

$$P-P. + P \rightarrow P-P-P. \quad \text{(trimer), etc.}$$

Roubal and Tappel (1966A, B) studied the reactions, at $37°C$, of aqueous solutions of various proteins and enzymes in the presence of peroxidizing fish oil fractions or linoleate. There was a loss of solubility and of enzymatic activity. Molecular weight determinations indicated the formation of polymers and aggregates, with a simultaneous loss of methionine, histidine, cysteine and lysine.

The measurement of lipids covalently incorporated into these protein polymers shows that polymerization mechanisms (17.1) and (17.2) occur simultaneously.

These various effects were found to be similar to those resulting from protein irradiation by γ rays.

In some conditions, especially in the dry state, protein free radicals may not be able to diffuse and react with each other. Scission, rather than polymerization may then occur in the protein chain (Zirlin and Karel 1969).

γ Irradiation of proteins in the dry state may also provoke chain scissions.

Carbonylamine Reactions.—Long or short chain aldehyde derivatives resulting from the breakdown of peroxides of unsaturated fatty acids bind to amino groups of proteins by Schiff base type reactions:

$$L - C = O + H_2 N - P \rightarrow L - C = N - P$$

Bifunctional aldehydes such as malonaldehyde,

$$O = CH - CH_2 - CH = O$$

known to form during the peroxidation of many unsaturated lipids, may cause the formation of protein intra- or intermolecular cross-links.

Thus Andrews *et al.* (1965) showed that insulin stored with oxidizing linoleate at 50°C in the dry state became more resistant to trypsin, and that the content of FDNB-reactive lysine decreased. These effects were inhibited in the presence of bisulfites. Braddock and Dugan (1973) incubated salmon myosin in a buffer with oxidizing linoleate. IR, UV, visible and fluorescence spectra indicated the presence of C = N groups (similar compounds were found in extracts from frozen stored salmon). There was a marked loss of histidine, lysine and methionine.

Chio and Tappel (1969) reacted ribonuclease with malonaldehyde and observed the appearance of fluorescence, some degree of cross-linking, and losses in lysine, histidine and tyrosine. The fluorescence was attributed to the formation of a 1-amino-3-imino-propene structure, proving a reaction of malonaldehyde with free amino groups in proteins.

$$P - NH - CH = CH - CH = N - P$$

Thus aldehydes from oxidized lipids appear to be able to react primarily with amino groups (carbonylamine reactions) but also to react with or attack other groups from various amino acid residues.

Such reactions are thought to play a role in the denaturation of protein (decrease in solubility and salt-extractibility; loss of water-binding capacity; hardening of texture) observed in frozen and freeze-dried fish products.

Roubal (1971) investigated the relative importance of free radical and carbonylamine reactions in protein damage. A dry solid model system of peroxidizing fish oils plus muscle or plasma proteins was analyzed after various periods of contact at 37°C under oxygen. Maximum free radical concentration (and no fluorescence) was observed at 14 hr, while at 72 hr there was maximum fluorescence and very few ESR signals. At 14 hr, there was already a 20–40% loss in methionine, cysteine, tyrosine, alanine and lysine, while at 72 hr the additional loss was only 15%. The author concludes that damage due to free radicals (in the first period) was more severe than than due to aldehydes from decomposing lipid peroxides.

Matsushita and co-workers (Matsushita 1975) were able to confirm the existence of both radical and carbonylamine reactions by reacting either purified linoleic acid peroxides or a mixture of secondary breakdown products of the peroxides with enzymes.

The exact nature of amino acid modifications (with or without protein cross-links) by mechanisms (1) and (2) is not known, although cysteine, lysine, histidine, and methionine are almost always damaged. Tyrosine, arginine, tryptophan, serine and proline are often touched.

Methionine and cysteine destruction may be due to oxidation (Tannenbaum *et al.* 1969) and does not imply any participation of cross-links. Cysteine however is known to react with lipid peroxides or with their breakdown products.

Some of the amino acid losses reported in the preceding pages may be due to the presence of lipids at the moment of hydrochloric acid hydrolysis.

The exact reactions of malonaldehyde and other lipid-derived aldehydes with side-groups of amino acid residues other than lysine are also little understood.

Nutritional Effects of Lipid-Protein Reactions.—The extensive studies of Yanagita and co-workers (see Yanagita and Sugano 1975) have shown that casein (or ovalbumin) reacted with oxidizing ethyl linoleate at 50–60°C in aqueous medium or in the dry state, had a much reduced nutritive value. Contents in various amino acids, digestibility *in vivo*, PER, and Biological Value were all repeatedly and markedly reduced.

In addition to these studies with model systems, the effects of lipid oxidation on the nutritive value of fish meal has been investigated. After storage in air at 25°C for 1 yr, herring meal with 15% lipid showed a small 9% decrease in FDNB-reactive lysine (and no decrease when storage was done under nitrogen or after extraction of the lipids (Lea *et al.* 1958). Oxidized fat does bind more lysine at higher temperatures. Commercial fish meals which have been overheated during storage as a result of exothermic lipid oxidation reactions are usually severely damaged, with a low content in available lysine (Carpenter *et al.* 1963; Bunyan and Woodham 1964).

Reaction with Aldehydes

Apart from reducing carbohydrates and from carbonyl compounds resulting from Maillard-type reactions or lipid oxidation, a number of aldehydes may find themselves in contact with food proteins during processing or storage.

Aldehyde Treatment of Feed Proteins.—Feed proteins given to ruminants (especially soluble casein or soybean meal) are extensively hydrolyzed and deaminated by the bacterial flora of the rumen. Some ammonia is converted into microbial proteins; these are further digested in the lower gut but are often of lower nutritive value than the feed proteins.

Postruminal infusion of feed proteins or amino acids, especially methionine, has been demonstrated to improve the production of wool, meat and milk. It is therefore of interest to ensure that feed proteins reach the intestinal sites of digestion and absorption without preliminary modification in the rumen (Broderick 1975; Clark 1975).

Treatment of milk proteins and various vegetable protein meals with formaldehyde, glyoxal or glutaraldehyde (by spraying) has recently been shown to reduce protein degradation and ammonia formation in the rumen, and sometimes to increase nitrogen retention and/or wool growth in sheep (Broderick 1975; Clark 1975). This is probably due to a reduction of protein solubility.

Treatment of casein or whey proteins with 1% formaldehyde gives good results, while 2–3% formaldehyde is generally used with oilseed meals and with grass silage. However, in many cases, no beneficial effects were noted. This is particularly the case when initially insoluble protein feeds (such as fish meal) are treated with formaldehyde. Moreover, overtreatment with aldehydes reduces the protein digestibility in the lower gut, resulting in protein loss in the feces. The best treatment is probably that which gives a maximum intestinal absorption of methionine and the few next nutritionally limiting amino acids (Broderick 1975; Clark 1975).

A formaldehyde-treated casein adequate for sheep was found not to support growth in rats unless given as 40% of the whole diet. This low nutritional value was due to a decreased protein digestibility and was not improved by lysine, histidine and tryptophan supplementation (Hove and Lohrey 1976).

A four to sixfold increase in the concentration of ϵ-N-methyllysine has been observed in sheep plasma after feeding formaldehyde-treated casein. This derivative was also identified in the treated casein, after acid hydrolysis (Reis and Tunks 1973). This shows that formaldehyde treatment induces the formation of some nonreversible links. It has however been proved that formaldehyde binding is partly reversed at an acid pH. This phenomenon appears to depend on the duration of aldehyde treatment.

Chemistry of Formaldehyde-Protein Interactions.—The chemistry of formaldehyde-treated proteins is not entirely elucidated. It is known that formaldehyde reacts with ϵ-amino groups of lysine. This reaction is thought to proceed more to the dihydroxymethyl derivative than to the Schiff base (Feeney $et\ al.$ 1975):

$$\text{RNH}_2 + \underset{\underset{\displaystyle \text{OH}}{|}}{\overset{\overset{\displaystyle \text{H}}{|}}{\text{H—C—OH}}} \overset{-\text{H}_2\text{O}}{\rightleftharpoons} \underset{\underset{\displaystyle \text{H H}}{}}{\overset{\overset{\displaystyle \text{H}}{|}}{\text{R—N—C—OH}}} \overset{-\text{H}_2\text{O}}{\rightleftharpoons} \text{R—N}=\text{C}\underset{\displaystyle \text{H}}{\overset{\displaystyle \text{H}}{\Big\langle}}$$

$$\underset{\text{hydrate}}{\text{formaldehyde}} \Updownarrow + \text{CH}_2\text{O} \qquad\qquad \text{Schiff base}$$

$$\underset{\displaystyle \text{CH}_2\text{OH}}{\overset{\displaystyle}{\text{R—N—CH}_2\text{OH}}} \quad \underset{\text{derivative}}{\text{dihydroxymethyl}}$$

These compounds are degraded during acid hydrolysis.

Residues of cysteine, histidine, arginine, tyrosine, methionine, tryptophan, glutamine, and asparagine may also react with formaldehyde at neutral or alkaline pH (Means and Feeney 1971; Feeney $et\ al.$ 1975).

It is likely that the decrease in solubility and digestibility caused by treatments of proteins with aldehydes is due to cross-linking between polypeptide chains rather than to simple substitution of ϵ-amino lysyl or other amino acid residues.

Data on the "tanning effect" of formaldehyde or glutaraldehyde on leather, and on the "firming" of fish meal by formaldehyde, support this hypothesis. A 0.14% formaldehyde treatment of fish immediately before fish meal processing had a greater biological effect than chemical binding effect on lysine (Carpenter and Opstvedt 1976).

In a study of the detoxification of microbial toxin proteins with formaldehyde (to produce immunologically active toxoids), lysine derivatives could be identified, after acid hydrolysis, which corresponded to the covalent cross-linking of one (or two) ϵ-amino groups of lysyl residues by a methylene bridge to a tyrosyl residue in position 3 (and 5) of the phenol ring (Mannich-type compounds) (Bizzini and Raynaud 1974).

The formation of these and possibly other cross-links should be further studied and correlated with protein digestibility and amino acid availability. The extent of cross-linking and its stability (to pH) probably depends on the prior degree of polymerization of the aldehyde, on the duration and pH of reaction, and on further treatments such as heating or drying.

Other Aldehyde-Protein Interactions.—The reactions of glutaraldehyde, pyruvaldehyde, malonaldehyde, etc., with proteins have also been investigated (Means and Feeney 1971; Feeney *et al.* 1975). The chemistry of glutaraldehyde-protein interactions has been the object of much study because this aldehyde is widely used for immobilizing enzymes and antibodies on various supports.

Hexamethylenetetramine, a bacteriostatic agent used in northern Europe for the preservation of pickled raw fish (pH 4.1–4.5), progressively decomposes into formaldehyde.

Wood smoke and more specifically aldehydic fractions of wood smoke (containing acetaldehyde, formaldehyde, etc.) have also been shown to react with ϵ-amino groups in proteins of cured meats or in model systems; they may reduce the availability of lysine and the NPU in animals (Chen and Issenberg 1972).

Gossypol, a polyphenol present in cotton seed, may react with the ϵ-amino groups of the seed proteins through its aldehyde groups.

Free gossypol is toxic to animals. Gossypol binding may occur during certain processing operations of cotton seed meals, especially at high temperatures. Bound gossypol is undigestible, and much less toxic. Gossypol binding, however, reduces the content of available lysine and often lowers the nutritional value of cotton seed proteins (Mauron 1972).

Reaction with Quinones

It appears that quinones formed by the enzymatic or alkaline oxidation of polyphenols present in sunflower seeds or grasses may react with amino acid residues and lower the nutritional value of protein meals or concentrates. It may therefore be useful to prevent polyphenol oxidation and browning during the drying of grasses and the extraction and purification of sunflower seed protein.

Quinones can react with amino groups of free amino acids, and provoke oxidative deamination. Benzoquinone can react with the thioether group of methionine (Vithayathil and Murthy 1972). Quinones can also react with amino and sulfhydryl groups of proteins (Mason and Peterson 1965; Pierpoint 1969A, B). These reactions may lead to the formation of protein polymers.

The existence of covalently linked phenolic acids in various plant proteins, including sunflower seed protein concentrate, is well documented (Van Sumere *et al.* 1975). However, the type of linkage between phenolic acid and amino acids is still poorly known.

Casein allowed to react (in the presence of *o*-diphenol oxidase) with caffeic acid, isochlorogenic acid, or other phenolic compounds has been found to be inferior to control casein in terms of digestibility and biological value in the rat, and of available lysine content (Horigome and Kandatsu 1968). Casein model systems incubated with caffeic acid at alkaline pH or in the presence of polyphenol oxidase were found to contain less lysine than control casein. The release of free lysine from such model systems by proteolytic action was also decreased (Guilleux 1975).

In a study of various nonlegume leaf protein concentrates, Allison (1971) found significant negative correlations between nonavailable lysine and nutritional value. "Chloroplastic" protein fractions of leaves had a higher content of nonavailable lysine and a lower nutritional value than "cytoplasmic" protein fractions. The latter are known to contain less polyphenol than chloroplastic fractions.

Heat processing (especially above 80°C) of leaf protein concentrates reduces the nitrogen digestibility and the nutritional value; it would be of interest to find out if the detrimental effects of overprocessing are due to reactions of quinones, or of carbohydrates, with ϵ-amino or sulfhydryl groups.

Reaction with Nitrites

The fate of nitrites added to meat or meat-curing model systems composed of myoglobin, nitrite and ascorbate has been studied, with the use of ^{15}N-nitrite (Cassens *et al.* 1974; Fujimaki *et al.* 1975). Nitrosomyoglobin formed in cured meat corresponds approximately to 15 ppm of nitrite (10–20% of added nitrite). The nitrate content increases markedly with the curing period and length of storage, while the residual nitrite content decreases.

Nitrites may react with secondary or tertiary amines initially present in foods, or formed through autolysis, bacterial action or cooking. Some free amino acids such as proline, tryptophan, arginine, histidine, and possibly the corresponding amino acid residues, may constitute such reactive amines. These reactions may take place during the cooking of foods, or even during digestion, at the low pH of the stomach. Some of the resulting nitrosamines or nitrosamides, especially dimethylnitrosamine, are known to be potent carcinogens. As an example, nitrosopyrrolidine was found to occur in heated bacon; it probably comes from nitrosation of free proline, followed by decarboxylation of nitrosoproline.

Studies of nitrite interactions with amino acid residues in peptides and proteins (Means and Feeney 1971; Kurosky and Hofmann 1972) have shown that α and ϵ-amino groups can be first nitrosated, then split from the molecule: deamination occurs with N_2 elimination (Van Slyke reaction). Nitric oxide (NO) appears to be the reactive compound in nitrite-protein interaction.

Nitrosation of tyrosine, histidine and tryptophan residues can also take place. Nitrosation of tryptophan leads to the formation of an N^1-nitrosamine. An N-acetyl tryptophan dipeptide has indeed been shown to give a nitrosamine on reaction with nitrites (Bonnet and Holleyhead 1974). These reactions are enhanced by heat and low pH; they may well take place at the moderately acid pH of most meats.

An *in vitro* study of nitrite interaction with bovine serum albumin has shown that nitrite reacts with lysine and tyrosine residues; after extensive proteolysis, 6-hydroxynorleucine (formed through ϵ-amino deamination of lysyl residues), 3,4-DOPA and 3-nitrotyrosine were identified. The latter two compounds are formed by the oxidation of 3-nitrosotyrosine (Knowles *et al.* 1974).

Nitrites can react quantitatively with sulfhydryl groups in meat at pH 2–3, to give S-nitrosothiols (Kubberød *et al.* 1974). At meat pH and high temperature, only 10–25% of added nitrites form nitrosothiols.

Nitrites are normally present in meats at too low a concentration to bring about a significant decrease in the content or availability of lysine, tryptophan and/or cysteine. The formation of nitrosamines is of much more significance.

Reaction with Sulfites

Sulfite ions react with disulfides to form S-substituted thiosulfates (also called S-sulfonates) and thiols (see Means and Feeney 1971):

$$P - S - S - P + SO_3^{-2} \;\rightleftharpoons\; P - S - SO_3^- + P - S^-$$
$$S\text{-sulfocysteine}$$

Sulfitolysis is enhanced at pH 7. Many cystine residues in proteins, however, especially those of intramolecular bonds, resist the action of sulfites. S-Sulfonates are unstable in strongly acid or alkaline solutions, and usually decompose to disulfides. Treatment with reducing agents yields cysteine residues.

No detrimental effect of sulfites on the nutritive value of proteins can be inferred from these data.

Reaction with Chlorinated Molecules

Trichloroethylene combines with the sulfhydryl groups in proteins under prolonged heating. The S-dichlorovinyl-L-cysteine thus formed appears to be the toxic factor in trichloroethylene-extracted soybean meal, producing aplastic anemia in calves (McKinney *et al.* 1959).

Lipid solvents such as dichloromethane, tetrachloroethylene, 1,1,1-trichloroethane and fluorochlorocarbons do not appear to react with proteins (Valle-Riestra 1974).

1,2-Dichloroethane was found to react with fish proteins, especially under slightly alkaline conditions, resulting in the partial destruction of histidine and cystine; the availability of cystine, histidine and methionine was reduced (Morrison and Munroe 1965; Morrison and McLaughlan 1972). These findings suggest that 1,2-dichloroethane may alkylate sulfhydryl groups of proteins, and even cross-link protein chains through protease-resistant thioether bonds. 1,2-Dichloroethane extracted fish protein concentrate was fed to young calves as a milk replacer; it was found to have a lower nutritional value than dried skim milk, even when methionine was added (Makdani *et al.* 1971).

Agene (nitrogen trichloride), formerly used as a maturing agent for flour, reacts with methionine residues in wheat proteins to form methionine sulfoximine, a toxic compound.

Peanut protein isolates have been treated with 0.3% sodium hypochlorite at pH 8–9, in order to destroy aflatoxins. The treatment caused a 50% reduction in tryptophan and tyrosine contents, probably through oxidation (Natarajan *et al.* 1975).

CONCLUSIONS

Although this review is by no means exhaustive, it is clear that numerous detrimental reactions can take place when food proteins are processed by physical or chemical means. Only a few of these reactions may lead to the formation of toxic derivatives; many, however, cause a reduction in nutritive value, often through the formation of covalent cross-links between polypeptide chains. This reduction has both biological and economic consequences.

An improved knowledge of these reactions will obviously help for a better control of food processing and storage. A limited degree of damage, however, appears unavoidable, and must be considered together with the beneficial effects of food processing on food safety, flavor, shelf-life and convenience.

ACKNOWLEDGMENTS

This study was supported in part by the Centre National de la Recherche Scientifique, Paris—(E.R.A. N° 614: Modifications biochimiques et nutritionnelles de protéines alimentaires).

BIBLIOGRAPHY

ADRIAN, J., FRANGNE, R., PETIT, L., GODON, B., and BARBIER, J. 1966. Nutritional effects of soluble products from Maillard reaction. Ann. Nutr. Alimentation *20*, 257–277. (French)

ADRIAN, J., AND SUSBIELLE, H. 1975. The Maillard reaction. 10. Effects of premelanoidins on reproduction in rats. Ann. Nutr. Alimentation *29*, 151–158. (French)

ALLISON, R. M. 1971. Factors influencing the availability of lysine in leaf protein. *In* Leaf Protein. N. W. Pirie (Editor). Blackwell Scientific Publications, Oxford.

ANANTHARAMAN, K., and CARPENTER, K. J. 1971. Effects of feed processing on the nutritional value of groundnut products. 2. Individual amino acids. J. Sci. Food Agr. *22*, 412–418.

ANDERSON, G. H., LI, G. S. K., JONES, J. D., and BENDER, F. 1975. Effect of hydrogen peroxide treatment on the nutritional quality of rapeseed flour fed to weanling rats. J. Nutr. *105*, 317–325.

ANDREWS, F., BJORKSTEN, J., TRENK, F. B., HENICK, A. S., and KOCH, R. B. 1965. The reaction of an autoxidized lipid with proteins. J. Am. Oil Chem. Soc. *42*, 779–786.

BENDER, A. E. 1970. Factors affecting the nutritive value of protein foods. *In* Evaluation of Novel Protein Products, A. E. Bender, R. Kihlberg, B. Löfqvist, and L. Munck (Editors). Pergamon Press, Oxford.

BENDER, A. E. 1972. Processing damage to protein foods. Protein Advisory Group Bull. *2*, 10–19.

BIZZINI, B., and RAYNAUD, M. 1974. The detoxification of protein toxins by formaldehyde. Possible underlying mechanisms and new developments. Biochimie *56*, 297–303. (French)

BJARNASON, J., and CARPENTER, K. J. 1969. Mechanisms of heat damage in proteins. 1. Models with acylated lysine units. Br. J. Nutr. *23*, 859–868.

BJARNASON, J., and CARPENTER, K. J. 1970. Mechanisms of heat damage in proteins. Chemical changes in pure proteins. Br. J. Nutr. *24*, 313–329.

BOCTOR, A. M., and HARPER, A. E. 1968. Measurement of available lysine in heated and unheated foodstuffs by chemical and biological methods. J. Nutr. *94*, 289–296.

BOHAK, Z. 1964. N^ϵ-(DL-2 amino-2-carboxyethyl)-L-lysine, a new amino acid formed on alkaline treatment of proteins. J. Biol. Chem. *239*, 2878–2887.

BONNETT, R., and HOLLEYHEAD, R. 1974. Reaction of tryptophan derivatives with nitrite. J. Chem. Soc. *Perkin I*, 961–964.

BOOTH, V. H. 1971. Problems in the determination of FDNB-available lysine. J. Sci. Food Agric. *22*, 658–665.

BOYNE, A. W., CARPENTER, K. J., and WOODHAM, A. A. 1961. Assessment of laboratory procedures suggested as indicators of protein quality in feeds. J. Sci. Food Agric. *12*, 832–848.

BRADDOCK, R. J. and DUGAN, L. R. 1973. Reaction of autoxidizing linoleate with coho salmon myosin. J. Am. Oil Chem. Soc. *50*, 343–347.

BRESSANI, R., ELIAS, L. G., GOMEZ BRENES, R. A. 1972. Improvement of protein quality by amino acid and protein supplementation. *In* Protein and Amino Acid Functions. E. J. Bigwood (Editor). Pergamon Press, Oxford.

BRESSANI, R., VITERI, F., ELIAS, L. G., DE ZAGHI, S., ALVARADO, J., and ODELL, A. D. 1967. Protein quality of a soybean protein textured food in experimental animals and children. J. Nutr. *93*, 349–360.

BRODERICK, G. A. 1975. Factors affecting ruminant response to protected amino acids and proteins. *In* Protein Nutritional Quality of Foods and Feeds. Part 2. M. Friedman (Editor). Marcel Dekker, Inc., New York.

BUNYAN, J., and WOODHAM, A. A. 1964. Protein quality of feeding-stuffs. 2. The comparative assessment of protein quality in three fish meals by microbiological and other laboratory tests, and by biological evaluation with chicks and rats. Br. J. Nutr. *18*, 537–544.

BURACZEWSKI, S., BURACZEWSKA, L., and FORD, J. E. 1967. The influence of heating of fish proteins on the course of their digestion. Acta Biochim. Polonica. *14*, 121–131.

CARPENTER, K. J. 1960. The estimation of the available lysine in animal-protein foods. Biochem. J. 77, 604–610.

CARPENTER, K. J. 1973. Damage to lysine in food processing: Its measurement and its significance. Nutr. Abst. *43*, 423–451.

CARPENTER, K. J., MARCH, B. E., MILNER, C. K., and CAMPBELL, R. C. 1963. A growth assay with chicks for the lysine content of protein concentrates. Br. J. Nutr. *17*, 309–323.

CARPENTER, K. J., MORGAN, C. B., LEA, C. H., and PARR, L. J. 1962. Chemical and nutritional changes in stored herring meal. 3. Effect of heating at controlled moisture contents on the binding of amino acids in freeze-dried herring press cake and in related model systems. Br. J. Nutr. *16*, 451–465.

CARPENTER, K. J., and OPSTVEDT, J. 1976. Application of chemical and biological assay procedures for lysine to fish meal. J. Agr. Food Chem. *24*, 389–393.

CASSENS, R. G., SEBRANEK, J. G., KUBBERRØD, G., and WOOLFORD, G. 1974. Where does the nitrite go? Food Prod. Dev. (Reprint)

CHEFTEL, C., CUQ, J. L., PROVANSAL, M., and BESANCON, P. 1976. Influence of processing on the composition and the nutritive value of protein foods. Rev. Fr. Corps Gras. *1*, 7–13. (French)

CHEN, L. B. and ISSENBERG, P. 1972. Interactions of some wood smoke components with ε-amino groups in proteins. J. Agr. Food Chem. *20*, 1113–1115.

CHIO, K. S., and TAPPEL, A. L. 1969. Inactivation of ribonuclease and other enzymes by peroxidizing lipids and by malonaldehyde. Biochemistry *8*, 2827–2832.

CLARK, J. H. 1975. Nitrogen metabolism in ruminants: protein solubility and rumen bypass of protein and amino acids. *In* Protein Nutritional Quality of Foods and Feeds. Part 2. M. Friedman (Editor). Marcel Dekker, Inc., New York.

CLARK, A. V. and TANNENBAUM, S. R. 1974. Isolation and characterization of pigments from protein-carbonyl browning systems. Models for two insulin-glucose pigments. J. Agr. Food Chem. *22*, 1089–1093.

CUQ, J. L. 1973. Oxidation of methionyl residues of casein; nutritional effects. Thesis. University of Sciences and Techniques, Montpellier. (French)

CUQ, J. L., PROVANSAL, M., CHARTIER, L., BESANCON, P., and CHEFTEL, C. 1974. Effects of oxidative and alkaline treatments of food proteins on the nutritional availability of methionine and lysine. Proceedings of the 4th International Congress of Food Science and Technology, Madrid. In press.

CUQ, J. L., PROVANSAL, M., GUILLEUX, F., and CHEFTEL, C. 1973. Oxidation of methionine residues of casein by hydrogen peroxide. J. Food Sci. *38*, 11–13.

DE GROOT, A. P. 1963. The influence of dehydration of foods on the digestibility and the biological value of the protein. Food Technol. *18*, 339–343.

DE GROOT, A. P., and SLUMP, P. 1969. Effects of severe alkali treatment of proteins on amino acid composition and nutritive value. J. Nutr. *98*, 45–56.

DE GROOT, A. P., SLUMP, P., VAN BEEK, L., and FERON, V. J. 1975. Severe alkali treatment of protein. (Abstr.) 35th Annual Meeting of the Institute of Food Technologists, Chicago.

DILLARD, C. J., and TAPPEL, A. L. 1976. Do carbonylamine browning reactions occur in vivo? J. Agr. Food Chem. *24*, 74–77.

DONOSO, G., LEWIS, O. A. M., MILLER, D. S., and PAYNE, P. R. 1962. Effect of heat treatment on the nutritive value of proteins: Chemical and balance studies. J. Sci. Food Agric. *13*, 192–196.

ERBERSDOBLER, H. 1973. The normal course of digestion of food proteins. *In* Proteins in Human Nutrition. J. W. G. Porter and B. A. Rolls (Editors). Academic Press, New York.

EVANS, R. J., BAUER, D. H., SISAK, K. A. and RYAN, P. A. 1974. The availability for the rat of methionine and cystine contained in dry bean seed (Phaseolus vulgaris). J. Agr. Food Chem. *22*, 130–133.

FEENEY, R. E., BLANKENHORN, G., and DIXON, H. B. F. 1975. Carbonylamine reactions in protein chemistry. Adv. Prot. Chem. *29*, 135–203.

FINOT, P. A. 1973. Nonenzymic browning. *In* Proteins in Human Nutrition. J. W. G. Porter and B. A. Rolls (Editors). Academic Press, New York.

FINOT, P. A. and MAURON, J. 1972. Damage to lysine by the Maillard reaction. 2. Chemical properties of the N-(-1-deoxy-D-fructosyl) and N-(-1-deoxy-D-lactulosyl) derivatives of lysine. Helv. Chim. Acta. *55*, 1153–1164.

FORD, J. E. 1973. Some effects of processing on nutritive value. *In* Proteins in Human Nutrition. J. W. G. Porter and B. A. Rolls (Editors). Academic Press, New York.

FORD, J. E., and SALTER, D. N. 1966. Analysis of enzymically digested food proteins by Sephadex-gel filtration. Br. J. Nutr. *20*, 843–860.

FORD, J. E., and SHORROCK, C. 1971. Metabolism of heat-damaged proteins in the rat. Influence of heat damage on the excretion of amino acids and peptides in the urine. Br. J. Nutr. *26*, 311–322.

FUJIMAKI, M., EMI, M., and OKITANI, A. 1975. Fate of nitrite in meat-curing model systems composed of myoglobin, nitrite and ascorbate. Agr. Biol. Chem. *39*, 371–379.

GOMYO, T., and FUJIMAKI, M. 1970. Studies on changes of protein by dye sensitized photooxidation. 3. On the photodecomposition products of lysozyme. Agr. Biol. Chem. *34*, 302–309.

GUILLEUX, F. 1975. Interactions between polyphenols and food proteins. Thesis, University of Sciences and Techniques, Montpellier. (French)

HANSEN, L. P., JOHNSTON, P. H., and FERREL, R. E. 1975. The assessment of thermal processing on wheat flour proteins by physical, chemical, and enzymatic methods. *In* Protein Nutritional Quality of Foods and Feeds. Part 2. M. Friedman (Editor). Marcel Dekker, Inc., New York.

HAYASE, F., KATO, H., and FUJIMAKI, M. 1973. Racemization of amino acid residues in proteins during roasting. Agr. Biol. Chem. *37*, 191–192.

HAYASE, F., KATO, H., and FUJIMAKI, M. 1975. Racemization of amino acid residues in proteins and poly (L-amino acids) during roasting. J. Agr. Food Chem. *23*, 491–494.

HORIGOME, T., and KANDATSU, M. 1968. Biological value of protein allowed to react with phenolic compounds in presence of o-diphenol oxidase. Agr. Biol. Chem. *32*, 1093–1102.

HORN, M. J., JONES, D. B., and RINGEL, S. J. 1941. Isolation of a new sulfur-containing amino acid (lanthionine) from sodium carbonate treated wool. J. Biol. Chem. *138*, 141–149.

HOVE, E. L., and LOHREY, E. 1976. The effect of formaldehyde on the nutritive value of casein and lactalbumin in the diet of rats. J. Nutr. *106*, 382–387.

HURRELL, R. F., and CARPENTER, K. J. 1974. Mechanisms of heat damage in proteins. 4. The reactive lysine content of heat-damaged material as measured in different ways. Br. J. Nutr. *32*, 589–604.

HURRELL, R. F., and CARPENTER, K. J. 1975. The use of three dye-binding procedures for the assessment of heat damage to food proteins. Br. J. Nutr. *33*, 101–115.

IRIARTE, B. J. R., and BARNES, R. H. 1966. The effects of overheating on certain nutritive properties of proteins of soybeans. Food Technol. *20*, 835–838.

JOSEFSSON, E. 1975. Effects of variation of heat treatment conditions on the nutritional value of low-glucosinolate rapeseed meal. J. Sci. Food Agric. *26*, 157–164.

KAREL, M. 1973. Protein-lipid interactions. J. Food Sci. *38*, 756–763.

KAREL, M., SCHAICH, K., and BOY, R. B. 1975. Interaction of peroxidizing methyl linoleate with some proteins and amino acids. J. Agr. Food Chem. *23*, 159–163.

KELLEY, J. J., and PRESSEY, R. 1966. Studies with soybean protein and fiber formation. Cereal Chem. *43*, 195–206.

KENNEDY, T. S., and LEY, F. J. 1971. Studies on the combined effect of gamma radiation and cooking on the nutritional value of fish. J. Sci. Fd. Agr. *22*, 146–148.

KIES, G., and FOX, H. M. 1971. Comparison of the nutritional value of TVP, methionine enriched TVP and beef at two levels of intake for human adults. J. Food Sci. *36*, 841–845.

KIES, C., FOX, H., and APRAHAMIAN, S. 1975. Comparative value of L-, DL-, and D-methionine supplementation of an oat-based diet for humans. J. Nutr. *105*, 809–814.

KNIPFEL, J. E., BOTTING, H. G. and McLAUGHLAN, J. M. 1975. Nutritional quality of several proteins as affected by heating in the presence of carbohydrates. *In* Protein Nutritional Quality of Foods and Feeds. Part 2. M. Friedman (Editor). Marcel Dekker, Inc., New York.

KNOWLES, M. E., McWEENY, D. J., COUCHMAN, L., and THOROGOOD, M. 1974. Interaction of nitrite with proteins at gastric pH. Nature *247*, 288–289.

KORSLUND, M., KIES, C., and FOX, H. M. 1973. Comparison of the protein nutritional value of TVP, methionine-enriched TVP and beef for adolescent boys. J. Food Sci. *38*, 637–638.

KUBBERØD, G., CASSENS, R. G., and GREASER, M. L. 1974. Reaction of nitrite with sulfhydryl groups of myosin. J. Food Sci. *39*, 1228–1230.

KUROSKY, A., and HOFMANN, T. 1972. Kinetics of the reaction of nitrous acid with model compounds and proteins, and the conformational state of N-terminal groups in the chymotrypsin family. Can. J. Biochem. *50*, 1282–1296.

LABUZA, T. P. 1971. Kinetics of lipid oxidation in foods. Critical Rev. Food Technol. *2*, 355–405.

LEA, C. H., PARR, L. J., and CARPENTER, K. J. 1958. Chemical and nutritional changes in stored herring meal. Br. J. Nutr. *12*, 297–312.

LEE, C. M., CHICHESTER, C. O., and LEE, T. C. 1974. Physiological consequences of browned food products. Proceedings of the 4th International Congress of Food Science and Technology, Madrid.

LEY, F. J. 1970. Safety of irradiated food: Biological aspects. *In* SOS/70 Proceedings, 3rd International Congress of Food Science and Technology, Institute of Food Technologists, Chicago.

LIENER, I. E. 1975A. Protease inhibitors and hemagglutinins of legumes. Symposium on Proteins for Human Consumption, 35th Annual Meeting of the Institute of Food Technologists, Chicago.

LIENER, I. E. 1975B. Effects of anti-nutritional and toxic factors on the quality and utilization of legume proteins. *In* Protein Nutritional Quality of Foods and Feeds. Part. 2. M. Friedman (Editor) Marcel Dekker, Inc. New York.

LIENER, I. E. 1976. Legume toxins in relation to protein digestibility—A review. J. Food Sci. *41*, 1076–1081.

LIPTON, S. H., and BODWELL, C. E. 1975. Chemical approaches for estimating nutritionally available methionine. *In* Protein Nutritional Quality of Foods and Feeds. Part 2. M. Friedman (Editor) Marcel Dekker, Inc., New York.

LIPTON, S. H., and BODWELL, C. E. 1976. Preparation and chemical cleavage of S-phenacyl-L-methioninesulfonium bromide. J. Agr. Food Chem. *24*, 32–36.

MAKDANI, A. B., HUBER, J. T., and MICHEL, R. L. 1971. Nutritional value of 1,2-dichloroethane-extracted fish protein concentrate for young calves fed milk replacer diets. J. Dairy Sci. *54*, 886–892.

MASON, H. S., and PETERSON, E. W. 1965. Melanoproteins. 1. Reactions between enzyme-generated quinones and amino-acids. Biochim. Biophys. Acta *111*, 134–146.

MATSUSHITA, S. 1975. Specific interactions of linoleic acid hydroperoxides and their secondary degraded products with enzyme proteins. J. Agr. Food Chem. *23*, 150–154.

MAURON, J. 1972. Influence of industrial and household handling on food protein quality. *In* Protein and Amino Acid Functions, E. J. Bigwood (Editor). Pergamon Press, Oxford.

MAURON, J. 1975. Nutritional evaluation of processed proteins. Deutsche Lebensm. Rundsch. *71*, 27–35. (German)

MCKINNEY, L. L., ELDRIDGE, A. C., and COWAN, J. C. 1959. Cysteine thioethers from chloroethylenes. J. Amer. Chem. Soc. *81*, 1423–1427.

MEANS, G. E., and FEENEY, R. E. 1971. Chemical Modification of Proteins. Holden-Day, Inc., San Francisco.

MILLER, E. L., CARPENTER, K. J., and MILNER, C. K. 1965A. Availability of sulphur amino acids in protein foods. Chemical and nutritional changes in heated cod muscle. Br. J. Nutr. *19*, 547–564.

MILLER, E. L., CARPENTER, K. J., MORGAN, C. B., and BOYNE, A. W. 1965B. Availability of sulphur amino acids in protein foods. Assessment of available methionine by chick and microbiological assays. Br. J. Nutr. *19*, 249–267.

MILLER, E. L., HARTLEY, A. W., and THOMAS, D. C. 1965C. Availability of sulphur amino acids in protein foods. Effect of heat treatment upon the total amino acid content of cod muscle. Br. J. Nutr. *19*, 565–573.

MILLER, D. S., and SAMUEL, P. D. 1970. Effects of the addition of sulphur compounds to the diet on the utilization of protein in young growing rats. J. Sci. Food Agric. *21*, 616–618.

MILLER, S. A., TANNENBAUM, S. R., and SEITZ, A. W. 1970. Utilization of L-methionine sulfoxide by the rat. J. Nutr. *100*, 909–915.

MILNER, C. K., and WESTGARTH, D. R. 1973. A collaborative study on the determination of available lysine in feeding stuffs. J. Sci. Food Agric. *24*, 873–882.

MORRISON, A. B., and McLAUGHLAN, J. M. 1972. Availability of amino acids in foods. *In* Protein and Amino Acid Functions. E. J. Bigwood (Editor). Pergamon Press, Oxford.

MORRISON, A. B., and MUNRO, I. C. 1965. Factors influencing the nutritional value of fish flour. 4. Reaction between 1,2-dichloroethane and protein. Can. J. Biochem. *43*, 33–40.

MOTTU, F., and MAURON, J. 1967. The differential determination of lysine in heated milk. 2. Comparison of the in vitro methods with the biological evaluation. J. Sci. Food Agric. *18*, 57–62.

MUNYUA, J. K. 1975. Hydrogen peroxide alteration of whey protein in skim milk and whey protein solutions. Milchwissenschaft. *30*, 730–734.

MUSTAKAS, G. C., ALBRECHT, W. J., BOOKWALTER, G. N., McGHEE, J. E., KWOLEK, W. F., and GRIFFIN, E. L., JR. 1970. Extruder-processing to improve nutritional quality, flavor, and keeping quality of full-fat soy flour. Food Technol. *20*, 1290–1296.

NAKAI, T., and OHTA, T. 1976. β-3-oxindolylalanine: The main intermediate in tryptophan degradation occurring in acid hydrolysis of proteins. Biochim. Biophys. Acta. *420*, 258–264.

NATARAJAN, K. R., RHEE, K. C., CATER, C. M., and MATTIL, K. F. 1975. Effect of sodium hypochlorite on peanut protein isolates. J. Food Sci. *40*, 1193–1198.

NEUCERE, N. J., CONKERTON, E. J., and BOOTH, A. N. 1972. Effect of heat on peanut proteins. 2. Variations in nutritional quality of the meals. J. Agr. Food Chem. *20*, 256–259.

NEUMANN, N. P. 1967. Analysis for methionine sulfoxides. *In* Methods in Enzymology, Vol. 11. S. P. Colowick and N. O. Kaplan (Editors). Academic Press, New York.

NJAA, L. R. 1962. Utilization of methionine sulphoxide and methionine sulphone by the young rat. Br. J. Nutr. *16*, 571–577.

PATCHORNIK, A., and SOKOLOVSKY, M. 1964. Chemical interactions between lysine and dehydroalanine in modified bovine pancreatic ribonuclease J. Am. Chem. Soc. *86*, 1860–1861.

PIENIAZEK, D., RAKOWSKA, M., and KUNACHOWICZ, H. 1975A. The participation of methionine and cysteine in the formation of bonds resistant to the action of proteolytic enzymes in heated casein. Br. J. Nutr. *34*, 163–173.

PIENIAZEK, D., RAKOWSKA, M., SZKILLADZIOWA, W., and GRABAREK, Z. 1975B. Estimation of available methionine and cysteine in proteins of food products by in vivo and in vitro methods. Br. J. Nutr. *34*, 175–190.

PIERPOINT, W. S. 1969A. O-Quinones formed in plant extracts. Their reactions with amino acids and peptides. Biochem. J. *112*, 609–616.

PIERPOINT, W. S. 1969B. O-Quinones formed in plant extracts. Their reaction with bovine serum albumin. Biochem. J. *112*, 619–629.

PISANO, J. J., FINLAYSON, J. S., PEYTON, M. P., and NAGAI, Y. 1971. ϵ-(γ-Glutamyl) lysine in fibrin: Lack of crosslink formation in factor XIII deficiency. Proc. Nat. Acad. Sci. *68*, 770–772.

PRONCZUK, A., PAWLOWSKA, D., and BARTNIK, J. 1973. Effect of heat treatment on the digestibility and utilization of proteins. Nutr. Metabol. *15*, 171–180.

PROVANSAL, M., CUQ, J. L., and CHEFTEL, C. 1975. Chemical and nutritional modifications of sunflower proteins due to alkaline processing. Formation of amino acids crosslinks and isomerization of lysine residues. J. Agr. Food. Chem. *23*, 938–943.

RACZYNSKI, G., SNOCHOWSKI, M., and BURACZEWSKI, S. 1975. Metabolism of ϵ-(γ-L-glutamyl)-L-lysine in the rat. Br. J. Nutr. *34*, 291–296.

REIS, P. J., and TUNKS, D. A. 1973. Influence of formaldehyde-treated casein supplements on the concentration of ϵ-*N*-methyllysine in sheep plasma. Australian J. Biol. Sci. *26*, 1127–1136.

ROUBAL, W. T. 1971. Free radicals, malonaldehyde and protein damage in lipid-protein systems. Lipids *6*, 62–64.

ROUBAL, W. T., and TAPPEL, A. L. 1966A. Damage to proteins, enzymes and amino acids by peroxidizing lipids. Arch. Biochim. Biophys. *113*, 5–8.

ROUBAL, W. T., and TAPPEL, A. L. 1966B. Polymerization of proteins induced by free-radical lipid peroxidation. Arch. Biochim. Biophys. *113*, 150–155.

SAVIGE, W. E., and MAC LAREN, J. A. 1966. Oxidation of disulfides with special reference to cystine. *In* Reactions of Organic Sulphur compounds, Vol. 2. N. Kharasch and C. Y. Meyers (Editors). Pergamon Press. Oxford.

SCHAICH, K. M., and KAREL, M. 1975. Free radicals in lysozyme reacted with peroxidizing methyl linoleate. J. Food Sci. *40*, 456–459.

SCHULTZ, H. W., DAY, E. A., and SINNHUBER, R. O. 1962. Lipids and their oxidation. The AVI Publishing Company, Inc., Westport, Connecticut.

SEIDL, D., JAFFE, M., and JAFFE, W. G. 1969. Digestibility and proteinase inhibitory action of a kidney bean globulin. J. Agr. Food Chem. *17*, 1218–1321.

SLUMP, P., and SCHREUDER, H. A. W. 1973. Oxidation of methionine and cystine in foods treated with hydrogen peroxide. J. Sci. Fd. Agric. *24*, 657–661.

SPINELLI, J., GRONINGER, H., KOURY, B., and MILLER, R. 1975. Functional protein isolates and derivatives from fish muscle. Process Biochem. *10*, 31–40.

STERNBERG, M., KIM, C. Y., and SCHWENDE, F. J. 1975A. Lysinoalanine: Presence in foods and food ingredients. Science *190*, 992–994.

STERNBERG, M., KIM, C. Y., and PLUNKETT, R. A. 1975B. Lysinoalanine determination in proteins. J. Food Sci. *40*, 1168–1170.

TANAKA, M., AMAYA, J., LEE, T. C., and CHICHESTER, C. O. 1974. Effects of the browning reaction on the quality of protein. Proceedings of the 4th International Congress of Food Science and Technology, Madrid.

TANNENBAUM, S. R., AHERN, M., and BATES, R. P. 1970. Solubilization of fish protein concentrate. An alkaline process. Food Technol. *24*, 96–99.

TANNENBAUM, S. R., BARTH, H., and LE ROUX, J. P. 1969. Loss of methionine in casein during storage with autoxidizing methyl linoleate. J. Agr. Food. Chem. *17*, 1353–1354.

TURK, R. E., CORNWELL, P. E., DIANNE BROOKS, M., and BUTTER-WORTH, C. E. 1973. Adequacy of spun-soy protein containing egg albumin for human nutrition. J. Am. Dietetic Assoc. *63*, 519–524.

VALLE-RIESTRA, J. F. 1974. Food processing with chlorinated solvents. Food Technol. *28*, 25–32.

VALLE-RIESTRA, J. F. and BARNES, R. H. 1970. Digestion of heat-damaged egg albumen by the rat., J. Nutr. *100*, 873–882.

VAN BEEK, L., FERON, V. J., and DE GROOT, A. P. 1974. Nutritional effects of alkali-treated soyprotein in rats. J. Nutr. *104*, 1630–1636.

VAN SUMERE, C. F., ALBRECHT, J., DEDONDER, A., DE POOTER, H., and PE, I. 1975. Plant proteins and phenolics. *In* The Chemistry and Biochemistry of Plant Proteins. J. B. Harborne and C. F. Van Sumere (Editors). Academic Press, New York.

VARNISH, S. A., and CARPENTER, K. J. 1975A. Mechanisms of heat damage in proteins. 5. The nutritional values of heat-damaged and propionylated proteins as sources of lysine, methionine and tryptophan. Br. J. Nutr. *34*, 325–337.

VARNISH, S. A., and CARPENTER, K. J. 1975B. Mechanisms of heat damage in proteins. 6. The digestibility of individual amino acids in heated and propionylated proteins. Br. J. Nutr. *34*, 339–349.

VITHAYATHIL, P. J., and MURTHY, G. S. 1972. New reaction of a benzoquinone at the thioether group of methionine. Nature *236*, 101–103.

WAIBEL, P. E., and CARPENTER, K. J. 1972. Mechanisms of heat damage in proteins 3. Studies with ϵ-(γ-L-glutamyl)-L-lysine. Br. J. Nutr. *27*, 509–515.

WALKER, H. G., JR., KOHLER, G. O., ZURMICKY, D. D., and WITT, S. C. 1975. Problems in analysis for sulfur amino acids in feeds and foods. *In* Protein Nutritional Quality of Foods and Feeds. Part. 2. M. Friedman (Editor). Marcel Dekker, Inc., New York.

WOODARD, J. C., and SHORT, D. D. 1973. Toxicity of alkali-treated soyprotein in rats. J. Nutr. *103*, 569–574.

WOODARD, J. C., SHORT, D. D., ALVAREZ, M. R., and REYNIERS, J. 1975. Biologic effects of $N\epsilon$-(DL-2-amino-2-carboxyethyl)-L-lysine, lysinoalanine. *In* Protein Nutritional Quality of Foods and Feeds. Part 2. M. Friedman (Editor). Marcel Dekker, Inc., New York.

YANAGITA, T., and SUGANO, M. 1975. Liver and plasma lipids in rats fed casein reacted with oxidized ethyl linoleate. Agr. Biol. Chem. *39*, 63–69.

ZIEGLER, K. L., MELCHERT, I., and LURKEN, C. 1967. $N\delta$-(2-amino-2-carboxyethyl)-ornithine, a new amino-acid from alkali-treated proteins. Nature *214*, 404–405.

ZIRLIN, A., and KAREL, M. 1969. Oxidation effects in a freeze-dried gelatin-methyl linoleate system. J. Food Sci. *34*, 160–164.

18

Effects of Lysine Modification on Chemical, Physical, Nutritive, and Functional Properties of Proteins

Mendel Friedman

The nutritive value of a protein depends not only on the proportions of its essential amino acids, but also on their physiological availability. Availability varies and depends on protein source, processing treatments (such as heating, storage, and exposure to acids and alkalis), interactions with other diet components, and on the consumer's physiological state.

In many foods, especially those of plant origin, the small content of the essential amino acid lysine limits nutritive value. This is particularly important for cereals, which are the main source of protein for much of the earth's population.

A characteristic that restricts the availability of dietary lysine is that its ϵ-amino group, under the influence of light, heat, alkali, and other factors, interacts with other food constituents to become nutritionally unavailable. These reactions include those with carbonyl groups of sugars to form Schiff's bases that are further transformed to products that are incompletely characterized (Tanaka *et al.* 1975, 1977; Finot *et al.* 1977; Erbersdobler 1977; Nakai *et al.* 1975; Knipfel *et al.* 1975; El-Nockrashy and Frampton 1967). Lysine ϵ-amino groups also react with carboxyl groups of aspartic

446

and glutamic acid and with primary amide groups of asparagine and glutamine to give cross-linked products with unnatural amide (isopeptide) bonds (Hurrell *et al.* 1976). Cross-links are also formed when ϵ-amino groups (as well as N-terminal α-amino groups) interact with dehydroalanine and methyldehydroalanine residues to form lysinoalanine and methyllysinoalanine derivatives (Friedman 1977A). Similar homologous cross-links can also form from amino groups of ornithine, derived from arginine residues under the influence of alkali, or possibly directly from arginine. Histidine and other amino acids may react similarly (Friedman 1977B; Finley and Friedman 1976; *cf.* also Koenig *et al.* 1973).

Such reactions impair protein quality, not only by reducing its content of essential amino acids, but also by stabilizing its structure and conformation so that its susceptibility to normal digestion is decreased. Because nutritive value must be conserved during processing and storage, an important objective of food science research is to understand and control chemical transformations of the ϵ-amino groups of free (Vaghefi *et al.* 1974) and protein-bound lysine side chains. In this paper, I examine such transformations, means for estimating them, and their chemical and nutritional consequences. Much of this work concerns chemical, physical, nutritive, and functional properties of proteins in which the ϵ-amino groups of lysine residues have been modified with various alkylating and acylating reagents. Related chemical and photochemical transformations of possible industrial interest are described elsewhere (Krull and Friedman 1967A; Wall *et al.* 1968; Friedman *et al.* 1971; Friedman 1973A).

DISCOVERY AND PROPERTIES OF LYSINE

Lysine was discovered in 1891 by E. Drechsel and collaborators (Drechsel 1891; Drechsel and Kruger 1892), who demonstrated its presence in proteins and characterized some derivatives. Vickery and Schmidt (1931) described its early history. The crystalline-free base was first described by Vickery and Leavenworth in 1928. Osborne and Mendel (1914) showed that lysine is required for normal growth and development; it is an essential amino acid.

Lysine is a dibasic amino acid with one asymmetric carbon atom (Lehninger 1970). The apparent pK_a values of the three functional groups of lysine are: carboxyl, 2.18; α-amino, 8.98; ϵ-amino, 10.53 (Greenstein and Winitz 1961). The basicity of the ϵ-amino group, which is reflected in its chemical behavior in proteins, is less when protein-bound than in free lysine and varies from protein to protein. The changing basicity is apparently due mainly to hydrogen-bonding interactions of free and protonated amino groups with adjacent electrophilic and nucleophilic sites. Lysine ϵ-amino groups are protonated at physiological pH.

Several lysine derivatives have also been reported to occur naturally in proteins from animal sources (Vickery 1972). These include hydroxylysine, ϵ-N-acetyllysine, ϵ-N-methyllysine, ϵ-N-dimethyllysine, and ϵ-N-trimethyllysine. Their importance in nutrition has not been determined.

CHEMICAL ASPECTS

Effect of pH

Cyanoethylation of amino acids as a function of pH was studied in the range pH 8–12 by Friedman and Wall (1964). Rates increase rapidly as the pH approaches the pK_a value of the amino group and, with further increase in pH, approach limiting value. The dependence of rates on pH can be ascribed to the effect of pH on the concentration of the various ionized states of the amino acids and their relative reactivities. Mathematical descriptions of the change in reaction rate with pH in terms of concentrations of the amino acid species can be derived from the ionization constants of the amino acids (Friedman and Wall 1964; Friedman and Williams 1977).

Generally, the reaction rate in aqueous solution increases in proportion to the concentration of unprotonated amino groups, either free or protein-bound. However, nonaqueous solvents and the electronic and steric microenvironments near protein amino groups profoundly alter their reactivities as discussed in the following sections with the aid of Fig. 18.1–18.5.

Application of theoretical predictions to experimental data from protein modification, *e.g.*, with ethyl vinyl sulfone (Friedman and Williams 1977) is approximate at best, because the protein lysine residues vary in reactivity, partly because of different (micro) pK values. Results for reaction of bovine serum albumin (BSA) with ethyl vinyl sulfone show that the lysine reactivities range widely in water, but are more closely bunched, although not identical, in 50% DMSO. The BSA is probably denatured in 50% DMSO so that the lysine groups are comparably accessible. If this is so, other protein denaturants should have similar effects.

If there is no interaction among reactive lysine side chains, reactivities for all of these can be totaled as described (Friedman and Williams 1977). See also discussion of effects of alkylation and acylation of amino groups on basicities of the modified amino groups under Physicochemical Aspects.

Solvent Effects

Dimethyl sulfoxide (DMSO) and other dipolar aprotic solvents strongly affect the reactivity of ionizable reactive sites in proteins.

Several effects contribute to this result. In particular, such solvents, because of differences in dielectric constant and hydrogen bonding properties, influence ionization equilibria, hydrogen-bonding and hydrophobic interactions, and the nucleophilic strengths of protein functional groups. The stabilities of ground and transition states are therefore modified and, consequently, chemical reactivities of protein functional groups, protein conformations, and, for instance, enzyme reactivities (Friedman 1967A, 1968; Krull and Friedman 1967A, B; Friedman and Krull 1970; Friedman and Koenig 1971; Whitfield and Friedman 1972; Chirgadze and Ovsepyan 1972; Koenig *et al.* 1973).

Protein-solvent interactions may be illustrated with DMSO in more detail. The following factors influence the reaction: (1) preferential solvation of positive charges by DMSO leaves negative charges free from the destabilizing influences of positive charges and ion pair aggregation; and (2) the strong dipole of the sulfoxide group in DMSO strongly influences physicochemical properties of proteins.

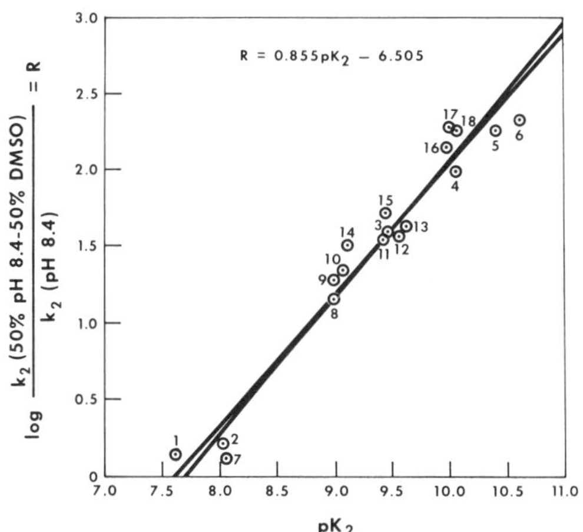

From Friedman (1967A)

FIG. 18.1. PLOT OF RATIOS OF SECOND-ORDER RATE CONSTANTS (k_2 X 10^4 LITERS/MOLE SEC) FOR THE REACTION OF AMINO GROUPS IN AMINO ACIDS AND PEPTIDES WITH ACRYLONITRILE AT 30°C

Note that because the ε-amino group in ε-aminocaproic acid (compound 6) has the highest pK value, it experienced the greatest rate enhancement. The rate constant for compound 6 is about 200 times greater than that for tetraglycine (compound 1). [1, Tetraglycine; 2, Digylcine; 3, Glycine; 4, β-Alanine; 5, γ-Aminobutyric acid; 6, ε-Aminocaproic acid; 7, L-Alanylglycine; 8, DL-Phenylalanine; 9, L-Tyrosine; 10, DL-Methionine; 11, L-Leucine; 12, DL-α-Alanine; 13, DL-Norleucine; 14, DL-α-Phenyl-α-alanine; 15, DL-α-Methylmethionine; 16, DL-Isovaline; 17, α-Aminoisobutyric acid; and 18, 1-Aminocyclopentane-1-carboxylic acid].

Friedman (1967A) measured reaction rates of amino groups in structurally different amino acids, peptides, and proteins with conjugated vinyl compounds in a medium consisting of 50% pH 8.4 borate buffer and 50% DMSO and compared them with analogous rates in an aqueous buffer. The results (Fig. 18.1) were described by the following equation:

$$\log \frac{k_2 \ (50\% \ \text{pH } 8.4 - 50\% \ \text{DMSO})}{k_2 \ (\text{pH } 8.4)} = R = 0.855 \ pK_2 - 6.505$$

In the mixed solvent, the rate increase (R), which ranged from 1.4 for tetraglycine to 209 for ϵ-aminocaproic acid, varied directly with the pK_a value of the amino group. This rate increase can be rationalized as follows. Water strongly solvates an amino acid anion of structure $NH_2 CH(R)COO^-$ as illustrated in I. In this case solvation decreases the electron density of nitrogen and, consequently, decreases the nucleophilicity of the amino group. Water can also hydrogen bond to the amino groups as shown in II. This solvation process increases the electron density on nitrogen and therefore increases the nucleophilicity of the amino group. Evidently, because water can act as both a hydrogen donor and acceptor, its effect results from a balance of these opposing actions. Because DMSO acts only as an acceptor, only the hydrogen-bonding interactions shown in III and IV are possible in this medium. These interactions strengthen both the basicity and nucleophilicity of the amino group. In addition, by hydrogen-bonding interactions shown in V–VI, DMSO can lower the basicity of amino groups as measured by pK.

Friedman (1967A) also examined the reaction of acrylonitrile with amino groups in three proteins (Fig. 18.2 and 18.3). The rate enhancement factor in the mixed solvent medium ranged from 20 for bovine serum albumin to 60 for polylysine. Although such rate studies suggest that proteins can be chemically modified more rapidly in mixed aqueous-nonaqueous media than in dilute aqueous solvents, it is difficult to ascertain how much of the rate enhancement by DMSO is due to the described direct influences of the nonaqueous compound on electrostatic and hydrogen-bonding environments of protein amino groups and how much to effects on protein conformation.

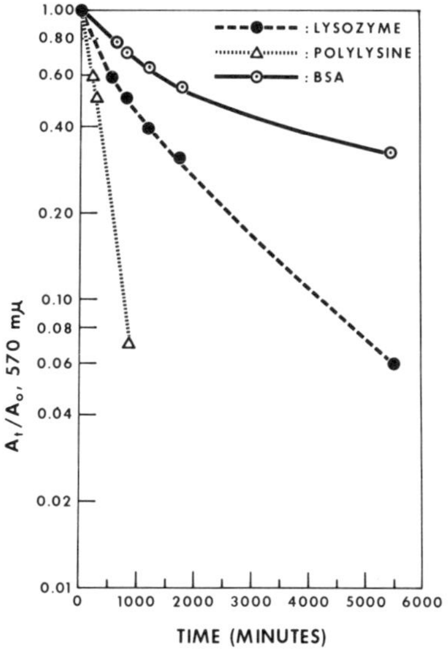

From Friedman (1967A)

FIG. 18.2. PLOT OF LOG A_1/A_0 VS TIME FOR REACTION OF PROTEIN (0.005 M IN NH_2 GROUPS) WITH ACRYLONITRILE (0.0825 M) IN pH 8.4 BORATE BUFFER MEDIUM AT 30°C

Ninhydrin was used to determine the amount of free amino groups left at various times. BSA stands for bovine serum albumin.

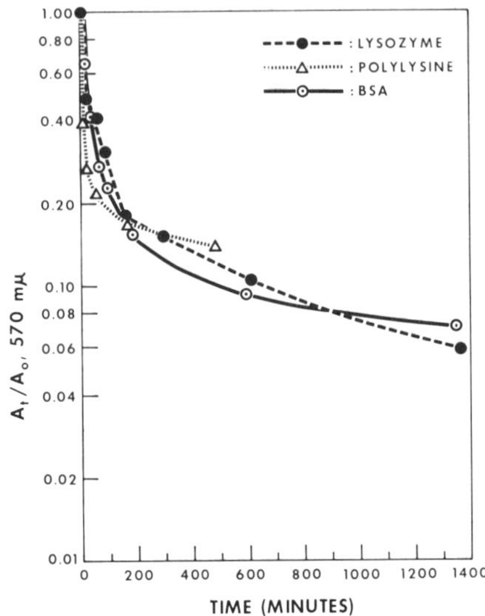

From Friedman (1967A)
FIG. 18.3. PLOT OF LOG A_t/A_o VS TIME
FOR REACTION OF PROTEINS (0.005 M
IN NH$_2$ GROUPS) WITH ACRY-
LONITRILE (0.0825 M) IN 50% pH 8.4
BUFFER-50% DMSO AT 30°C
BSA stands for bovine serum albumin.

Structure-Reactivity Correlations

Analysis of reaction rates of acrylonitrile with amino groups in three steric series of amino acids and peptides (in which the amino groups are attached to primary, secondary, and tertiary carbon atoms, respectively, and one protein) shows that both polar and steric factors affect reactivities (Fig. 18.4 and 18.5). These rates can be described in terms of linear free energy relationships that specify observed reactivities of amino groups in amino acids, peptides, and proteins in terms of polar, steric, electrophilic, and nucleophilic parameters (Friedman and Wall 1964, 1966; Friedman *et al.* 1965).

Analogous linear free energy relationships have also been developed by correlating reaction rates of sulfhydryl groups in addition reactions with vinyl compounds such as acrylonitrile and dehydroalanine (Friedman *et al.* 1965; Friedman 1972; 1973B; Snow *et al.* 1975A, B), and similarly for reactions of amino groups with vinyl compounds, in which structural variations have been imposed on both reactants (Fig. 18.2) (Friedman and Wall 1966), for reactions of amino groups with ninhydrin (Friedman and Sigel 1966), for the addition of amines to double bonds (Shenhav *et al.* 1970), and for the reaction of amino acids with insecticides and with fluorodinitrobenzene (Burchfield and Storrs 1957, 1958; Bunnett and Hermann 1970).

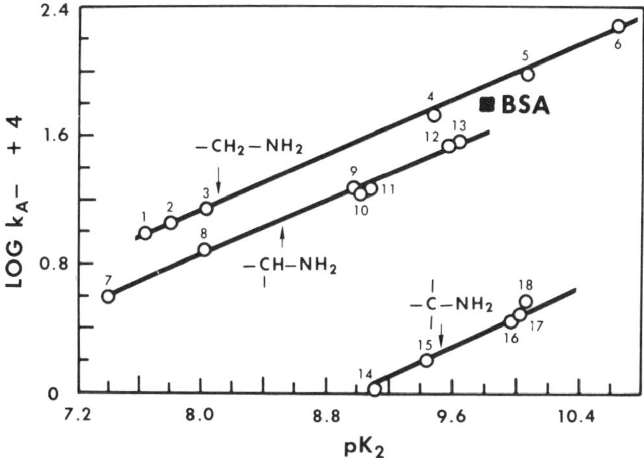

FIG. 18.4. VARIATION OF LOG (k_A-) WITH pK_2 OF THE AMINO GROUPS IN VARIOUS AMINO ACIDS AT 30°C
Numbers correspond to compounds listed in Fig. 18.1. Note that the second-order anion (inherent) rate constants (k_A-) can be determined directly at a pH about two units above the pK_2 values of the protonated amino groups or calculated from the observed second-order rate constants (k_2), determined at any pH, by means of the following equation (Friedman and Wall 1964): k_A- = k_2 [1 + (H$^+$/K_2]. Comparisons of relative nucleophilic reactivities of ionized amino groups in nucleophilic addition and displacement reactions are valid only when done in terms of anion (and not observed) rate constants. Note also that ϵ-aminocaproic acid, compound 6, is a model for lysine and that the calculated rate constant for BSA lies on the top line of the figure directly above the BSA point. The proximity of predicted and determined rate constants indicates that the free energy relationships derived for amino acids and peptides may be used to predict rates for proteins within an order of magnitude.

These relationships are fundamental contributions to bio-organic chemical theory. They also have practical value for assessing relative contributions of polar and steric factors to observed reactivities of amino groups and for calculating the estimated rate constants of amino groups with a large number of reactants. Furthermore, knowledge of the chemical reactivities of protein functional groups in terms of fundamental free-energy parameters is invaluable for planning and carrying out selective and nonselective chemical modifications of proteins for various end uses. A few examples will be cited to show the interplay of various influences on reactivity.

In examining the reaction of p-nitrophenyl acetate with ϵ-acetyllysine, α-acetyllysine, lysine, and polylysine, Kristol et $al.$ (1975) noted that rates of ammonolysis increased with pH. This fact suggests that the unprotonated amino group is the reactive species. They also found that the inherent second-order rate constants do not vary consistently with the pK_a values of the various amino groups in the four compounds. Therefore, the observed rates for polylysine are

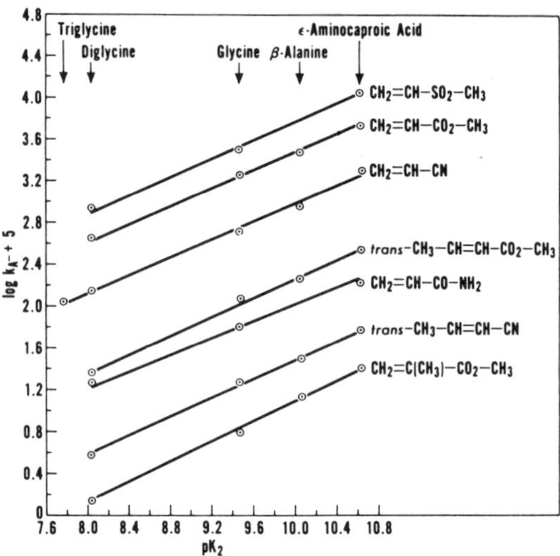

From Friedman and Wall (1966)

FIG. 18.5. RATES OF ALKYLATION OF AMINO ACIDS AS A FUNCTION OF pK AT 30°C

Note the dramatic influence of both steric and electronic effects on the observed reactivities of the vinyl compounds with the amino groups.

deduced to be governed by combined effects of basicity and nucleophilicity and by steric factors. Steric factors are influenced by conformational changes of the protein as the pH of the reaction medium is changed.

In studies of reactions of the amino groups of ribonuclease A with trinitrobenzenesulfonic acid (TNBS), Goldfarb and collaborators (Goldfarb 1974; Goldfarb *et al.* 1974) observed that the enzyme contains four subsets of reactive amino groups, each with a different reactivity towards TNBS and a different pK_a. In triethanolamine buffer, the first and second subsets include most of the ε-amino groups of the enzyme; the third subset includes only the α-amino group of lys-1; the fourth subset only the ε-amino group of lys-41. The number and type of amino group in each subset was markedly changed when phosphate was used instead of triethanolamine. Thus the nature of the buffer as well as the pH of the medium influences rates.

Kinetic analysis of the data (Goldfarb 1974) also suggests that a substituent in one position in a protein chain can influence reactivity at another position; the observed decrease in reaction rate with increasing protein concentration may be due to protein-protein interactions that hinder accessability of the reagent to some of the amino groups.

These and related studies (Benesch and Benesch 1958; Goldfarb 1966, 1970; Friedman 1967A; Haynes *et al.* 1967; Nakaya *et al.* 1967; Freedman and Radda 1968; Scheele and Lauffer 1969; Muhlrad *et al.* 1970; Fields 1971; Koenig *et al.* 1973; Friedman *et al.* 1974; Dopheide 1975; Duggleby and Kaplan 1975; Gilbert and O'Leary 1975; Re and Kaper 1975; Goldfarb and Martin 1976A, B; Friedman and Williams 1977; Chauffe and Friedman 1977) all suggest that kinetic analysis of the reactivity of protein amino groups with various reagents is inherently difficult. Rates vary not only among proteins, but even in a single protein chain. Thus, the relative contributions of the various governing parameters are difficult to assign.

ANALYTICAL, PROCESSING, AND NUTRITIONAL ASPECTS

Available and Unavailable Lysine

Several methods have been proposed for estimating chemically and, presumably, nutritionally available lysine (Carpenter 1960; Blom *et al.* 1967; Roach *et al.* 1967; Matheson 1968; Kakade and Liener 1969; Ostrowski *et al.* 1970; Booth 1971; Allison *et al.* 1973; Ford 1973; Hall *et al.* 1973; Lein *et al.* 1973; Milner and Westgarth 1973; Okumura and Tasaki 1973; Bailey 1974; Bhatty and Wu 1974; Hurrell and Carpenter 1974; Hussain 1974; Sandler and Warren 1974; Concon 1975; Ellwell and Soares 1975; Hansen 1975; Holsinger and Posati 1975; Rhodes 1975; Sarwar *et al.* 1975; Synge 1975, 1976; Paulis and Wall 1975; Osner 1975; Eklund 1976; DellaMonica *et al.* 1976). None of these methods appears entirely satisfactory with respect to sensitivity, reproducibility, time required, convenience, and correlation with biological methods. To avoid the difficulties of available procedures, we devised a new method based on the alkylation of ϵ-amino groups of proteins with methyl acrylate (Finley and Friedman 1973; Cavins and Friedman 1967, 1968A, B). The chemically unavailable or unreacted lysine was estimated by chromatographic amino acid analysis. Available lysine was measured by comparing lysine content of hydrolysates of alkylated and native proteins. The decrease in lysine content measures chemically available lysine (Tables 18.1, 18.2).

The products of the ϵ-amino group modification, however, could not be measured directly with the amino acid analyzer because of overlapping peaks. We, therefore, continued to search for reagents that would form lysine derivatives stable to acid hydrolysis and resolvable on the basic column of an amino acid analyzer. Such a reagent would permit simultaneous assay of chemically available (reactive) and unavailable (unreactive) lysine (lysine kept from

TABLE 18.1
COMPARISON OF THREE METHODS FOR
DETERMINING AVAILABLE LYSINE[1]

Protein	Methyl Acrylate	FDNB	TNBS
	(g/100 g protein)		
Bovine serum albumin	12.82	12.85	12.86
Lysozyme	5.64	5.60	5.78
Egg albumin	5.08	5.09	5.10
Ribonuclease	10.60	10.18	10.34
Gluten	1.24	1.26	1.24
Casein	8.08	8.17	8.06

[1] Finley and Friedman (1973).

TABLE 18.2
APPARENT AVAILABLE LYSINE IN HEAT-TREATED SAMPLES[1]

Protein	Time at 205°C (min)	Total Lysine	Lysine After Alkylation (g/100 g material)	Available Lysine by Methyl Acrylate
Casein	0	8.31	0.23	8.08
Casein + 20% dextrose	30		1.97	6.34
Casein + 20% dextrose	60		3.27	0.04
Gluten	0	1.25	0.06	1.19
Gluten + 20% dextrose	30		0.27	0.98
Gluten + 20% dextrose	60		1.22	0.03

[1] Finley and Friedman (1973); treated at 30% moisture (205°C).

reacting by its physical environment or chemical combination, for example, through hydrolyzable isopeptide bonds or Schiff's base reaction products). Previous rate studies of amino groups with methyl vinyl sulfone (Friedman and Wall 1966) suggested that alkyl vinyl sulfones would be suitable reagents for the simultaneous assay.

Our results (Friedman and Finley 1975A, B) show that ethyl vinyl sulfone (EVS) can alkylate ϵ-amino groups in proteins to give two derivatives, ϵ-N-(ethylsulfonylethyl) lysine and ϵ,ϵ-N,N-bis(ethylsulfonylethyl) lysine (Fig. 18.6). The new amino acids were eluted from the long basic (physiological) column of an amino acid analyzer at about 56 and 113 min, respectively, and were well resolved from the other amino acids (Fig. 18.7, 18.8). Rate studies showed that as reaction progressed, the amount of the faster peak increased at the

**REACTION OF LYSINE $\mathcal{E}$–NH$_2$ GROUPS
WITH ETHYL VINYL SULFONE**

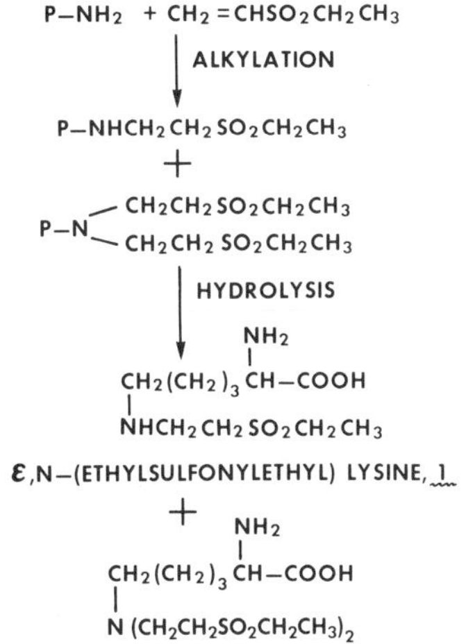

FIG. 18.6. REACTION OF LYSINE $\mathcal{E}$-AMINO
GROUPS WITH ETHYL VINYL SULFONE

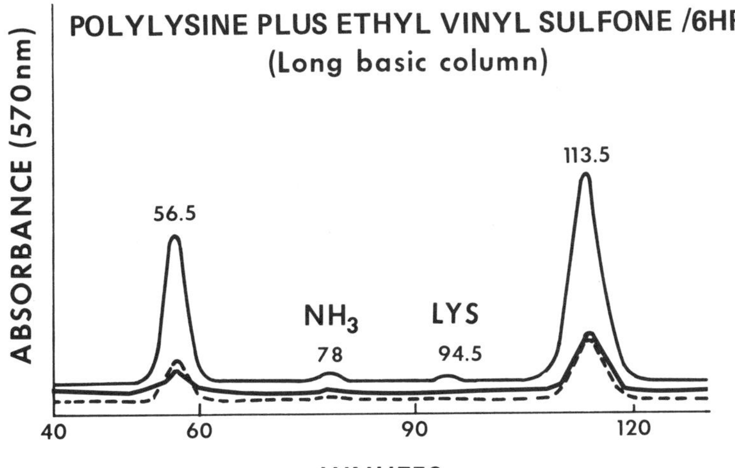

From Friedman and Finley (1975A, B)
FIG. 18.7. AMINO ACID CHROMATOGRAM (LONG
BASIC COLUMN) OF A HYDROLYZATE OF EVS-AL-
KYLATED POLYLYSINE
The compound eluted at 113.5 min is the monosubstituted
lysine derivative and that at 56.6 min, the disubstituted one

expense of the slower, implying that consecutive reactions take place
in which the monosubstituted derivative is transformed to the
disubstituted one.

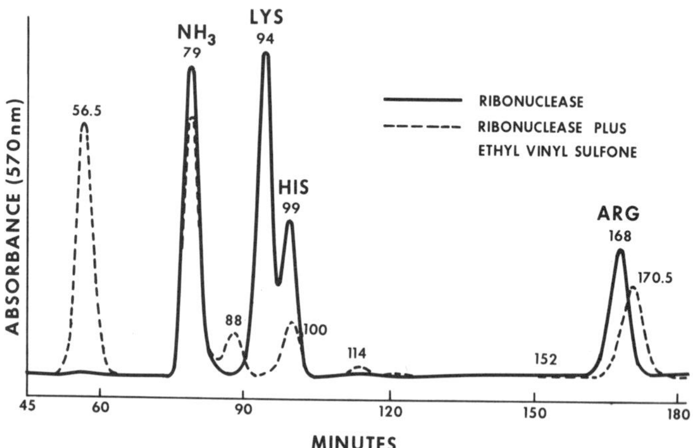

From Friedman and Finley (1975A, B)

FIG. 18.8. AMINO ACID CHROMATOGRAM (LONG BASIC COL-
UMN) OF HYDROLYZATES OF UNTREATED AND EVS-TREATED
RIBONUCLEASE

TABLE 18.3

AMINO ACID COMPOSITION (m moles/100 g) OF PROTEIN
HEATED AT 205°C WITH GLUCOSE FOR 30 MIN AND
THEN TREATED WITH ETHYL VINYL SULFONE[1]

Protein	Dextrose (%)	Lysine (A)	62 Unk[2] (B)	117 Unk[3] (C)	A + B + C	Loss of Lysine on Heating (%)
Casein	0	56.4	0	0	56.4	
Casein	1	2.20	34.75	2.95	39.9	29.3
Casein	5	2.25	28.04	11.22	40.5	28.2
Casein	10	2.96	12.71	19.26	34.9	38.2
Gluten	0	12.75	0	0	12.75	
Gluten	1	trace	4.95	trace	5.0	60.8

[1] Heated proteins were treated with ethyl vinyl sulfone as described (Friedman and Finley 1975B).

[2] ϵ,ϵ-N,N-*bis*(ethylsulfonylethyl)lysine, Structure 2 in Fig. 18.6.

[3] ϵ,N-(ethylsulfonylethyl)lysine, Structure 1 in Fig. 18.6.

Preliminary studies indicated that the EVS method can, indeed, be used to measure both available and unavailable lysine in a single determination (Tables 18.3, 18.4). Chemically unavailable lysine is estimated from the lysine content of the EVS-treated protein, while the sum of the EVS derivatives measures the available lysine. Any difference in these two measurements as a result of heating or other treatment of a food protein is interpretable in terms of products that do or do not yield lysine after acid hydrolysis. The treatment is simple to carry out and reproducible. The relationships of this EVS method to biological assays based on feeding tests and to other chemical assays for available lysine remain to be established.

TABLE 18.4
EFFECTS OF HEATING[1] ON APPARENT AVAILABLE LYSINE[2]
IN ACYLATED AND NONACYLATED PROTEINS[3,4]

| | Acylating Agent | | | |
	None	Succinic Anhydride	Maleic Anhydride	Diketene
Protein	Loss of Available Lysine (%)			
Casein	2.8	3.2	2.7	2.7
Casein + 1% glucose	29.3	4.3	4.2	4.4
Casein + 5% glucose	28.2	5.1	4.2	3.4
Casein + 10% glucose	38.2	6.9	7.1	6.8
Casein + 20% glucose	76.2	13.8	11.1	11.5
Gluten	4.8	6.2	7.1	7.3
Gluten + 15% glucose	60.8	13.1	9.5	10.2

[1] Heat treatments were at 205°C for 30 min.

[2] Available lysine was estimated by the ethyl vinyl sulfone (EVS) method after removing acyl groups.

[3] Acylation and deacylation of proteins was carried out as described by Means and Feeney (1971).

[4] Friedman and Finley (1975B).

X-Ray Photoelectron Spectroscopy

Recent studies (Friedman and Millard 1976; Millard and Friedman 1976) indicate that X-ray photoelectron spectroscopy (XPS), also called "electron spectroscopy for chemical analysis" (ESCA), can be used to estimate the extent of chemical modification of proteins, such as BSA, with ethyl vinyl sulfone. The analysis is based on the difference in binding energy (BE) of 2p electrons, from sulfur in the native protein at 163 electron volts (eV), and the corresponding binding energy in the sulfur of the introduced sulfone group at 168 eV. The nitrogen to sulfur (N/S) atom ratios, determined by XPS, for a series of BSA samples modified to different degrees agree with corresponding values from elemental analyses and confirmed results from amino acid analysis. For example, a plot of N/S ratio (Y), determined by XPS, against the extent of lysine modification (X) in five BSA derivatives is linear and is described by the following equation:

$$Y = 146.3 - (9.55 \pm 0.47)X$$

With this equation, the extent of lysine modification can be calculated from N/S ratios obtained by XPS; protein hydrolysis and analysis of amino acids are not required. Analogous equations should be applicable to other proteins. Analysis by XPS, which takes only a few minutes, is potentially useful for determining the extent of chemical modification of many modified proteins and should find application in studies of protein structure and reactivity (*cf.* also Walker *et al.* 1975).

Reaction of Amino Groups with Ninhydrin

In principle, the number of available lysine amino groups can be determined in a protein by reaction with ninhydrin and measurement of the liberated color without having to resort to hydrolysis and chromatographic analysis. Reaction of ninhydrin with amino groups in amino acids and related compounds to form a chromophore that absorbs at 570 nm is the basis of sensitive specific methods for analysis of amino groups. The reaction of ninhydrin with amino acids has been extensively studied (Friedman 1967A; Friedman and Williams 1973, 1974), but the ninhydrin assay of protein amino groups has been less satisfactory. The ninhydrin color yield of several proteins is about two-thirds of the theoretical value (Friedman and Williams 1974). Because the ninhydrin color from proteins is derived largely from ϵ-amino groups of lysine side chains, the question arises as to the fate of the lysine residues not accounted for.

During studies designed to elucidate factors that influence the ninhydrin reaction in amino acids, peptides, proteins, and related compounds, we noted (Friedman and Williams 1974) that the reaction of protein amino groups continued slowly even after the usual 15-min reaction time and that not all protein amino groups

TABLE 18.5
NINHYDRIN COLOR YIELDS AND LYSINE
CONTENT OF NINHYDRIN-TREATED PROTEINS[1]

Protein	Reaction Time (min)	NH_2 Group Content (A) from Ninhydrin Reaction	Lysine (B) from Amino Acid Analysis	A + B
Polylysine	30	269 (m-moles/100 g)	128	407
Lysozyme	0		30.3	30.3
	30	27.9	5.1	33.0
Wool	0		19.2	19.2
	5	8.7	10.6	19.3
	51	15.6	5.2	20.8

[1] Friedman and Williams (1974).

appeared to react with ninhydrin (Table 18.5). Longer heating times, however, usually increased color yield only slightly. (Wool is an exception.) Apparently, any unreacted free amino groups are unaffected by chemical modification of the undisrupted peptide chain. The continued slow color production is presumably due to slow hydrolysis of peptide bonds and reaction of the resulting amino groups with ninhydrin or to denaturation leading to exposure of buried amino groups.

The best solvent for the ninhydrin reaction of proteins was DMSO-water in the range 40–60% by volume (Friedman 1971).

Under optimum conditions, the ninhydrin color yield of protein ϵ-amino groups is equivalent to a 60–70% yield of Ruhemann's purple after a correction for α-amino groups is made. Friedman and Williams (1973) suggested that free amino groups in proteins can be estimated by the ninhydrin method if color yields at long reaction times are extrapolated back to zero time to eliminate contributions by the slow secondary reactions and if a correction is made for the 60–70% color yield from ϵ-NH$_2$ groups.

The ninhydrin reaction of amino groups was adapted to estimate lysine directly in corn meal (Beckwith *et al.* 1975), to measure, in cereal grains, free amino acids that correlate with lysine content (Mertz *et al.* 1975; Mertz 1975; Misra *et al.* 1975; Rhodes 1975), to assess the extent of modification due to processing of food proteins and the effect of that modification on protein utilization by ruminants (Broderick 1977; Friedman and Broderick 1977), and to follow reaction rates of amino groups in amino acids, peptides, and proteins (Friedman and Wall 1964; Friedman 1967A; Ansari *et al.* 1975). Ninhydrin also reacts with aromatic amines (Friedman 1967B).

Nutritional Availability

For nutritional assimilation, lysine must be liberated from a food protein by digestion. It is, therefore, theoretically and practically important to establish the factors that affect peptide bond cleavage of lysine derivatives as well as the nutritional availability and safety of such derivatives. Only a few such studies are recorded (Carpenter and Bjarnason 1968).

Neuberger and Sanger (1944), who evaluated the availability of lysine and lysine derivatives for rat growth, concluded that, unlike The L-isomer, ϵ-N-acetyl-D-lysine cannot replace lysine in the diet. They also found that ϵ-N-methyl-DL-lysine promoted growth of rats as effectively as DL-lysine. (Only the L-isomer was effective in each case.)

Seeley and Benoiton (1969, 1970) examined the hydrolysis of various derivatives of lysine and lysine homologues by trypsin and carboxypeptidase B. (Trypsin catalyzes the hydrolysis of primary amide, secondary amide (peptide), and ester bonds of carboxyl groups of lysine and arginine. Carboxypeptidase B liberates the C-terminal amino acid of a peptide chain ending with lysine or arginine). They reported, first, that susceptibility to trypsin is limited to diamino acids in which either three or four methylene groups separate the two carbon atoms that bear the amino groups. Second, of the three possible ϵ-N-methylated lysines (mono-, di-, and tri-), only derivatives of monomethyl-L-lysine are cleaved by trypsin.

Third, ϵ-*N*-formyllysine amides and esters resist hydrolysis by trypsin. Finally, carboxypeptidase B readily hydrolyzes poly- ϵ-*N*-methyl-L-lysine and also poly-ϵ-*N*-dimethyl-L-lysine, the latter more slowly than the monomethylated polypeptide.

These and related studies (Weil and Telka 1957; Elmore *et al.* 1967; Wang and Carpenter 1968; Gertler 1971A, B; Zoltobrocki *et al.* 1974; Jering *et al.* 1974) show that charge, basicity, and steric bulk of the substituted amino groups strongly influence the susceptibility of lysine derivatives to enzymatic hydrolysis.

Although little information is available about other enzymes that specifically act on lysine derivatives, Leclerc and Benoiton (1968) show that the enzyme ϵ-*N*-lysine acylase, present in kidneys of rats and hogs, can cleave the acyl groups from ϵ-*N*-formyl and ϵ-*N*-acetyllysines. Although this enzyme did not hydrolyze analogous ornithine, homolysine, and diaminobutyric acid derivatives, a similar enzyme preparation from chicken and pigeon kidneys was less specific; it hydrolyzed not only those compounds but also ϵ-*N*-propionyllysine.

Finally, it is important that whereas lysine in ϵ-glutamyllysine residues, formed in proteins subjected to alkali and heat, is nutritionally available (Carpenter and Booth 1973; Bjarnason and Carpenter 1969; Waibel and Carpenter 1972; Mauron *et al.* 1975; Hurrell *et al.* 1976; Hurrell and Carpenter 1974, 1977), carbohydrate derivatives of lysine formed during browning reactions are not (Carpenter and Booth 1973).

Protected Proteins in Animal Nutrition

Chemical treatment of proteins in feeds, can, by cross-linking and modifying amino and other functional groups in protein chains, decrease their solubility and microbial degradation in the rumen (Broderick 1975; Clark 1975; Friedman and Broderick 1977). Formaldehyde treatment of protein-containing rations increased nitrogen retention and body weight of cattle and sheep and wool growth of sheep. However, results of feeding proteins treated with formaldehyde and related compounds are not always consistent. Feeds in which proteins have been reliably modified with inexpensive reagents under readily controlled conditions need to be evaluated for possible beneficial effects on meat, milk, and wool production.

We are approaching this objective by systematically evaluating many derivatives of casein and other proteins made with various acylating and alkylating reagents (Friedman and Broderick 1977). Preliminary results (Table 18.6) indicate that all such treatments decrease protein digestion by rumen microorganisms. Some of the treatments gave equal or better ruminal protection than for-

TABLE 18.6
RATE OF RUMINAL *IN VITRO* DEGRADATION OF
CASEIN PREPARATIONS[1]

Treated with	Release Rate[2] μmole/hr/ml RF[4]	%	Release Rate[3] μmole/hr/ml/RF	%
Control	5.52	100.0	4.17	100.0
Formaldehyde[5]	1.28	23.2	0.901	21.6
N-Methyloylacrylamide	1.17	21.2	0.886	21.2
Divinylsulfone	0.798	14.5	0.578	13.9
Hydroxyethyl acrylate	2.67	48.4	2.04	48.9
N-Vinyl-2-pyrrolidinone	3.26	59.1	2.40	57.6

[1] Friedman and Broderick (1977); 180 mg of each preparation incubated in 15 ml of medium containing 5 ml strained ruminal fluid.

[2] Medium contained no hydrazine sulfate.

[3] Medium contained 1 m-mole hydrazine sulfate.

[4] Ruminal fluid.

[5] All treatments were carried out for 2 hr.

maldehyde. The nutritive value and safety of the modified proteins is still to be determined. We hope that systematic *in vitro* screening for digestibility of these protein derivatives may disclose useful new feed components.

Lysinoalanine Formation and Inhibition

Alkali and heat may convert a fraction of the ϵ-amino groups of lysine side chains to lysinoalanine residues. Lysinoalanine, either free or protein-bound, has been implicated as a possible cause of kidney lesions when fed to rats (Woodard *et al.* 1975; Gould and MacGregor 1977).

Lysinoalanine and related cross-linked amino acids may be derived from reaction of lysine with dehydroalanine residues formed from substituted serine, cystine, and cysteine residues in proteins. Substituted threonine residue reacts similarly to form methylated homologues. The double bond of dehydroalanine, which is part of an activated, conjugated double-bond system, readily reacts with SH and NH_2 groups of cysteine and lysine to form lanthionine and lysinoalanine, respectively (Fig. 18.9, 18.10). The relative rates of these reactions appear to determine the nature of heated or alkali-treated food proteins. To gain insight into the factors that govern these reactions, we are measuring the reactions of various functional groups in amino acids and proteins with dehydroalanine and dehydroalanine methyl ester (Snow *et al.* 1975A, B; Finley *et al.* 1976A, B; Friedman *et al.* 1977).

When proteins were treated with dehydroalanine esters at pH 10, then hydrolyzed and analyzed, varying amounts of lysine residues were found to have been transformed to lysinoalanine side chains.

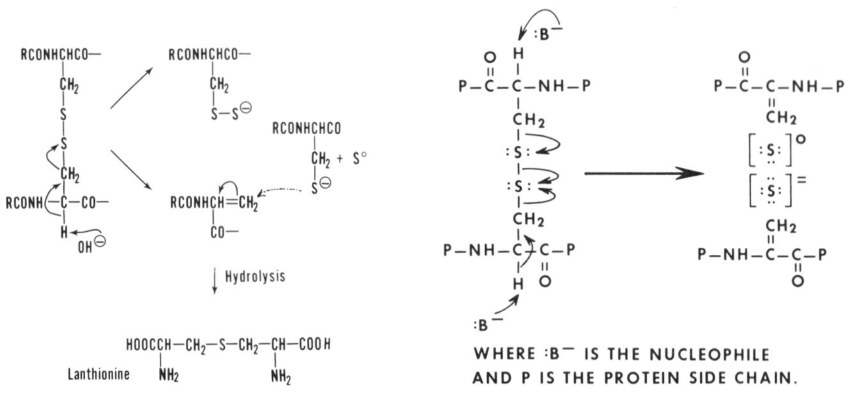

FIG. 18.9. TRANSFORMATION OF A PROTEIN DISULFIDE BOND TO DEHYDROALANINE RESIDUES

A disulfide bond can give rise to one (left-hand side) or two (right-hand side) dehydroalanine residues. The persulfide anion (RSS⁻) shown on the left-hand side can produce another dehydroalanine residue or undergo several other transformations (Fletcher *et al.* 1963).

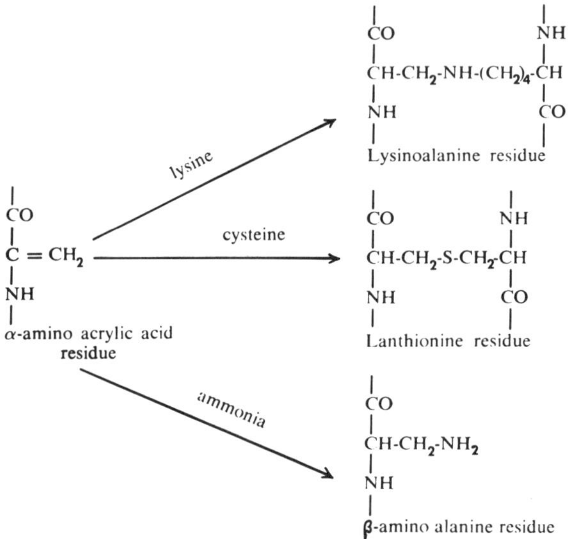

FIG. 18.10. ADDITION REACTIONS OF A DEHYDROALANINE RESIDUE IN PROTEIN

The nature and concentrations of reaction products formed is a direct function of the nucleophilic reactivities of protein or added cuntional groups towards the double bond of dehydroalanine. Since SH groups are much more reactive than NH₂ groups, addition of cysteine or another thiol will minimize formation of lysinoalanine at the expense of other adducts.

TABLE 18.7

RATES OF REACTION OF AMINO AND MERCAPTIDE GROUPS WITH ACRYLONITRILE[1]

(AT pH 8.1 AND 30°C; $\mu = 0.3$)

Compound	$k_2(S^-)$, 1/mole/sec	pK_1 (SH)	$k_2(NH_2) \times 10^4$ 1/mole/sec	pK_2 $(NH_3{}^+)$	Ratio of Observed Rates (S^-/NH_2)
Mercaptoacetic acid	0.0390	10.20			
β-Mercaptopropionic acid	0.0270	10.05			
N-Acetylcysteine	0.0573	9.52			
DL-Homocysteine	0.120	8.70	4.20	10.46	285
Glutathione	0.173	8.56	3.68	9.57	470
Cysteine	0.134	8.15	4.40	10.37	304
Cysteine ethyl ester	0.183	6.53	1.42	9.05	1290
Penicillamine	0.00650	7.90	9.44	10.42	6.4

[1] Friedman *et al.* (1965).

Complete transformation, which might provide a convenient method for determining available lysine, was not feasible at pH 10 because the esters hydrolyzed to less reactive acrylic acid derivatives.

Analogous studies of the reaction of dehydroalanine with SH groups in reduced proteins were carried out at pH 7.5 (Friedman *et al.* 1977). The SH groups were completely alkylated to form lanthionine derivatives; thus, the cysteine and (half)-cystine content of proteins could be estimated. Moreover, since SH groups react much more rapidly than amino groups with dehydroalanine and related vinyl compounds (Table 18.7), addition of thiols, such as cysteine, during alkali treatment of proteins should trap the residues of dehydroalanine as transient products and/or eliminate potential precursors for dehydroalanine as described below. This competitive reaction should minimize lysinoalanine formation. We demonstrated that reaction (Finley *et al.* 1976A, B).

Inhibition of lysinoalanine formation by added thiols or other nucleophiles such as sulfite ions can occur by at least three distinct mechanisms. First, by direct competition the added nucleophile can trap dehydroalanine residues derived from protein amino acid side chains as mentioned earlier (Eq 18.1, 18.2). Second, the added nucleophile can cleave protein disulfide bonds (Eq 18.3, 18.4) and thus generate free SH groups, which may, in turn, combine with dehydroalanine residues as illustrated in Eq (18.5) (indirect competition). Third, the added nucleophiles by cleaving disulfide bonds can diminish a potential source for dehydroalanine inasmuch as cystine residues are known to undergo β-elimination reactions to form dehydroalanine more readily than negatively charged $P\text{-}S^-$ and $P\text{-}S\text{-}SO_3^-$/protein side chains (suppression of dehydroalanine formation mechanism).

It may be possible to distinguish among these mechanisms. With radioactive-labeled cysteine as the added nucleophile, for example, direct competition will give rise to labeled lanthionine, secondary competition to unlabeled lanthionine, while the suppression mechanism, to the extent it is successful, to no lanthionine. More than one of the cited pathways can of course operate simultaneously. We are currently attempting to decide among these possibilities.

Direct Competition.—

$$R\text{-}SH \rightleftharpoons R\text{-}S^- + H^+ \qquad\qquad 18.1a$$

$$R\text{-}S^- + CH_2{=}C\text{-}(NHCOP)\text{-}P \rightleftharpoons R\text{-}S\text{-}CH_2\text{-}\overline{C}\text{-}(NHCOP)\text{-}P \quad 18.1b$$

$$H^+ + R\text{-}S\text{-}CH_2\text{-}\overline{C}\text{-}(NHCOP)\text{-}P \rightleftharpoons R\text{-}S\text{-}CH_2\text{-}CH\text{-}(NHCOP)\text{-}P \quad 18.1c$$

$$SO_3^- + CH_2{=}C\text{-}(NHCOP)\text{-}P \rightleftharpoons {}^-O_3S\text{-}CH_2\text{-}\overline{C}\text{-}(NHCOP)\text{-}P \quad 18.2a$$

$$H^+ + {}^-O_3S\text{-}CH_2\text{-}\overline{C}\text{-}(NHCOP)\text{-}P \rightleftharpoons O_3S\text{-}CH_2\text{-}CH\text{-}(NHCOP)\text{-}P \quad 18.2c$$

Suppression and First Step of Indirect Competition.—

$$2R\text{-}S^- + P\text{-}S\text{-}S\text{-}P \rightleftharpoons 2P\text{-}S^- + R\text{-}S\text{-}S\text{-}R \quad 18.3$$

$$SO_3^{--} + \quad P\text{-}S\text{-}S\text{-}P \quad \rightleftharpoons \quad P\text{-}S^- + P\text{-}S\text{-}SO_3^- \qquad 18.4$$

Indirect competition.—

$$P\text{-}S^- + \quad CH_2 {=} C\text{-}(NHCOP)\text{-}P \quad \rightleftharpoons \quad P\text{-}S\text{-}CH_2\text{-}\overline{C}\text{-}(NHCOP)\text{-}P \quad 18.5$$

$$H^+ + \quad P\text{-}S\text{-}CH_2\text{-}\overline{C}\text{-}(NHCOP)\text{-}P \rightleftharpoons P\text{-}S\text{-}CH_2\text{-}CH\text{-}(NHCOP)\text{-}P \quad 18.6$$

Lysinoalanine may also form via substitution reactions, as appears to be the case when wool is treated with some phosphate salts (Touloupis and Vassiliadis 1977), rather than by the described elimination-addition reactions (*cf.* also Friedman 1977).

Several largely speculative points are noteworthy. First, dehydroalanine may react with essential SH and NH_2 groups of proteins and nucleic acids *in vivo*. Such reactions have been postulated for peptide antibiotics that contain dehydroalanine and methyldehydroalanine (Gross 1971, 1974, 1977). Second, lysinoalanine may exert its pharmacological effect on kidneys by undergoing a reverse-Michael reaction *in situ* to reform dehydroalanine as the reactive alkylating agent. Lysinoalanine in effect may act as a carrier of dehydroalanine or similar alkylating agent to kidney target sites (*cf.* MacGregor and Clarkson 1974; Kupchan 1974). Third, transition metal ions also show strong affinity for SH and NH_2 groups (Friedman 1974; Friedman and Masri 1973). Therefore, they may compete with dehydroalanine for such biological nucleophiles, thus minimizing or preventing the formation of lanthionine and lysinoalanine. These possibilities and the possible role of lysinoalanine in metal binding *in vivo* deserve careful study. Fourth, lysinoalanine may compete with lysine or other amino acid(s) during protein biosynthesis, *i.e.*, it may act as a competitive inhibitor of protein biosynthesis. In this connection, it is noteworthy that the antimicrobial action of lysine analogs and derivatives has been ascribed to competitive inhibition of lysine incorporation into cell walls of bacteria (Perlman 1975) and to the release of cellular constituents, respectively (Nakamiya *et al.* 1976). Fifth, lysinoalanine may competitively inhibit lysine catabolism by reacting with α-ketoglutarate to prevent the analogous transformation of lysine to saccharopine, a key intermediate of lysine metabolism (Hutzler and Dancis 1975; Higashino *et al.* 1971; Chang 1976). Note the structural resemblance between lysine, lysinoalanine, and saccharopine.

<table>
<tr><td></td><td></td><td>COOH</td><td></td><td>COOH</td></tr>
<tr><td>CH₂-NH₂</td><td>CH₂-NH-CH₂-CH</td><td></td><td>CH₂-NH-CH</td><td></td></tr>
<tr><td>(CH₂)₃</td><td>(CH₂)₃</td><td>NH₂</td><td>(CH₂)₃</td><td>(CH₂)₂</td></tr>
<tr><td>CHNH₂</td><td>CHNH₂</td><td></td><td>CHNH₂</td><td>COOH</td></tr>
<tr><td>COOH</td><td>COOH</td><td></td><td>COOH</td><td></td></tr>
<tr><td>Lysine</td><td>Lysinoalanine</td><td></td><td>Saccharopine</td><td></td></tr>
</table>

Sixth, the secondary amino group of lysinoalanine may interact with nitrites both *in vitro* and *in vivo* to form a toxic *N*-nitroso derivative. Seventh, lysinoalanine may act as a local irritant or allergen of kidney tissue. Finally, lysinoalanine may exert its effect by inhibiting or affecting normal reabsorption of amino acids by kidney tubules.

PHYSICOCHEMICAL ASPECTS

Friedman and Romersberger (1968) observed that chemical modification of amino groups with conjugated vinyl compounds

TABLE 18.8
pK_2 VALUES OF AMINO GROUPS IN AMINO ACIDS
AND DERIVATIVES AT 30°C[1]

Compound	pK_2	ΔpK_2
DL-Phenylalanine	9.00	
N-Methylcarbonylethyl DL-phenylalanine	8.60	0.40
N-Carbamidoethyl-DL-phenylalanine	8.10	0.90
N-2-Bis(β'-chloroethyl)phosphonylethyl-DL-phenylalanine	7.30	1.70
N-Cyanoethyl-DL-phenylalanine	6.60	2.40
N-Methylsulfonylethyl-DL-phenylalanine	5.87	3.13
N-2-Cyanopropyl-DL-phenylalanine	6.35	2.65
ϵ-Aminocaproic acid	10.62	
N-Cyanoethyl-ϵ-aminocaproic acid	8.15	2.47
N-Methylsulfonylethyl-ϵ-aminocaproic acid	7.80	2.82
N-2-Cyanopropyl-ϵ-aminocaproic acid	8.25	2.37
β-Alanine	10.06	
N-Cyanoethyl-β-alanine	7.85	2.21
Asparagine	8.72	
α-*N*-Cyanoethylasparagine	6.46	2.26
Tyrosine	9.00	
N-Cyanoethyltyrosine	6.50	2.50
N,*N*-Dicyanoethyltyrosine	4.13	4.87

[1] Friedman and Romersberger (1968); Friedman (1966).

significantly decreases amino group basicity (Table 18.8). This decrease ranges from 0.4 pK_a units, for the methyl vinyl ketone derivative of phenylalanine, to 3.13 pK_a units, for the methyl vinyl sulfone derivative of phenylalanine. The decrease seems to be an additive function of the number of introduced subsituents, since, for example, the decrease for *N*-cyanoethyltyrosine is about one-half that for *N*,*N*-dicyanoethyltyrosine (Friedman 1966). Although one would expect differences in pK_a values in going from a primary to a secondary and then to a tertiary amino group, the exact differences need to be determined in each case.

The basicity of the amino group decreases greatly on cyanoethylation, in marked contrast to propylation, for example (Friedman and Wall 1964). The pK_a values for the amino groups in propylamine, dipropylamine, and tripropylamine are 10.67, 11.01, and 10.74, respectively. Thus, adding a second propyl group to propylamine increases its basicity, presumably because of an inductive effect increasing the electron density near the nitrogen atom. However, introduction of a third propyl group results in a slight decrease in the pK_a. This decrease is ascribed to steric effects that hinder proton binding. The remarkable effect of cyanoethylation in hindering proton binding—for example, in the following equilibria—can be explained in part as a result of the bulky substituent, but more specifically as an inductive electron-withdrawing effect of the cyano group. Simple alkyl substituents such as propyl lack this mechanism.

These results imply that analogous derivatives of proteins, especially those in which amino group hydrogens have been replaced by electron-withdrawing substituents will have greatly altered acid-base equilibria, isoelectric properties, charge distributions, and conformations. These effects allow corresponding purposeful changes in protein functional properties (*cf.* Wu *et al.* 1971).

In a related study, Baxter and Byvoet (1975) observed a progressive decrease in the charge density of the ϵ-amino group of lysine with increasing methyl substitution. Analysis of carbon-13 NMR spectra of ϵ-N-methyllysine, and the corresponding di- and trimethyl analogues revealed that progressive methyl substitution progressively shifts the signals downfield from those for ϵ-CH_2-N and N-methyl carbon atoms (Brown *et al.* 1976). This result implies that progressive N-methyl substitution of lysine residues leads to a progressive charge decrease on the CH_2-N and CH_3-N carbon atoms. That is, the charge densities at *both* the side-chain nitrogen atoms and the next-neighbor carbon atoms decrease regularly with increasing methyl substitution. A second implication is that the greater the number of ϵ-N-methyl groups on protein basic amino acid residues, the greater the affinity of these residues for anionic sites. However, the situation is really more complex than that since, in contrast to a primary amino group, a quaternary amino group cannot participate in hydrogen-bonding interactions which are paramount in protein structure and conformation.

The influence of charge on the amino group is strikingly illustrated by the effect of maleylation of the ϵ-amino group in the proteolytic enzyme elastase (Gertler 1971A, B); elastolytic activity is significantly decreased, but esterase activity is changed only slightly. Gertler attributed this behavior to the hypothesis that elastase requires the basic amino groups for activity, because the enzyme has to form salt linkages between its positively charged amino groups and the carboxyl groups of elastin before it can act. Maleylation not only eliminates the positive charges of the amino groups, but introduces a negative charge for each positive charge that is eliminated, thus preventing primary binding between enzyme and substrate. The enzyme subtilisin behaved similarly (*cf.* Elbein and Pan 1977).

Maleylation, as well as trifluoroacetylation, has been used to acylate NH_2 groups reversibly (Fig. 18.11). Protein NH_2 groups

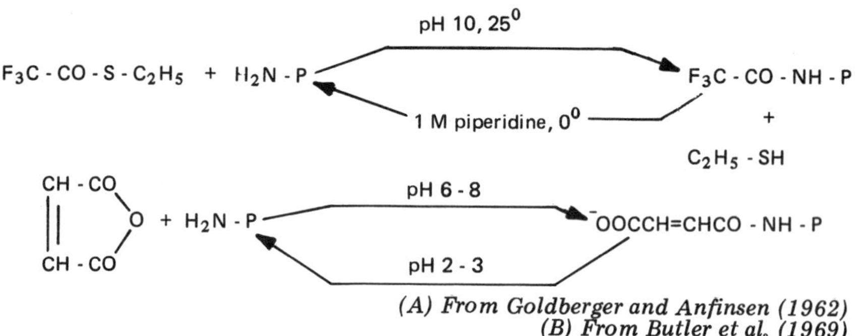

(A) From Goldberger and Anfinsen (1962)
(B) From Butler et al. (1969)

FIG. 18.11. REVERSIBLE ACYLATIONS OF PROTEIN AMINO GROUPS
(A) Trifluoroacetylation; (B) Maleylation

could, in principle, also add across the double bonds of maleic anhydride and the maleylated protein. However, these additions would be expected to proceed much more slowly than maleylation (*Cf.* Friedman and Wall 1964, 1966).

FUNCTIONAL PROPERTIES

Because chemical modification of lysine amino groups with various reagents causes major changes in physicochemical properties of proteins, it is not surprising that such modifications also change various functional properties. Although the relation between physicochemical and functional properties is a relatively unexplored subject in food science, it is necessary for practical understanding of protein properties, and for the creation or improvement of desirable functional properties and the suppression of undesirable ones. Modification of food proteins should, however, be approached with caution. Chemical modification can impair nutritional quality besides changing functional behavior. Therefore, both effects should be measured (Melnick and Oser 1949). Some modifications may well form protein derivatives that are toxic or have other undesirable properties. Current efforts are designed to modify the functionality of food proteins so that new and improved food products can be developed; a few examples follow.

Cottonseed Flour

Childs and Park (1976) evaluated functional properties of glandless cottonseed flour acylated with succinic and acetic anhydrides. Their results (Table 18.9) show that (a) oil-holding and

TABLE 18.9
FUNCTIONALITY OF ACYLATED COTTONSEED PRODUCTS[1]

Capacity	Glandless	Acetylated	Succinylated
Water-holding	3.50 (ml/g)	7.87 (ml/g)	5.5 (ml/g)
Oil-holding	2.60	8.8	19.60
Emulsifying	456.7	416.7	776.67
Foam	126.7	250.00	150.0

[1] Childs and Park (1976).

water-holding capacities were markedly increased by both reagents; (b) emulsifying capacity was increased by succinylation, but not by acetylation; (c) foam capacity was increased (doubled) by acetylation, but only by about 10% by succinylation; and (d) the specific viscosity of the flour was increased markedly by succinylation but only slightly by acetylation. (Data in Table 18.9 do not illustrate this last point.)

The different effects of acetic and succinic anhydrides apparently reflect the respective abilities of these reagents to modify both the shape and charge of the food protein molecule. Thus, modification with acetic anhydride transforms basic amino groups to neutral amide groups and introduces a two-carbon fragment (the acetyl group) that can participate in both hydrogen-bonding and hydrophobic interactions. The corresponding modification with succinic anhydride introduces a four-carbon fragment that, in addition to a carbonyl group, also contains a carboxyl side chain. The carboxyl group participates in acid-base equilibria (Fig. 18.12); its pK_a value is about 4.6. According to the pH of the medium, the newly introduced carboxyl group contributes varying negative charge to the protein and strongly influences physicochemical and functional properties. For example, electrostatic forces are increased; such

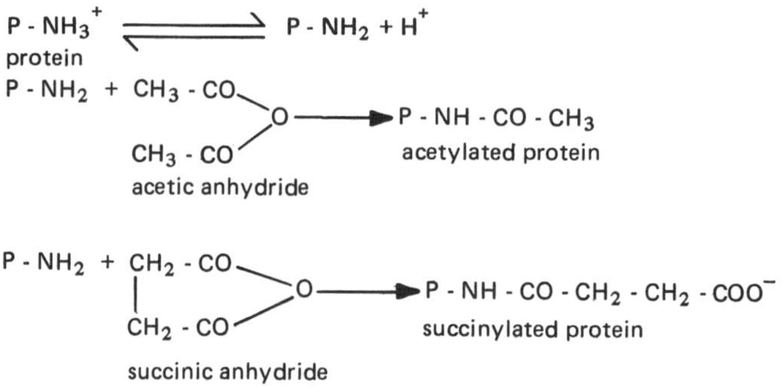

$$P\text{-}NH_3^+ \rightleftharpoons P\text{-}NH_2 + H^+$$

protein

$$P\text{-}NH_2 + \begin{matrix} CH_3\text{-}CO \\ \\ CH_3\text{-}CO \end{matrix}\!\!\!> O \longrightarrow P\text{-}NH\text{-}CO\text{-}CH_3$$

acetic anhydride   acetylated protein

$$P\text{-}NH_2 + \begin{matrix} CH_2\text{-}CO \\ | \\ CH_2\text{-}CO \end{matrix}\!\!\!> O \longrightarrow P\text{-}NH\text{-}CO\text{-}CH_2\text{-}CH_2\text{-}COO^-$$

succinic anhydride   succinylated protein

FIG. 18.12. ACETYLATION AND SUCCINYLATION OF A
PROTEIN AMINO GROUP

charges are no doubt responsible for the increased specific viscosity of the modified proteins (Carpenter and Saffle 1965; Habeeb *et al.* 1958).

The toxic pigment gossypol, which is present in normal cottonseed, interacts with cottonseed proteins, presumably to form Schiff's bases which can then undergo further transformations. This interaction shifts the pH of minimum solubility and the apparent isoelectric points of cottonseed protein fractions to lower values (Damaty and Hudson 1975). Thus, the presence of gossypol may have effects similar to those of acylation.

Single-Cell Proteins

McElwain *et al.* (1975) examined the solubility, viscosity, emulsifying capacity, and digestibility of protein concentrates extracted from *Candida utilis* and subsequently succinylated so as to modify 84% of the amino groups. Both native and modified protein isolates had similar solubilities at the isoelectric point (pH 4); the succinylated protein was essentially insoluble below that pH. That effect apparently was due to elimination of most positive charges on the amino groups, with accompanying decreases in acid-binding and hydrophilic properties. (Note that succinylation lowers the isoelectric point of the starting protein.) Succinylation increased emulsion viscosity, but decreased emulsion stability. The authors ascribed the increased viscosity to the observed threadlike, unfolded structure of the modified protein compared to the globular, interwoven structure of the starting material. Decreased stability was attributed to the increased negative charge due to succinylation. Digestibility, as measured by enzymatic release of free lysine, was 3.3% for the succinylated derivative.

Fish Proteins

Miller and Groninger (1976), who studied functional properties of enzyme-modified, acetylated and succinylated fish protein derivatives, reported that emulsifying activity and capacity, gelation, water sorption, aeration, and foam stability all increased with acylation until 43–59% of the amino groups were acylated. Higher degrees of modification did not significantly change these properties. They suggested that the desirable changes in the acylated and bromelain-modified fish proteins should find application in foods.

In related studies Chen *et al.* (1975) observed that succinylation of proteins from fish protein concentrates gave derivatives with improved solubility and emulsifying and curd-forming properties. The modified protein was, however, less digestible than the original.

Analogous changes in functional and physicochemical properties following chemical, heat, and enzyme treatments have been observed with egg white (Gandhi *et al.* 1968), myosin (Oppenheimer *et al.* 1967; Ito and Ando 1974, 1975), casein (Clarke and Naki 1972; Osner 1975), and soy proteins (Hashizume *et al.* 1975; Puski 1975; Terrell and Staniec 1975). Effects of enzymes on nutritional quality and availability of proteins was examined in detail by Schwimmer (1975).

Niwa and Kitayama (1975) discovered that addition of lysine and related diamines increased the jelly-strength of fish gel from dolphins. This effect is presumably due to the ability of these positively charged, basic compounds to form cross-linked salt bridges with the fish proteins.

In conclusion, I would stress that understanding of the complete interplay of chemical composition and physical and chemical properties of native and modified proteins must be a prime target of research to guide improvement in nutritive and functional values of proteins. In such studies, all amino acid residues are important, but lysine is especially so. Because it is a highly reactive essential amino acid, its chemical changes are vitally important in designing new food products and processes.

ACKNOWLEDGMENT

It is a pleasure to thank my colleagues whose names are listed in the cited publications as authors and co-authors for excellent scientific collaboration.

BIBLIOGRAPHY

ALLISON, R. M., LAIRD, W. M., and SYNGE, R. L. M. 1973. Notes on a deamination method proposed for determining chemically available lysine in proteins. Br. J. Nutr. *29*, 51–55.

ANSARI, A., KIDWAL, S. A., and SALAHUDDIN, A. 1975. Acetylation of amino groups and its effect on the conformation and immunological activity of ovalbumin. J. Biol. Chem. *250*, 1625–1632.

BAILEY, J. B. 1974. Automated analysis of available lysine and tyrosine in foodstuffs. J. Sci. Agri. *25*, 1007–1014.

BAXTER, C. S., and BYVOET, P. 1975. CMR studies of protein modification. Progressive decrease in charge density at the ϵ-amino function of lysine with increasing methyl substitution. Biochem. Biophys. Res. Commun. 64, 514–518.

BECKWITH, A. C., PAULIS, J. W., and WALL, J. S. 1975. Direct estimation of lysine in corn by the ninhydrin color reaction. J. Agr. Food Chem. 23, 194–196.

BENESCH, R., and BENESCH, R. E. 1958. Thiolation of proteins. Proc. Natl. Acad. Sci. 44, 848–853.

BHATTY, R. S., and WU, K. K. 1974. Lysine screening in barley with a modified Udy dye-binding method. Can. J. Plant. Sci. 55, 685–689.

BJARNASON, J., and CARPENTER, K. J. 1969. Mechanism of heat damage to proteins. I. Models with acylated lysine units. British J. of Nutr. 23, 859–868.

BLOM, L., HENDRICKS, P., and CARIS, J. 1967. Determination of the available lysine in foods. Anal. Biochem. 21, 382–400.

BOOTH, V. H. 1971. Problems in the determination of FDNB-available lysine. J. Sci. Food. Agr. 22, 658–664.

BRODERICK, G. A. 1975. Factors affecting ruminant responses to protected amino acids and proteins. In Protein Nutritional Quality of Foods and Feeds, Part 2, M. Friedman (Editor). Marcel Dekker, New York.

BRODERICK, G. A. 1977. Effect of processing on protein utilization by ruminants. In Protein Crosslinking: Nutritional and Medical Consequences. M. Friedman (Editor). Plenum Press, New York. 172nd Am. Chem. Soc. Meet. 1976.

BROWN, L. R., DEMARCO, A., WAGNER, G., and WUTHRICH, K. 1976. A study of the lysyl residues in the basic pancreatic trypsin inhibitor using ^{1}H NMR at 360 MHz. Eur. J. Biochem. 62, 103–107.

BUNNETT, J. F., and HERMANN, D. H. 1970. Kinetics of dinitrophenylation of amino acids. Biochemistry 9, 816–822.

BURCHFIELD, H. P., and STORRS, E. E. 1957. Chemical structures and dissociation constants of amino acids, peptides, and proteins in relation to their reaction rates with 2,4-dichloro-6-(o-chloroanilino)-s-triazine. Boyce Thompson Institute Contrib. 18, 395–418.

BURCHFIELD, H. P., and STORRS, E. E. 1958. Relative reactivities of 1-fluoro-2,4-dinitrobenzene and 2,4-dichloro-6-(o-chloroanilino)-s-triazine with metabolites containing various functional groups. Boyce Thompson Institute Contrib. 19, 169–176.

BUTLER, P. J. G., HARRIS, J. I., HARTLEY, B. S., and LEBERMAN, R. 1969. The use of maleic anhydride for the reversible blocking of amino groups in polypeptide chains. Biochem. J. 112, 679–689.

CARPENTER, J. A., and SAFFLE, R. L. 1965. Some physical and chemical factors affecting emulsifying capacity of meat protein extracts. Food Technol. 19, 111–115.

CARPENTER, K. J. 1960. The estimation of available lysine in animal protein foods. Biochem. J. 77, 604–610.

CARPENTER, K. J., and BJARNASON, J. 1968. Nutritional evaluation of proteins by chemical methods. In Evaluation of Novel Protein Products, A. E. Bender, R. Kihlberg, B. Lofquist, and L. Munck (Editors). Pergamon Press, New York.

CARPENTER, K. J., and BOOTH, V. H. 1973. Damage to lysine in food processing: Its measurement and its significance. Nutr. Abst. Rev. 43, 424–451.

CAVINS, J. F., and FRIEDMAN, M. 1967. New amino acids derived from reactions of ϵ-amino groups in proteins with α,β-unsaturated compounds. Biochemistry 6, 3766–3770.

CAVINS, J. F., and FRIEDMAN, M. 1968A. Specific modification of protein sulfhydryl groups with α,β-unsaturated compounds. J. Biol. Chem. 243, 3357–3360.

CAVINS, J. F., and FRIEDMAN, M. 1968B. Automatic integration and computation of amino acid analyses. Cereal Chem. 45, 172–176.

CHANG, Y. F. 1976. Pipecolic acid pathway: The major lysine metabolic route in the rat brain. Biochem. Biophys. Res. Commun. 69, 174–180.

CHAUFFE, L., and Friedman, M. 1977. Factors which affect the cyanoborohydride reduction of protein Schiff's bases derived from aromatic aldehydes. In Protein Crosslinking: Biochemical and Molecular Aspects. M. Friedman (Editor). Plenum Press, New York.

CHEN, L. F., RICHARDSON, T., and AMUNDSON, C. H. 1975. Some functional properties of succinylated proteins from fish protein concentrate. J. Milk Food Technol. 38, 89–93.

CHILDS, E. A., and PARK, K. K. 1976. Functional properties of acylated glandless cottonseed flour. J. Food Sci. 41, 713–714.

CHIRGADZE, Y. N., and OVSEPYAN, A. M. 1972. Hydration mobility in peptide structures. Biopolymers 11, 2179–2186.

CLARK, J. H. 1975. Nitrogen metabolism in ruminants: Protein solubility and rumen by-pass of protein and amino acids. In Protein Nutritional Quality of Foods and Feeds, Part 2, M. Friedman (Editor). Marcel Dekker, New York.

CLARKE, R. F. L., and NAKI, S. 1972. Effect of modification of carboxyl and ϵ-amino groups on the structure and stabilizing ability of κ-casein. Biochim. Biophys. Acta 257, 70–75.

CONCON, J. M. 1975. Chemical estimation of critical amino acids in cereal grains and other products. In Protein Nutritional Quality of Foods and Feeds, Part 1, M. Friedman (Editor). Marcel Dekker, New York.

DAMATY, S., and HUDSON, B. J. F. 1975. Interaction of gossypol with cottonseed protein: Potentiometric studies. J. Sci. Food Agric. 26, 1667–1672.

DELLA MONICA, E. S., STROLLE, E. O., and McDOWELL, P. E. 1976. A modified method for determining available lysine in protein recovered from heat treated potato juice. Anal. Biochem. 73, 274–279.

DOPHEIDE, A. A. 1975. Reactivity and ionization constants of the lysine residues in apovitellin I of emu egg yolk low-density lipoprotein by competitive labeling. Aust. J. Biol. Sci. 28, 433–437.

DRECHSEL, E. 1891. Abbau der Eiweisstoffe. Arch. Anat. Physiol. Abt., 248–278.

DRECHSEL, E., and KRUGER, R. R. 1892. Zur Kenntniss des Lysins. Ber. 25, 2454–2457.

DUGGLEBY, R. G., and KAPLAN, H. 1975. A competitive labeling method for the determination of the chemical properties of solitary functional groups in proteins. Biochemistry 14, 5168–5175.

EKLUND, A. 1976. The determination of available lysine in casein and rapeseed protein concentrates using TNBS as a reagent for free epsilon amino group of lysine. Anal. Biochem. 70, 434–439.

ELBEIN, A. D., and PAN, Y. T. 1977. Protein: polyanion interactions. Studies on the trehalose-P synthetase as a model system. In Protein Crosslinking: Biochemical and Molecular Aspects, M. Friedman (Editor). Plenum Press, New York.

ELLWELL, D., and SOARES, J. H., JR. 1975. Amino acid bio-availability. A comparative evaluation of several assay techniques. Poultry Sci. *54*, 78–85.

ELMORE, D. T., ROBERTS, D. V., and SMYTH, J. J. 1967. The kinetics of hydrolysis of derivatives of L-lysine and S-(β-aminoethyl) L-cysteine (thialysine) by bovine trypsin. Biochem. J. *102*, 728–734.

EL-NOCKRASHY, A. S., and FRAMPTON, V. L. 1967. Destruction of lysine by nonreducing sugars. Biochem. Biophys. Res. Commun. *28*, 675–681.

ERBERSDOBLER, H. F. 1977. The biological significance of carbohydrate lysine crosslinking during heat-treatment of food proteins. *In* Protein Crosslinking: Nutritional and Medical Consequences, M. Friedman (Editor). Plenum Press, New York. 172nd Am. Chem. Soc. Meet., 1976.

FIELDS, R. 1971. The measurement of amino groups in proteins and peptides. Biochem. J. *124*, 581–590.

FINLEY, J. W., and FRIEDMAN, M. 1973. Chemical methods for available lysine. Cereal Chem. *50*, 101–105.

FINLEY, J. W., and FRIEDMAN, M. 1976. Products derived from the alkali treatment of proteins. 172nd Am. Chem. Soc. Meet., 1976.

FINLEY, J. W., SNOW, J. T., JOHNSTON, P. H., and FRIEDMAN, M. 1976A. Inhibition of lysinoalanine formation during alkali treatment of proteins. (Submitted for publication.)

FINLEY, J. W., SNOW, J. T., JOHNSTON, P. H., and FRIEDMAN, M. 1976B. Inhibition of lysinoalanine formation in foods. 36th Annu. Meet. Inst. Food Technol., Anaheim, Calif., June 5–9, Abstract No. 236.

FINOT, P. A., BUJARD, E., MOTTU, F., and MAURON, J. 1977. Availability of the true Schiff's bases of lysine. Chemical evaluation of the Schiff's base between lysine and lactose in milk. *In* Protein Crosslinking: Nutritional and Medical Consequences, M. Friedman (Editor). Plenum Press, New York. 172nd Am. Chem. Soc. Meet., 1976.

FLETCHER, J. C., ROBSON, A., and TODD, S. 1963. The sulfur balance of wool. Biochem. J. *87*, 560–567.

FORD, J. E. 1973. Some effects of processing on nutritive value. *In* Proteins in Human Nutrition, J. W. G. Porter and B. A. Rolls (Editors). Academic Press, London and New York.

FREEDMAN, R. B., and RADDA, G. K. 1968. The reaction of 2,4,6-trinitrobenzenesulphonic acid with amino acids, peptides and proteins. Biochem. J. *108*, 383–391.

FRIEDMAN, M. 1966. A novel differential titration to determine pK values of phenolic groups in tyrosine and related aminophenols. Biochem. Biophys. Res. Commun. *23*, 626–632.

FRIEDMAN, M. 1967A. Solvent effects in reactions of amino groups in amino acids, peptides, and proteins with α,β-unsaturated compounds. J. Amer. Chem. Soc. *89*, 4709–4713.

FRIEDMAN, M. 1967B. Mechanism of the ninhydrin reaction II. Preparation and spectral properties of reaction products from primary aromatic amines and ninhydrin hydrate. Canad. J. Chem. *45*, 2271–2275.

FRIEDMAN, M. 1968. Solvent effects in reactions of protein functional groups. Q. Rep. Sulfur Chem. *3*, 125–144.

FRIEDMAN, M. 1971. Solvent effects in absorption spectra of the ninhydrin chromophore. Microchem. J. *16*, 204–209.

FRIEDMAN, M. 1972. Selective chemical modification of protein sulfhydryl groups. Intra-Science Chem. Rep. *6*, No. 4, 23–34.

FRIEDMAN, M. 1973A. Reactions of cereal proteins with vinyl compounds. *In* Industrial Uses of Cereal Grains, Y. Pomeranz (Editor). Am. Assoc. Cereal Chem., Minneapolis, Minn.

FRIEDMAN, M. 1973B. The Chemistry and Biochemistry of the Sulfhydryl Group in Amino Acids, Peptides, and Proteins. Pergamon Press, Oxford, England and Elmsford, New York.

FRIEDMAN, M. 1974. Protein-Metal Interactions, Plenum Press, New York.

FRIEDMAN, M. 1977A. Protein Crosslinking: Biochemical and Molecular Aspects. (Part A). Protein Crosslinking: Nutritional and Medical Consequences. (Part B). Plenum Press, New York.

FRIEDMAN, M. 1977B. Crosslinked amino acids—Stereochemistry and nomenclature. *In* Protein Crosslinking: Nutritional and Medical Consequences, M. Friedman (Editor). Plenum Press, New York.

FRIEDMAN, M., and BRODERICK, G. A. 1977. Protected proteins in animal nutrition. *In vitro* evaluation of casein derivatives. *In* Protein Crosslinking: Nutritional and Medical Consequences, M. Friedman (Editor). Plenum Press, New York. 172nd Am. Chem. Soc. Meet., 1976.

FRIEDMAN, M., CAVINS, J. F., and WALL, J. S. 1965. Relative nucleophilic reactivities of amino groups and mercaptide ions with α,β-unsaturated compounds. J. Amer. Chem. Soc. *87*, 3672–3682.

FRIEDMAN, M., and FINLEY, J. W. 1975A. Reactions of proteins with ethyl vinyl sulfone. Int. J. Peptide and Protein Res. *7*, 481–486.

FRIEDMAN, M., and FINLEY, J. W. 1975B. Vinyl compounds as reagents for available lysine in proteins. *In* Protein Nutritional Quality of Foods and Feeds, Part 1. M. Friedman (Editor). Marcel Dekker, New York.

FRIEDMAN, M., FINLEY, J. W., and YEH, L-S. 1977. Reactions of proteins with dehydroalanine. *In* Protein Crosslinking: Nutritional and Medical Consequences, M. Friedman (Editor). Plenum Press, New York. 172nd Am. Chem. Soc. Meet., 1976.

FRIEDMAN, M., and KOENIG, N. H. 1971. Effect of dimethyl sulfoxide on chemical and physical properties of wool. Text. Res. J. *41*, 605–609.

FRIEDMAN, M., and KRULL, L. H. 1970. N- and C-alkylation of proteins in dimethyl sulfoxide. Biochim. Biophys. Acta *297*, 361–364.

FRIEDMAN, M., KRULL, L. H., and ESKINS, K. 1971. Protein chemistry and photochemistry in dimethyl sulfoxide. J. Appl. Polym. Sci. Appl. Polym. Symp. *18*, 297–306.

FRIEDMAN, M., and MASRI, M. S. 1973. Sorption behaviour of mercuric salts on modified wool and polyamino acids. J. Applied Polym. Sci. *17*, 2179–2186.

FRIEDMAN, M., and MILLARD, M. M. 1976. Radiation effects in protein analysis by X-ray photoelectron spectroscopy. Fed. Proc. *35*, 1456 (Abstract).

FRIEDMAN, M., and ROMERSBERGER, J. A. 1968. Relative influences of electron-withdrawing functional groups on basicities of amino acid derivatives. J. Org. Chem. *33*, 154–157.

FRIEDMAN, M., and SIGEL, C. W. 1966. A kinetic study of the ninhydrin reaction. Biochemistry *5*, 478–484.

FRIEDMAN, M., and WALL, J. S. 1964. Application of a Hammett-Taft relation to kinetics of alkylation of amino acid and peptide model compounds with acrylonitrile. J. Amer. Chem. Soc. *86*, 3735–3741.

FRIEDMAN, M., and WALL, J. S. 1966. Additive linear free energy relationships in reaction kinetics of amino groups with α,β-unsaturated compounds. J. Org. Chem. *31*, 2888–2894.

FRIEDMAN, M., and WILLIAMS, L. D. 1973. The ninhydrin reaction. VI. The reaction of ninhydrin with keratin proteins. Anal. Biochem. *54*, 333–345.

FRIEDMAN, M., and WILLIAMS, L. D. 1974. The ninhydrin reaction. VII.

Stoichiometry of formation of Ruhemann's purple in the ninhydrin reaction. Bioorganic Chem. *3*, 267–280.

FRIEDMAN, M., and WILLIAMS, L. D. 1977. A mathematical analysis of consecutive, competitive reactions of amino groups. *In* Protein Crosslinking: Biochemical and Molecular Aspects. M. Friedman (Editor). Plenum Press, New York.

FRIEDMAN, M., WILLIAMS, L. D., and MASRI, M. S. 1974. Reductive alkylation of proteins with aromatic aldehydes and sodium cyanoborohydride. Int. J. Peptide Protein Res. *6*, 183–185.

GANDHI, S. K., SCHULTZ, J. R., BOUGHEY, F. W., and FORSYTHE, R. H. 1968. Chemical modification of egg white with 3,3-dimethylglutaric anhydride. J. Food Sci. *33*, 163–169.

GERTLER, A. 1971A. Selective, reversible loss of elastolytic activity of elastase and subtilisin resulting from electrostatic changes due to maleylation. Eur. J. Biochem. *23*, 36–40.

GERTLER, A. 1971B. The non-specific electrostatic nature of the adsorption of elastase and other basic proteins on elastin. Eur. J. Biochem. *20*, 541–546.

GILBERT, H. F., and O'LEARY, M. H. 1975. Modification of arginine and lysine in proteins with 2,4-pentanedione. Biochemistry *14*, 5194–5199.

GOLDBERGER, R. F., and ANFINSEN, C. B. 1962. The reversible masking of amino groups in ribonuclease and its possible usefulness in the synthesis of the protein. Biochemistry *1*, 401–405.

GOLDFARB, R. A. 1966. A kinetic study of the reactions of amino acids and peptides with trinitrobenzenesulfonic acid. Biochem. J. *5*, 2570–2578.

GOLDFARB, R. A. 1970. Reactivity of amino groups in proteins. Biochim. Biophys. Acta *200*, 1–8.

GOLDFARB, R. A. 1974. Reactions of amino groups in ribonuclease A. 1. Kinetic studies. Bioorganic Chem. *3*, 249–259.

GOLDFARB, R. A., BUCHANAN, D. N., and SUTTON, D. E. 1974. Reactions of amino groups in ribonuclease A. 2. Identification of reactive amino groups of ribonuclease. Bioorganic Chem. *3*, 260–266.

GOLDFARB, A. R., and MARTIN, P. D. 1976A. Reactions of amino groups of RNAse A. 3. The existence of pH-dependent values of pK. Bioorganic Chem. *5*, 137–145.

GOLDFARB, A. R., and MARTIN, P. D. 1976B. Reactions of amino groups of RNAse A. IV. The effects of protein concentration. Bioorganic Chem. *5*, 147–153.

GOULD, D. H., and MACGREGOR, J. T. 1977. Biological effects of alkali-treated protein and lysinoalanine: an overview. *In* Protein Crosslinking: Nutritional and Medical Consequences, M. Friedman (Editor). Plenum Press, New York.

GREENSTEIN, J. P., and WINITZ, M. 1961. Chemistry of the Amino Acids. John Wiley and Sons, New York.

GROSS, E. 1971. Structure and function of peptides with α,β-unsaturated amino acids. Intra-Science Chem. Repts. *5*, 405–408.

GROSS, E., 1974. α,β-Unsaturated amino acids in peptides and proteins: Formation, chemistry and biological role. 168th Meet. Am. Chem. Soc., Atlantic City, New Jersey.

GROSS, E. 1977. α,β-Unsaturated and related amino acids in peptides and proteins. *In* Protein Crosslinking: Nutritional and Medical Consequences, M. Friedman (Editor). Plenum Press, New York. 172nd Am. Chem. Soc. Meet., 1976.

HABEEB, A. F. S. A., CASSIDY, H. G., and SINGER, S. J. 1958. Molecular structural effects produced in proteins by reaction with succinic anhydride. Arch. Biochem. *29*, 587–593.

HALL, R. J., TRINDER, N., and GIVERNS, D. I. 1973. Use of 2,4,6-trinitro-benzenesulfonic acid for the determination of available lysine in animal protein concentrates. Analyst *98*, 673–686.

HANSEN, N. G. 1975. Comparison of chemical methods of protein quality evaluation with biological values determined on rats. Z. Tierphysiol. Tiernaehr. Futtermitt. *35*, 302–310.

HASHIZUME, K., NAKAMURA, N., and WATANABE, T. 1975. Influence of ionic strength and conformation changes of soybean proteins caused by heating and relationship of its conformation changes to gel formation. Agr. Biol. Chem. *39*, 1339–1347.

HAYNES, R., OSUGA, D. T., and FEENEY, R. E. 1967. Modification of amino groups in inhibitors of proteolytic enzymes. Biochemistry *6*, 541–547.

HIGASHINO, K., FUJIOKA, M., and YAMAMURA, Y. 1971. The conversion of L-lysine to saccharopine and α-aminoadipate in mouse. Arch. Biochem. Biophys. *142*, 606–614.

HOLSINGER, V. H., and POSATI, L. P. 1975. Chemical estimation of available lysine in dehydrated dairy products. *In* Protein Nutritional Quality of Foods and Feeds, Part 1, M. Friedman (Editor). Marcel Dekker, New York.

HURRELL, R. F., and CARPENTER, K. J. 1974. Mechanisms of heat damage in proteins. 4. The reactive lysine content of heat-damaged material as measured by different ways. Brit. J. Nutr. *32*, 589–604.

HURRELL, R. F., and CARPENTER, K. J. 1977. Nutritional significance of cross-link formation in food processing. *In* Protein Crosslinking: Nutritional and Medical Consequences, M. Friedman (Editor), Plenum Press, New York. 172nd Am. Chem. Soc. Meet., 1976.

HURRELL, R. F., CARPENTER, K. J., SINCLAIR, W. J., OTTERBURN, M. S., and ASQUITH, R. S. 1976. Mechanism of heat damage in proteins. 7. The significance of lysine-containing isopeptides and of lanthionine in heated proteins. Brit. J. Nutr. *35*, 383–395.

HUSSAIN, L. A. 1974. Comparison of methods for the determination of available lysine value in animal protein concentrate. J. Sci. Food Agr. *25*, 117–120.

HUTZLER, J., AND DANCIS, J. 1975. Lysine-ketoglutarate reductase in human tissues. Biochim. Biophys. Acta *377*, 42–51.

ITO, T., and ANDO, N. 1974. The effect of the modification of amino groups with β-naphthoquinone-4-sulfonic acid on some physicochemical properties. Agr. Biol. Chem. *38*, 1423–1432.

ITO, T., and ANDO, N. 1975. The effect of the modification of amino groups with β-naphthoquinone-4-sulfonic acid on heat denaturation of myosin. Agr. Biol. Chem. *39*, 575–587.

JERING, H., SCHORP, G., and TSCHESCHE, H. 1974. Enzymic cleavage of the N-ϵ-acetyl protecting group from lysine in proteins (Kunitz trypsin-kallikrein inhibitor) in vitro and in vivo. Z. Physiol. Chem. *355*, 1129–1134.

KAKADE, M. L., and LIENER, I. E. 1969. Determination of available lysine in foods. Anal. Biochem. *27*, 237–280.

KNIPFEL, J. E., BOTTING, H. G., and MCLAUGHLAN, J. M. 1975. Nutritional quality of several proteins as affected by heating in the presence of carbohydrates. *In* Protein Nutritional Quality of Foods and Feeds, Part 2, M. Friedman (Editor). Marcel Dekker, New York.

KOENIG, N. H., MUIR, M. W., and FRIEDMAN, M. 1973. Properties of wool modified with vinyl compounds. Textile Res. J. *43*, 682–688.

KRISTOL, D. S., KRAUTHEIM, P., STANLEY, S., and PARKER, R. C. 1975. The reaction of p-nitrophenyl acetate with lysine and lysine derivatives. Bioorganic Chem. *4*, 299–304.

KRULL, H., and FRIEDMAN, M. 1967A. Anionic polymerization of methyl acrylate to protein functional groups. J. Polym. Sci. *A-1*, *5*, 2535–2546.

KRULL, L. H., and FRIEDMAN, M. 1967B. Ion-exchange separation and quantitiative determination of dimethyl sulfoxide. J. Chromatog. *26*, 336–338.

KUPCHAN, S. M. 1974. Selective alkylation: A mechanism of tumor inhibition. Intra-Science Chem. Rep. *8*, 57–66. Fed. Proc. *33*, 2288–2295.

LECLERC, J., and BENOITON, L. 1968. Further studies on ϵ-lysine acylase. The ϵ-N-acyl-diamino acid hydrolase activity of avian kidney. Canad. J. Biochem. *46*, 471–475.

LEHNINGER, A. L. 1970. Biochemistry. Worth Publishing Co., New York.

LEIN, K-A., SCHON, W. J., and BRUNCKHORST, K. 1973. The quality of cereal proteins from the plantbreeder's viewpoint. Qual. Plant. *23*, 311–324.

MCELWAIN, M. D., RICHARDSON, R., and AMUNDSON, C. H. 1975. Some functional properties of succinylated single cell protein concentrate. J. Milk Food Technol. *38*, 521–526.

MACGREGOR, J. T., and CLARKSON, T. W. 1974. Distribution, tissue binding and toxicity of mercurials. *In* Protein-Metal Interactions, M. Friedman (Editor). Plenum Press, New York.

MATHESON, N. A. 1968. Available lysine. 2. Determination of available lysine in feedingstuffs by dinitrophenylation. J. Sci. Food Agr. *19*, 496–502.

MAURON, J., FINOT, P. A., and MOTTU, F. 1975. Treatment of foods poor in lysine. Ger. Offen. 2,423,089. Chem. Abst. *84*, 15898g.

MEANS, G. E., and FEENEY, R. E. 1971. Chemical Modification of Proteins. Holden Day, Inc., San Francisco, Calif.

MELNICK, D., and OSER, B. L. 1949. The influence of heat-processing on the functional and nutritive properties of protein. Food Technol. *3*, 57–71.

MERTZ, E. T. 1975. Breeding for improved nutritional value of cereals. *In* Protein Nutritional Quality of Foods and Feeds, Part 2, M. Friedman (Editor). Marcel Dekker, New York.

MERTZ, E. T., MISRA, P. S., and JAMBUNATHAN, R. 1975. Rapid ninhydrin color test for screening high-lysine mutants of maize, sorghum, barley, and other cereals. Cereal Chem. *51*, 304–307.

MILLARD, M. M., and FRIEDMAN, M. 1976. X-ray photoelectron spectroscopy of BSA and ethyl vinyl sulfone modified BSA. Biochem. Biophys. Res. Commun. *70*, 451–458.

MILLER, R., and GRONINGER, H. S., JR. 1976. Functional properties of enzyme-modified acylated fish protein derivatives. J. Food Sci. *41*, 268–272.

MILNER, C. K., and WESTGARTH, D. R. 1973. A collaborative study on the determination of available lysine in feedingstuffs. J. Sci. Food Agr. *24*, 873–882.

MISRA, P. S., MERTZ, E. T., and GLOVER, D. V. 1975. Studies of corn proteins. 8. Free amino acid content of opaque-2 double mutants. Cereal Chem. *52*, 844–848.

MUHLRAD, A., AJTAI, K., and FABIAN, F. 1970. Reaction of myosin with salicylaldehyde. Biochim. Biophys. Acta *205*, 342–354.

NAKAI, T., OHTA, T., HATSUMI, M., and HORIKISHI, S. 1975. Isolation of N-ϵ-(2-furoylmethyl)-L-lysine (furosine) from an acid hydrolyzate of casein heated with glucose. Agr. Biol. Chem. *39*, 2421–2422.

NAKAMIYA, T., MIZUNO, H., MEGURO, T., RYONO, H., and TAKINAMI, K. 1976. Antibacterial activity of lauryl ester of DL-lysine. J. Ferment. Technol. 54, 369–373.

NAKAYA, K., HORINISHI, H., and SHIBATA, K. 1967. Monochlorotrifluoroquinone as a new reagent for discriminating of amino groups. J. Biochem. 61, 337–344.

NEUBERGER, A., and SANGER, F. 1944. The availability of ϵ-acetyl-D-lysine and ϵ-methyl-DL-lysine for growth. Biochem. J. 38, 125–129.

NIWA, E., and KITAYAMA, M. 1975. Studies on the mechanism of strengthening fish gel by basic amino acids. Agr. Biol. Chem. 39, 2075–2076.

OKUMURA, J., and TASAKI, I. 1973. Digestibility, biological value and 'available' lysine content of some protein concentrates for poultry. Japan Poultry Sci. 10, No. 1, 37–40.

OPPENHEIMER, H., BARANY, K., HAMOIR, G., and FENTON, J. 1967. Succinylation of myosin. Arch. Biochem. Biophys. 120, 108–118.

OSBORNE, T. B., and MENDEL, L. B. 1914. Amino acids in nutrition and growth. J. Biol. Chem. 17, 325–349.

OSNER, R. C. 1975. Nutritional and chemical changes in heated casein. 3. A comparison of the alkali-soluble and insoluble fractions. J. Food Sci. 10, 533–539.

OSTROWSKI, H., JONES, A. S., and CADENHEAD, A. 1970. Availability of lysine in protein concentrates using Carpenter's method and a modified Silcock method. J. Sci. Food Agr. 21, 103–107.

PAULIS, J. W., and WALL, J. S. 1975. Protein quality in cereals evaluated by rapid estimation of specific protein fractions. In Protein Nutrition Quality of Foods and Feeds, Part 1, M. Friedman (Editor). Marcel Dekker, New York.

PERLMAN, D. 1975. Lysine and lysine analog potentiation of antibiotic and antimicrobial activity. J. Antibiotics 2, 997–999.

PUSKI, G. 1975. Modification of functional properties of soy proteins by proteolytic enzyme treatment. Cereal Chem. 52, 655–664.

RE, G. G., and KAPER, J. M. 1975. Chemical accessability of tyrosyl and lysyl residues in turnip yellow mosaic virus capsids. Biochemistry 14, 4492–4497.

RHODES, A. P. 1975. A comparison of two rapid screening methods for the selection of high-lysine barleys. J. Sci. Food Agr. 26, 1793–1805.

ROACH, A. G., SANDERSON, P., and WILLIAMS, D. R. 1967. Comparison of methods for the determination of available lysine value in animal and vegetable protein sources. J. Sci. Food Agr. 18, 247–279.

SANDLER, L., and WARREN, F. L. 1974. Effect of ethyl chloroformate on the dye binding capacity of protein. Anal. Chem. 46, 1870–1872.

SARWAR, G., SHANNON, D. W. F., and BOWLAND, J. P. 1975. Effects of processing conditions on the availability of amino acids in soybean and rapeseed proteins when fed to rats. J. Inst. Can. Sci. Technol. Aliment. 8, 137–141.

SCHEELE, R. B., and LAUFFER, M. A. 1969. Restricted reactivity of the ϵ-amino groups of tobacco mosaic virus protein toward trinitrobenzenesulfonic acid. Biochemistry 8, 3597–3603.

SCHWIMMER, S. 1975. Effects of enzymes on nutritional quality and availability of proteins. In Protein Nutritional Quality of Foods and Feeds, Part 2, M. Friedman (Editor). Marcel Dekker, New York.

SEELEY, J. H., and BENOITON, N. L. 1969. The carboxypeptidase β-catalyzed hydrolysis of poly-ϵ-N-methyl-L-lysine and poly-ϵ,N,ϵ-N-dimethyl-L-lysine. Biochem. Biophys. Res. Commun. 37, 771–776.

SEELEY, J. H., and BENOITON, N. L. 1970. Effect of N-methylation and chain length on kinetic constants of trypsin substrates. ϵ-*N*-Methyl-lysine and homolysine derivatives as substrates. Can. J. Biochem. *48*, 1122–1131.

SHENHAV, H., RAPPOPORT, Z., and PATAI, S. 1970. Nucleophilic attacks on carbon-carbon double bonds. Part 12. Addition of amines to electrophilic olefins and reactivity order of the activating group. J. Chem. Soc. *(B)*, 469–476.

SNOW, J. T., FINLEY, J. W., and FRIEDMAN, M. 1975A. A kinetic study of the hydrolysis of *N*-acetyldehydroalanine and *N*-acetyldehydroalanine methyl ester. Int. J. Peptide Protein Res. *7*, 461–466.

SNOW, J. T., FINLEY, J. W., and FRIEDMAN, M. 1975B. Relative reactivities of sulfhydryl groups with *N*-acetyldehydroalanine and *N*-acetyldehydro-alanine methyl ester. Int. J. Peptide Protein Res. *8*, 57–64.

SYNGE, R. L. M. 1975. Interactions of polyphenols with proteins and plant products. Qual. Plant. *24*, 337–350.

SYNGE, R. L. M. 1976. Damage to nutritional value of plant proteins by chemical reactions during storage and processing. Qual. Plant, In press.

TANAKA, M., KIMIAGAR, M., LEE, T. C., and CHICHESTER, C. O. 1977. Effect of the Maillard browning reaction on nutritional quality of protein. *In* Protein Crosslinking: Nutritional and Medical Consequences, M. Friedman (Editor). Plenum Press, New York. 172nd Am. Chem. Soc. Meet., 1976.

TANAKA, M., LEE, T. C., and CHICHESTER, C. O. 1975. Nutritional consequences of the Maillard reaction. J. Nutr. *105*, 989–994, and references cited therein.

TERRELL, R. N., and STANIEC, W. P. 1975. Comparative functionality of soy proteins used in commercial meat food products. J. Am. Oil Chem. Soc. *52*, 263A–266A.

TOULOUPIS, C., and VASSILIADIS, A. 1977. Lysinoalanine formation in wool after treatments with some phosphate salts. *In* Protein Crosslinking: Nutritional and Medical Consequences, M. Friedman (Editor). Plenum Press. New York.

VAGHEFI, S. B., MAKDANI, D. D., and MICKELSEN, O. 1974. Lysine supplementation of wheat proteins. A review. Am. J. Clin. Nutr. *27*, 1231–1246.

VICKERY, H. B. 1972. The history of the discovery of the amino acids. 2. A review of amino acids described since 1931 as components of native proteins. Adv. Prot. Chem. *26*, 81–170.

VICKERY, H. B., and LEAVENWORTH, C. S. 1928. A note on the crystalliza-tion of free lysine. J. Biol. Chem. *76*, 437–443.

VICKERY, H. B., and SCHMIDT, C. L. A. 1931. The history of the discovery of the amino acids. Chem. Rev. *9*, 169–318.

WAIBEL, P. E., and CARPENTER, K. J. 1972. Mechanism of heat damage in proteins. 3. Studies with ϵ-(γ-L-glutamyl)-L-lysine. British J. Nutr. *27*, 509–515.

WALKER, H. G., KOHLER, G. O., KUZMICKY, D. D., and WITT, S. C. 1975. Problems in analysis for sulfur amino acids in foods and feeds. *In* Protein Nutritional Quality of Foods and Feeds. Part 1. M. Friedman (Editor). Marcel Dekker, New York.

WALL, J. S., FRIEDMAN, M., BECKWITH, A. C., KRULL, L. H., and CAVINS, J. F. 1968. Chemical modification of wheat gluten proteins and related compounds. J. Polym. Sci. *C, 24*, 147–161.

WANG, S. S., and CARPENTER, F. H. 1968. Kinetic studies at high pH of the trypsin-catalyzed hydrolysis of *N*-ϵ-benzoyl derivatives of L-arginamide, L-lysinamide, and S-2-aminoethyl-L-cysteinamide and related compounds. J. Biol. Chem. *243*, 3702–3710.

WEIL, L., and TELKA, M. 1957. Tryptic digestion of native and chemically modified α-lactalbumin. Arch. Biochem. Biophys. *71*, 473–478.

WHITFIELD, R. E., and FRIEDMAN, M. 1972. Chemical modification of wool with dicarbonyl compounds in dimethyl sulfoxide. Text. Res. J., *42*, 344–347.

WOODARD, C. J., SHORT, D. D., ALVAREZ, M. R., and REYNIERS, J. 1975. Biologic effects of N,ϵ-(DL-2-amino-2-carboxyethyl)-L-lysine, lysinoalanine. *In* Protein Nutritional Quality of Foods and Feeds, Part 2. M. Friedman (Editor). Marcel Dekker, New York.

WU, Y. V., CLUSKEY, J. E., KRULL, L. H., and FRIEDMAN, M. 1971. Some optical properties of S-β-(4-pyridylethyl)-L-cysteine and its wheat gluten and serum albumin derivatives. Canad. J. Biochem. *49*, 1042–1049.

ZOLTOBROCKI, M., KIM, J. C., and PLAPP, B. V. 1974. Activity of liver alcohol dehydrogenase with various substituents on the amino groups. Biochemistry *13*, 899–903.

19

Protein Texturization

F. E. Horan

In a treatise on the "fundamental aspects of proteins basic to foods," it is understandable that the majority of contributions would come from the areas where the bulk of the fundamental scientific investigations is being carried out, namely, in the academic institutions and the government laboratories. This volume attests to that fact.

Still, there might be little justification in continuing these investigations if history does not record that in some way and at some time they were beneficial to mankind. Therefore, the approach of the present discussion is to take the liberty of broadening the scope of the topic to introduce a viewpoint from industry which may be basic to the total concept of developing protein foods.

Emphasis is on Foods

One particular word is extremely important and it is "foods." In a recent publication Anton (1976) stated the situation well: "But the demand is not for more protein at lower cost; accurately, it is a demand for traditional foods, which contain the protein, at lower cost." Over the last several years there have been countless meetings, discussions and publications, both in the scientific realm as well as in the lay world, in which proteins are featured, but sometimes the essential consideration of proteins as *part* of a completely acceptable food system is lost. To explain, there is little value in referring to the pounds of protein per acre from any source unless this protein is ready to be eaten as a food.

To be sure, most of the fundamental aspects of the major types of food proteins, from a scientific point of view, are covered in this monograph in considerable depth. From an industrial point of view, it is almost axiomatic that much reliance must be placed upon the results of basic scientific studies carried out in the university and government laboratories. The "fundamental" aspect for industry is to promote the technology from the laboratories into commercially accepted foodstuffs in the marketplace. This presentation concerns itself with some of the factors involved in developing so-called textured protein foods.

Meaning of Texture

This word "texturization" needs clarification. Texture is a distinctive and identifying characteristic of foods, but it is one about which the consumer generally exhibits very little spontaneous awareness. Texture for the most part is taken for granted, being considered an integral part of a food. As pointed out by Szczesniak (1971), "an average consumer may have difficulty in visualizing the concept of texture *per se*." Likewise Corey (1970) has observed: "We 'know' without words all the nuances of texture. Difficulties arise only when we wish to describe these forces (and feelings other than flavor sensed in the mouth while chewing a piece of food) further either in qualitative or in quantitative terms."

In recent years the food technologist has brought into common usage the term "textured vegetable protein." It is now difficult to ascribe the origin of this term to any one investigator or to any one event. One is abruptly faced with the need for exactness in definition when trying to explain this concept to a person from another country in the language of that country, *e.g.*, to a Brazilian in Portuguese; or to the appropriate body in the Food and Drug Administration in an attempt to establish a Standard of Identity for this class of food products.

Even the patent lawyers struggle within their vocabulary to protect the positions of inventors in this field. In a commercially important patent (Atkinson 1970) an approach is to invent an entirely new word, *i.e.*, "plexilamellar" to attempt to delineate the uniqueness of a "textured vegetable protein" product. Quoting from this patent: "The protein-containing product of *plexilamellar* structure is a tough, resilient, dry to semi-dry, open celled, funicular structure made up of interlaced, interconnected funiculi of varying width and thickness." It is questionable whether this legal terminology offers the food scientist a greater insight into the concept of "textured vegetable protein."

It is generally conceded among food scientists that there is no accepted definition of texture. Szczesniak (1963) has developed a system for classification of textural characteristics of foods based on fundamental rheological principles. These characteristics are defined and classified into mechanical and geometrical qualities, as well as those related to the moisture and fat content of the food. Mechanical properties are related to the reaction of the food to stress, and include such characteristics as hardness, chewiness, and adhesiveness. The geometrical characteristics refer more specifically to the arrangement of the physical constituents within the food and deal with particle size, shape, and orientation; some examples are the descriptive terms of chalky, lumpy, gritty, flaky, pulpy, or crystalline. In line with this classification Bourne (1975) has introduced the term "textural properties" which emphasizes that texture cannot really be described completely by the observation of only one sensory characteristic.

In a review on the sensory assessment of food texture Abbott (1972) describes a three-phase texture profile involved with the mechanical properties in the ingestion of food: (1) the first few chews encounter the properties of hardness, brittleness, and viscosity; (2) in the subsequent masticatory stage up to the time of swallowing the characteristics of gumminess, chewiness, and viscosity appear; (3) in the third or residual stage the condition of the residue is associated with general mouthfeel. In addition to the mechanical properties, some of the geometrical properties of food particle size and shape may exist during each of the previous three stages. Still other characteristics affecting mouthfeel are the moisture and fat content of food.

The description of the textural properties of a food becomes a communication problem in which words must be selected which have the same meaning to all individuals who are interested in being conscious of the textural characteristics of a food. The ideal would be to minimize routine subjective assessment in terms of an instrument which would produce numbers capable of being quantified in an objective manner. Bourne (1975) has summarized many of the instrumental measuring devices which are employed in the food industry today, including penetrometers, compressimeters, viscometers, consistometers, shearing devices, and masticometers. Ultimately, however, the machine must be calibrated against human organoleptic opinions relating word descriptions to mouthfeel, and therein lies the inherent difficulty in obtaining accurate and meaningful assessments of food texture.

In this connection Corey (1970) raises a very fundamental question: "Can texture in fact be measured external to the human being?" Does texture have an objective existence or is it the result of the interaction of man with the mechanical and geometrical properties of the food?

Upon reflection one realizes that the textural properties of foods are primarily related to the internal structure of these foods. Even though natural foods such as meat, fruit, and vegetables have predetermined internal structures which govern their textural properties, it is true that man has historically modified the structures of many foods to create more desirable textures and thus make the foods more palatable.

The structure of a raw egg is changed to perhaps a more desirable one by boiling the egg in water. Wheat flour is consumed more readily in the form of bread, crackers, cookies, or pasta products. Corn meal undergoes a structural change when transformed into a tortilla. The great variety of dry breakfast cereals in the marketplace today gives forceful testimony to the importance of structure, and the resulting texture, in having the commodities of corn, wheat, oats, and rice consumed as human foods. It is highly doubtful that a beef steak would merit a top billing on a restaurant's menu if after having been broiled properly it were freeze-dried and ground to a fine powder and then served to the customer as a high quality protein food. When the food scientist talks about texturizing foods today, he is not embarking on a revolutionary journey. His path is evolutionary in that he is merely extending the practice of changing the structure of foods, a practice which actually began with the origin of man himself.

Why Protein Foods?

In the realm of the so-called protein foods textural changes have been made as mankind has proceeded from one generation into the next. One need only consider the number of recipes available for preparing eggs in a variety of forms which are differentiated by structure and texture. Milk has been transformed into cheeses. Beef and pork have been reconstituted into an assortment of patties, loaves and sausages, all of which have different textural characteristics from the source material. The question then arises: Why is there at this time so much interest in the texturization of vegetable protein? To attempt to answer this question prompts one to delve more deeply into the protein food situation in the world today in an attempt to envisage the years ahead.

The world protein situation can oftentimes be blurred with statistics. The recommended daily dietary allowance (RDA) for protein, as established by the National Academy of Sciences (1974), is 56 g (based on a high percentage of animal protein) which should be adequate for the normal, healthy, human adult. Statistical data (United Nations 1974) exist showing that for the world the per capita food energy supply is 100% of the requirement, and the per capita protein available is 69 g per day. Of course, the food supply which is theoretically available and the actual amounts consumed on an individualistic basis do not balance because of unequal distribution of the food. It is a fact that a significant percentage of the world's population is underfed resulting in malnutrition, which to a large extent is clearly linked to low incomes and poverty.

In many of the developing countries, low incomes necessitate diets composed largely of cereals. When considered in terms of income elasticity of demand[1] (USDA Economic Research Service 1974), as incomes rise from very low levels the consumption of grain increases rapidly, giving a high positive income elasticity. As contrasted with North America, where incomes are relatively high, the elasticity of demand for grain for direct consumption is negative, and the per capita consumption of cereals actually decreases when incomes rise. However, income elasticity of demand for meat in the developed countries is relatively high; comparatively speaking, it is still higher in the developing regions, but low-income levels place these animal protein foods out of reach for most consumers.

Throughout most of the world diets are such that the percent of calories from protein ranges between 9-12%. But the percent of protein from animal sources may vary from a low of about 10% in some of the developing countries to a high of about 70% in the most developed countries. In comparing the average available supplies of edible proteins throughout the world one observes a distinct difference in the ratio of vegetable proteins to animal proteins in the developing and the more developed countries, as indicated in Table 19.1 (Autret 1970).

In view of the fact that animal protein sources are relatively small in many of the developing regions of the world, the food scientist is prompted to conclude that there has to be a growing need for foods

TABLE 19.1
PROTEIN SUPPLIES THROUGHOUT THE WORLD (1963–65)

Source	Percentage of Total Protein	
	Vegetable	Animal
Developing regions	81	19
Developed regions	46	54
World	68	32

[1] The income elasticity of demand is a measure of the percentage increase in the quantity demanded of a commodity associated with a one percent increase in consumer income.

such as textured vegetable protein products. No doubt, the need does exist, but a sizeable market cannot be there until the income level and the purchasing power of the individual consumer is increased manyfold over what it is today.

In a developed country it is difficult to build a strong case for the textured vegetable protein products solely in terms of an urgent need for more protein in the diet. The level of protein available from the food supply in the U.S. has always been abundant in relation to the nutritional needs of the population. The protein situation is summarized in Table 19.2 (Friend and Marston 1975) and Table 19.3 (USDA 1975A).

TABLE 19.2

PROTEINS AVAILABLE FOR U.S. CIVILIAN CONSUMPTION

(PER CAPITA PER DAY)

Year	Grams	Percentage Animal
1957–59	95	67
1974	99	69

TABLE 19.3

ANNUAL U.S. PER CAPITA CONSUMPTION OF SOME MEATS

(POUNDS)

Year	Beef	Pork	Chicken and Turkey
1965	100	59	41
1974	117	67	50

In general, the protein resources available to the individual American consumer for the daily diet consisted of furnishing the following average percentages of the total protein in 1974: Meat and poultry, 38%, of which beef was 18%; dairy products, 22%; flour and cereal products, 18%. It is interesting to note that in 1974 the percentage of dietary protein obtained from beef was about the same as that derived from flour and cereal products. A significant shift in the composition of the American diet is observed by referring back to 1909–13 when only 10% of the protein came from beef and about 36% was furnished by flour and the cereal products. These data point out the increasing, strong preference of the U.S. consumer for foods from animal sources. This trend is represented in Fig. 19.1 (USDA 1975B).

If the supply of animal and vegetable protein foods is adequate and these foods are available to the consumer, then where does a product such as textured vegetable protein fit in the marketplace? It appears that from a nutritional viewpoint the typical American

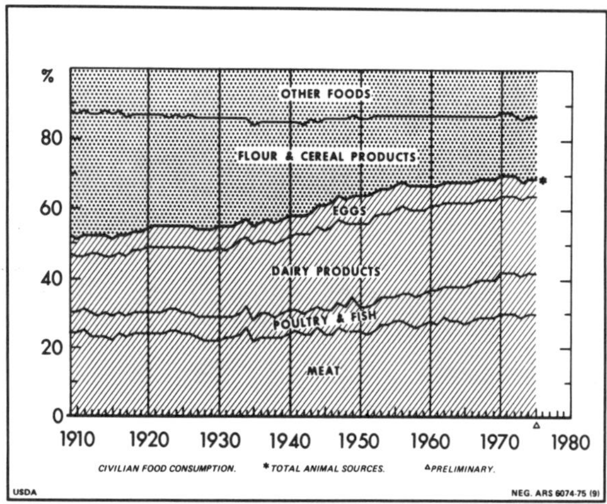

FIG. 19.1. SOURCES OF PROTEIN

consumer is not lacking protein in the daily diet. The factor which could change this situation is that the animal protein might become too expensive. Insofar as textured vegetable protein products are to be used as meat extenders, the economics of cutting costs in preparing daily meals becomes the overriding force.

In many of the developing countries at least 50% of the personal income is spent on food. In the U.S., the amount of the disposal income spent on foods historically has been under 20%. The breakdown of disposable income in Fig. 19.2 (USDA 1975B)

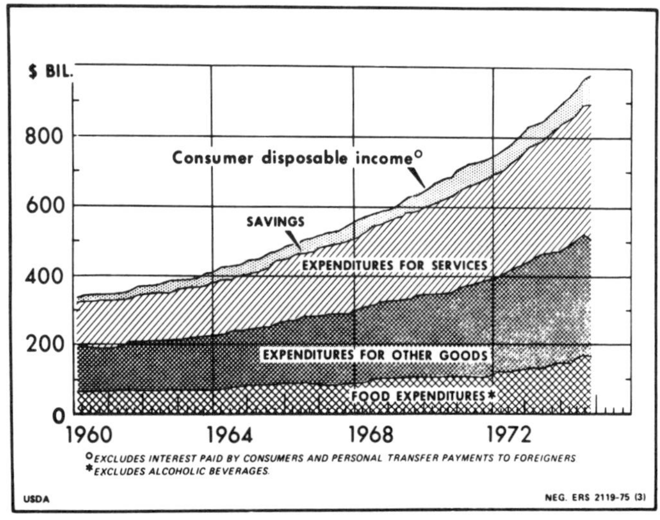

FIG. 19.2. INCOME AND EXPENDITURES

illustrates how additional income is disposed of in the American way of life, and how the expenditure for food is held at a comparatively low percentage level.

The facts at hand are that the U.S. consumer continues to increase the annual per capita consumption of beef, *i.e.*, from 100 lb in 1965 to 117 lb in 1974. The retail price of beef continues to increase: from $0.85/lb in 1968 to $1.46 in 1975. The production of beef likewise continues to grow (USDA 1975A): from 15 billion pounds in 1960 to 23 billion in 1974. Paralleling these changes has been a similar increase in the consumer's disposable income. Nevertheless, when all of these factors are taken into consideration it can be shown that there has been little change in the percentage of income spent for beef over the last 20 yr. This is summarized in Table 19.4 (USDA 1976).

TABLE 19.4
DISPOSABLE INCOME SPENT FOR BEEF

Year	Disposable Income	(%)	Consumption (lb) (Per Capita)
1955	1,666	2.57	70
1960	1,937	2.66	85
1965	2,436	2.42	100
1970	3,376	2.46	114
1972	3,843	2.54	116
1973	4,295	2.56	110
1974	4,623	2.59	117
1975	5,040	2.57	119

An important observation can be made about the time interval 1972–74. In 1972, the average per capita consumption of beef was 116 lb, the average retail price of hamburger was $0.75/lb, and the proportion of disposable income spent on beef was 2.54%.

In 1973 the average retail price of hamburger increased sharply to $1.00/lb. This resulted in a decrease in the per capita consumption of beef to 110 lb, but the percentage of disposable income allocated to beef did not decrease; it actually increased slightly to 2.56.

Then in 1974 the price of hamburger receded to $0.85/lb, the per capita consumption of beef returned to its upward path and reached 117 lb, and the percentage of disposable income increased somewhat more to 2.59.

It was during the high beef price period of 1973 that blends of textured soy protein and ground beef entered the retail food market. Gallimore (1976) collected and analyzed data dealing with the market penetration of soy-beef blend in three cooperating food chains over a period of 46 weeks, beginning in May, 1973. The blend's market share was about 24% of all ground beef sold during this period, amounting to at least 15 million pounds of the dry textured soy protein material. This study showed that consumers would purchase a meat product containing soy protein if there were sufficient price differential between the all-meat and the blend product.

However, it also indicated that the soy-beef blend sales were very responsive to changes in the price of regular ground beef. With more plentiful supplies of beef and related lower prices after the latter half of 1974, demand for the soy-beef blends in the retail market declined to the extent that the market share for the blends was only about 8–10% of the total ground beef sales in November 1975. This corresponded to a price differential of less than 10 cents per pound, as compared with 20 cents per pound during the high-price period for beef in 1973.

With beef at a low retail price there is little incentive for the consumer to use the soy-beef blend. However, the food chains feel that these blends are now established items and that sales will increase when retail beef prices increase. If and when meat supplies are restricted, the soy-meat blends provide a way of stretching the meat which is available. Likewise, if and when the price of meat products resumes its upward trend in the years ahead, budgetary pressures will prevail so that the average consumer will turn to the blended products in attempting to keep the percentage of disposal income spent for meat under control. In other words, the experiences gained over the past few years demonstrate that soy-meat blends are viable products in the marketplace.

EVOLUTION OF TEXTURED PROTEIN FOODS

Most human beings during their first days, months, or even sometimes years have been sustained on a food protein system provided by their mothers, *i.e.*, human milk. As compared with the development of new foods by the food technologist, human milk eliminates all of the consumer testing, marketing, or distribution problems, and in addition to its very important nutritional characteristics, human milk furnishes anti-infective, contraceptive, and economic advantages which render it a precious natural resource (Jelliffe and Jelliffe 1975). Several thousands of years ago mankind learned to supplement human milk with cow's milk, even though subtle textural differences in the form of mouthfeel exist, and this change in eating habits has been accepted as a sound evolutionary step in the dietary development of the human in most parts of the world.

The First Protein Food—Human Milk

If one desires to envisage what the next step might be if mammalian milks were in short supply, interesting speculations can be drawn by comparing the proximate chemical compositions of material from other natural resources, such as soybeans. As listed in Table 19.5 (Watt and Merrill 1963) one might hastily conclude that

TABLE 19.5
PROXIMATE COMPOSITION OF MILKS AND SOYBEANS

(Percentages on Moisture-Free Basis)

Material	Proteins	Fats	Carbohydrates
Human milk	8	27	64
Whole cow's milk	28	28	39
Soybeans	40	22	32

the soybean could conceivably be a better nutrient than human milk, in that it affords about five times as much protein.

Since protein plays a critical role in the nutriture of the infant, then why have soybeans not gained a greater stature in the direct feeding of the human infant? Of course, the answer is that the proximate analyses present too simple a picture, and we must consider further the quality of the protein, the nature of the fat, the possible disadvantages of the carbohydrates, as well as many other unidentified growth factors and delicate textural properties present in mammalian milks.

After the infant is weaned from its mother there is a transition to the partaking of a mixture of foods rather than having the relatively simple human milk as the sole source of human food. This transition proceeds from milk to gruels to pureed vegetables, meat and fruit and eventually to a great variety of foods with various types and degrees of texture. Most human beings are constantly striving for variety in their diets, and the gratification of this desire is generally directly proportional to the personal disposable income. Likewise, in the infant the craving for greater variety is probably related to the appearance of teeth, whereas in the aged the spectrum of food texture narrows due in part to the disappearance of teeth.

One does not normally eat just one particular protein material day in and day out during an entire life span, but does experience many food mixtures. Therefore, even if one protein source may be somewhat deficient in a specific amino acid, it is conceivable that a mixture of foods in the diet may result in mutual complementation of the available proteins to yield a result with higher protein quality.

An outstanding example of this type of nutritionally blended food is Wheat-Soy Blend (WSB), produced commercially for the U.S. Government PL-480 Title II program to assist in feeding the undernourished in developing countries (Horan 1973). WSB is a blended flour product consisted primarily of wheat (wheat protein concentrate and bulgur) and defatted soy flour. An examination of the limiting amino acids in wheat protein (lysine) and in soy protein (methionine plus cystine) illustrates the complementary effects possible in blending these materials, as indicated in Table 19.6. The

TABLE 19.6
LYSINE AND METHIONINE PLUS CYSTINE IN WHEAT AND SOY
(PERCENTAGE OF FAO REFERENCE)

Amino Acid	Bulgur	Wheat Protein Concentrate	Defatted Soy Flour
Lysine	63	94	152
Methionine plus cystine	101	88	50

soy flour improves the wheat by supplying additional lysine, and the wheat assists the soy flour in overcoming its limitations with the sulfur amino acids.

The functional properties of WSB are such that it is limited primarily to a food preparation in the form of a gruel. Other types of nutritionally blended cereal foods have appeared on the scene in recent years, and some of these are listed in Fig. 19.3, showing a variety of textural properties.

An extremely interesting manner to improve protein nutrition in the diet without causing any change in eating habits is the addition of soy flour to bakery products. Technological progress now permits the addition of soy flour up to 12% in white bread. This not only increases the quantity of protein in the bread 50% (from 8% to 12%) but also increases the nutritional quality of the protein about 100% (from a PER of 1.00 to 1.95).

Transforming a powdery protein material (a flour) into one which has texture, described as "chewiness" and having a fibrous character, adds attractiveness to it as a food. The most significant innovation in this area in recent years is that of textured vegetable protein products, which are finding applications primarily in food systems based on meat. At this point it might be worthwhile to review some of the highlights in the historical development of these products.

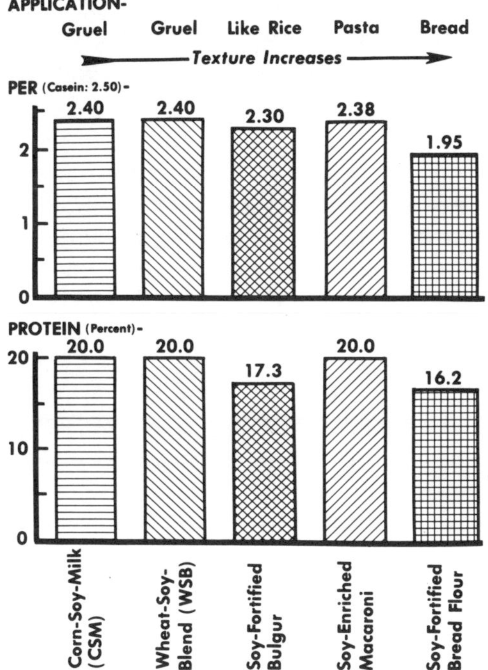

FIG. 19.3. COMMERCIAL CEREAL-SOY BLENDED PRODUCTS

Early Achievements with Textured Protein Products

In America the technology of textured vegetable protein products probably had its beginning with groups of people who chose to have nonflesh diets, commonly known as vegetarian. For the most part this mode of living was prompted by specific religious beliefs. One such group is the Seventh-Day Adventist denomination which in 1866 established the forerunner of the Battle Creek Sanitarium in Michigan under the administration of Dr. John H. Kellogg, who with his brother originated the breakfast cereal industry (Hardinge and Crooks 1963). One of the early patents in the field was issued to Kellogg (1907) in which the object of the invention was "to provide an improved food product which is very palatable and nourishing and one adapted for use as a meat substitute." The process consisted essentially of mixing wet wheat gluten and casein and vegetable oil in a shredding machine and then cooking the mix in a can under retorting conditions.

Two commercial firms have supplied the food demands of the Seventh-Day Adventists over the last 75 years: Loma Linda Foods in California, and Worthington Foods in Ohio (a division of Miles Laboratories, Inc.). These early protein foods were based largely on wheat gluten as the primary ingredient and took the form of patties or small cutlets, which were either canned in a broth or ground into a solid-pack "burger-type" product (Hartman 1966).

Some 25 years ago Wrenshall (1951) was awarded a patent which employed nonfat dry milk solids as the basic ingredient in making meatless food products closely simulating commercial comminuted meat preparations such as country sausage, bologna and luncheon loaves. The process consisted in heating an aqueous mixture of the milk solids and a texturizing agent (starchy materials), along with flavors and colors, to a temperature sufficiently high to coagulate and set the milk solids.

During the 1950s numerous publications appeared showing the emergence of the trend toward using soy protein in human food products. A detailed study by Frank and Circle (1959) pointed out the possibility of making nonmeat products resembling the frank-furter and bologna types from edible isolated soy protein. It was during this same period that Anson and Pader were issued a series of patents dealing with the utilization of soy protein in simulating meat products. One of their first patents (Anson and Pader 1957) introduced the concept of a gel precursor, which consisted of an aqueous paste solubilized out of isolated soy protein. This paste was then blended with a small amount of wheat flour dough, an oil emulsion, colors, and flavors and then heated to convert the mass into a chewy protein gel simulating meat products of varying kinds.

Further progress in the development of nonmeat luncheon loaves is described by Durst (1963) who employed mixtures of vegetable oil, wheat germ, yeast and egg white. At about the same time a patent was issued to MacAllister and Finucane (1963) which discloses a preparation of hamburger-like granules from mixtures of wheat gluten, soy flour and egg albumen. In an effort to expand the fabrication of meat-like products beyond the types represented by spreads and loaves, Rusoff *et al.* (1962) developed a relatively simple method for preparing fibrous protein materials having shred-like texture from soy protein; this method comprises rapidly heating the undenatured hydrated protein material to 300–400°F, with vigorous agitation, to orient and coagulate the protein and then cooling the resultant shred-like material as rapidly as possible.

More Recent Achievements with Textured Protein Products

The textured products designed for the limited "vegetarian" market for the most part were unknown to the majority of the U.S. population for many years. Advances in the technology for fabricating these nonmeat food products were made, and gradually the food industry began to take notice. A significant breakthrough was reported in a patent issued to Boyer (1954) in which he described a method of processing edible soy protein, first by solubilizing and then utilizing textile spinning techniques to orient the molecules into continuous filaments. Many modifications have been made on this process commercially, but in general the state of the art as summarized concisely by Rosenfield (1974) is an extension of the original work by Boyer. Some misconceptions have arisen with spun protein fibers in that it is not always recognized that these fibers do not serve as a principal ingredient in fabricated foods, but make up only 20–50% of the finished food products. An edible binder, which is heat coagulable, is essential to the fabrication of these spunfiber products. Fibers *per se* are not foods.

Over the past decade commercial products resembling bacon bits, ham and chicken chunks, an assortment of fish types, and breakfast links, patties, and slices have appeared in the marketplace. Perhaps the most sophisticated product, from a technological point of view is that of the bacon-slice analog, containing both the red and white parallel strips resembling the lean and fatty portions of the typical animal product. The fabrication steps consist of preparing separately a red mixture and a white mixture, which are alternately layered in a suitable container to form a loaf of distinct red and white layers to simulate the stratification of bacon. The loaf is then heat-treated, preferably in an autoclave, to set the binders, refrigerated, sliced, and packaged. A typical formulation is given in Table 19.7 (Corliss and

TABLE 19.7
COMPOSITION OF BACON ANALOG

Ingredient	Red (%)	White (%)
Spun fiber	18.0	1.5
Egg albumen	10.0	8.2
Tapioca starch	7.5	5.8
Water	42.5	42.3
Corn oil	6.7	25.7
Soy isolate	3.4	2.6
Carrageenan	0.5	0.2
Sodium caseinate	none	5.2
Colors, flavorings, seasonings	11.4	8.5

Furgol 1975). Other formulations can be found in the patent literature (Hartman 1967; General Foods Corp. 1975).

In addition to the spun-fiber types of textured protein products, which serve primarily as complete meat analogs, another class of materials is firmly established commercially, and these are known as thermoplastic extruded protein products which function chiefly as meat extenders. These products have been described voluminously in the literature over the past several years, but the essential elements of the general process are covered in detail by Flier (1976) and the chief characteristics of the products are recorded by Atkinson (1970).

ASSESSMENT OF TEXTURED PROTEIN FOODS

Scientific and Technological Aspects

As is often the case with new commercial developments in the food industry, the technological achievements surpass the scientific ones. This is true for the textured protein products: the spun-fiber as well as the thermally extruded types.

Nevertheless, the scientific studies of Kelley and Pressey (1966) serve as a significant attempt to understand the molecular changes involved in spinning soy protein into fibers. It is generally recognized that soy protein consists of four principal ultracentrifugal components, designated as 2, 7, 11, and 15S. In observing the change in these components after dissolving the protein at high pH and then spinning with coagulation into an acid bath, an hypothesis has been proposed to explain the spinning process from a molecular viewpoint. At first, the native globular protein is converted to unfolded polypeptide chains, along with undergoing some dissociation to material of lower molecular weight, as evidenced by the appearance of new ultracentrifugal components, essentially 3S. Secondly, the alkaline conditions favor sulfhydryl-disulfide interchange reactions in which new disulfide bonds can be formed. While these changes are occurring the fibers are stretched to cause molecular orientation, which promote additional hydrogen and ionic bonding, all of which eventually controls the structure and textural characteristics of the fiber.

The more recent work of Chiang and Sternberg (1974) gives further evidence that the soy protein structure in spun fibers is held together by hydrogen bonds as well as disulfide bonds. During storage of fibers, the amounts of free sulfhydryl groups (SH) decrease to correspond with an increase in disulfide bonds, and these molecular changes can be related to the fact that aged fibers retain less water, are more tough, and suffer a loss of elasticity. Disc electrophoresis measurements also corroborate this shift of sulfhydryl groups to disulfide bonds during the aging of the fibers.

The importance of the disulfide bonds in determining the final texture in spun fibers has been suggested by Stanley *et al.* (1972) as a result of their investigations in comparing textural properties of cooked meat with spun soy fibers by means of an Instron Universal

Testing Machine.[2] They found that spun soy fibers have a much higher breaking strength and greater break elongation compared to cooked meat. Nevertheless, cooked meat exhibited less stress relaxation than the soy fibers, which indicates that the meat is more viscoelastic, and that the soy fibers do not have the equivalent masticatory properties as the meat. This is probably related to the connective tissues found in meat. By learning more about the molecular interactions involved in spinning the soy protein, the food technologist should be able to fabricate products which will meet the specifications desired in the finished meat analog.

The technology involved in thermoplastic extrusion is even more of an art than that of the spinning technology. The state of the art of cooking extruders has been treated well by Harper and Harmann (1973). As the spinning process is related to technology developed in the textile industry, the thermoplastic extrustion process is widely employed in the plastics industry, and considerable extrusion theory has been developed for the simpler system where Newtonian flow exists. For the more complicated food systems such as vegetable proteins, however, the process operations are developed empirically.

Most cooking extruders use a single screw to feed, cook, and meter or pump the proteinaceous material. As indicated in Fig. 19.4,

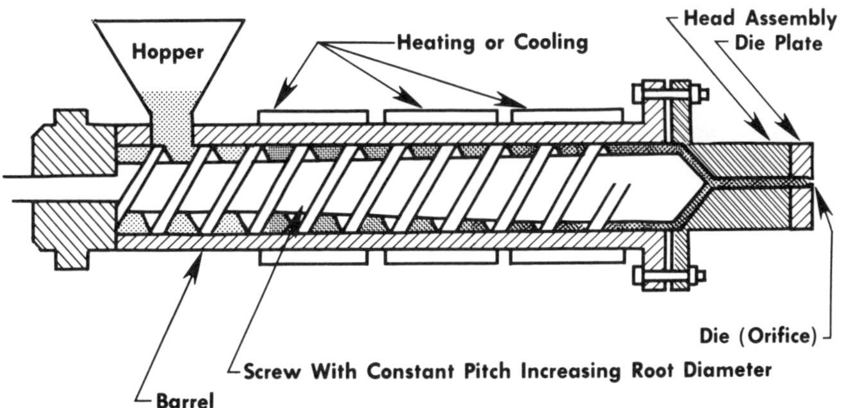

FIG. 19.4. DIAGRAM OF A TYPICAL COOKER-EXTRUDER

granular material is first tempered with water or steam, and then enters the feed section where the flights of the screw convey it down the extruder channel and begin to work it into a homogeneous dough. As the material moves through the transition section, it is thoroughly worked into a partially plastic dough by partially cooking under conditions of elevated temperature and pressure. In the pumping zone the material becomes completely plastic by the time it

---

[2]Instron Engineering Corp., 2500 Washington Ave., Canton, Mass.

reaches the exit die, at which point the pressure is rapidly released, and the material expands into irreversible structures upon cooling. The extruded product may then pass through a drier to further reduce its moisture content.

The food technologist is concerned with the processing variables which affect the quality of the finished product including density, amount of expansion, stability of the structure, and degree of toughness upon rehydration. In the final analysis, the production rate is essential in determining the economic viability of a particular product. For the most part the optimum selection of variables has to be made by cut-and-try methods. To illustrate this point in terms of relating the design of an extruder screw to rate of production, it is found that productivity is a function of the number of flights in the pump zone, as well as the depth of the flight (Fig. 19.5). Further

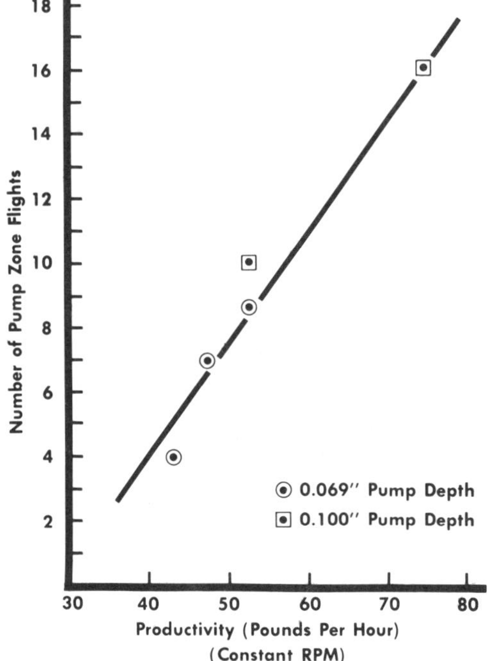

FIG. 19.5. PRODUCTIVITY AS A FUNCTION
OF AN EXTRUDER PUMP ZONE
(1.75 in. Prodex extruder)

experimental variables are discussed by Taranto *et al.* (1975) in a study concerning the extrusion texturization of cottonseed meal.

From a practical point of view a number of different types of commercial extruders may be investigated. The essential characteristics of a few typical ones are summarized in Fig. 19.6. A more

Manley

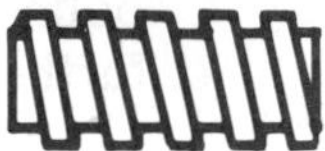

Constant pitch flights—deep & close
Root diameter increases slightly
Moderate to high RPM
Barrel with opposing threads
L:D ratio small (approx. 4)

Prodex

Constant pitch flights
Root diameter increases 4:1
Slow RPM
Smooth barrel
L:D = 24:1

Wenger

Decreasing pitch flights
Constant root diameter
Moderate RPM
Smooth or fluted barrel
L:D = 12:1

Anderson

Constant pitch interrupted flights
Constant root diameter
Moderate RPM
Barrel with "mixing pins"
L:D = 12:1

FIG. 19.6. SOME TYPES OF EXTRUDER SCREWS

generalized treatment of food extrusion is given by Rossen and Miller (1973).

The heat required for extrusion necessarily affects the molecular make-up of the vegetable protein involved. A study by Cumming *et al.* (1973) focused on the changes in the ultracentrifugal components of soy protein as a result of extruding defatted soy flour over a range of elevated temperatures. The results indicated that for good fiberization to occur, it is necessary to employ a temperature high enough to cause a redistribution of the water soluble protein moieties with a subsequent overall insolubilization of the protein.

Electrophoretic patterns showed that with increasing temperature there is a significant decrease of the 15S and 11S components accompanied by a noticeable increase in breakdown products of lower molecular weights. As a result of texturization the product shows only 27% as much soluble protein as the unprocessed soy flour, and it is believed that this lowering of solubility is related to changes occurring with disulfide bonds.

Cumming *et al.* (1972) have also employed various physical testing methods to gain further insight into the structure of textured soy protein products. Examination of products made at different elevated temperatures with shear force and work measurements allowed these investigators to conclude that upon increased thermal processing, the textured soy protein products became more difficult to shear and easier to pull apart because of lateral fissuring, as measured by breaking strength with an Instron Machine. The fact that break elongation and stress relaxation, however, appeared to be independent of processing temperature may be that these parameters are more directly related to the strength of inter-peptide bonds, rather than the secondary structure related to the aligning of fibers.

Additional work with the Instron Machine has been done at the University of Minnesota in an attempt to shed more light on the relationship of objective physical measurements to actual sensory evaluations in textured soy protein products, alone and also when employed as extenders in ground beef patties. A report by Breene and Barker (1975) introduced the concept of identifying portions of the Instron force-distance curve with textural properties, namely, hardness, cohesiveness, extrudability, chewiness, maximum force and average maximum force. This system of evaluation is known as the Minnesota Texture Method, and has been employed by Loh (1975) to evaluate several types of textured soy protein products in meat patties at 25% and 50% levels of meat replacement. With the beef-soy mixtures it was possible to correlate the instrumental parameters with sensory tenderness, chewiness and springiness. This method also shows promise of differentiating texturally among various textured vegetable protein materials.

In the development of textured protein products to meet particular specifications, the instrumental methods can be useful in giving guidance to the direction of the technical work. This is illustrated in Fig. 19.7 from which it is seen that a product with a stronger internal structure can be made by using a 70% protein product (soy protein concentrate) in place of the usual 50% protein soy flour. Likewise, the curves indicate that the products tested retain their structure under retorting conditions, a useful characteris-

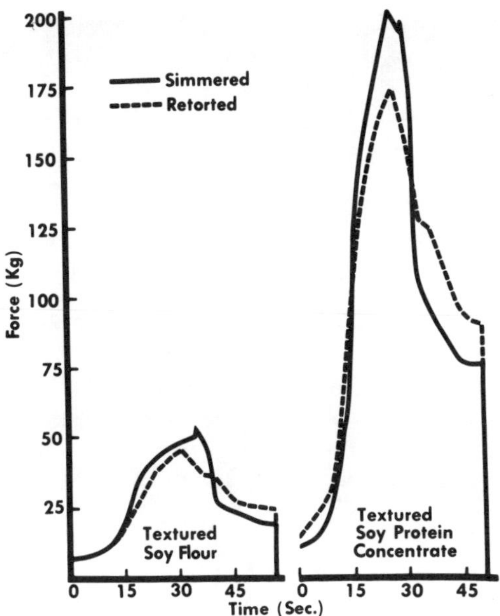

FIG. 19.7. INSTRON MEASUREMENTS ON TEXTURED SOY FLOUR AND TEXTURED SOY PROTEIN CONCENTRATE

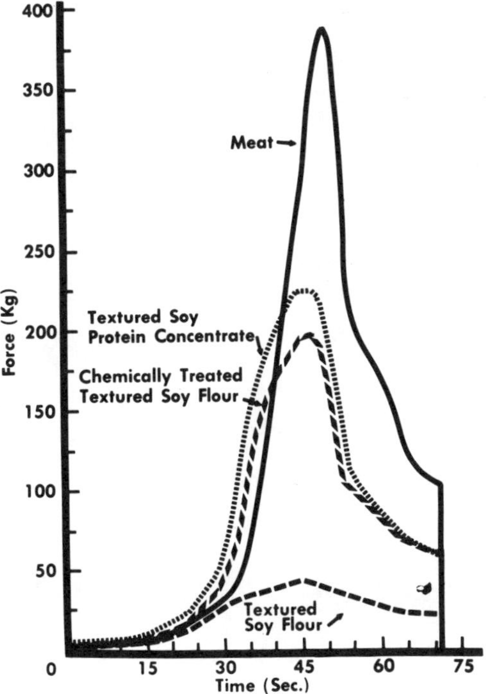

FIG. 19.8. INSTRON MEASUREMENTS ON TEXTURED SOY PRODUCTS AND MEAT

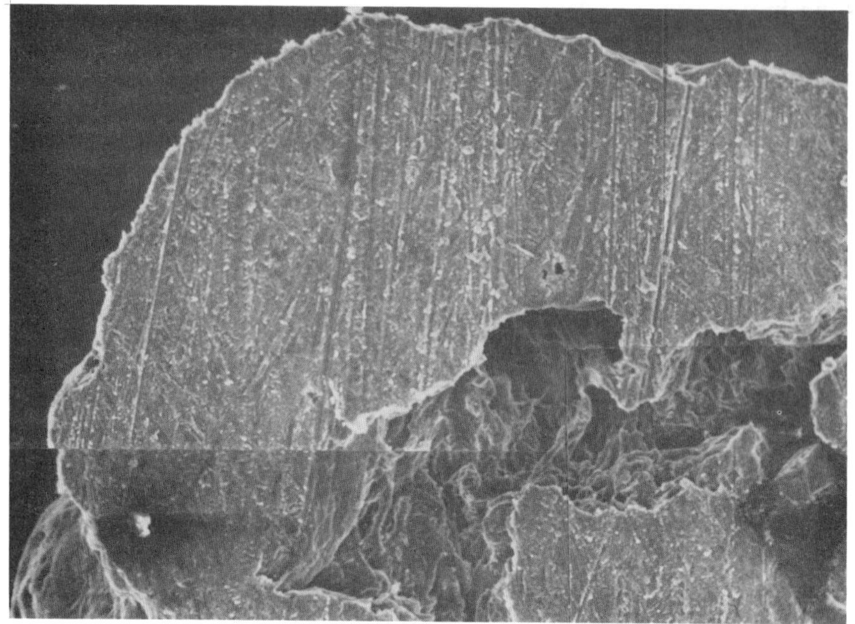

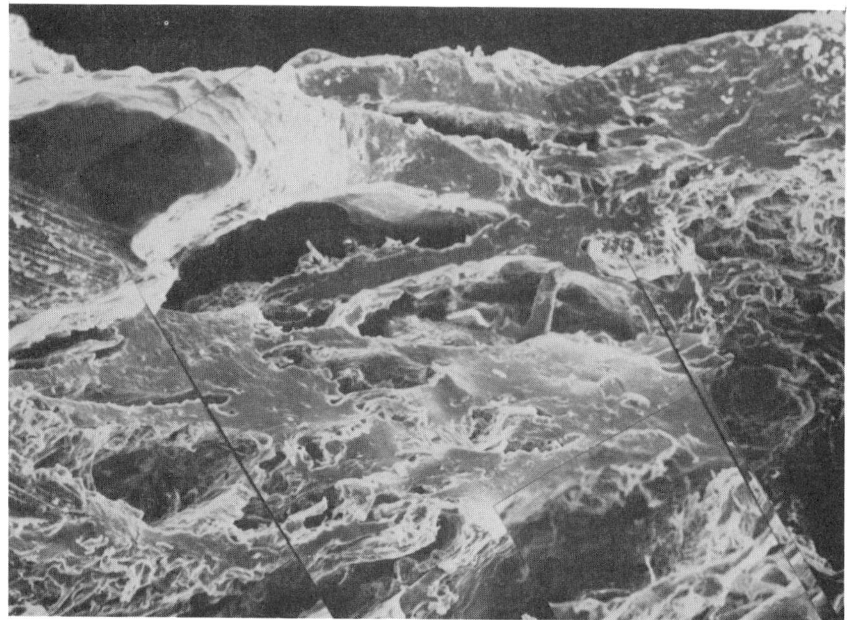

FIG. 19.9. SCANNING ELECTRON MICROSCOPE PHOTOGRAPHS
(Perpendicular to Direction of Flow): Upper—Nonexpanded (soy isolate); Lower—Expanded (soy flour)

tic in canned food products. In an effort to approach the sensory properties existing in a beef chunk, the curves in Fig. 19.8 indicate that changing from 50% protein to 70% is probably a step in the right direction. Visual structural characteristics suggestive of particular textural properties are given in Fig. 19.9, in the form of scanning electron microscope photographs, supplementing the report of deMan (1976).

Marketing Aspects

Textured vegetable protein products are not marketable unless they are acceptable to the consumer. Acceptability refers to organoleptic satisfaction, convenience in preparation, nutritional adequacy or even advantages, favorable costs, among a number of other related factors. Not all of these components are required for all applications. Still, if the product is not organoleptically acceptable the other aspects are meaningless.

As reported by Telser and Stermer (1976) a majority of the consumers tested eat margarine instead of butter, because margarine is usually less costly, has less cholesterol, is considered to be more healthful, and out of habit the family likes it; those who continue to eat butter do so because they claim that it tastes better. This study showed that the nutritional and health factors can be pushed aside if the price differential between margarine and butter changes significantly, as was evidenced in the recent past. In 1973 when the price spread favored margarine by 49¢, the sales of margarine increased 4% over the previous year and butter decreased 3%. However, in 1974 when butter was only 32¢ more expensive than margarine, the sales of margarine dropped 1% and butter increased 8%. In some respects the consumer behaves similarly with the use of textured vegetable protein products.

Psychological factors also play a role in the consumer's attitude to new products. A study by Weimer (1976) consisted of feeding volunteers four hamburger-type products, one of which was all-beef and the other three were commercial beef-soy blends. Half of the panelists were informed about the nature of the products before eating; the other half were not. Analysis of the data showed that without prior knowledge of the product content, there was no significant difference in preference among the products tasted. However, prior knowledge of the products resulted in the all-meat product being preferred significantly over each of the beef-soy blends. From a general marketing point of view, it is believed that the more acceptable products are those in which the textured protein cannot be easily identified.

The outstanding exception to this assumption is the situation of the spun-fiber products which are positioned in the marketplace as analogs, *i.e.*, a complete replacement of specific forms of meat products. For the most part these analogs are not priced lower than that of the comparable meat item, and therefore cost is not an overriding factor. If a product has convenience value it can be a more easily accepted substitute. For instance, the bacon bit products designed for condiment use do not require cooking or refrigeration. In the case of the breakfast links and related products made from vegetable proteins, the consumer purchases them primarily because of a nutritional consideration, namely, "no cholesterol," even though they may be more costly than the comparable all-meat product. The fact that the spun-fiber products are inherently more costly to manufacture requires that they carry a uniqueness which can be promoted in the marketplace.

The thermoplastic extruded products, however, stand out prominently when contrasted economically with the spun-fiber products. The basic consideration here is that these extruded products are usually hydrated with approximately two parts of water when used as meat extenders, resulting in a product having a protein content equivalent to that of ground meat but with a cost usually under 10 cents per pound, as indicated in Fig. 19.10. A typical example of savings to the consumer is shown in Table 19.8 (Adolphson and Horan 1974).

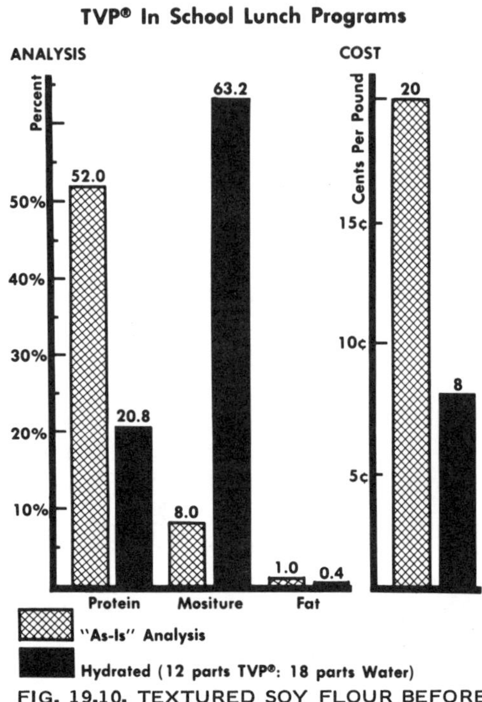

FIG. 19.10. TEXTURED SOY FLOUR BEFORE AND AFTER HYDRATION

TABLE 19.8
CONSUMER'S COST COMPARISON OF MEAT AND MEAT-SOY BLEND[1]

	Regular Ground Beef	Beef-Soy Blend[2]	Savings (%)
$/lb	0.75	0.59	21
Raw weight	16 oz	16 oz	
Cooked weight	11.2 oz	12.8 oz	
$/lb cooked	1.07	0.74	31

[1] Average retail prices in the Midwest, August, 1974.
[2] 25% replacement of beef with hydrated extruded textured vegetable protein (TVP®).

TABLE 19.9
FAT AND WATER RELATIONSHIPS IN EXTENDED BEEF PATTIES

		Lost (%)	
Product	Cooked Yield (%)	Fat	Water
Ground beef (GB)	70	35	35
Extended GB[1]	73	44	28

[1] 25% replacement of beef with hydrated extruded textured vegetable protein.

Another important advantage in employing these textured soy products as extenders in beef patties is that upon cooking, either in an oven or on a grill, the yield of product is higher and the fat content is lower for the extended product. This is summarized in Table 19.9 (Anderson and Lind 1975) with beef patties containing about 25% fat before cooking.

Historically, the extruded textured soy products appeared first in the marketplace as extenders for patties and fresh ground beef but applications are numerous as shown in Fig. 19.11. Of more recent interest is their use in comminuted cooked or cured meat food products, as described in considerable detail by Terrell and Staniec (1975). The same general economic advantages for extended frankfurters prevail as is illustrated in Table 19.10.

The commercial successes achieved in recent years with soy proteins have prompted many speculations about their expanded use in the meat industry in the years ahead. The U.S. Department of Agriculture (1972) has projected for 1980 a low level of substitution in all processed beef and pork items at 10% and a high level at 21%, representing 1.8 and 3.8 billion pounds of meat respectively. For the thermoplastic extruded products, estimates of 117 million pounds have been made for 1975, increasing steadily to about one billion pounds in 1980 (Iammartino 1974).

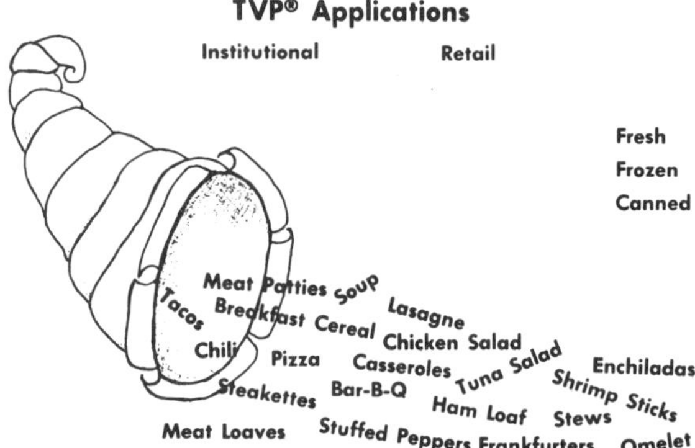

FIG. 19.11. INSTITUTIONAL AND RETAIL APPLICA-
TIONS OF TEXTURED SOY PROTEIN

TABLE 19.10
ECONOMIC BASIS FOR TVP® IN FRANKFURTERS

Material	All-Meat $/100 lb	30% TVP® (Hydrated) $/100 lb
Beef trim (10% fat)[1]	$36.72	25.64
Plates (50% fat)[1]	16.23	11.33
TVP®[1]	—	1.11
Seasonings	0.95	0.95
Water	(10.0%)	(29.0%)
Sub-total	53.90	39.03
Total (yield correction)[2]	57.65	42.89
Savings	—	25.60%

[1] Lean beef trimmings @ $0.844/lb; beef plates @ $0.373/lb; TVP® @ $0.16/lb.
[2] 93.5% yield for all-meat frankfurter; 91.0% for beef-TVP®.

To focus on one segment of the current market activity Shelef and
Morton (1976) investigated in 1974 institutional food services as
good potential users of soy protein. Food services in the metro-
politan Detroit area were investigated, including schools, hospitals,
nursing and convalescent centers, and industrial services. At this
beginning era of growing commercial acceptance of the soy protein
product, only 15 out of 100 services indicated that they used soy
protein in their feeding programs and 6 were schools representing
50% of all the schools contracted. The schools served between 2,300
and 5,000 meals per day, using soy in about 50% of the meals. The
number of meals prepared per day in the other food services was
between 200 and 15,000. It is not difficult to grasp from the limited
study that a tremendous potential exists for the expanded use of soy
products in the years ahead.

OUTLOOK FOR THE FUTURE

The foregoing discussion was designed to indicate that the age of using textured vegetable protein products has just begun. Much still remains to be done, both in the areas of technology and marketing. In looking towards the future, the food scientist, technologist, and marketer might be advised to give serious thought to the following considerations:

(1) From a market point of view one of the major obstacles confronting a more rapid acceptance of textured vegetable proteins is a lack of clear definition or a lack of uniformity with respect to government regulatory guidelines. It is true that the U.S. Department of Agriculture gave an important boost to the large scale introduction of textured soy protein in the school lunch program (USDA 1971). Still much regulatory confusion prevails: the FDA has issued a proposal (U.S. Govt., 1974A) on a "common or usual name for plant protein products;" the Federal Trade Commission has issued a proposed regulation (U.S. Govt. 1975) on "Advertising and labeling of protein supplements", the USDA Food and Nutrition Service has revised its regulations (U.S. Govt. 1974B) for textured vegetable protein products in the school lunch program; the Animal and Plant Health Inspection Service (APHIS) of the USDA has published a proposal (U.S. Govt. 1973) on "textured vegetable products in meat products;" on top of all of this APHIS has made another proposal (U.S. Govt. 1976) which broadens the definition of "meat," and this could affect the extending of meat products with nonmeat protein products. Outside of the U.S., Canada has issued detailed regulations (Canada 1975) which place emphasis on nutrient composition. The Food Standards Committee of England (Ministry of Agriculture, Fisheries, and Food 1974) has issued regulations to be in effect for 5 years on safety, labeling, nutrition, and composition. To add to the complexity, each of the Western European countries has its own specific regulations concerning the use of vegetable protein materials (van Gils 1975). It is evident that more cooperative progress is needed to unify regulations both in the U.S. and worldwide, but this will require considerable dedication on the parts of all the industrial and governmental organizations involved.

(2) Some of the difficulty encountered with governmental regulations is due in part to a lack of accurate definition of terms. Just what is meant by "textured vegetable protein?" More in-depth research is needed to characterize the essential structural parameters related to textural properties. Continued investigations with the scanning electron microscope, and with force-deformation curves obtained with the Instron Testing Machine (Peleg 1976) need to be encouraged.

Once clear and explicit regulations are established, a function of the regulatory agencies is to have available reliable methods of analysis so that meat products blended with nonmeat proteins appearing in the marketplace can be monitored from time to time. This presents an important challenge to the analytical chemist as discussed by Smith (1975); promising approaches to the solution of this problem, reported by Lee *et al.* (1975, 1976), will undoubtedly be of real value in the final development of workable regulatory guidelines.

(3) More knowledge is needed with respect to understanding more clearly the mechanism of thermoplastic extrusion from a molecular point of view. Present evidence indicates that disulfide bonds play an important role in the depolymerization and repolymerization of the protein components involved. An interesting practical discovery has been made (Jenkins 1970), in which it was determined empirically that a small amount of free sulfur or sulfur compounds added to defatted soy flour before extrusion caused the proteinaceous material to expand more readily and more uniformly. A further study of this effect could possibly lead to a more thorough understanding of the extrusion process.

(4) Many forecasts for the textured vegetable protein market have been made, and estimates as high as 5 billion pounds annually by the year 2000 have been given. When one considers that a large extruder can produce at best about 5000 lb of textured product per hour, amounting to some 40 million pounds per year, it is clear that a minimum of 100 extruders would be required to meet the demand. This type of installation entails the commitment of many millions of dollars in capital investment. For this reason it is important to study the fundamentals of extrusion in more detail so that greater efficiencies can be achieved. A better understanding of the variables involved and their control to obtain increased production are worthy of further study. Steps in this direction have been taken by Aguilera and Kosikowski (1976) in which the technique of Response Surface Analysis was employed to investigate the variables of process temperature, feed moisture content, and screw speed in making a soybean extruded product.

The magnitude of the growth potential for textured vegetable protein products encourages other approaches in addition to the usual commercial cooker-extruders. One such process for producing expanded textured protein products is referred to as steam texturization and has recently achieved commercial significance (Strommer and Beck 1973).

(5) Even though a preponderance of the protein extrusion technology has been concerned with defatted soy flour as a starting ingredient, the state of the art has now progressed to the point where

other materials are being studied. The extrusion of the 70%-protein soy concentrate will undoubtedly generate a new family of textured products characterized by the lack of sugars, better flavor, and different degrees of structural integrity.

For nutritional considerations it is advantageous to be able to increase the PER value of soy protein to that of casein. One approach is to use mixtures of soy and other nonmeat proteins to effect a complementary balance of the amino acids present. Some progress has been made in employing mixtures of soy flour and wheat gluten (Odell 1975).

As other nonmeat proteins appear on the commercial scene attempts will be made to apply the existing technology to them. A notable example in recent years is the texturization of single-cell protein (Tannenbaum 1975).

(6) Perhaps the most difficult problem encountered technologically in obtaining complete acceptance of textured vegetable protein products is connected with the incorporation of artificial meat-like flavors which bear a close resemblance to those of the animal product when prepared for eating in a number of different ways. The commercial development of these textured products in recent years has stimulated accelerated research efforts on meat flavors, as reviewed by Hashida (1974). New successes in flavor development and application in textured products will certainly assist in the total market growth of these products in the years ahead.

(7) In view of the fact that soy protein in the defatted 50%-protein form is probably the lowest cost source of protein in the world, and also that textured soy protein products are dramatically less costly than meat, the question is frequently asked: Why is not the market for textured protein in the developing countries growing more rapidly than in the U.S.? As has been pointed out by Caruso and Moore (1976), "systematic marketing research is not a luxury that can be bypassed in formulating low-cost, high-protein foods in developing nations—use of marketing research techniques is mandatory for product success." A study in Ecuador proceeded in five consecutive stages to finally reach a decision that a strawberry-flavored soy milk held promise for being accepted in a specific market segment of children six years of age and younger. It is extremely important that products be specifically formulated for a local population's taste and that each country's marketing system be considered fully.

(8) To date the extruded textured soy products require a defatted material for their manufacture. In developing countries this restriction imposes severe limitations in establishing production facilities,

because of the relatively high capital requirement in installing an oil extraction plant. A more direct use of the soybean into a textured product would seem desirable. A breakthrough in this area has been reported by Berry *et al.* (1975) who describe a process for making an expanded edible protein product having up to 35% of fatty glyceride material; the other critical ingredients are free sulfur and a filler such as cellulose or silicon dioxide. No doubt this work will be followed by additional innovations.

(9) In looking ahead for the next 10 to 20 years one searches for trends which seem to be appearing on the horizon related to the eating habits of the U.S. consumer. Many changes are constantly occurring, some of which have been presented by Becker (1976). The consumer today is interested in time-saving insofar as food preparation is concerned; this may be a result of the fact that the percentage of women in the work force is increasing. There is a greater demand for fast service institutional types of foods. The sales of microwave ovens are increasing more rapidly than the established conventional types and this will require some changes in food formulations. The number of households is increasing, but the number in each is decreasing, which brings greater significance to portion-control foods. The U.S. consumer is becoming more health conscious, particularly with respect to the daily food intake, with a watchful eye on calories, cholesterol and dietary fiber. With an awareness of these changes it behooves the food scientist and technologist to realize that the trend is not for the U.S. consumer to continually increase his or her daily per capita intake of food, but to be more discriminating about what is eaten. Thus, for textured vegetable protein products to succeed in the marketplace they will have to offer something of real value over the conventional types of foods which are being replaced.

BIBLIOGRAPHY

ABBOTT, J. A. 1972. Sensory assessment of food texture. Food Technol. *26*, No. 1, 41–49.

ADOLPHSON, L. C., and HORAN, F. E. 1974. Textured vegetable protein products as meat extenders. Cereal Sci. Today *19*, 441–446.

AGUILERA, J. M., and KOSIKOWSKI, F. V. 1976. Soybean extruded product: A Response Surface Analysis. J. Food Sci. *41*, 647–650.

ANDERSON, R. H., and LIND, K. D. 1975. Retention of water and fat in cooked patties of beef and of beef extended with textured vegetable protein. Food Technol. *29*, No. 2, 44–45.

ANSON, M. L., and PADER, M. 1957. Protein food product and process. U.S. Patent 2,802,737.

ANTON, J. J. 1976. Soy-protein of the decade. Chemtech. Feb., 90–91.

ATKINSON, W. T. 1970. Meat-like protein food product. U.S. Patent 3,488,770.

AUTRET, M. 1970. World protein supplies and needs. *In* Proteins as Human Food, R. A. Lawrie (Editor). AVI Publishing Co., Westport, Conn.

BECKER, J. C. 1976. A presentation given at "Birth of a New Product," 17th Annu. Symp., Central States Section, Am. Assoc. Cereal Chem., St. Louis, Mo.

BERRY, M. F., REESMAN, S. H., and SMITH, M. L. 1975. Expanded protein product and manufacture thereof. Canadian Pat. 977606.

BOURNE, M. C. 1975. Texture properties and evaluations of fabricated foods. *In* Fabricated Foods, G. E. Inglett (Editor). AVI Publishing Co., Westport, Conn.

BOYER, R. A. 1954. High protein food product and process for its preparation. U.S. Patent 2,682,466.

BREENE, W. M., and BARKER, T. G. 1975. Development and application of a texture measurement procedure for textured vegetable protein. J. Texture Studies 6, 459-472.

CANADA. 1975. Can. Gaz. Part II, 208.

CARUSO, R. V., and MOORE, R. J. 1976. Marketing research for high-protein foods for developing countries. Food Product Dev. 10, 54-58.

CHIANG, J. P. C., and STERNBERG, M. 1974. Physical and chemical changes in spun soy fibers during storage. Cereal Chem. 51, 465-471.

COREY, H. 1970. Texture in foodstuffs. CRC Crit. Rev. Food Technol. 1, 161-198.

CORLISS, G. A., and FURGAL, H. P. 1975. Simulated bacon product and process therefor. U.S. Patent 3,930,033.

CUMMING, D. B., STANLEY, D. W., and DEMAN, J. M. 1972. Texture—structure relationships in texturized soy protein. 2. Textural properties and ultrastructure of an extruded soybean product. J. Inst. Can. Sci. Technol. Aliment. 5, 124-128.

CUMMING, D. B., STANLEY, D. W., and DEMAN, J. M. 1973. Fate of water soluble protein during thermoplastic extrusion. J. Food Sci. 38, 320-323.

deMAN, J. M. 1976. Texture—structure relationships. Cereal Foods World 21, 10-13.

DURST, J. R. 1963. Meat-like product and process. U.S. Patent 3,108,873.

FLIER, R. J. 1976. Protein product and method for forming same. U.S. Patent 3,940,495.

FRANK, S. S., and CIRCLE, S. J. 1959. The use of isolated soybean protein for nonmeat, simulated sausage products, frankfurter and bologna types. Food Technol. 13, No. 6, 307-313.

FRIEND, B., and MARSTON, R. 1975. Nutritional Review. U.S. Dep. Agric., ARS. NFS-154, Nov. 1975, 26-32.

GALLIMORE, W. W. 1976. Estimated sale and impact of soy-beef blends in grocery stores. U.S. Dep. Agr. NFS-155, Feb. 37-44.

GENERAL FOODS CORPORATION. 1975. Meat analog and manufacture thereof. Brit. Patent Specification 1,379,411.

HARDINGE, M. G., and CROOKS, H. 1963. Nonflesh dietaries. 1. Historical background. J. Amer. Dietetic Assoc. 43, 545-549.

HARPER, J. M., and HARMANN, D. V. 1973. Research needs in extrusion cooking and forming. Trans. ASAE, 941-943.

HARTMAN, W. E. 1966. Vegetarian protein foods. Food Technol. 20, 39-40.

HARTMAN, W. E. 1967. Vegetable base high protein food product. U.S. Patent 3,320,070.

HASHIDA, W. 1974. Flavour potentiation in meat analogues. Food Trade Rev. 44, 21-31.

HORAN, F. E. 1973. Wheat-soy blends—high quality protein products. Cereal Sci. Today *18*, 11–14.

IAMMARTINO, N. R. 1974. Fabricated protein foods. Chem. Eng. *81*, 50–54.

JELLIFFE, D. B., and JELLIFFE, E. F. P. 1975. Human milk, nutrition, and the world resource crisis. Science *188*, 557–561.

JENKINS, S. L. 1970. Method for preparing a protein product. U.S. Patent 3,496,585.

KELLEY, J. J., and PRESSEY, R. 1966. Studies with soybean protein and fiber formation. Cereal Chem. *43*, 195–206.

KELLOGG, J. H. 1907. Food product. U.S. Patent 869,371.

LEE, Y. B., RICKANSRUD, D. A., HAGBERG, E. C., and BRISKEY, E. J. 1975. Quantitative determination of soybean protein in fresh and cooked meat-soy blends. J. Food Sci. *40*, 380–383.

LEE, Y. B., RICKANSRUD, D. A., HAGBERG, E. C., and FORSYTHE, R. H. 1976. Detection of various nonmeat extenders in meat products, J. Food Sci. *41*, 589–593.

LOH, J. 1975. Texture analysis of soy protein products: correlation of instrumental and sensory evaluations. Thesis. University of Minnesota, Minneapolis.

MACALLISTER, R. V., and FINUCANE, T. P. 1963. High protein food granules. U.S. Patent 3,102,031.

MINISTRY OF AGRICULTURE, FISHERIES, AND FOOD. 1974. FSC/REP/62. London.

NATIONAL ACADEMY OF SCIENCES. 1974. Recommended Dietary Allowances, 8th Edition. Washington, D.C.

ODELL, A. D. 1975. Private communication. Industrial Grain Products, Ltd., Montreal, Canada.

PELEG, M. 1976. Texture profile analysis parameters obtained by an Instron Universal Testing Machine. J. Food Sci. *41*, 721–722.

ROSENFIELD, D. 1974. Spun-fiber vegetable protein products. Chemtech *4*, 352–355.

ROSSEN, J. L., and MILLER, R. C. 1973. Food extrusion. Food Technol. *27*, No. 8, 47–54.

RUSOFF, I. I., OHAN, W. J., and LONG, C. L. 1962. Protein food product and process. U.S. Patent 3,047,395.

SHELEF, L. A., and MORTON, L. R. 1976. Soybean protein foods. Food Technol. *30*, No. 4, 44–49.

SMITH, P. S. 1975. Detection and estimation of vegetable protein in the presence of meat protein. Inst. of Food Sci. and Technol. (UK) Proceedings, *8*, No. 4, 154–162.

STANLEY, D. W., CUMMING, D. B., and DEMAN, J. M. 1972. Texture—structure relationships in texturized soy protein. 1. Textural properties and ultrastructure of rehydrated spun soy fibers. Inst. Can. Sci. Technol. Aliment. *5*, 118–123.

STROMMER, P. K., and BECK, C. I. 1973. Method for texturizing protein material. U.S. Patent 3,754,926.

SZCZESNIAK, A. S. 1963. Classification of textural characteristics. J. Food Sci. *28*, 385–389.

SZCZESNIAK, A. S. 1971. Consumer awareness of and attitudes to food texture. 1: Adults. J. Texture Studies *2*, 280–295.

TANNENBAUM, S. R. 1975. Texturizing process for single cell protein containing protein mixtures. U.S. Patent 3,925,562.

TARANTO, M. V., MEINKE, W. W., CATER, C. M., and MATTIL, K. F. 1975. Parameters affecting production and character of extrusion texturized defatted glandless cottonseed meal. J. Food Sci. *40*, 1264–1269.

TELSER, E., and STERMER, J. 1976. Nutrition: Consumer beliefs and actions. Paper presented at 1976 Nutrition Symposium, April 21, Chicago, Ill. A. C. Nielsen Company, Northbrook, Ill.

TERRELL, R. N., and STANIEC, W. P. 1975. Comparative functionality of soy proteins used in commercial meat food products. J. Am. Oil Chemists' Soc. *52*, 263A–266A.

UNITED NATIONS WORLD FOOD CONFERENCE. 1974. Assessment of the World Food Situation, Rome. Nov. 5–16, 58.

U.S. DEP. AGRIC. 1971. FNS Notice 219, Feb. 22, 1–2.

U.S. DEP. AGRIC. 1972. Synthetics and substitutes for agricultural products. Projections for 1980. ERS, Marketing Research Report No. 947.

U.S. DEP. AGRIC. 1974. The world food situation and prospects to 1985. Economic Research Service, Foreign Agricultural Economics Report No. 98, 48.

U.S. DEP. AGRIC. 1975A. Agricultural Statistics. Washington, U.S. Govt. Printing Office, 349.

U.S. DEP. AGRIC. 1975B. Handbook of Agricultural Charts. Agricultural Handbook No. *971*.

U.S. DEP. AGRIC. 1976. Livestock and meat situation. Feb.

U.S. GOVT. 1973. Textured vegetable products in meat products. Federal Register *38*, No. 86, 11090–11094.

U.S. GOVT. 1974A. Common or usual name for plant protein products; withdrawal of textured protein products proposal. Federal Register *39*, No. 116, 20892–20895.

U.S. GOVT. 1974B. National school lunch program; special food service program for children. Federal Register *39*, No. 60, 11296–11298.

U.S. GOVT. 1975. Advertising and labeling of protein supplements. Federal Register *40*, No. 173, 41144–41148.

U.S. GOVT. 1976. Definition of meat and classes of meat, permitted uses, and labeling requirements. Federal Register *41*, No. 82, 17560–17566.

VANGILS, H. W. 1975. European regulations and laws governing the use of soya protein in food products. Paper presented at Conference on Food Regulations at the American Embassy in London (Nov.).

WATT, B. K., and MERRILL, B. K. 1963. Composition of Foods. Agricultural Handbook No. *8*, U.S. Govt. Printing Office, Washington, D.C.

WEIMER, J. 1976. Taste preference for hamburger containing textured vegetable protein. U.S. Dep. Agric., NFS–155, Feb., 45–46.

WRENSHALL, C. L. 1951. Meat substitute and process of making same. U.S. Patent 2,560,621.

20

Plant Protein Sources

G. O. Kohler
C. K. Lyon

During the past 10 years much consideration has been given to population growth in relation to the present and potential food supplies (Economic Research Service 1974; Pimentel *et al.* 1975; United Nations 1974; Univ. of Calif. 1974). Even though there is some question as to the degree of protein deficit as compared with caloric deficit, there is no doubt that much larger supplies of protein will be needed in the future. Appreciation of this problem has stimulated research on improving the yields and the quality of conventional protein rich foods and on the development of novel, unconventional protein sources.

To arbitrarily divide discussion of such novel sources on the basis of end use (food versus feed) is unnecessary because protein products that are nutritionally suitable for monogastric animals are potentially usable as food or food ingredients if the special requirements are met of palatability, functional properties, and a higher degree of toxicological safety. Further, the use of new protein resources by animals should free materials currently being used in feeds for use in human diets.

In this discussion, we shall compare the basic physical and chemical characteristics of protein rich tissues from various types of plants and show how these characteristics determine the methods used to produce protein products from them. The discussion will be

TABLE 20.1
PLANT PROTEIN SOURCES

A. Nonendospermous "Seeds" — Usually Achenes
Soybean	Sesame seed
Cottonseed	Castor seed
Sunflower seed	Rapeseed
Safflower seed	Peanuts
	Linseed

B. Endospermous "Seeds" — Usually Caryopses
Wheat bran	Triticale
Rice bran	Oats
Triticale bran	Coconut

C. Leafy Plants
Alfalfa	Vegetable wastes
Clovers	Aquatic plants
Grasses	Miscellaneous

limited to plant materials including seeds and leaves (Table 20.1). The seeds may be subdivided into oilseeds and starchy seeds. In many starchy seeds (*e.g.*, cereals) the endosperm comprises the major portion of the seed (endospermous types). In many of the oilseeds, the cotyledons make up the bulk of the seed (nonendospermous types) (McMasters *et al.* 1971; Weiss 1971; Winton and Winton 1932; Wolf and Cowan 1971).

THE OILSEEDS

Research on oilseeds as unconventional protein sources includes work on sunflower, safflower, cottonseed, sesame, castor, rapeseed, and many other seeds. Soybeans could be included but for purposes of this discussion are considered conventional protein sources. In the nonendospermous types, the cotyledons make up 80–90% of the seed in many cases. The cotyledons are rich in oil (*e.g.*, 20–65%) and contain protein as the next most prevalent component. Many of the so-called "seeds," such as safflower seeds, are achenes, rather than true seeds, and contain an easily detachable pericarp in addition to the seed coat. In the remainder of this paper the word seed will be used in the looser sense to include achenes. The methods of processing the other oilseeds follow the same general schemes as have been developed for soybeans which are not achenial. Differences in the physical nature of the different seeds require specific modifications in processing procedures.

It will not be possible to cover all oilseeds in detail. We shall, therefore, use safflower as an example and mention others only where certain unusual properties are important.

Safflower

Safflower differs markedly from soybeans in several respects. The hull fraction of safflower, which includes the seed coat plus the pericarpal layers of the achene, amounts to about 40% of the weight of the conventional type seed. Much research has been done to reduce the hull level by breeding and some success has been achieved (Ebert and Knowles 1968; Guggolz *et al.* 1968; Weiss 1971). In contrast to safflower hull, the soybean hull is essentially all seed coat and amounts to only about 8–10% of the seed. The pericarp of soybeans and other legumes is the pod which is removed by the combine in the field. A further important difference is that the safflower kernel is extremely high in oil (*e.g.*, 60%) and is mushy in nature. The soybean kernel is firm and contains much less oil (*e.g.*, 22%).

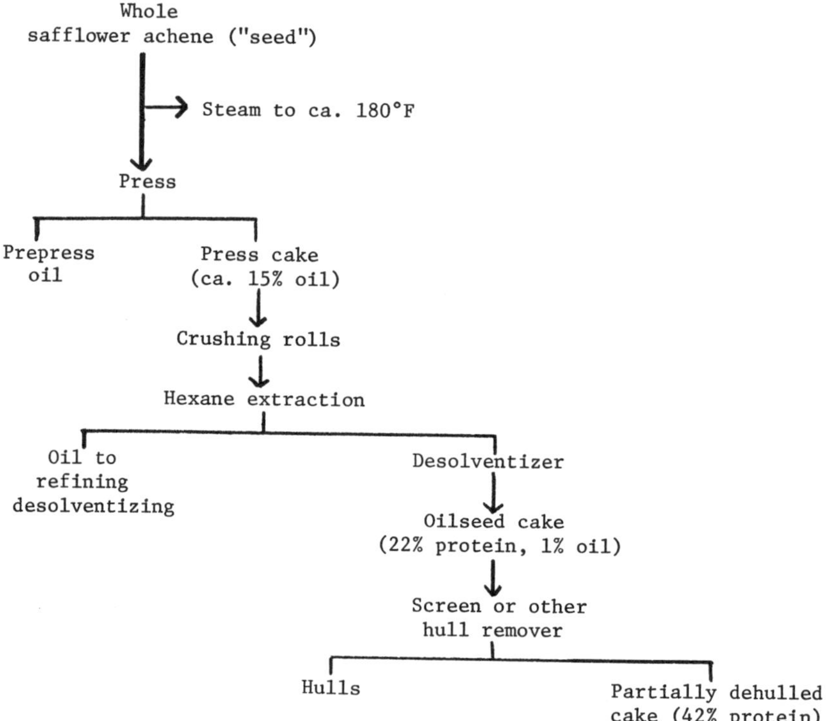

FIG. 20.1. GENERAL SCHEME FOR PROCESSING OILSEEDS AS ADAPTED FOR SAFFLOWER SEED

Figure 20.1 shows a commercial procedure for oil extraction and oilseed cake production from safflower seed (Purdy 1976). The whole seed is conditioned by steaming to: (1) partially denature the protein so as to permit more facile removal of the oil; (2) inactivate enzymes; (3) raise the temperature so as to reduce oil viscosity; and (4) adjust the moisture content for optimum oil removal. The conditioned seed is then pressed to remove a part of the oil. Safflower press cake usually retains 12–15% oil (in contrast to about 5% for soybean expeller meal). The safflower cake is next put through rollers to crack any remaining unbroken seeds and is then extracted with hexane to remove the residual oil. Solvent is removed from the cake by means of injected steam. A higher quality meal may be separated by screening and/or air separation to give hull fraction and a 42–44% protein product (Goodban 1967; Huff and Robe 1973; Kopas and Kneeland 1966).

This meal still contains about 15–17% crude fiber. If all of the hull is removed, as by dissection (Guggolz *et al.* 1968), the extracted kernel contains about 60% protein. A product almost this good also has been obtained by dehulling prior to oil extraction as is done in the case of soybeans. However, the mushy nature of the safflower kernel and its high oil content, together with the toughness of the hull, make it virtually impossible to carry out such a dehulling operation without incurring large oil losses in the hull fraction. Further, some fiber is necessary in the kernel to get good prepress operation. Some progress has been made toward producing a hull-free flour (Goodban 1967; Goodban and Kohler 1970), but even if a low cost, 60% protein dehulled product were produced commercially, it would still be unusable as a food product (Agren 1966; Betschart *et al.* 1975; Kohler 1966; Weiss 1971) since it has a strong bitter flavor and a mild laxative principle(s).

These problems have been under investigation at our laboratory. A bitter principle, matairesinol β-D-glucoside (Palter and Lundin 1970), and a laxative principle, 2-hydroxyarcitiin, have been identified (Palter *et al.* 1972). These and other bitter or laxative components which may be present are removed from the oilseed meal by extraction with aqueous alcohol, acetone or water (Kohler 1966, 1969; Lyon and Saunders 1975). The extracted meals are bland and produce little or no laxation when fed to rats at fairly high levels. Thus, the same procedures used for making soy protein concentrates (Smith and Circle 1972) may be used to prepare safflower protein concentrates. In both cases, deleterious components are removed and high quality edible products containing 60–80% protein are obtained.

Similarly, safflower protein isolates have been prepared on a laboratory scale (Betschart 1975) by modifying only slightly the procedure used for producing soy isolates. We shall review the development of such a safflower isolate process, an example of the approaches used with unconventional oilseeds. In order to gain information on classes of protein present, unheated, hexane extracted, partially decorticated meal (47% protein, 6.3% fat, 13% crude fiber) was serially extracted with water, 1 N NaCl, 70% EtOH, and 0.1 N NaOH. The extracts were dialyzed and the retentates were freeze-dried. The procedure was essentially that of Osborne and Mendel (1914). Table 20.2 shows that about 18% of the protein of

TABLE 20.2
PROTEIN FRACTIONS FROM SAFFLOWER[1]

Extractant	% of Total Nondialyzable Protein Extracted	% Lysine in Protein (g/16 g N)
Water	18.1	4.55
1 N NaCl	41.5	2.17
70% Ethanol	0.2	—
0.1 N NaOH	39.1	1.97

[1] Lysine content of original safflower meal and extracted residue: 2.8, 2.62, respectively; % total nitrogen extracted, 91.2%.

safflower was extracted by water and that this fraction was significantly higher than the other fractions in lysine, the first limiting amino acid of unfractionated safflower protein (Kohler 1966). The bulk of the protein was in the salt and alkali soluble fractions, and these proteins were low in lysine.

The next stage in development of the process was determination of the effect of pH on extraction of nitrogen from the meal. As shown in Fig. 20.2, poor solubilization of safflower meal nitrogen occurred in acid or neutral solutions. In order to obtain good yields, it was necessary to extract at pH 8.5 or over. The nitrogen of the hexane extracted (cold) expeller press cake is as easily extractable as that of laboratory prepared, unheated meal so that it could provide protein isolates in good yields. The commercial desolventizing process, on the other hand, was severe enough (*e.g.*, 104–110°C) to cause a marked degree of insolubilization of nitrogen which is evident at all pH values. Further work showed that the solubility profile of the precipitated isolate had a much narrower valley near the isoelectric region, due undoubtedly to the absence of impurities which adversely affect solubility.

Sunflower Meal

Sunflower meal (Gheyasuddin *et al.* 1970A) shows greater nitrogen extractability than safflower at lower pH values and lesser extractability at high pH values. Other protein sources vary in their

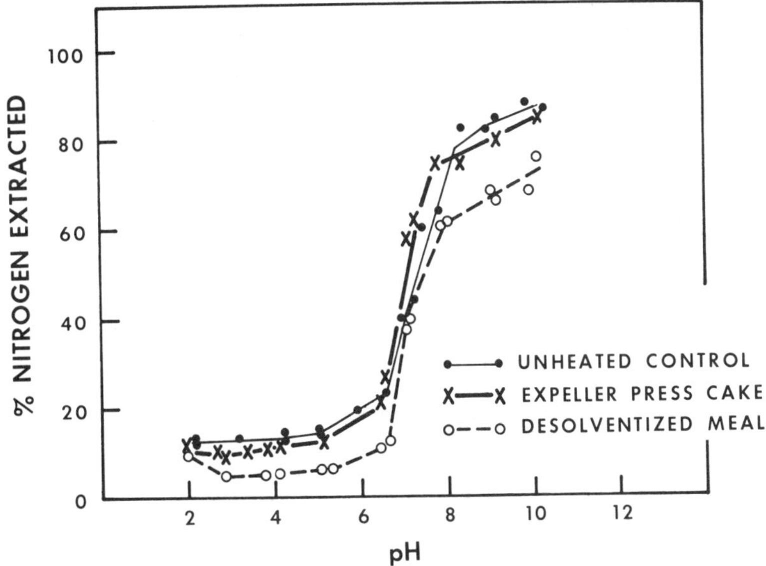

FIG. 20.2. EXTRACTION OF SAFFLOWER MEAL

pH extractability profiles which are characteristic of the raw materials in question (Mattil 1971). Several factors that affect extraction, in addition to such obvious factors as extraction temperature, solvent-to-meal ratio, and previous heat treatments are the presence of salts and phenolic compounds (*e.g.*, chlorogenic acid) (Betschart *et al.* 1975; Cater *et al.* 1972; Gheyasuddin *et al.* 1970A). It is thus apparent that the physical and morphological nature of the specific oilseeds determine differences in the conditions used, but, in general, approaches are quite similar.

The presence of undesirable components in safflower has been mentioned. The problem is a broadly occurring one among plant materials and will be discussed further. Table 20.3 shows some of the undesirable components that have been found in various oilseeds. The wide array of such undesirable components in soybeans (Liener 1975; Mickelsen and Yang 1966; National Academy of Sciences 1973; Rackis 1972; Wolf and Cowan 1971) provides the classical example. Raw soybeans are poorly digestible, low quality feed. Over 30 years of research has led to development of highly nutritious, high protein feed products from soybeans by processes based largely on controlled heating of the meal. Heating destroys the proteinase inhibitor, urease, and hemagglutinin and reduces the beany and bitter flavors of the meal. Overheating destroys lysine so that careful process control is necessary. The flatulence factor(s), the goitrogen, the estrogens, and the saponins are reduced only by extraction

TABLE 20.3

UNDESIRABLE COMPONENTS OF SEVERAL OILSEEDS

	Principle or Effect	Identified Component	Means for Elimination or Amelioration of Effect
Soybean	Proteinase inhibitor	Protein	Heat (autoclaving, toasting)
	Urease	Protein	Heat (autoclaving, toasting)
	Saponins	Triterpenoid glycosides	—
	Estrogens	Isoflavone glycosides	—
	Flatulence factors	Oligosaccharides + ?	Concentrate or isolate preparation
	Bitter and beany flavors	Phenolic compounds	Heat or extraction (concentrates and isolates)
	Hemagglutinin	Protein	Heat (autoclaving, toasting)
	Goitrogen	Sulfur containing glycoside	—
Safflower	Bitterness	Matairesinol monoglucoside	Extraction-concentrates, isolates
	Laxative	2-Hydroxyarctiin	Extraction-concentrates, isolates
Rapeseed	Goitrogen, growth inhibitor	Glucosinolate and cleavage products	Control of enzyme activity extraction, breeding
Peanuts	Hemolytic factor	Saponin	—
	Goitrogen	—	—
	Trypsin inhibitor	Protein	Heat
Cottonseed	Toxic, reduces nutritive value of protein	Gossypol	Breeding (glandless seed), heat gravity separation (liquid cyclone)
	Dark egg yolk factor		
	Pink egg white factor	Cyclopropene fatty acids	Through fat extraction
	Allergen	Protein	Alkali treatment
Sunflower	Green color development	Chlorogenic acid, etc.	Diffuse out of seed or extract with polar solvents
Castor	Toxin	Ricin (protein)	Heat
	Toxin	Ricinine (alkaloid)	Water extraction
	Allergen	Protein-polysaccharide	Lime or ammonia with heat
	Hemagglutinin	Protein	Heat

during preparation of concentrates or isolates. The remaining residues of these factors in the concentrates and isolates are low and probably of no practical significance.

The problems of safflower and their solutions have already been discussed briefly. It appears that the components that affect humans adversely are of importance (Agren 1966; Weiss 1971) and must be coped with before edible products become a reality. However, this is not the case in poultry nutrition, since simple amino acid supplementation (lysine, methionine) of safflower-meal based diets produced higher growth rates in chicks than optimally treated and supplemented soybean meal (Kuzmicky and Kohler 1968). The problems posed by rapeseed glucosinolate has been the subject of much research (Kozlowska *et al.* 1972; National Academy of Sciences 1973; Ohlson 1972; Ohlson and Sepp 1975). A promising approach lies in prevention of enzymatic cleavage of the glucosinolate present and leaching of the glucoside from the protein. Peanuts contain saponins, goitrogens and trypsin inhibitors (Dieckert *et al.* 1959; Liener 1975; Oke *et al.* 1975; Srinivasan *et al.* 1957), but at the levels present, they do not seem to pose serious problems. Cottonseed meal contains a potent allergen as well as the cyclopropene fatty acids and gossypol. Special procedures have been formulated to eliminate these materials (especially the gossypol), but commercial application has been difficult and time consuming (Martinez and Hopkins 1975; Mickelsen and Yang 1966; National Academy of Sciences 1973; Ridlehuber and Gardner 1974).

Sunflower meal also has its unique array of problems that center largely around its high content of chlorogenic acid. The general problem of protein-phenolic interactions has been reviewed recently by Van Sumere *et al.* (1975). Polyphenolic compounds react with peptide groups through hydrogen bonding at acid pH values and reduce the solubility of the protein. Upon oxidation of ortho-polyphenolic compounds such as chlorogenic acid the orthoquinone formed may react to form more stable compounds through covalent bonds with numerous functional groups of the protein. Further, the quinones may polymerize to form the dark green colored compounds from which chlorogenic acid derives its name (Van Sumere *et al.* 1975). Research has been done by several workers to eliminate chlorogenic acid (Betschart *et al.* 1975; Gheyasuddin *et al.* 1970B; O'Connor 1971; Sosulski *et al.* 1972), but thus far we know of no commercial application of the results.

Castor Meal

The problem of deleterious components is particularly severe in the case of castor meal (Bolley and Holmes 1958; Fuller *et al.* 1971;

Jenkins 1963; Weiss 1971). It contains an extremely potent array of allergens, a very highly toxic protein, ricin, a somewhat toxic alkaloid, ricinine, and a potent hemagglutinin. The most effective way to make the castor meal usable as a feed seems to be treatment with heat and lime, alkali, or ammonia (Betschart *et al.* 1975; Fuller *et al.* 1971; Mottola *et al.* 1971, 1972). These treatments are similar to, but more rigorous than, the treatments developed for inactivation of aflatoxin in oilseed meals (Goldblatt 1971; Masri *et al.* 1969).

Thus, in addition to the physical characteristics of the seed and the extraction characteristics of the protein (especially for isolates), a further important determinant for process development is the presence in the raw material of deleterious components that might accompany the protein into the final product. Each raw material has a unique array of such components. This means that in many cases uniquely adapted processes must be developed. The treatments that have been found useful for a number of different components are (a) heat treatment, (b) aqueous alcohol or isoelectric water extraction, (c) alkali treatment, and (d) mild alkaline extraction followed by acid and/or heat precipitation of the extracted protein (*i.e.*, isolate preparation).

STARCHY SEEDS

Cereal Grains

Let us next consider the starchy type of seeds. In the case of the cereal grains, the so-called "seed" is more properly called a caryopsis. The caryopsis is a seed with a tightly adhering seed coat and pericarp. The latter cannot be removed by a simple dehulling procedure such as that used for soybeans. In the cases of rice and barley, the pericarp, seedcoat and outer portion (aleurone layer) of the endosperm are separated from the intact starchy endosperm by grinding off the surface layers of the grain with special mills (Houston 1972; Pomeranz *et al.* 1971). In the case of wheat, the seed has a deep crease in it so that surface milling techniques are not very effective. This problem has been solved by the development of a complicated system of roll grinding and sieving. Cereal technology has evolved over thousands of years and has been described in detail in a multitude of books and papers (Inglett and Anderson 1973; Lockwood 1960; Ziegler and Greer 1971). The main focus of these technologies has been recovery of the starchy endosperm for food, but our interest in this paper is in the utilization of the outer layers of the grains (*i.e.*, wheat bran, wheat shorts, rice bran, etc.) as unconventional protein sources. These materials are used primarily in animal feeds for ruminants and are designated by the general term "millfeeds" (Millers' National Federation 1972).

Amino acid composition data show that the proteins of millfeeds are considerably better balanced in essential amino acids (Kasarda *et al.* 1971; Miladi *et al.* 1972) than the protein of wheat flour or polished rice, for examples. However, their high fiber content, strong flavor, and susceptibility to rancidity have severely limited their use in foods. These considerations have led to research to produce protein concentrates from cereal millfeeds by further processing.

In this connection, let us first consider the microscopic structure of the parts of the wheat kernel (Fig. 20.3). The starchy endosperm

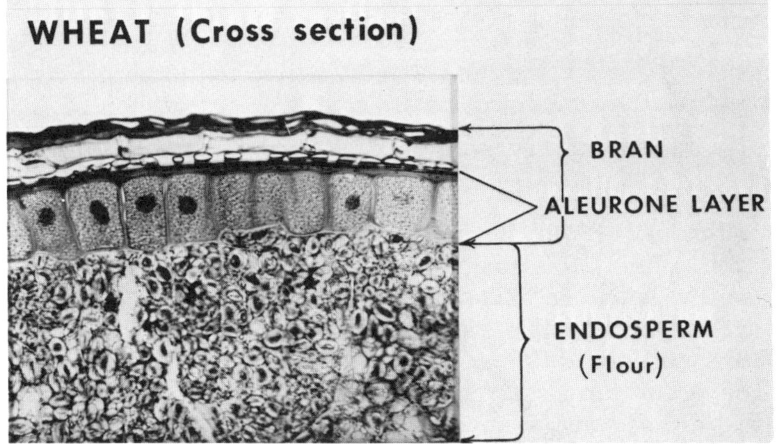

FIG. 20.3. CROSS SECTION OF PART OF WHEAT KERNEL

constitutes about 85% of the weight of the kernel; the aleurone layer of the endosperm, 7%; the seed coat and nucellar layer, 2.2%; the pericarp, about 6%, and the germ, about 2–3%. The protein is concentrated in the aleurone layer and the germ (not shown in Fig. 20.3). Commercial milling yields only about 72% of the weight of the wheat kernel as flour. Thus, the millfeed fraction contains considerable starchy endosperm material in addition to the components mentioned above. Commercial wheat millfeed contains about 14–18% crude protein (N × 6.25), 5% crude fat, 12% fiber and 15% starch (all on a 14% moisture basis). A dry milling procedure has been developed for producing a wheat protein concentrate (WPC) (Fellers *et al.* 1968; Fellers 1973), a material which has been utilized to a substantial degree in wheat products for donation programs (*e.g.*, Flour Blend A and wheat-soy blend, WSB). However, the level of protein which can be economically attained in these dry milled WPC products is only in the range of 20–22%, and their range of application is severely limited by poor functionality.

A second approach being taken is to produce cereal protein concentrates from millfeeds by wet alkaline extraction and recovery of a protein-lipid concentrate by heat or acid precipitation (Saunders and Kohler 1975). As a basis for this work, let us consider the protein fractions obtained on the basis of solubilities shown in Table 20.4 (Teller and Teller 1932). About 20% is in the albumin, 14% in

TABLE 20.4
BRAN PROTEINS[1]

	% of Protein
Albumin	20.5
Globulins	13.8
Prolamines	13.8
Glutelins	22.4
Unclassified	19.2

[1] Total protein N $\times$ 6.25 = 14.6; lysine, 3.87.

the globulin, 14% in the prolamine, and about 22% in the glutelin fraction. The 19% unaccounted for probably includes some physically bound in intact aleurone cells and a fraction (ca 10%) which is not digestible even by ruminant animals (Saunders *et al.* 1969; Saunders and Kohler 1972).

The pH-nitrogen extraction profile of wheat bran is quite similar to that of safflower, although, as would be deduced from Table 20.4, poorer extraction is obtained. Once the protein has been extracted and reprecipitated the solubility profile of the product shows a much narrower range of insolubility as was found in the case of oilseed protein isolates as well (Saunders *et al.* 1975A, B).

While the basic process closely resembles the safflower isolate process, most of the fat and the starch of the bran is extracted along with the protein. When the extract is acidified to pH 4 or heated to 85°C at pH 6, the precipitate obtained is a protein-starch-lipid isolate rather than a conventional protein isolate. For some end uses this is desirable. For other uses, such as in a milk replacer for lambs, the starch must be removed because it is poorly digested by young lambs and causes severe digestive disturbances. The starch may be simply removed during the processing by inserting a centrifuge in the line prior to protein coagulation.

Table 20.5 shows the effects of drying method on the composition and lipid extractability of a protein concentrate from wheat millrun (bulk of the starch removed) (Saunders *et al.* 1975A, B; 1976). These products contain about 55–60% protein, 1% fiber, 5–8% starch, and 4–6% ash. It will be noted that ether extractable lipids decreased sharply as more vigorous drying conditions were used. Total lipids, however, were readily extractable with chloroform-methanol (2:1).

TABLE 20.5

COMPOSITION OF PROTEIN CONCENTRATES PREPARED
BY ALKALINE EXTRACTION[1]

Raw Material	Drying Method	Protein %	Crude Fat (ether) (%)	Total Lipid ($CHCl_3$-MeOH) (%)
Wheat millrun	Freeze-dried	59	16	22
	Spray-dried	57	12	21
	Drum-dried	56	11	22
Rice bran	Freeze-dried	38	51	—
Triticale bran	Freeze-dried	71	29	—

[1] Centrifuged to remove part of starch.

Also included in Table 20.5 are data on concentrates (Chen and Houston 1970; Saunders *et al.* 1974, 1975B) from rice bran and triticale. Due to the high oil content of rice bran, the concentrate contains 50% lipid, making it especially valuable for high energy formulations (*e.g.*, milk replacers). The triticale bran concentrate is outstanding for its very high protein content. Protein concentrates have been prepared similarly from oats by extraction with alkali, followed by precipitation with acid (Cluskey *et al.* 1973; Wu *et al.* 1973).

LEAFY GREEN PLANTS

Let us turn our attention to the third category of unconventional protein sources—leafy plants. Their potential as a possible food protein source was pointed out over 50 years ago by Osborne *et al.* (1921). A great deal of research has been carried out in England over the past 35 years with the primary objective of developing a simple low cost process for use in developing countries (Pirie 1942, 1971). At present, however, there are no commercial plants in operation although prototype plants are being used in Nigeria and India to produce leaf protein concentrates for feed experiments with children (Olatunbosun *et al.* 1972; Kamalanathan and Devadas 1971).

Green leaves constitute the basic photosynthetic organ of higher plants which are in turn the basis of all animal life. Cultivated forage crops are the most efficient leafy crops from the standpoint of nutrient production per acre per year. Since alfalfa (*Medicago sativa*) is the best of the forage crops for large areas of the world, it will be used as the example for this discussion, with the stipulation that most leafy crops present the same problems when it comes to preparation of protein concentrates for foods or feed. The chief exceptions are leafy crops which contain strongly acid juice or large amounts of gum in the juice. The agronomy and technology of alfalfa have been thoroughly reviewed in a recent monograph (Hanson 1972). Alfalfa is a perennial legume, which may be harvested throughout the growing season (*e.g.*, 30-day cutting cycle) and which produces consistently high quality forage. On a dry basis the whole plant, as harvested, contains 15–25% crude protein, 20–30% crude fiber, and 10% ash. The lipid fraction amounts to only 4–8%, but includes important vitamins (tocopherol, vitamin K, carotene) and xanthophylls, (primarily lutein, an economically important poultry pigmenter).

About 50% of the dry weight and 75% of the total crude protein of alfalfa as conventionally harvested (about 2 in. above the ground) is in the leaves (Chrisman *et al.* 1971). Experimentally, enriched leafy fractions have been obtained by use of a leaf stripper (Currence and Buchele 1967) or by a dual cutting system, whereby the leafy tops of the plants are harvested separately from the bottoms (Kohler *et al.* 1973; Ogden and Kehr 1968). Although these leafy fractions are superior raw materials for leaf protein (LPC) production, commercial harvesters to produce them are not yet available.

The morphology and histochemistry of leaves have probably been studied more intensively than those of other plant tissues because of the overriding importance of the photosynthetic process. In general, the leaves of succulent plants are composed of an outer skin (epidermis) which encloses the internal tissue or mesophyll (Fig. 20.4). There are specialized structures within the leaf for nutrient

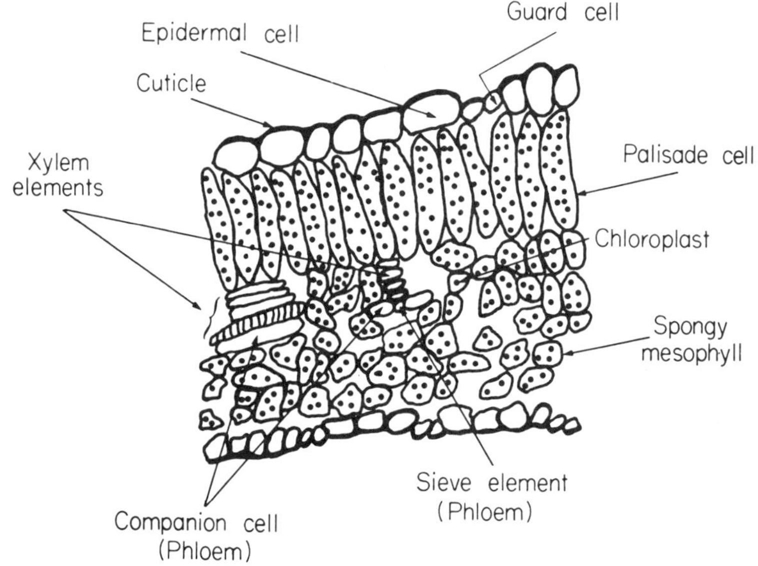

From Zelitch (1975)

FIG. 20.4. CROSS SECTION OF A LEAF SHOWING THE RELATION OF THE CHLOROPLAST-CONTAINING PALISADE AND MESOPHYLL CELLS
A portion of the xylem and phloem is shown in longitudinal section.

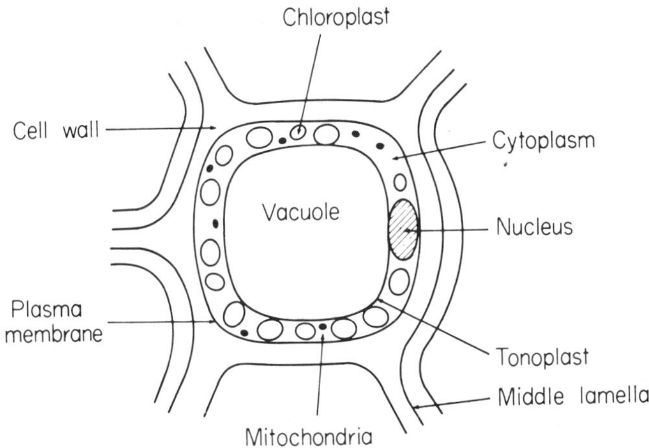

From Zelitch (1975)

FIG. 20.5. REPRESENTATION OF A LEAF PARENCHYMA CELL SHOWING THE RELATION OF THE CYTOPLASM AND ITS ORGANELLES TO THE REMAINDER OF THE CELL

transport, transpiration control, etc., but the primary functions of the leaf (photosynthesis and biochemical synthesis of proteins, carbohydrates, lipids, and intermediates) are carried out in the mesophyll cells (Fig. 20.5). In addition to a nucleus, many chloroplasts (over 100 per cell) and thousands of other much smaller bodies, the mitochondria and spherosomes, are present. All of these bodies (organelles) are embedded in a protein gel, the cytoplasm, which adheres to the outer cellulosic cell wall and surrounds a central vacuole. This vacuole, which is full of water, occupies 60–80% of the total cell volume. The layer of cytoplasm with its organelles contains more than 95% of the total protein of the leaves but only 20% of the leaf water (Kohler *et al.*, MIT Report, in press).

The outer membrane of the cytoplasm maintains the turgidity of the cell. If its permeability is reduced as by treatment with ether-saturated water, some cytoplasmic protein passes through the cell walls and can be recovered (Chibnall 1939). However, this fraction amounts to less than 20% of the total leaf protein. The proteins of the nucleus and chloroplast remain in the cell. The turgidity of the cells may also be destroyed by heating the leaf tissue to 60–80°C. This denatures and insolubilizes the proteins and thus fixes them within the cells. The juice, or diffusate, from heated leaves is composed largely of the vacuole contents.

It is thus apparent that the first requirement for extraction of large amounts of protein from leaves is to rupture the walls of unheated mesophyll cells (Osborne *et al.* 1921; Pirie 1942). Commercial grinders that effectively rupture cells are available.

Studies into the physical characteristics of leaf cells at the University of Wisconsin have led to the approach of forcing the plant material through a die in order to rupture cells (Koegel *et al.* 1973). This approach may lead to lower energy requirements for this unit operation. A second requirement for the extraction step of a leaf protein process is based on the fact that rupturing the cell walls leads simultaneously to a mixing of enzymes with substrates. Two specific examples of problems that result from this mixing are the rapid loss of protein through protease activity in ground alfalfa or alfalfa juice (de Fremery *et al.* 1972) and the rapid loss of carotenoids due to lipoxygenase, possibly through coupled oxidation with fatty acids (Grossman *et al.* 1969). These enzymatic reactions may be slowed or stopped by raising the pH to over 8.0, or by refrigeration. If the harvesting, transportation, and processing are carried out very rapidly in a continuous process, the damage done by enzyme action may not be serious, but under practical conditions it is extremely difficult to reduce postharvest, preprocess delays to a satisfactory degree. Thus, treatment to stop enzyme action should be done as soon as possible (*e.g.*, at the time of harvest).

A third consideration during extraction, imposed by the basic structure of leaves, is that when the cell walls are broken, the contents of the vacuole are mixed with the cytoplasm and its organelles. The vacuole is a repository for waste cell metabolites some of which (*e.g.*, phenolic compounds) react with the proteins to render them insoluble. It is, therefore, important to prevent such reactions. In our experience, the addition of sodium metabisulfite at grinding step substantially inhibits oxidation of polyphenols to quinones, which react with protein and may polmerize to give "off colors" (Bickoff and Kohler 1974).

The above considerations have thus provided the basis for a process (Pro-Xan process) for extraction and recovery of feed grade, leaf protein concentrate (LPC) from alfalfa which utilizes commercially available equipment, gives satisfactory yields and appears to be economically feasible. A full-scale plant (40 tons/hr) based largely on USDA technology has been constructed in France and will begin production in 1976. The economic feasibility of the process is strongly affected by the fact that drying the products requires only about half as much energy as does alfalfa dehydration. The company installing the plant is the world's largest producer of dehydrated alfalfa. Fig. 20.6 shows a flow diagram of the USDA process (*i.e.*, the Pro-Xan process). The alfalfa is harvested and chopped in the field to lengths of 1/2–2 in. It is rapidly transported to the plant where it is ground to a pulp in a nonclogging hammer mill. After milling, the material enters a twin-screw press where 55–65% of the weight of the

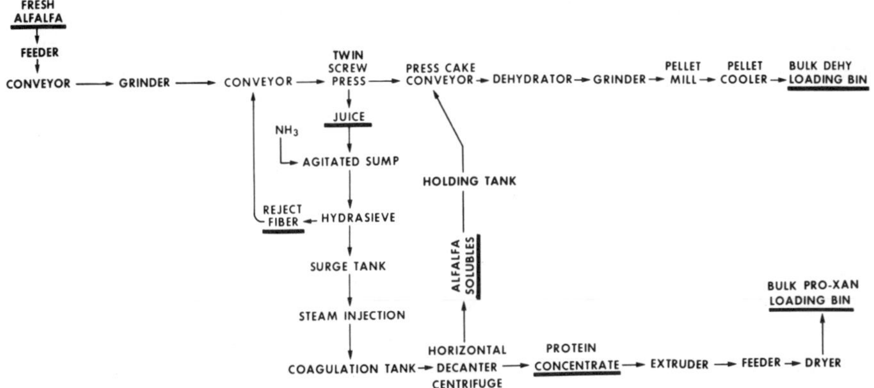

FIG. 20.6. PRODUCTION OF PRO-XAN FROM ALFALFA

fresh feed is recovered as an opaque green juice. This juice is immediately treated with ammonia to raise the pH to 8.5. The juice is then heated to 85–95°C by direct steam injection and is pumped into a decanter-type centrifuge which separates the green protein curd from the clear brown "whey." The separated curd at about 40–50% solids is dried either directly or after extruding to form noodle-like pieces (Miller *et al.* 1972) using a rotary dehydrator, a tunnel drier, or a fluidized bed drier. During drying, the product temperature should not exceed 60–80°C (Arkcoll 1969) to avoid heat damage. Current research shows that addition of ammonia or sodium hydroxide to the alfalfa during harvesting is preferred (rather than to the juice as shown) since considerable proteolysis occurs in the chopped alfalfa unless rapid transport and processing can be achieved. In the flow diagram we show that the clear brown "whey" is added back to the extracted alfalfa prior to dehydration.

Economic feasibility studies (Vosloh *et al.* 1976) have shown that when the price of natural gas is increased from the present 67¢ to $2.00 per 10⁶ BTU it will be more economical to concentrate the whey to 50–60% solids by means of a multiple effect evaporator with return of the concentrate to the press cake prior to dehydration.

Because it was felt that the whole green product (Pro-Xan) had a flavor and color that essentially ruled it out as a human food in developed countries, a program was initiated to take further advantage of the basic nature of the leaf to produce a white protein product. Solvent extraction of the coagulated protein or the juice has been used but many unsolved problems remain in its application (Huang *et al.* 1971; Wilson and Tilley 1965; Hartman *et al.* 1967). Another approach is to separate the juice protein into white and green fractions. As noted above almost all of the protein of the leaf is in the cytoplasm and its organelles; about 80% of the total resides in the chloroplasts where about half is associated with lipoprotein membranes (Wildman and Bonner 1947). These membranes contain the chlorophyll, carotenoids, and other lipid material of the chloroplast. The other half of the chloroplast protein is in the stroma between the membranes (Kawashima and Wildman 1970). In the living plant it is in the form of a gel. When the cell is subjected to physical rupture, the chloroplasts are broken up unless great care is taken to prevent this occurrence. The forces created within the cell cause the gel to become a sol, which becomes mixed with the cytoplasm. Neither the cytoplasm, nor the stroma, nor the organelles other than the chloroplasts contain chlorophyll or carotenoids. Thus, if one could separate the soluble proteins from the insoluble proteins (lipoproteins) of the pressed juice, one would have a white protein fraction desirable as a food, and a green protein fraction even richer in carotenoids than the whole leaf protein. This separation can readily be made in the laboratory using the ultracentrifuge at 75,000 $\times$ G. However, there is no large scale commercial equipment available that can accomplish this separation. One approach, which is in the pilot plant stage at the USDA's Western Regional Research Center, is fractionation by differential heat treatments (Rouelle 1773; Henry and Ford 1965; Byers 1967; de Fremery *et al.* 1973; Edwards *et al.* 1975). The pressed juice is heated to 60–65°C and pumped to a continuous centrifuge. The rapid, controlled heat treatment causes the chloroplast fragments to agglomerate. The larger particles formed are centrifugable in commercial equipment. The supernatant liquid is filtered to remove residual green particles. The clear filtered juice contains most of the soluble protein which is recoverable by heat coagulation.

Typical yields of solids and protein of both whole and fractionated LPCs are shown in Table 20.6. The yield of whole leaf protein product (*i.e.*, Pro-Xan) containing 57% protein (dry basis) is about 15% of the dry weight of the alfalfa. The crude protein recovered in the product is about 42% of the crude protein of the original alfalfa and enough protein is left in the combined press cake plus whey fraction to make it an acceptable ruminant feed product. The xanthophyll (lutein) content of the Pro-Xan is about three to four times that of the original alfalfa and makes the product of especial value in poultry rations. The recoveries shown here (Edwards *et al.* 1976) are about 60% higher than we obtained a year ago. The improvement is largely due to improvement in cell rupture achieved.

Since the heat coagulation step as described here caused denaturation and loss of functional properties, other methods of protein recovery have been considered. Early studies on acid

TABLE 20.6
DISTRIBUTION OF SOLIDS IN THE PRO-XAN SYSTEM

	Dry Wt %	Crude Protein (% of Original)	Soluble Protein (% of Original)
Raw material			
Solids	100		
Crude protein	21	100	
Soluble protein	7	33	100
Recovery in Pro-Xan process products			
Whole LPC	15	42	
Press cake	68	43	
Brown juice	17	15	
Recovery in Pro-Xan II products			
White LPC	3.3	16	43
Green LPC	11	26	

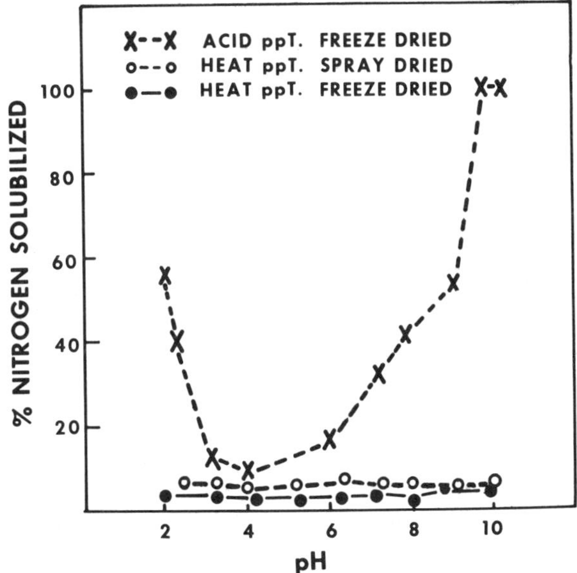

FIG. 20.7. NITROGEN SOLUBILITY OF HEAT AND ACID PRECIPITATED ALFALFA PROTEIN ISOLATES AS A FUNCTION OF pH

precipitation (Betschart and Kohler 1975) (Fig. 20.7) showed that the recovered protein was more soluble than the heat coagulated protein, but was not as soluble (*e.g.*, at pH 7) as the protein in the starting material. It was found, however, that if acid precipitation were carried out at reduced temperatures (Fig. 20.8) the precipitated protein was almost totally soluble at pH 7 (Miller *et al.* 1975). The pH solubility profile (Fig. 20.9) shows that even though damage was

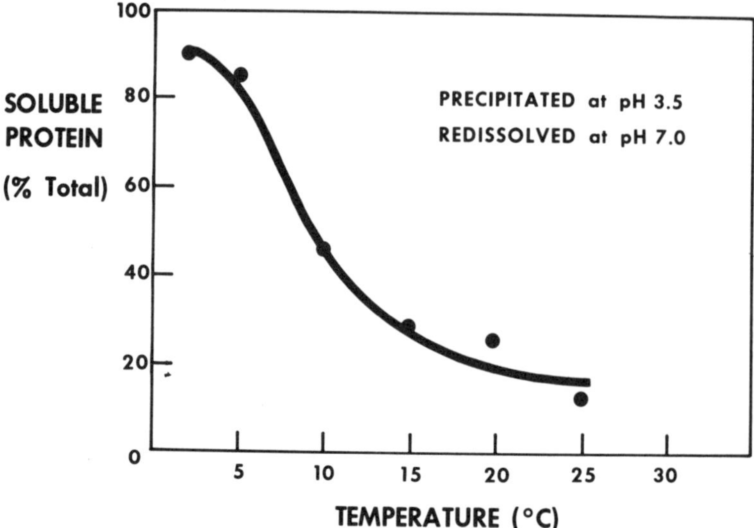

FIG. 20.8. SOLUBILITY OF ALFALFA PROTEIN PRECIPITATED AT pH 3.5
AND REDISSOLVED AT pH 7.0
Temperature maintained at indicated level during precipitation, washing, and
redissolving.

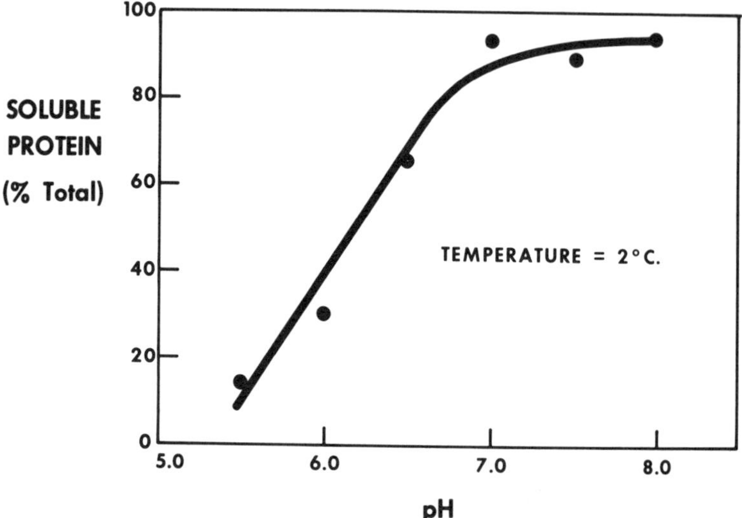

FIG. 20.9. SOLUBILITY OF ALFALFA PROTEIN PRECIPITATED AT pH 3.5
AND REDISSOLVED AT VARIOUS pH LEVELS
Temperature maintained at 2°C during precipitation, washing, and redissolving.

reduced at low precipitation temperature, the product still shows reduced solubility at pH 6, the original pH of the juice.

Ultrafiltration of the chlorophyll-free juice (Knuckles *et al.* 1975) yielded a protein with enhanced solubility (Fig. 20.10). The solid dot

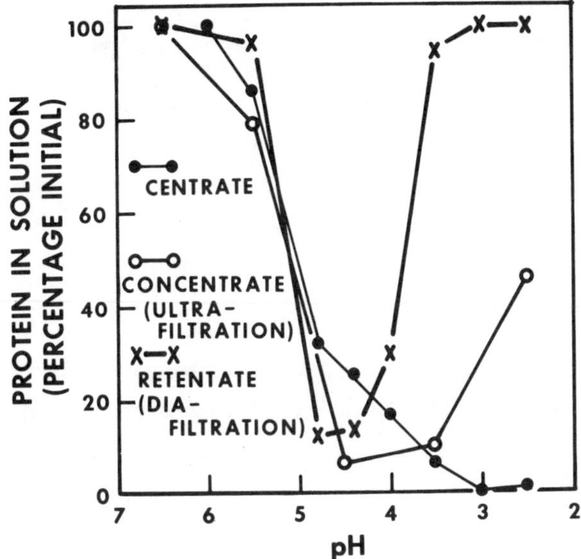

FIG. 20.10. EFFECT OF ULTRAFILTRATION AND DIAFILTRATION ON ACID SOLUBILITY OF ALFALFA PROTEIN

curve shows the raw material (centrate) solubility profile. At acid pHs very low solubility was observed. The ultrafilter concentrate showed some increase in acid solubility. The diafiltration retentate, obtained by continuously adding wash water to the ultrafilter showed good solubility at low pH values (*e.g.*, 3.5). All curves show high solubility of the nitrogen at pH 5.5 and above. These observations have provided the basis for pilot plant studies to produce soluble forms of "white" alfalfa protein.

Leaves, like seed sources of unconventional proteins, contain materials that are undesirable in a food. These deleterious substances are extremely varied in chemical nature. Examples are cyanogenetic glycosides (sudan grass, sorghum, rye grass), alkaloids (ryegrass, lupines), glucosinolates (various crucifer leaves), estrogens (legumes including clover, alfalfa, birdsfoot, trefoil, soybean). In the case of the alfalfa protein processes, the saponins and coumestrol (estrogen) tend to appear in the protein concentrates. However, special washing procedures are being developed to reduce the levels of these components to a minimum even though the levels present in normal alfalfa are not deleterious to animals. In the case of saponins, low saponin cultivars are available which yield very low saponin protein products.

In conclusion, it may be stated that the fundamental physical and chemical nature of the raw materials determine the unit operations needed and their sequence in the overall process. In the case of leaf protein, this relationship is even more critical than for oilseeds and cereal products because leaves constitute an extremely dynamic system of interacting components and opportunity for spoilage through enzymatic action or microbial attack.

BIBLIOGRAPHY

AGREN, G. 1966. Cathartic activity of safflower flour in feeding tests with children. Personal communication. Uppsala, Sweden.

ARKCOLL, D. B. 1969. Preservation of leaf protein preparation by air drying, J. Sci. Food Agr. *20*, 600-602.

BETSCHART, A. A. 1975. Factors influencing the extractability of safflower protein. J. Food Sci. *40*, 1010-1013.

BETSCHART, A. A., and KOHLER, G. O. 1975. Certain functional properties of alfalfa protein isolates prepared and dried by various methods. *In* Proceedings, 12th Technical Alfalfa Conf. 115-122. Overland Park, Kansas, Nov. 1974.

BETSCHART, A. A., LYON, C. K., and KOHLER, G. O. 1975. Sunflower, safflower, sesame and castor protein. *In* Food Protein Sources, International Biological Programme, Vol. 4, N. W. Pirie (Editor). Cambridge Univ. Press, Cambridge.

BICKOFF, E. M., and KOHLER, G. O. 1974. Preparation of edible protein from leafy green crops such as alfalfa. U.S. Pat. 3,823,128, July 9.

BOLLEY, D. S., and HOLMES, R. L. 1958. Inedible oilseed meals. *In* Processed Plant Protein Foodstuffs. A. M. Altschul (Editor). Academic Press, N.Y.

BYERS, M. 1967. *In vitro* hydrolysis of leaf proteins. 2. The action of papain on protein concentrates extracted from leaves of different species. J. Sci. Food Agric. *18*, 33-38.

CATER, C. M., GHEYASUDDIN, S., and MATIL, K. F. 1972. The effect of chlorogenic, quinic and caffeic acids on the solubility and color of protein isolates, especially from sunflower seed. Cereal Chem. *49*, 508-514.

CHEN, L., and HOUSTON, D. F. 1970. Solubilization and recovery of protein from defatted rice bran. Cereal Chem. *47*, 72-79.

CHIBNALL, A. C. 1939. Protein Metabolism in the Plant. Yale Univ. Press, New Haven, Conn.

CHRISMAN, J., KOHLER, G. O., MOTTOLA, A. C., and NELSON, J. W. 1971. High and low protein fractions by separation milling of alfalfa. U.S. Dep. Agric. ARS 74-57.

CLUSKEY, J. E., WU, Y. V., WALL, J. S., and INGLETT, G. E. 1973. Oat protein concentrate from a wet milling process: Preparation. Cereal Chem. *50*, 475-481.

CURRENCE, H. D., and BUCHELE, W. F. 1967. Leaf-strip harvester for alfalfa. J. Agric. Eng. *48*, 20-23.

DE FREMERY, D., BICKOFF, E. M., and KOHLER, G. O. 1972. PRO-XAN process: Stability of proteins and carotenoid pigments in freshly expressed alfalfa juice. J. Agr. Food Chem. *20*, 1155-1158.

DE FREMERY, D., MILLER, R. E., EDWARDS, R. H., KNUCKLES, B. E., BICKOFF, E. M., and KOHLER, G. O. 1973. Centrifugal separation of white and green protein fractions from alfalfa juice following controlled heating. J. Agr. Food Chem. *21*, 886-889.

DIECKERT, J. W., MORRIS, N. J., and MASON, A. F. 1959. Saponins of the peanut: Isolation of some peanut sapogenins and their comparison with the soya sapogenols by glass-paper chromatography. Arch. Biochem. Biophys. *82*, 220–228.

EBERT, W. F., and KNOWLES, P. F. 1968. Development and anatomical characteristics of thin-hull mutants of *Carthamus tinctorius*. Am. J. Bot. *55*, 421–430.

ECONOMIC RESEARCH SERVICE, USDA. 1974. The world food situation and prospects to 1985. Foreign Agricultural Economic Report No. 98.

EDWARDS, R. H., DE FREMERY, D., MILLER, R. E., and KOHLER, G. O. 1976. Pilot plant production of alfalfa leaf protein concentrate (PRO-XAN). Presented at 81st National Meeting, Amer. Inst. Chem. Eng., Kansas City, Missouri, April 1976.

EDWARDS, R. H., MILLER, R. E., DE FREMERY, D., KNUCKLES, B. E., BICKOFF, E. M., and KOHLER, G. O. 1975. Pilot plant production of an edible white fraction leaf protein concentrate from alfalfa. J. Agric. Food Chem. *23*, 620–626.

FELLERS, D. A. 1973. U.S. wheat products in world feeding programs. *In* Wheat: Production and Utilization. G. E. Inglett (Editor). AVI Publishing Co., Westport, Conn.

FELLERS, D. A., SHEPHERD, A. D., BELLARD, N. J., MOSSMAN, A. P., JOHNSTON, P. H., and WASSERMAN, T. 1968. Protein concentrates by dry milling wheat millfeeds. II. Compositional aspects. Cereal Chem. *45*, 520–529.

FULLER, G., WALKER, H. G., JR., MOTTOLA, A. C., KUZMICKY, D. D., KOHLER, G. O. and VOHRA, P. 1971. Potential for detoxified castor meal. J. Am. Oil Chem. Soc. *48*, 616–618.

GHEYASUDDIN, S., CATER, C. M., and MATIL, K. F. 1970A. Effect of several variables on the extractability of sunflower seed proteins. J. Food Sci. *35*, 453–456.

GHEYASUDDIN, S., CATER, C. M., and MATIL, K. F. 1970B. Preparation of a colorless sunflower protein isolate. Food Technol. *24*, 242–243.

GOLDBLATT, L. A. 1971. Control and removal of aflatoxin. J. Am. Oil Chem. Soc. *48*, 605–610.

GOODBAN, A. E. 1967. Safflower protein products for food use. First Research Conference on the Utilization of Safflower. May 25–26, 1967, Albany, Calif. Agricultural Research Service ARS-74-43.

GOODBAN, A. E., and KOHLER, G. O. 1970. Preparation of high protein products from safflower. U.S. Pat. 3,542,559. Nov. 24.

GROSSMAN, S., BEN AZIZ, A., BUDOWSKI, P., ASCORELLI, I., GERTLER, A., BIRK, Y., and BONDI, A. 1969. Enzymic oxidation of carotene and linoleate by alfalfa: Extraction and separation of active fractions Phytochemistry *8*, 2287–2293.

GUGGOLZ, J., RUBIS, D. D., HERRING, V. V., PALTER, R., and KOHLER, G. O. 1968. Composition of several types of safflower seed. J. Am. Oil Chem. Soc. *45*, 689–693.

HANSON, C. H. (Editor). 1972. Alfalfa Science and Technology. Amer. Soc. Agronomy, Inc., Madison, Wisc.

HARTMAN, G. H., AKESON, W. R., and STAHMANN, M. A. 1967. Leaf protein concentrate prepared by spray-drying. J. Agric. Food Chem. *15*, 74–79.

HENRY, K. M., and FORD, J. E. 1965. Nutritive value of leaf protein concentrates determined in biological tests with rats and by microbiological methods. J. Sci. Food Agric. *16*, 425-432.

HOUSTON, D. F. 1972. Rice: Chemistry and Technology. Am. Assoc. of Cereal Chemists, St. Paul, Minn.

HUANG, K. H., TAO, M. C., BOULET, M., RIEL, R. R., JULIEN, J. P., and BRISSON, G. J. 1971. A process for the preparation of leaf protein concentrates based on the treatment of leaf juices with polar solvents. Can. Ins. Food Tech. J. *4*, 85-90.

HUFF, R. F., and ROBE, K. 1973. Precision gyratory screening controls protein content. Food Processing *34*, No. 5, 46-48.

INGLETT, G. E., and ANDERSON, R. A. 1973. Flour milling. *In* Wheat: Production and Utilization, G. E. Inglett (Editor). AVI Publishing Co., Westport, Conn.

JENKINS, F. P. 1963. Allergenic and toxic components of castor meal: Review of the literature and studies of the inactivation of these components. J. Sci. Food Agric. *14*, 773-780.

KAMALANATHAN, G., and DEVADAS, R. P. 1971. Acceptability of food preparations containing leaf protein concentrates. *In* Leaf Protein: Its Agronomy, Preparation, Quality and Use. IBP Handbook 20. N. W. Pirie (Editor). Blackwell Scientific Publications, Oxford, England.

KASARDA, D. D., NIMMO, C. C., and KOHLER, G. O. 1971. Proteins and the amino acid composition of wheat fractions. *In* Wheat Chemistry and Technology. Y. Pomeranz (Editor). Amer. Assoc. Cereal Chem., Inc., St. Paul, Minn.

KAWASHIMA, N., and WILDMAN, S. G. 1970. Fraction I protein. Ann. Rev. Plant Physiology *21*, 325-358.

KNUCKLES, B. E., de FREMERY, D., BICKOFF, E. M., and KOHLER, G. O. 1975. Soluble protein from alfalfa juice by membrane filtration. J. Agric. Food Chem. *23*, 209-212.

KOEGEL, R. G., FOMIN, V. I., and BRUHN, H. D. 1973. Cell rupture properties of alfalfa. Trans. Amer. Soc. Agr. Eng. *16*, 712-716.

KOHLER, G. O. 1966. Safflower, a potential source of protein for human food. *In* World Protein Resources, R. F. Gould (Editor). Advances in Chemistry Series, No. 57. American Chemical Society, Washington, D.C.

KOHLER, G. O. 1969. Utilization prospects of safflower. Third Safflower Research Conference, May 7-8, 1969, Davis, Calif.

KOHLER, G. O., CHRISMAN, J., and BICKOFF, E. M. 1973. Separation of protein from fiber in forage crops. *In* Alternative Sources of Protein for Animal Production. National Academy of Sciences, Washington, D.C.

KOHLER, G. O., WILDMAN, S. G., JORGENSEN, N. A., ENOCHIAN, R. V., and BRAY, W. J. Leaf protein in relation to forage crop production and utilization. Vol. 3 of comp. study. *In* Protein Research and Technology: Status and Research Needs. Principal Investigators: N. S. Scrimshaw and D. I. C. Wang. Study Coordinator: M. Milner, Pub. by Govt. Print. Office (In press).

KOPAS, G. A., and KNEELAND, J. A. 1966. Process for preparing feed from undecorticated oil free safflower seed residue. U.S. Pat. 3,271,160, Sept. 6.

KOZLOWSKA, H., SOSULSKI, F. W., and YOUNGS, C. G. 1972. Extraction of glucosinolates from rapeseed. Can. Inst. Food Sci. Technol. J. *5*, 149-154.

KUZMICKY, D. D., and KOHLER, G. O. 1968. Safflower meal-utilization as a protein source for broiler rations. Poultry Sci. *47*, 1266-1270.

LIENER, I. E. 1975. Effects of anti-nutritional and toxic factors on the quality and utilization of legume proteins. *In* Protein Nutritional Quality of Foods and Feeds, Part 2, M. Friedman (Editor). Marcel Dekker, Inc., New York.

LOCKWOOD, J. 1960. Flour Milling. 4th Edition. Northern Publishing Co., Liverpool.

LYON, C. K., and SAUNDERS, R. M. 1975. Aqueous extraction of safflower meals: Removal of cathartic and bitter components. Abstracts of 49th Annual Fall Meeting. *In* J. Am. Oil Chem. Soc. *52*, 441A.

MARTINEZ, W., and HOPKINS, D. T. 1975. Cottonseed protein products: Variation in protein quality with product and process. *In* Food Protein Sources, International Biological Programme, Vol. 4, N. W. Pirie (Editor). Cambridge Univ. Press, Cambridge, England.

MASRI, M. S., VIX, H. L. E., and GOLDBLATT, L. A. 1969. Process for detoxifying substances contaminated with aflatoxin. U.S. Pat. 3,429,709, Feb. 25.

MATTIL, K. F. 1971. The functional requirements of proteins for foods. J. Am. Oil Chem. Soc. *48*, 477-480.

MCMASTERS, M. M., HINTON, J. J. C., and BRADBURY, D. 1971. Microscopic structure and composition of the wheat kernel. *In* Wheat Chemistry and Technology, Y. Pomeranz (Editor). Am. Assoc. of Cereal Chemists, St. Paul, Minn.

MICKELSEN, O., and YANG, M. G. 1966. Naturally occurring toxicants in foods. Fed. Proc. *25*, 104-123.

MILADI, S., HEGSTED, D. M., SAUNDERS, R. M., and KOHLER, G. O. 1972. The relative nutritive value, amino acid content and digestibility of the proteins of wheat mill fractions. Cereal Chem. *49*, 119-127.

MILLER, R. E., EDWARDS, R. H., LAZAR, M. E., BICKOFF, E. M., and KOHLER, G. O. 1972. PRO-XAN process. Air drying of alfalfa leaf protein concentrate. J. Agric. Food Chem. *20*, 1151-1154.

MILLER, R. E., DE FREMERY, D., BICKOFF, E. M., and KOHLER, G. O. 1975. Soluble protein concentrate from alfalfa by low-temperature acid precipitation. J. Agric. Food Chem. *23*, 1177-1179.

MILLERS' NATIONAL FEDERATION. 1972. Millfeed Manual. Millers' National Federation, Chicago, Ill.

MOTTOLA, A. C., MACKEY, B., and HERRING, V. 1971. Castor meal antigen deactivation—pilot plant steam process. J. Am. Oil Chem. Soc. *48*, 510-513.

MOTTOLA, A. C., MACKEY, B., HERRING, V., and KOHLER, G. 1972. Castor meal antigen deactivation—pilot plant ammonia process. J. Am. Oil Chem. Soc. *49*, 101-105.

NATIONAL ACADEMY OF SCIENCES. 1973. Toxicants Occurring Naturally in Foods, 2nd Edition. National Academy of Sciences, Washington, D.C.

O'CONNOR, D. E. 1971. Preparing light-colored protein isolate from sunflower meal by alkali extraction under inert gas blanket followed by membrane ultrafiltration. U.S. Pat. 3,622,556, Nov. 23.

OGDEN, R. L., and KEHR, W. R. 1968. Field management for dehydration and hay production, U.S. Dep. Agric. ARS 74-46, p. 23-37.

OHLSON, R. 1972. Projection and prospects for rapeseed and mustard seed. J. Am. Oil Chem. Soc. *49*, 522A-526A.

OHLSON, R., and SEPP, R. 1975. Rapeseed and other crucifers. *In* Food Protein Sources, International Biological Programme, Vol. 4, N. W. Pirie (Editor). Cambridge Univ. Press, Cambridge, England.

OKE, O. L., SMITH, R. H., and WOODHAM, A. A. 1975. Groundnut. *In* Food Protein Sources, International Biological Programme, Vol. 4, N. W. Pirie (Editor). Cambridge Univ. Press, Cambridge, England.

OLATUNBOSUN, D. A., ADADEVOH, B. K., and OKE, O. L. 1972. Leaf protein: A new protein source for the management of protein calorie malnutrition in Nigeria. Nig. Med. J. 2, 195-199.

OSBORNE, T. B., and MENDEL, L. B. 1914. Nutritive properties of proteins of the maize kernel. J. Biol. Chem. 18, 1-16.

OSBORNE, T. B., WAKEMAN, A. J., and LEAVENWORTH, C. S. 1921. The proteins of the alfalfa plant. J. Biol. Chem. 49, 63-91.

PALTER, R., and LUNDIN, R. E. 1970. A bitter principle of safflower: Matairesinol monoglucoside. Phytochemistry 9, 2407-2409.

PALTER, R., LUNDIN, R. E., and HADDON, W. F. 1972. A cathartic lignan glycoside isolated from carthamus tinctorius. Phytochemistry 11, 2871-2874.

PIMENTEL, D., DRITSCHILO, W., KRUMMEL, J., and KUTZMAN, J. 1975. Energy and land constraints in food protein production. Science 190, 754-761.

PIRIE, N. W. 1942. Some practical aspects of leaf protein manufacture. Food Manufacture 17, 283-286.

PIRIE, N. W. 1971. Leaf Protein: Its Agronomy, Preparation, Quality, and Use. IBP Handbook 20. Blackwell Scientific Publications, Oxford, England.

POMERANZ, Y., KE, H., and WARD, A. B. 1971. Composition and use of milled barley products. 1. Gross composition of roller-milled and air-separated fraction. Cereal Chem. 48, 47-58.

PURDY, R. H. 1976. Safflower processing. Personal communication. Richmond, Calif.

RACKIS, J. J. 1972. Biologically active components of soybeans. In Soybeans: Chemistry and Technology, Vol. 1, Proteins, A. K. Smith and S. J. Circle (Editors). AVI Publishing Co., Westport, Conn.

RIDLEHUBER, J. M., and GARDNER, H. K. JR. 1974. Production of food-grade cottonseed protein by the liquid cyclone process. J. Am. Oil Chem. Soc. 51, 153-157.

ROUELLE, H. M. 1773. Observations on the sediments or green parts of plants and on glutinous or vegeto-animal matter. (J. Med. Chir. Pharm. 40, 59.) Translation In Leaf Protein: Its Agronomy, Preparation, Quality and Use. IBF Handbook 20. 1971. N. W. Pirie (Editor). Blackwell Scientific Publications, Oxford, England.

SAUNDERS, R. M., BETSCHART, A. A., EDWARDS, R. H., and KOHLER, G. O. 1976. Nutritive assessment and potential food applications of protein concentrates prepared by alkaline extraction of wheat millfeeds. In Proc. of 9th National Conf. on Wheat Utilization Research. Agricultural Research Service, ARS-NC-40, 9-22. Seattle, Washington, Oct. 1975.

SAUNDERS, R. M., BETSCHART, A. A., and KOHLER, G. O. 1975B. Preparation of cereal protein concentrates. Bakers Digest 49, 49-52.

SAUNDERS, R. M., CONNOR, M. A., EDWARDS, R. H., and KOHLER, G. O. 1975A. Preparation of protein concentrates from wheat shorts and wheat millrun by a wet alkaline process. Cereal Chem. 52, 93-101.

SAUNDERS, R. M., and KOHLER, G. O. 1972. In vitro determination of protein digestibility in wheat millfeeds for monogastric animals. Cereal Chem. 49, 98-103.

SAUNDERS, R. M., and KOHLER, G. O. 1975. Concentrates by wet and dry processing of cereals. In Food Protein Sources. International Biological Programme, Vol. 4. N. W. Pirie (Editor). Cambridge Univ. Press, Cambridge, England.

SAUNDERS, R. M. WALKER, H. G., JR., and KOHLER, G. O. 1969. Aleurone cells and the digestibility of wheat mill feeds. Poultry Sci. 48, 1497-1503.

SAUNDERS, R. M., *et al.* 1974. Preparation and properties of triticale protein concentrates prepared by wet-processing of triticale bran. *In* Triticale: First Man-Made Cereal. C. C. Tsen (Editor). Amer. Assoc. Cereal Chem., Inc., St. Paul, Minn.

SMITH, A. K., and CIRCLE, S. J. (Editors) 1972. Soybeans: Chemistry and Technology, Vol. 1, Proteins, AVI Publishing Co., Westport, Conn.

SOSULSKI, F. W., McCLEARY, C. W., and SOLIMAN, F. S. 1972. Diffusion extraction of chlorogenic acid from sunflower kernels. J. Food Sci. *37*, 253–256.

SRINIVASAN, V., MONDGAL, N. R., and SARMA, P. S. 1957. Studies on goitrogenic agents in food. 1. Goitrogenic action of groundnut. J. Nutr. *61*, 87–95.

TELLER, G. L., and TELLER, W. K. 1932. A study of proteins of wheat bran. Cereal Chem. *9*, 560–572.

UNITED NATIONS. 1974. Assessment of the world food situation, present and future, Rome. United Nations Bull.

UNIVERSITY OF CALIFORNIA. 1974. A Hungry World: The Challenge to Agriculture. Univ. of California, Div. Agricultural Sciences.

VAN SUMERE, C. F., ALBRECHT, J., DEDONDER, A., dePOOTER, H., and PE, I. 1975. Plant proteins and phenolics. *In* Chemistry and Biochemistry of Plant Proteins, J. B. Harborne and C. F. Van Sumere (Editors). Academic Press, New York.

VOSLOH, C. J., JR., EDWARDS, R. H., ENOCHIAN, R. V., KUZMICKY, D. D., and KOHLER, G. O. 1976. Leaf protein concentrate (Pro-Xan) from alfalfa: An economic evaluation. (Accepted for publication as an Agr. Econ. Rept. by ERS, USDA.)

WEISS, E. A. 1971. Castor, Sesame, and Safflower. Barnes & Noble, Inc., New York.

WILDMAN, S. G., and BONNER, J. 1947. The proteins of green leaves. I. Isolation, enzymatic properties and auxin content of spinach cytoplasmic proteins. Arch. Biochem. *14*, 381–413.

WILSON, R. F., and TILLEY, J. M. A. 1965. Amino-acid composition of lucerne and of lucerne and grass protein preparations. J. Sci. Food Agric. *16*, 173–178.

WINTON, A. L., and WINTON, K. B. 1932. Structure and Compositon of Foods: Vol. 1, Cereals, Nuts, Oilseeds. John Wiley & Sons, New York.

WOLF, W. J., and COWAN, J. C. 1971. Soybeans as a Food Source. The Chemical Rubber Co., Clevelend, Ohio.

WU, Y. V., CLUSKEY, J. E., WALL, J. S., and INGLETT, G. E. 1973. Oat protein concentrates from a wet milling process: Composition and properties. Cereal Chem. *50*, 481–488.

ZELITCH, I. 1975. Photosynthesis, Photorespiration, and Plant Productivity. Academic Press, New York.

ZIEGLER, E., and GREER, E. N. 1971. Principles of milling. *In* Wheat Chemistry and Technology. Y. Pomeranz (Editor). Amer. Assoc. Cereal Chem., Inc., St. Paul, Minn.

21

Food Protein Production: Land, Energy, and Economics

David Pimentel
Marcia Pimentel

The nutrient protein is but one part of the total food component in the food/energy/population equation that the world is trying to balance. Although food energy is of prime importance, protein—especially as produced and consumed in this country—deserves serious scrutiny. This is because protein consumption per capita in the United States is one of the highest in the world. The current average consumption of 99 g per day per capita is considered generous by all standards (USDA 1975A). This is about twice the 56 g protein suggested in Recommended Dietary Allowances (NAS 1974) and 41 g protein suggested by the United Nations Food and Agriculture Organization and World Health Organization (FAO/WHO Expert Group in Protein-Caloric Requirements 1940) as the daily allowance for a 71 kg adult male. This allowance of 41 g is for the average adult male, but varies according to age, sex, activity and condition of the individual.

Furthermore, of the 99 g of protein consumed per capita per day in the U.S., about 66 g or 2/3 is of animal origin (USDA 1975A). This is in sharp contrast to the 56 g of protein consumed per capita in the Far East, of which 8 g is animal protein (PSAC 1967).

In the U.S. the major source of protein is from meat, poultry, and fish. Beef provides 19% of the total protein; pork, 7%; poultry, 8%;

and fish, 4%. Eggs and dairy products contribute 5 and 12%, respectively (USDA 1975A).

In contrast plant proteins account for less than 1/3 of the total protein consumed. Flour and cereal products contribute about 18% of the total protein; dry beans, peas, nuts, soy flour and corn grits, 5.6%; sweet and white potatoes, 2.4%; and vegetables and fruits, 5.1% (USDA 1975A).

To produce these large amounts of protein, considerable resources of land, labor and fossil fuel energy are required. Each of these factors is interrelated and, fortunately, within limits can be substituted for one another. For example, fossil energy that is used to make fertilizers and run machinery can be used to reduce the labor input and vice versa. Also increasing the intensity of land management through the use of fertilizers and tractors can reduce the land area needed in production. The reverse is also possible when the amount of available land is not a constraint (Pimentel *et al.* 1975).

Energy, land, and labor all can be evaluated in economic terms of price and also in terms of energy values (kcal). Use of economic terms facilitates comparisons of relative value on the marketplace and is effective in communicating policy decisions to the public. Unfortunately price values change depending on supply and demand of the resources, whereas energy values measured in kcal remain constant. Therefore, information about both price per se and kcal as quantitative measures of energy, land, and man-hours of labor is needed for sound decision-making about food production. Furthermore, in assessing the relative efficiency of protein production a simple comparison of protein input (kg) to protein output is a helpful guide.

COSTS OF LIVESTOCK PRODUCTION

At present, animal protein production in the United States requires tremendous inputs of energy, land, and labor. The use of feed grains and concentrates in feeding animals is in large measure responsible for this. During the last few decades reliance on use of feed grains and other concentrates in livestock production has increased. At present in the U.S., 91% (24.6 million metric tons) of the estimated 27.1 million metric tons of cereal, legume, and plant protein suitable for human consumption is fed to livestock. In turn these livestock produce only 5.3 million metric tons of protein that is consumed by humans (Table 21.1). Thus, in general, for every 1 kg of animal protein produced, about 5 kg of plant protein is fed to the livestock. These costs of livestock production include not only the protein and calorie inputs necessary to feed the livestock used for food production, but also the protein and calorie inputs used to maintain the breeding herd population.

TABLE 21.1
ESTIMATED PLANT AND ANIMAL PROTEIN PRODUCED,
FED TO LIVESTOCK AND AVAILABLE TO HUMANS FOR 1975[1]

Plant/Animal Proteins	Protein Produced	Protein Fed to Livestock	Estimated Protein Consumed by Humans
Cereals	17.0[2]	15.5[3]	1.5[4]
Legumes	9.3[2]	9.0[2]	0.3[4]
Other vegetable protein	0.8[2]	0.1[5]	0.7[4]
Livestock	6.0[2]	0.7[6]	5.3[4]
Fish	1.0[2]	0.8[2]	0.2[4]
Total protein produced	34.1	26.1	8.0

[1] After Pimentel *et al.* (1975).
[2] Data from USDA (1971; 1973).
[3] Data from USDA (1974).
[4] Data from USDA (1975C).
[5] Estimated as to the amount of apple pomace, sugar beet, and other vegetable matter fed to cattle, sheep, and hogs.
[6] Estimated as meat and milk by-products fed to livestock.

Data on specific livestock systems in various regions of the U.S. have been gathered and the protein production efficiencies estimated (Table 21.2). Of the livestock systems analyzed, milk production is the most efficient in converting plant protein into animal protein. For this system, about 31% of the plant protein fed is converted into milk protein. Or, put another way, 188 kg of plant protein are fed a dairy cow to produce about 59 kg of milk protein. Of the total plant protein fed, however, 98 kg is grain protein utilizable by humans and the remaining 90 kg is from forages.

Now consider the energy inputs for this same system. About 7 million kcal (188 kg feed) of plant protein energy are consumed by a dairy cow to produce 236,000 kcal (59 kg) of milk protein. This means 30 kcal of plant protein energy are fed for each kcal of milk protein produced (Table 21.2). This ratio of 30:1 substantiates that of Reid (1970) who reported a ratio of 28:1 for milk production. The ratio does not take into account the fossil energy input for feed production nor the energy necessary to operate the equipment used in dairy systems. The fossil energy costs for equipment and other miscellaneous production inputs are estimated to be 36 kcal of energy for 1 kcal of milk protein (Table 21.2). This ratio for the U.S. is somewhat higher than the ratio of 21:1 estimated by Leach (1975) for the United Kingdom. When this energy input is added to plant protein energy input the value is much less favorable, or 60:1.

Egg protein is the next most efficiently produced protein in terms of converting plant protein into animal protein. About 27% of the plant protein fed chickens is converted into egg protein (Table 21.2).

TABLE 21.2

ANIMAL PROTEIN (kg) PRODUCED PER HECTARE IN THE UNITED STATES
WITH VARIOUS INPUTS OF FEED, LABOR, AND ENERGY[1]

| Animal Product | Feed Protein Yield (kg) | Animal Protein Yield (kg) | Feed Energy Input (10^3 kcal) | Fossil Energy Input (10^3) kcal) for the Production of: | | | Kilocalorie Ratio | | Man-hours Labor |
				Feed	Animal Husbandry	Feed and Animal Husbandry	Feed Input/ Protein Output[2]	Fossil Energy Input/ Protein Output[3]	
Milk	188	59	6,963	2,328	6,179	8,561	30	35.9	23
Eggs	672	182	14,406	6,070	3,490	9,560	20	13.1	174
Broilers	651	116	8,886	6,446	3,787	10,233	19	22.1	38
Pork	689	65	17,021	6,774	2,438	9,212	65	35.4	28
Beef (feedlot)	786	51	24,952	7,129	8,716	15,845	122	77.7	31
Beef (rangeland)	33	2.2	1,420	0	89	89	164	10.1	1

[1] After Pimentel *et al.* (1975).
[2] Animal protein yield is converted into kilocalories and divided into feed energy input.
[3] Animal protein yield is converted into kilocalories and divided into fossil energy input for production of feed and animal husbandry.

Unlike the plant protein fed dairy cattle, however, most of the plant protein fed to chickens is also suitable for human consumption. Broiler production is somewhat similar to egg production but the efficiency of protein conversion drops to 18% (Table 21.2). Whenever consideration is given to plant proteins suitable for human consumption, conversion efficiency is reduced even more.

Swine production is similar to poultry because hogs are also fed plant proteins suitable for human consumption. Based on total plant protein fed, the conversion rate is only 6% (Table 21.2). Looking at energy input, the plant protein energy converted into pork protein energy is about 65:1, or about twice that of milk production. This ratio of 65:1 is somewhat greater than Reid's (1970) estimate of 51:1, but both indicate a conversion substantially less than that for milk.

Feedlot beef, with an efficiency conversion of only 6%, is the least effective converter of plant to animal protein (Table 21.2). Their usual plant protein feed consists of about 42% forage and 58% grains and other feed materials. Likewise the energy inputs are high compared to yield. An estimated 25 million kcal of plant energy must be fed to a beef animal to produce 51 kg of beef protein (204,000 kcal) or about 122 kcal of feed energy is fed per kcal of beef protein produced (Table 21.2). This result agrees favorably with Reid (1970) who reported a ratio of 123:1. In addition, about 78 kcal of fossil fuel energy is required per kcal of beef protein produced (Table 21.2). Also, an important factor, as Reid (1970) points out, is that protein production with beef cattle is energetically expensive "mainly because of the cost of maintaining the breeding herd."

Note that the operation of all livestock systems is "costly" in several ways other than just inefficient conversion of plant to animal protein. When livestock consume large quantities of grains that are suitable for human consumption, food supplies available to humans are being used up.

In addition to extensive land use, substantial fossil fuel energy inputs are necessary in the animal husbandry industry to grow and process feed grains. Futhermore, animal proteins require a product distribution system which most often requires processing and refrigeration. This too costs both in terms of fossil energy and money expended.

FORAGE-FED VERSUS FEEDLOT BEEF

At the present time forages and pasture grasses that are not useful to man account for about 60% of the animal protein produced. Obviously, when livestock are fed on forage, any competition with man for food is eliminated. This practice also lessens competition for the land that could be used for cultivation of grain crops suitable for human consumption. Therefore, the conversion of forage grasses and shrubs on pastureland and rangeland into animal protein (milk and meat) by cattle, sheep, and goats is extremely valuable.

Pastoral systems vary in their effectiveness to support animal protein production. For example under exceptionally good rangeland conditions in Texas where the rainfall is about 70 cm per year, approximately 2.2 kg of beef protein is produced per hectare (Table 21.2). The feed input to protein output as measured in kcal is about 164:1. Under these Texas rangeland conditions, the fossil energy input expended mainly for truck and machinery used in herding and herd management is about 4 kcal per 1 kcal of beef protein produced. Even under less optimum conditions, the energy ratio will be significantly lower than the 78:1 ratio for production of feedlot beef.

During the last two decades forage utilization increased slightly while feed grain utilization increased significantly. The increased use of feed grains was feasible because of the abundance of grain, especially corn and sorghum which were produced at relatively low costs. This change in feed practices was reflected in the fact that food energy from forage-fed beef cattle declined 10% between 1950 and 1972 (Reid 1975). Similarly, between 1960 and 1973, food energy from hay-fed dairy cattle declined 24%; and from pasture-fed, 32% (Reid 1975). As mentioned, this shift occurred because surplus grain was available. Thus, energy-intensive but profitable feed production systems evolved (Reid 1975).

After 1972, however, the sudden increase in grain exports and the disappearance of U.S. grain surpluses radically altered the picture. Grain prices soared from $50 to between $115 and $160 per ton (USDA 1970; 1975B). Because feed costs are the single greatest cost of production in beef cattle (2/3 of total production cost) and dairy cattle (1/2 of production cost), the price of meat and milk began to rise (Reid 1975). Concurrently, fossil fuel energy costs escalated and also contributed to the price increases of these animal protein foods.

PLANT PROTEIN PRODUCTION

The analyses for plant production such as those listed in Table 21.3 can be made for only a relatively few crops and represent average values, because data on crop production from different regions in the U.S. is scant and incomplete. Based on a general comparison of inputs of labor, land and fossil energy, however, plant protein is much more efficiently produced compared to cycling plant protein through animals to produce animal protein. The kg of plant

TABLE 21.3

PLANT PROTEIN PRODUCTION PER HECTARE FOR VARIOUS CROPS
IN THE UNITED STATES REQUIRING DIFFERENT AMOUNTS OF
LABOR AND ENERGY[1]

Crop	Crop Yield in Protein (kg)	Crop Yield (kg)	Crop Yield in Food Energy (10^6 kcal)	Fossil Energy Input for Production (10^6 kcal)	kcal Fossil Energy Input/ kcal Protein Output	Labor (Man-hours)
Alfalfa	710	6,451 (dry)	11.4	2.694	0.95	9
Soybeans	640	1,882	7.6	5.285	2.06	15
Brussels sprouts	604	12,320	5.5	8.492	3.51	60
Potatoes	524	26,208	20.2	8.907	4.25	60
Corn	457	5,080	17.9	6.644	3.63	22
Rice	388	5,796	21.0	15.536	10.01	30
Dry beans	325	1,457	5.0	4.478	3.44	15
Oats	276	1,900	7.4	2.978	2.70	6
Wheat	274	2,284	7.5	3.770	3.44	7

[1] After Pimentel *et al.* (1975).

protein produced per hectare varies with the crop, but generally is eight times the yield of animal protein per hectare.

The average yield of wheat per hectare is 2,284 kg and equates to 274 kg protein and 1.1 million kcal. Both corn and rice outproduce wheat in terms of quantities and kcal of protein per hectare. Corn yields 457 kg protein and 1,828,000 kcal; rice, 328 kg and 1,552,000 kcal. Of the legumes, soybeans yield the largest amount of protein, or 640 kg and 2.6 million kcal. Note especially the excellent yield of potatoes, 524 kg of protein per hectare.

When crops are compared on the basis of kcal fossil energy input to kcal protein output, soybeans are highly efficient (Table 21.3). Next come oats, dry beans, wheat, potatoes, and corn in order of decreasing efficiency. Rice is relatively inefficient mainly because of the high energy inputs required in irrigation.

The labor resource input per kg of protein varies greatly in kinds of plant and animal protein products produced (Tables 21.2 and 21.3). In general, about 30 kg of plant protein are produced per man-hour of labor, compared to about 2 kg of animal protein. Thus considering both grams of protein and energy used, plant protein production is much more efficient than that of animal protein.

TABLE 21.4

THE COST OF 20 g OF PROTEIN AND THE PRICE PER POUND

IN ITHACA, NEW YORK, MAY, 1976

Food Item	Price per Pound[1]	Cost of 20 g of Protein[1]
Soybeans	$0.49	$0.06
White, wheat flour	0.16	0.07
Rye flour	0.20	0.07
Red beans	0.39	0.08
Navy beans	0.48	0.09
Cornmeal	0.24	0.12
Oats (meal)		0.12
Peanut butter	0.77	0.12
White bread	0.40	0.20
Hamburger	0.92	0.23
Eggs, Grade A large	0.66	0.25
Rice	0.39	0.26
Milk	0.20	0.26
Tuna	1.47	0.27
Potatoes	0.14	0.29
Turbot/cod	1.23	0.31
Chicken breast	1.14	0.40
Haddock	2.39	0.58

[1] U.S. dollars.

PROTEIN PRICES

The consumer-market price of proteins reflects the costs of production as well as supply and demand (Table 21.4). In general, the price of plant protein costs about 1/3–1/2 that of animal protein for an equivalent of 20 g (Table 21.4). Of the plant protein, soybeans, wheat flour, and dried red beans cost the least ($0.06–0.08 for 20 g), whereas rice and potatoes cost the most ($0.26–0.29).

RESEARCH NEEDS FOR THE NEXT 25 YEARS

Pressure for energy, food, land, water, and other resources is significant even today. Therefore, with the U.S. population, and especially the world population, projected to increase greatly during the next 25 years, food shortages and energy, water, and land limitations will become critical. Hence, it is important that research on various aspects of the food production problem, and in particular protein supply, be initiated. The following is a list of five priority areas:

Grass-Fed Livestock

The feasibility of maintaining various kinds of livestock production in the future will depend upon the extent to which the animals compete with man for food grains and land, plus the economic pressures resulting from a depletion of nonrenewable energy supplies, especially fossil fuel.

Much data support the fact that feeding large amounts of grain to animals is a costly method of producing protein. The use of grains that can be consumed by man compounds the inefficiency. In addition, Reid (1970) predicts that little grain concentrate will be available for animal production systems by 2000 A.D., so alternate procedures for protein production will probably have to be developed.

Could the problem be alleviated by changing the production of animal protein from grain-fed to forage-fed? A preliminary estimate suggests that if this were possible, animal protein production would decline from 6 to 2 million metric tons (Pimentel *et al.* 1975). This estimate assumes a shift in animal protein production from swine and poultry, which compete with man for plant protein to species like beef, lamb and goat that can utilize forage.

A thorough investigation into this possibility is needed. Included also would be ways to improve forage production on our pasturelands and to reduce overgrazing.

Public Acceptance of Modified Protein Intake

Obviously if a substantial reduction in available supplies of protein occurs and population continues to increase, food consumption patterns will have to be modified. Whether the changes will be reflected in a decrease in total protein or amount of animal protein—or both—remains to be seen.

In the United States a decrease in per capita consumption of animal protein could be compensated for by an increase in consumption of plant protein.

One problem to be faced in such a change is the fact that animal proteins are of higher nutritional quality than plant proteins because they contain the eight essential amino acids and in the amounts required by man (Burton 1965). For this reason, the public will have to be instructed on how to complement plant proteins to ensure that the essential amino acids will measure up to those of animal proteins. Fortunately, protein combinations such as grains and legumes (*i.e.*, rice and beans) not only achieve a satisfactory amino acid composition but also make a palatable food (FAO 1970).

The fuller utilization of plant protein foods holds much promise in the future because they are economical and energy-efficient protein foods. In addition compared to animal proteins, they are low in saturated fat and cholesterol.

Eating patterns in all cultures are slow to change. Because the U.S. public is unfamiliar with the vast variety of plant proteins, much effort will have to be directed toward developing palatable ways to serve them. Thus, the ultimate question to be answered is how can a population that is used to eating a high total protein-high animal protein diet be convinced to substitute a lower protein, lower animal protein diet? Granted, supply and price alter buying and consumption patterns, but the basic need is to explore other factors that would successfully motivate this type of major change in our eating patterns.

Degradation of Agricultural Land

Cropland is one of our most valuable resources and will be even more so as we try to feed our increasing population, whether we produce plant or animal protein. Attention must be given these lands for they are deteriorating at an alarming rate. For instance, during the past years, highway and urban construction has removed about 36 million acres of excellent cropland from production (Pimentel *et al.* 1976). At the same time both water and wind have eroded the soil from about 200 million acres, and thereby have ruined the land for crop production (Bennett 1939). At present an estimated 460 million acres of cropland are left in the U.S. and much of this land already has lost about 1/3 of its topsoil (NAS 1970).

Thus, a careful study of the rate of cropland loss is needed. Even more vital is the development of a sound program to protect and conserve our remaining valuable cropland.

Relocation of Agricultural Production in the United States

During the past three decades agricultural production has shifted from local production to more regional production. For example, in the past most of the fresh vegetables consumed by cities and towns were produced within a short distance of the city and town. Most of this type of production was also seasonal.

Gradually, this local seasonal production has shifted to certain regional production. Now the produce grown in an area is transported to the cities and towns throughout the United States. For example, most of the broccoli (95%), lettuce (80%), and processing tomatoes (80%) is now produced in California and shipped to various localities in the U.S. About 96% of the rice is produced in Arkansas, California, Louisiana, and Texas; and 76% of the dried beans are produced in Michigan, California, Idaho, Nebraska, and Colorado.

An analysis is needed to determine the current fossil energy inputs required for production of vegetables in California using heavy irrigation and then transporting them to eastern cities versus producing vegetables locally near the eastern cities. Production would, of course, be seasonal, and a variety of fresh produce would be limited. However, as the price of fossil energy increases, more localized production may become necessary.

Energy Costs for Packaging and Processing

Large quantities of fossil energy are used in processing and packaging foods. For example, more than three times as much energy is required to either freeze or can corn than is required to produce the quantity of corn put in either package (Brown and Batty 1975). Also large amounts of energy are expended to package small amounts of fresh fruits and vegetables for the consumer. Thus, the quantities of energy and other resources that are inputs in processing and packaging should be assessed. In this way the most efficient technology in processing and packaging can be developed and used to keep energy expenditures at a minimum. The factor of energy-efficient packaging is especially important if we continue to maintain regional production of food crops.

BIBLIOGRAPHY

BENNETT, H. H. 1939. Soil Conservation. McGraw-Hill, New York.
BROWN, S. J., and BATTY, J. C. 1976. Another perspective of energy and America's food system. Trans. Amer. Soc. Agr. Eng. (In press).
BURTON, B. T. 1976. The Heinz Handbook of Nutrition. McGraw-Hill, New York.

FAO. 1970. Amino acid content of foods and biological data on proteins. Food and Agriculture Organization Studies, No. 24, Rome.

LEACH, G. 1975. Energy and Food Production. International Institute for Environment and Development, London.

NAS. 1970. Biology and the Future of Man., P. Handler, (Editor). National Academy of Sciences, Oxford Univ. Press. New York.

NAS. 1974. Food and Nutrition Board: Recommended Dietary Allowances, 8th Edition. National Academy of Sciences, Washington, D.C.

PIMENTEL, D., DRITSCHILO, W., KRUMMEL, J., and KUTZMAN, J. 1975. Energy and land constraints in food-protein production. Science *190*, 754–761.

PIMENTEL, D., TERHUNE, E. C., DYSON-HUDSON, R., ROCHEREAU, S., SAMIS, R., SMITH, E., DENMAN, D., REIFSCHNEIDER, D., and SHEPARD, M. 1976. Environmental effects of land degradation on food and energy resources. Science *194*, 149–155.

PSAC. 1967. The World Food Problem. Report of Panel on the World Food Supply, Pres. Sci. Advis. Comm., The White House. Vol. 1, 2, 3. U.S. Govt. Printing Office, Washington, D.C.

REID, J. T. 1970. Will meat, milk and egg production be possible in the future? Proc. Cornell Nutr. Conf. for Feed Manufacturers, Buffalo, N.Y., Nov.

REID, J. T. 1975. A re-evaluation of the importance of forages. Dep. Animal Sciences, Cornell University, Ithaca, N.Y. October. Unpublished mimeo.

USDA. 1970. Agricultural Statistics, 1970. U.S. Dep. Agric., Washington, D.C.

USDA. 1971. World demand prospects for grain in 1980. Econ. Res. Serv., Foreign Agric. Econ. Rep. No. 75.

USDA. 1973. Agricultural Statistics, 1973, U.S. Dep. Agric., Washington, D.C.

USDA. 1974. The world food situation and prospects to 1985. Foreign Agric. Econ. Rep. No. 98.

USDA. 1975A. National food situation. Econ. Res. Serv. NFS-154.

USDA. 1975B. Agricultural Statistics, 1975, U.S. Dep. Agric., Washington, D.C.

USDA. 1975C. National food situation. Econ. Res. Serv. NFS-151.

22

Outlook for Alternative Protein Sources in Processed Foods

R. H. Forsythe
E. J. Briskey

Several previous authors in this book have thoroughly reviewed the classical food protein systems. Indeed, some new protein sources have also been discussed. By our own definition, we have included in the discussion to follow some proteins which are in substantial commercial production and therefore probably not considered "new" by some standards. We have limited our discussion to those proteins reasonably expected to be used in the United States for food processing functions and which, we believe, offer significant growth opportunities for the future.

We have chosen to use the term "alternative proteins" rather than new or unconventional, as we believe it is the simplest description of the departure from traditional proteins such as milk, meat, and eggs. While traditional proteins have become accepted, both from the standpoint of desirable organoleptic properties and safety for human consumption over the centuries, alternative foods are new, or relatively new, most having been introduced within the past 25 years and do not enjoy those advantages. We will discuss currently used alternative proteins as well as those still in the research or conceptual phase.

The most obvious questions, often left unasked, are "Why does a food processor have an interest in alternative proteins for processed foods?" and "Why are food ingredient manufacturers in the U.S.

investing large sums of money developing alternative resources?" Even though accurate investment data are not made public, large sums are being invested by at least 20 major food companies in this area. If the sketchy reports available are reasonably accurate, the returns to date have been largely disappointing. There are, however, sound reasons for these investments which include:

(1) In the long run, conventional animal proteins will be in short supply and higher in price (Scrimshaw *et al.* 1975; Willett 1974; Wittwer 1975).

(2) On the international level, significant increases in protein consumption will be required to supply adequate nutrition.

(3) Improved functional performance will be required for many new fabricated foods.

While the contributions to many companies' total profits from new protein sources will be minimal for the next five years, Hale (1975) predicts the long-term profits will look better by 1985.

ALTERNATIVE PROTEINS

Commercial Alternative Proteins

A number of alternative proteins are now in commercial use as additives in a wide variety of food products. The product categories and the amount of added protein are shown in Table 22.1. From these estimates, it is apparent that the growth of alternative protein ingredients is expected to be in the manufacture of fabricated foods, pasta and bakery products.

The alternative protein additives now being used and their relative amounts projected to 1980 are shown in Table 22.2. It is obvious that whey, casein and soy-based proteins are expected to dominate the market through the decade of the seventies (Muller 1974). Single-cell and other little recognized proteins will show the greatest growth rates, but still will constitute less than two percent of the total market for protein additives (Lipinsky and Litchfield 1974; MacLaren 1975).

New Alternative Proteins

A large number of proteins presently used for animal feed or not yet being produced commercially appear to offer growth possibilities as technology improves and economics force the industry to continue its search for less expensive ingredients. Some new potential protein sources for human food are given in Table 22.3. While considerable research is under way on several of these proteins, only cottonseed protein, fish meal, legume protein, triticale, mechanically deboned meat and single cell protein are expected to experience

TABLE 22.1

ESTIMATED U.S. PRODUCTION OF SELECTED FOODS CONTAINING PROTEIN

(MILLION POUNDS)

Product Category	1975		1980	
	Est. Prod.	Added Protein	Est. Prod.	Added Protein
Meat, processed meats including poultry	13,000	94	15,600	116
Baked goods, pasta	19,400	210	20,500	225
Cereals	2,000	44	2,000	44
Snack foods	6,600	66	8,100	82
Convenience foods[1]	6,400	92	7,600	136
Primary fabricated foods[2]	6,700	350	11,200	600
Total	54,100	856	65,000	1,203

Source: Adapted from Stone (1976).

[1] Includes dried and frozen dinners, instant breakfasts, etc.
[2] Includes nondairy toppings, coffee whiteners, high protein snacks, meat and poultry analogs.

TABLE 22.2
QUANTITY OF ALTERNATIVE PROTEIN USAGE
FOR HUMAN DIETS, 1975–1980
(MILLION POUNDS)

Protein	1975	1980
Whey	495	632
Casein (Caseinates)	120	202
Soy-based		
Flour/grits	160	185
Concentrates	35	42
Isolate	45	137
Textured	30	75
Single-cell protein	1	25
Other	5	

Source: Adapted from Hale (1975).

TABLE 22.3
FUTURE HUMAN FOOD PROTEINS

Near Term	Future
Cottonseed meal	Alfalfa meal
Fish meal	Okra seed
Legume protein	Fungal protein
Triticale	Bacterial protein
Mechanically deboned meat	Algal protein
Yeast protein	Blood
	Insects

much growth within the next ten years (Anon. 1975; Mulhern 1976; Bressani and Elias 1974; Finch 1970; Gray 1970; Halliday 1975; Karatoltsidis and Constantinides 1974; Lipinsky and Litchfield 1970; Oki and Kitai 1974; Parrish *et al.* 1974; Reed 1974; Seeley 1975; Waslien 1975; Watanabe *et al.* 1974).

Functional Properties

The functional properties expected of proteins in food systems have been reviewed in this volume. It should be emphasized, however, that at least in the United States, the primary reason for investigating these new food ingredients is to obtain better functionality at lower costs. Nutritional considerations, of most importance in the developing nations, should be considered but are currently of minor importance in the protein-rich countries.

CONSTRAINTS TO ALTERNATIVE PROTEIN USAGE

Richardson (1975) in a Symposium on Unconventional Sources of Protein held at the University of Wisconsin in April of 1975

TABLE 22.4

CONSTRAINTS ON USE OF ALTERNATIVE PROTEINS

Nutritional considerations
Toxic factors
Gastrointestinal factors
Organoleptic factors
Legal aspects
Costs

discussed several constraints to increased utilization of alternative proteins in the human diet and for food processing. Some of these are listed in Table 22.4.

Nutritional Considerations

Nutritional considerations, as pointed out earlier, are of lesser importance when using alternative protein additives in conventional food systems since the level of usage is usually low; however, when meat, milk or egg analogs are considered, the usage level increases significantly and nutritional concerns surface. Aspects of the nutritional problem include:

(1) The lack of essential amino acids in some proteins;
(2) The possibility that as yet unidentified micronutrients traditionally supplied with animal proteins, may not be present in the alternative source either because of their absence in the raw material or removal or destruction during processing; and
(3) Availability of nutrients is lowered due to lack of digestibility or inhibitors.

The total protein content of alternate proteins varies widely due to source and degree of processing. Several alternative proteins are compared with traditional proteins in Table 22.5.

Table 22.6 illustrates some of the amino acid differences in alternative proteins. It should be pointed out that these are average or typical values and the data will vary according to source, variety or processing and storage conditions. The primary limiting amino acids in soy and single cell protein are the sulfur-containing amino acids, methionine/cystine.

Lachance (1974) reviewed the limiting amino acids in a number of proteins and suggested fortification steps that might be taken. Wheat is deficient in lysine which can be accommodated by the addition of 5% fish protein concentrate. Other protein blends offer significant improvements in nutritive quality. Rice is also low in lysine and threonine. The Ajinomoto Company has developed a fortification program based on the addition of these amino acids in a "synthetic" rice kernel. Wheat and rice utilized in school feeding programs in Japan are fortified with lysine.

TABLE 22.5
PROTEIN CONTENT OF FOOD PROTEINS[1]

Source	% Protein
Beef	19.5
Whole Milk	3.5
Eggs	12.8
Casein	99.0
Rice	6.7
Soy, extruded	52.5
Soy Concentrate	70.0
Soy Isolate	95.0
Wheat	12.2
Cottonseed Flour	55.0
SCP, Yeast	45–55

[1] Adapted from Hale (1975).

TABLE 22.6
AMINO ACID CONTENT OF ALTERNATIVE PROTEINS

	WHO-FOA Profile	Whey	Casein	Soy	SCP	Hen Egg
PER		3.2	2.5	2.2	1.7	3.2
			(g/100g)			
Isoleucine	4.0	5.1	6.6	4.5	6.6	6.9
Leucine	7.0	7.7	10.1	7.1	8.0	9.4
Lysine	5.5	8.7	8.2	6.7	7.4	6.9
Methionine	3.5	1.9	3.3	1.5	1.7	3.3
Phenylalanine	6.0	3.8	5.8	5.1	5.3	5.8
Threonine	4.0	6.7	4.5	3.6	5.4	5.0
Tryptophan	1.0	3.3	1.5	1.0	1.2	1.6
Valine	5.0	5.9	7.4	4.5	6.2	7.4

A fortified corn product, Incaparina, has been used in Central America since 1957. A new Incaparina contains 70% rice and 30% soy flour. Peanut protein is deficient in three amino acids—lysine, methionine and threonine. Rosenfield and Hornstein (1974) contend that if population growth can be stabilized, man's protein needs can be met by fortification with amino acids and protein blends. The introduction of higher nutritive value varieties will be of considerable significance. In addition to amino acid deficiencies and imbalances, it is recognized that amino acid toxicities may result from an excess of an amino acid in the diet (Pike and Brown 1967). It is unlikely that this will become a serious problem in processed foods where the total alternative protein content will be lower.

In addition to the known effects of deficiencies of certain nutrients and the effects of factors reducing the bioavailability of alternative proteins, one must consider that extensive replacement of traditional proteins may result in other unidentified problems. Traditional proteins probably supply a number of trace elements and/or growth factors for which requirements have not been established. Long-term reduction of these micronutrients must be considered by food scientists developing foods based on alternative protein ingredients.

Toxic Factors

Rackis (1972), Richardson (1975) and the National Academy of Sciences (1973) have reviewed potential toxicological hazards associated with a number of alternative protein sources. These include allergens, amino acid antagonists, enzyme inhibitors, antivitamins, mycotoxins, trace elements and nucleic acid.

Allergens.—According to Jaffe (1973), many people show reactions to certain foods. While the nature of these reactions is not well understood, the intensity is usually much less than with true food toxins. The allergens thus probably should not be considered as food toxicants because the reason for the reaction rests with the individual and not the food. Most allergens are protein and have been observed in grains, milk, eggs, fish, crustaceans, tomatoes, strawberries and many other foods. Very few allergic reactions have been associated with soy, according to Rackis (1972) and, in fact, soy milk has been used for infants allergic to human and cow's milk. High thermostability is a characteristic of many plant allergens. Heating at 180° F for 30 min may be required to inactivate allergens as compared to the much lower thermostability of the antinutritive factors.

The ingestion of wheat proteins, principally gluten, results in a malabsorption syndrome in some individuals. This syndrome, called celiac disease, can result in severe malnutrition due to interference with absorption of amino acids, glucose and certain vitamins. Rye, barley and oats also result in celiac disease in sensitive individuals.

Amino Acid Antagonists.—Pike and Brown (1967) and Roe (1968) have described amino acid antagonism as the syndrome resulting from an amino acid imbalance causing an accumulation of the amino acid or its metabolites in the body. This problem is probably of limited significance except in the case of disease. The accumulation of phenylalanine and its metabolities in phenylketonuria is an example. It is also known, for example, that an excess of lysine can cause depressed utilization of arginine.

Enzyme Inhibitors.—The bioavailability of other nutrients may be adversely affected by several plant proteins. Rackis (1972, 1974) has extensively reviewed the biological and physiological factors in soybeans. Heating, particularly in the presence of moisture, reduces or eliminates many of these factors, but the demand for a nearly native protein for further processing emphasizes the need for consideration of these factors. Raw soy flours inhibit growth, depress fat absorption, reduce protein digestibility, modify enzyme secretions and reduce amino acid, vitamin and mineral availability. Intestinal microflora appear to intensify the growth inhibitory properties of raw meal. Soy isolates also create deficiency symptoms associated with decreased availability of calcium, magnesium, manganese, zinc and other trace elements. Trypsin inhibitor activity is effectively reduced by heating under alkaline conditions.

Whitaker and Feeney (1973) have also described a number of enzyme inhibitors in foods. A wide variety of beans inhibit trypsin and chymotrypsin. Amylase inhibitors are found in wheat, beans and a variety of fruits. Many plants contain significant quantities of acetyl cholinesterase inhibitors, the most potent being found in the potato. A powerful inhibitor, physostigimine, is found in the West African calabar bean. The carbamate insecticides are based on analogs of this compound.

While the largest body of research in this area has been conducted with soybeans, it is anticipated that as other alternative proteins are developed, these problems must receive added attention.

Antivitamins.—Somogyi (1973) defines an antivitamin as a compound that diminishes or abolishes the effect of a vitamin in a specific way. Probably the first such substance recognized was the antivitamin D substance, phytic acid from cereals. The antagonistic action of avidin on biotin has also been well documented. Thiamin antagonists have been noted in a variety of plants, including beans, mustard, cottonseed and flax. A number of bacteria also depress the utilization of thiamin. Corn and wheat bran have been observed to be antagonistic toward niacin. It is not possible to discuss here all of the antivitamin substances; our purpose is rather to indicate how these factors may serve as constraints to the utilization of alternative protein sources.

Mycotoxins.—The most common toxicant associated with alternative proteins is probably the broad class of toxins developed by fungi (Wilson and Hayes 1973). Mycotoxins are normally considered to be those associated with the filamentous fungi, and not those typically associated with mushroom poisoning. Except for fungi to be used directly as food, mycotoxins are not truly protein toxicants but result from poor handling of plant proteins during harvest, storage and processing. The primary concern has been with the ingestion of mycotoxins by animals but evidence that the human population, particularly in the undeveloped areas, is subjected to these hazards is increasing. The most common sources include peanuts, corn, grains and rice. Aflatoxins from aspergillus species are the most common, but fusarium toxins are being encountered increasingly.

Gray (1970) has discussed the use of fungi as food, the most common of which is, of course, the mushroom. The protein content of mushroom is in excess of 25% (dry basis). Fungi have long been used in food processing, for example, cheeses, a number of oriental foods such as tempeh and in the aging and flavoring of meat. Gray suggests that the fungi can be used to replace more traditional foods to supply future protein requirements with minimum land usage. Few molds actually produce toxins and culture conditions can be controlled to reduce or eliminate this hazard. He contends that almost unlimited quantities of protein can be produced by culturing fungal mycelia in submerged culture.

Trace Elements.—Many trace elements are associated with foods from all sources. In the case of plants, these are greatly dependent on the soil in which they are grown and in the case of animals, on the plants which they consume. This is true for either land or aquatic based species. Underwood (1973) considered the most toxic trace elements to be arsenic, lead, cadmium and mercury. Probably all trace elements could be classified as toxic if fed in high enough doses and/or for long periods of time. Selenium, for example, can result in either a deficiency or a toxicity, depending on the species and the levels ingested. Selenium toxicity, as with several other toxic substances, is affected by other dietary substances. Ten parts per million selenium is highly toxic to rats fed a 10% protein diet but is harmless in a 20% protein diet. Five parts per million arsenic in drinking water prevents selenosis in rats. Chromium has also been implicated in food toxicoses but information is meager at present. Fluorine and iodine must also be considered as possible toxicants for at least a portion of the population.

Possibly of equal concern is the reduction or elimination of a number of trace elements by the departure from traditional sources of protein in the diet. Unidentified growth factors in traditional foods may not manifest their effects for long periods of time in test animals and are quite unpredictable in the human population.

Nucleic Acid.—In the case of the yeasts, extremely high nucleic acid levels place rather severe restrictions on the amount that can be consumed. Typically, yeasts contain approximately 10% nucleic acid. Levels of over 2–3 g per day are considered unacceptable for humans. Higher levels can lead to increased formation of the relatively insoluble urates which have been associated with gouty arthritis (Edozien *et al.* 1970; Waslien *et al.* 1968). A number of processes have been developed to selectively hydrolyze the nucleic acids. These include acid hydrolysis, base hydrolysis, and endogenous and exogenous nucleases (Maul *et al.* 1970; Castro *et al.* 1971; Hedenskog and Ebbinghaus 1972).

Miscellaneous Toxicants.—A large number of potential food toxicants are reviewed in the National Academy of Sciences publication, *Toxicants Occurring Naturally in Foods* (1973) and Rackis (1974). These include cyanogenetic glycosides, plant phenolics, oxalates, phytohemagglutinins, saponins and sterols. All of these must be considered by future protein technologists.

Organoleptic Factors

Probably the most serious constraint on the utilization of alternative proteins in the next decade is that group of properties which determines the acceptability of the products to potential users. Any departure from the traditional is difficult and the use of alternative forms of protein is no exception. Surprisingly little information is available in the scientific literature on the taste, odor, texture and appearance of new proteins. This may be because the developers were first interested in animal consumption; currently, low levels are used in processed foods, and partly because maximum effort has been directed toward attempts to mask the undesirable flavors found in many products. This may also be due to the observation that most alternative proteins are being developed by biochemists, microbiologists, agronomists and chemical engineers who do not fully understand that no food becomes a nutrient until it is consumed.

It has long been recognized (Tappel 1962) that a wide variety of plants contain lipoxygenase enzymes which catalyze the peroxidation of polyunsaturated fatty acids. These reactions are responsible for a spectrum of desirable and undesirable flavor compounds. Heat destroys these enzymes but also denatures the protein, reducing its functionality and may reduce the nutritive value through nonenzymatic browning. Many of the commercial processes under development for the modification of plant proteins have ignored this basic biochemistry.

Gastrointestinal Factors

The primary gastrointestinal factor is the development of gas, commonly referred to as flatulence, which may lead to nausea, cramps, diarrhea, pain and social discomfort (Rackis 1972). This is primarily caused by the fermentation of dietary carbohydrates in the lower intestine. The amount of gas is determined by the type of the intestinal microflora and the ingested carbohydrate. In soy, the high flatulence fractions are alcohol-soluble oligosaccharides, primarily sucrose, raffinose and stachyose. Little activity is found in the hulls, fat or protein fractions. Soy isolates contain little flatus activity. Unprocessed navy beans and wheat cereal have high activity.

Legal Constraints

McNamara (1975) has outlined some of the legal aspects of introducing new foods. A manufacturer of a nutritious food ingredient is not free to just put it in the marketplace. Many new food ingredients are likely to be called "food additives" within the meaning of the law. This requires that the FDA must approve the additives use *before* they can be included as a food ingredient. Even if the necessary proof can be supplied, which if not available may take years to obtain, the manufacturer will still face stiff label declarations.

A food that has been used for years in other countries could not automatically be considered GRAS if introduced into the United States.

It may be necessary to specify, for example, the exact species involved in the manufacture of plant proteins in the future. Minimum nutritional standards may also be required. Regardless of a new food ingredient's nutritional qualities, at present it must be labeled as "Imitation" if it closely resembles a traditional food. This implied connotation of inferiority could seriously retard marketing efforts.

Another serious constraint is the aesthetic consideration. The FDA Act provides that a food shall be deemed to be adulterated "if it consists in whole or in part of any filthy substance, or it if is otherwise unfit for food." Whole fish protein concentrate is a good example of the exclusion of a wholesome alternate protein. It would appear that in the future, we may have to modify our traditional concepts of "filth" and "unfit for food."

The recent USDA proposals (Mulhern 1976) permitting a variety of mechanically deboned and low temperature rendered meat products to be declared as "meat" on labels should encourage the development of these products if approved.

It is impossible to discuss even a small segment of the legal problems that can be anticipated in the introduction of alternative proteins. This will be a prime concern of tomorrow's food technologists.

Cost

While one of the most severe restraints to the use of alternative proteins will be the cost as compared to traditional proteins, this is also the most difficult to project. Comparable costs, based on 1974 data, per pound of protein are shown in Table 22.7.

The use of alternative ingredients in processed food is clearly dependent on functionality as well as protein content, since small quantities may significantly improve the final fabricated food.

TABLE 22.7
RELATIVE COSTS OF PROTEIN

Protein Source	Cost $/lb	Cost of Utilizable Protein $/lb
Beef	1.39	9.29
Milk	0.08	2.90
Eggs	0.32	2.67
Casein	0.82	1.15
MDM	0.25	1.75
Rice	0.13	2.43
Soy, extruded	0.57	1.87
Soy Isolate (90%)	0.64	0.75
Wheat	0.07	0.88
Cottonseed Flour	0.15	0.42
SCP	0.35	0.70

New protein sources such as alfalfa and peanuts should be cheaper than soybeans, based on raw material cost estimates and assuming similar processing costs. It has been estimated that peanut flour (60% protein) at 35¢/lb could replace NFDM at 65–70¢/lb and caseinate at $1.20.

Raw material and processing costs for alternative proteins are subject to the same influences as traditional sources, so current differences can be expected to be maintained over the next decade (Chancellor and Gross 1976). Probably the largest single factor to reduce costs of the newer proteins will be the savings that can result from increased manufacturing volume with the concurrent reductions in unit cost. Until general acceptance permits this type of expansion, unit costs can be expected to remain high. This will be particularly true for single-cell protein and other technology intensive products.

The greatest cost reductions can be expected in products developed from by-products or waste materials (Halliday 1975; Tybor 1975). The increased utilization of mechanically deboned meat from bones previously used only for rendering or bone meal is a good example. The recovery of beef tissue from trimmings by low temperature rendering yields a beef protein of a quality comparable to ground beef for use by food processors (Mulhern 1976). The utilization of blood protein, while not aesthetically acceptable at present, would offer a high quality, low cost and highly functional protein (Young and Lawrie 1974).

Costs will vary significantly, depending on the degree of processing, raw materials and demand. Major suppliers of proteins

TABLE 22.8
MAJOR SUPPLIERS OF ALTERNATIVE PROTEINS[1]

Company	Type of Protein
Amoco	SCP (yeast)
Anderson-Clayton	Soy
Anheuser-Busch	Yeast
Archer Daniels Midland	Soy
Cargill, Inc.	Soy
Central Soya	Soy, dairy replacer
DuPont	Research
Far-Mar Co.	Soy
Foremost-McKesson	Whey
General Mills	Soy
Griffith Laboratories	Soy
Hercules	Wheat gluten
Kractco	Soy, casein
Land O'Lakes	Whey
Miles Laboratories	Soy
Nabisco	Soy, rice
Nestle	SCP
Ralston Purina	Soy
A. E. Staley	Soy
Standard Brands	Yeast
Stauffer Chemical	Whey
Swift	Soy
Unilever, Ltd.	Soy

[1] Hale (1975).

who will probably continue to be the innovators for new products are shown in Table 22.8.

ALTERNATIVE TECHNOLOGY

It is anticipated that technological procedures will need to be improved to produce acceptable alternate proteins as well as to utilize them in traditional and new food products.

For the development of proteins, these will include (1) removal of toxicants, (2) improvement in flavor, (3) improvement in color, (4) improvement in texture, and (5) nutrient fortification.

The removal of toxic or potentially toxic substances will need to be given a higher priority by the food scientist if widespread use of alternative proteins is to expand. A variety of extraction, purification, fermentation, hydrolytic and osmotic technologies are currently available. The reduction of nucleic acid in yeast is an example of this requirement.

The flavor of many new proteins is not optimun. The first step is to produce as bland a flavor as possible and the second to incorporate stable natural or artificial flavors into the structure of the protein. This may be done during extrusion or reforming or by the development of encapsulated flavors into the protein mass.

In some protein systems, color is a problem and will need to be modified or removed. With present concern for artificial colors, more emphasis must be placed on natural products.

Much has already been reported here on the development of texture in proteins designed for use in analog products (Jaynes and Chou 1975; Smith 1975). In many systems, texture will be related to other functional components of the fabricated food and will be less critical.

The fortification of protein foods with the nutrients expected from their traditional counterparts is of equal importance. Fortification of products to be consumed as complete foods "as prepared" can be accomplished by direct addition or through encapsulation. Products where nutrient losses can be anticipated in cooking water offer greater challenges and may require chemical bonding of the nutrients to the protein.

NEW FOOD FORMS

While it is anticipated the largest use of alternative proteins in the next decade wlll be traditional food forms, or imitations thereof, we anticipate that a number of new or modified foods will result from the development of economical, functional and nutritious protein sources. These may include intermediate moisture foods, high protein "curds," new textured forms, compressed, dehydrated, pasta, and drinks.

The limits on these developments depend only on the imagination of the food scientists and technologists of the future.

SUMMARY

If the constraints to the expanded use of alternative proteins considered here—nutrition, toxicity, flavor, regulations and price— can be overcome through additional research, through relaxation of government attitudes and through economics of scale cost reductions, we anticipate a significant growth of these products in food processing.

Hale (1975) concludes that proteins for which major growth can be expected are those now in some commercial use with textured soy and soy isolates increasing by 20–25% per year for the next 5 years. Casein and caseinates will double and whey will continue to hold its top position with a growth of 125% by 1980. Single-cell protein, starting from a small base, is expected to grow by 30% per year but will still contribute less than 2% of the protein additive market.

A considerable increase in research and marketing activity will be necessary before several proteins such as cottonseed, fish protein concentrate, triticale and mechanically deboned meat will achieve a substantial portion of the market, partly because of the constraints discussed above, but primarily because of cost. The more exotic proteins from fungi, bacteria, leaves and insects will probably not be viable until at least 2000.

The alternative proteins will find their greatest usage as economical substitutes for higher priced traditional protein in convenience and fabricated foods. As functionality improves and costs are reduced, these alternative sources will be a valuable ingredient in new food concepts and forms. A variety of fortified blends to improve nutritional quality is expected during the next decade for marketing in protein-poor countries. The successful marketing efforts will be carried out primarily by private industry with significant financial support by governments. A high level of technological and management innovation will be required.

BIBLIOGRAPHY

ANON. 1975. Leaf protein: Scant research despite economic potential. Food Prod. Develop. 9, No. 9, 62, 64.

BALDWIN, A. R. 1974. Summary of the world soy protein conference. J. Amer. Oil chem. Soc. 51, 181A–184A.

BRESSANI, R., and ELIAS, L. G. 1974. Legume foods. In New Protein Foods, Vol. 1A, A. M. Altschul (Editor). Academic Press, Inc., New York.

CASTRO, A. C., SINSKEY, A. J., and TANNENBAUM, S. R. 1971. Reduction of nucleic acid content in Candida yeast cells by bovine pancreatic ribonuclease A treatment. Appl. Microbiol. 22, 422–427.

CHANCELLOR, W. J., and GROSS, J. R. 1976. Balancing energy and food production, 1975–2000. Science 192, 213–218.

DECAREAU, R. V. 1970. Microwave energy in food processing applications. Crit. Rev. Food Technol. 1, 199–224.

EDOZIEN, J. C., UDO, U. U., YOUNG, V. R., and SCRIMSHAW, N. S. 1970. Effects of high levels of yeast feeding on uric acid metabolism of young men. Nature 228, 180.

FINCH, R. 1970. Fish protein for human foods. Crit. Rev. Food Technol. 1, 519–580.

GRAY, W. D. 1970. The use of fungi as food and in food processing. Crit. Rev. Food Technol. 1, 225–329.

HALE, W. C. 1975. Impact of Technology on the Food Supply: Alternative Food Ingredients. Arthur D. Little, Inc., Cambridge, Mass.

HALLIDAY, D. 1975. Blood—A source of proteins. Process Biochem. 10, No. 4, 11–12.

HEDENSKOG, G. and EBBINGHAUS, L. 1972. Reduction of nucleic acid content of single cell protein concentrates. Biotechnol. Bioeng. 47, 447–457.

JAFFE, W. G. 1973. Toxic proteins and peptides. In Toxicants Occurring Naturally in Foods, 2nd Edition. National Academy of Sciences, Washington, D.C.

JAYNES, H. O., and CHOU, W. N. 1975. New method to produce soy protein-lipid films. Food Prod. Develop. *9*, No. 4, 86, 90.

KARAKOLTSIDIS, P. A., and CONSTANTINIDES, S. M. 1974. A new source of protein: The okra seed. 4. International Congress of Food Science and Technology 8a, 14.

LACHANCE, P. A. 1974. Protein, new sources. *In* Encyclopedia of Food Technology, A. H. Johnson and M. S. Peterson (Editors). AVI Publishing Co., Westport, Conn.

LIPINSKY, E. S., and LITCHFIELD, J. H. 1970. Algae, bacteria, and yeasts as food or feed. Crit. Rev. Food Technol. *1*, 581–618.

LIPINSKY, E. S., and LITCHFIELD, J. H. 1974. Single-cell protein in perspective. Food Technol. *28*, No. 5, 16, 18, 20, 22, 24, 40.

MACLAREN, D. D. 1975. Single cell protein: New processes open wider food uses. Food Prod. Develop. *9*, No. 6, 26–27, 30, 32.

MCNAMARA, S. H. 1975. Some legal aspects of providing a sufficient food supply for a hungry population. Food Drug Cosmetic Law J. *30*, 527–535.

MAUL, S. B., SINSKEY, A. J., and TANNENBAUM, S. R. 1970. New process for reducing the nucleic acid content of yeast. Nature *228*, 181.

MULHERN, F. J. 1976. Definition of meat and classes of meat, permitted uses and labeling requirements. Federal Register *41*, 17560–17566.

MULLER, L. L. 1974. Milk proteins in food. Process Biochem. *9*, No. 4, 7–9.

NATIONAL ACADEMY OF SCIENCES. 1973. Toxicants occurring naturally in foods. National Academy of Sciences, Washington, D.C.

OKI, T., and KITAI, A. 1974. Fermentation products from methanol. Process Biochem. *9*, No. 9, 31–32.

PARRISH, G. K., KROGER, M., and WEAVER, J. C. 1974. The prospects of leaf protein as a human food—and a close look at alfalfa. Crit. Rev. Food Technol. *5*, 1–13.

PIKE, R. L., and BROWN M. L. 1967. Nutrition: An Integrated Approach. John Wiley & Sons, New York.

RACKIS, J. J. 1972. Biologically active components. *In* Soybeans: Chemistry and Technology, Vol. 1 Proteins, A. K. Smith and S. J. Circle (Editors). AVI Publishing Co., Westport, Conn.

RACKIS, J. J. 1974. Biological and physiological factors in soybeans. J. Am. Oil Chemists Soc. *51*, 161A–174A.

REED, G. 1974. Comparison of the use of commercial yeasts. Process Biochem. *9*, No. 9, 11–12, 32.

RICHARDSON, T. 1975. Utilization of novel proteins in human foods. *In* Workshop on Unconventional Sources of Protein. University of Wisconsin, Madison.

ROE, D. A. 1968. Dietary inter-relationships. *In* Modern Nutrition in Health and Disease. M. G. Wohl and R. S. Goodhart (Editors). Lea & Febiger, Philadelphia.

ROSENFIELD, D., and HORNSTEIN, I. 1974. Protein foods from plant sources. *In* Encyclopedia of Food Technology, A. H. Johnson and M. S. Peterson (Editors). AVI Publishing Co., Westport, Conn.

SCRIMSHAW, N. S., WANG, D. I. C., and MILNER, M. 1975. Protein resources and technology: Status and research needs (NSF-RA-T-75-037) National Science Foundation, Washington, D.C.

SEELEY, R. D. 1975. Functional aspects of SCP are a key to potential markets. Food Prod. Develop. *9*, No. 7, 46, 51.

SMITH, O. B. 1975. Textures by extrusion processing. *In* Fabricated Foods, G. E. Inglett (Editor). AVI Publishing Co., Westport, Conn.

SOMOGYI, J. C. 1973. Antivitamins. *In* Toxicants Occurring Naturally in Foods, 2nd Edition. National Academy of Sciences, Washington, D.C.

STONE, H. 1976. Novel proteins—an alternate food source. Food Prod. Develop. *10*, No. 1, 64, 66, 68.

TAPPEL, A. L. 1962. Hematin compounds and lipoxidase as biocatalysts. *In* Lipids and Their Oxidatin. H. W. Schultz (Editor). AVI Publishing Co., Westport, Conn.

TYBOR, P. T., DILL, C. W., and LANDMANN, W. A. 1975. Functional properties of proteins isolated from bovine blood by a continuous pilot process. J. Food Sci. *40*, 155–159.

UNDERWOOD, E. J. 1973. Trace elements. *In* Toxicants Occurring Naturally in Foods, 2nd Edition. National Academy of Sciences, Washington, D.C.

WASLIEN, C. I. 1975. Unusual sources of protein for man. Crit. Rev. Food Technol. *6*, 77–151.

WASLIEN, C. I., CALLOWAY, D. H., and MARGEN, S. 1968. Uric acid production of men fed graded amounts of egg protein and yeast nucleic acid. Amer. J. Clin. Nutr. *21*, 892–897.

WATANABE, T., EBINE, H., and OKADA, M. 1974. New protein food technologies in Japan. *In* New Protein Foods, Vol. 1A. A. M. Altschul (Editor). Academic Press, Inc., New York.

WHITAKER, J. R., and FEENEY, R. E. 1973. Enzyme inhibitors in foods. *In* Toxicants Occurring Naturally in Foods, 2nd Edition. National Academy of Sciences, Washington, D.C.

WILLETT, J. W. 1974. The World Food Situation and Prospects to 1985. Foreign Agricultural Economic Report No. 98, ERS, USDA, Washington, D.C.

WILSON, B. J., and HAYES, A. W. 1973. Microbial toxins. *In* Toxicants Occurring Naturally in Foods, 2nd Edition. National Academy of Sciences, Washington, D.C.

WITTWER, S. H. 1975. World Food and Nutrition Study. Enhancement of food production for the United States. National Academy of Sciences, Washington, D.C.

YOUNG, R. H., and LAWRIE, R. A. 1974. Utilization of edible protein from meat industry by-products and waste. 2. The spinning of blood plasma proteins. J. Food Technol. *9*, 171–177.

23

Research Needs in Protein Resources

Max Milner
N.S. Scrimshaw
D.I.C. Wang

Recent Problems in World Protein Supply

Since 1972, the world protein supply has become a matter of more than usual international concern. On the one hand, the increasing population of many developing countries has been straining their food production capacities, producing a growing threat of malnutrition, and in some cases widespread starvation. On the other hand, several industrialized countries, such as Japan, the Soviet Union and the nations of Western Europe which depend on world markets for a considerable portion of their food supply, have been increasing their levels of animal protein consumption largely on the basis of importation of large amounts of feed grains and protein concentrates.

For almost 20 years prior to 1972, world grain surpluses, produced principally in North America, proved adequate to make up food grain shortages in the increasingly food-deficient developing countries and they were also sufficient to supply the feed grain and protein concentrate demands of wealthier countries which were increasing their consumption of meat and poultry products.

The sudden reversal in the world excess food supply, which occurred in 1972 due to widespread drought and other causes, is still very fresh in our minds. There is general international concern that this development may at last represent the realization, after a 200 year delay, of the dire Malthusian prophecy. Serious international discussion has resulted, such as that at the World Food Conference in 1974, of what must be done internationally and by individual countries to stimulate food production. Primary emphasis was given to initiatives to be taken by the developing countries themselves which have rapidly growing food gaps in the face of major population increases.

Even more recently, during the past 2 years, some lessening in international food demand has occurred due largely to the economic recession and the excellent 1975 grain and oilseed crop in North America, which may offset grain shortages elsewhere.

Nevertheless, it is clear that without positive steps to increase world supplies of grain and protein resources there can be little expectation for establishing the comfortable supply/demand situation that existed in the 1960s.

As we know, the U.S. is the largest international supplier of surplus grain including wheat, maize and soybeans. Indeed in recent years, it has exported 75% of its wheat crop and 50% of its soybean production. Nevertheless, the U.S. consumer has not been insulated from the sharp price escalation which began in 1972, coincident with the very large grain purchases by the Soviet Union, aggravated in 1974 by a setback in the U.S. to domestic maize and soybean crops. For these reasons questions have been raised about the longer term capability of U.S. food and protein production capacities to fill domestic needs, to sustain major foreign export demand, and to respond to urgent international emergency feeding programs.

Interest in Unconventional Proteins

It is important to recall that, notwithstanding the ample U.S. food supply and the frequency of surpluses prior to the dislocations of the 1970s, a great deal of scientific and technological interest had already developed in a variety of unconventional protein resources including not only edible soy products but also concentrates from fish, leaves, and microorganisms cultured on substrates such as hydrocarbons and waste cellulose. These research initiatives, as well as the growing concern for future world and domestic protein supplies, appear to have convinced the U.S. National Science Foundation that it must respond substantively to increasing requests for research support not only in unconventional sources of protein, but also for problems whose solution would enhance the productivity and quality of traditional protein sources as well. This realization led to initiation of an NSF project coordinated by the Department of Nutrition and Food Science at MIT whose objective it has been to analyze in some detail the status and potential of all relevant protein resources, and to provide recommendations to NSF for carefully selected projects and priorities worthy of research support. The emphasis of the study was on U.S. rather than global needs. However, the fact that U.S. priorities require concern for maximizing its export potentials to offset increasing energy imports, as well as for the U.S. commitment to provide large quantities of food to food-deficient Third World countries, makes the study relevant as well to the world situation.

This study recognized that many aspects of such research are already receiving attention from various governmental, academic and industrial institutions. Nevertheless, we were convinced that important areas of research vital to strengthening the U.S. food and protein productivity are not only poorly supported, but some may be entirely neglected in terms of their significance. These are the primary areas which this study has identified and recommended for NSF research emphasis (Scrimshaw *et al.* 1975).

SIGNIFICANT DETERMINANTS OF RESEARCH RECOMMENDATIONS

Review of the extensive documentation prepared for this study suggested that there were a number of significant conclusions or hypotheses which had considerable relevance when identifying research recommendations and priorities. The following are pertinent:

(1) U.S. agriculture will continue in the foreseeable future to supply in abundance the protein foods and feeds to which this country has become accustomed, but it will obviously do so with rising costs and increasing pressure on our land, energy and environmental resources.

(2) Appropriate research and technological development must be pursued more vigorously to achieve increases in productivity, protein quality, and protein quantity of primary food crops and aquatic resources, to meet both domestic demand and export opportunities.

(3) A strengthening of U.S. food and agricultural research capabilities will be necessary to ensure continued growth of food export, a major contributor to the maintenance of favorable U.S. trade balances. With adequate research support, and if other appropriate policies are adopted, U.S. food export capacity at least until 1985 should increase at a rate following the trend established prior to 1972.

(4) The food and feed demands of affluent industrial countries will continue to increase and will exert growing pressure on food resources available in international markets.

(5) Even with probable improvements, the agricultural production capacities of some major food deficit developing countries will not

soon be adequate to feed their growing populations. They will need to import food for many years to come and the U.S. will play a leading role in supplying these needs, largely on regular commercial terms.

(6) When food-deficit countries improve their own agriculture and food production capacities, not only will the nutritional status of their populations be increased, but, experience has shown, they will also tend to increase their commercial food purchases in the U.S.

(7) In the foreseeable future, the relative cost of proteins (*e.g.*, meat, soybeans) for human and animal feeding is likely to increase more rapidly than the cost for staple sources of food or feed energy (*e.g.*, corn, wheat).

(8) The foreseeable domestic and world demand for U.S. soybeans as a primary protein resource, notwithstanding increasing production in other countries such as Brazil, will grow faster than for other food or feed crops. While a major breakthrough in soybean yields may occur, it is not likely to be effective in the next ten years. This will stimulate interest in alternative protein sources of both conventional and unconventional types.

(9) It is likely that new and more productive grain crops will be developed, particularly from intergeneric crossing of cereals, of which triticale is a prototype.

(10) The marketplace must reflect the increased value of crops with improved protein quality and quantity if farmers are to be persuaded to produce them. Incentives may also be needed to encourage commercial initiatives by the seed industry and others directed toward more rapid development of improved food crop varieties.

(11) The increasing costs of agricultural food production may eventually encourage the use of unconventional techniques for protein production which are not now commercially competitive.

(12) The reclaiming of organic wastes of all kinds for food and feed use is a major priority.

(13) The development and application of innovative food processing techniques and of new science and technology to the production of protein foods will require fair and objective regulatory standards and procedures on the part of the Food and Drug Administration, the U.S. Department of Agriculture and the Environmental Protection Agency.

(14) Acceptance by the U.S. consumer of novel protein foods to any significant extent will require innovative marketing and promotional approaches which, to be effective, must go beyond emphasis on nutritional benefits.

(15) Adoption and enforcement by many countries of a 200-mile

offshore fishing limit will remove some prime fishing areas from free international exploitation and thus may affect protein supplies and prices in international markets. Over 30 countries including the U.S. have already adopted this limit.

COMMON PROBLEMS AND ISSUES REQUIRING RESEARCH

Nutrition

Adequate assessment of the protein problem is handicapped by lack of knowledge in two fundamental areas: (1) human requirements for protein at different ages and physiological states; and (2) the evaluation of protein quality of foods. These issues are further complicated by increasing evidence that both are greatly affected by caloric excess as well as caloric deficit (FAO/WHO 1973).

It is apparent that the present FAO/WHO recommendations for protein requirements and, similarly, those of the National Academy of Sciences/National Research Council's Food and Nutrition Board do not provide an adequate basis for determination of a safe practical allowance for either individuals or populations, especially under adverse environmental circumstances.

There is an urgent need to develop and apply improved methods of protein quality evaluation which are sufficiently rapid and inexpensive to be used by the food industry, by regulatory agencies, and by plant breeders seeking to develop new crop varieties of high nutritional value. These methods should be feasible with the extremely small samples likely to be available. For novel protein sources in general, and in particular for novel or conventional sources subject to extensive processing, evaluation of protein quality is especially important (PAG 1975).

Toxicology

Toxic factors, which even in small amounts may pose a threat to animals and humans, occur in many widely used foods unless the foods are specially processed, or restricted in use. A group of inhibitors of protein digesting enzymes—such as trypsin inhibitors as well as the hemagglutinins (lectins), which are themselves protein in nature—occur widely in food legumes (peas, beans) and oilseeds (soybeans) (Liener 1969). Fortunately, most of these inhibitors are deactivated by heat during processing. Cyanogens (glucosides containing hydrocyanic acid) occur in significant quantities in foods as diverse as lima beans, almonds, cassava, and sorghum. Saponins are toxins occurring in over 400 species of the plant kingdom—including sugar beets, spinach, and asparagus. Gossypol is a toxic pigment found in cottonseed (*Gossypium*). Favism is a genetically linked disease caused by ingestion of a toxin in broad beans (*Vicia faba*). Lathyrism can be severely debilitating and even fatal in humans ingesting the legumes of the *Lathyrus* (sweet pea) species. Goitrogens, usually in the form of thioglucosides, are present in the *Brassica* species (cabbage, turnips, and mustard). A variety of antivitamins also occur in plant foods (Liener 1969).

In addition to natural food toxins, substances which are potentially hazardous to man and animals may be added to foods as preservatives or coloring and flavoring agents. Some toxicants may be introduced unintentionally through the use of pesticides or fungicides. Food-safety problems associated with incomplete removal of mycotoxins and other microbial toxins are well-recognized (NAS/NRC 1973).

In many cases, studies on experimental animals alone are not sufficient to eliminate all possibilities of adverse reactions in man, since symptoms which may appear in human trials may not arise in experimental animal tests.

The need for toxicity studies is particularly evident with such proposed new protein sources as single-cell protein from both photosynthetic and nonphotosynthetic organisms, forage and leaf protein concentrates, and new legume and oilseed sources (PAG 1974A, B).

Innovative Technology for Protein Utilization

Traditional techniques to isolate proteins for processing into food are based mainly on extraction of a primary product, *e.g.*, oil from oilseeds, with little concern for the integrity of the protein. As a result, the by-product proteins may become unsuitable for many useful functional applications. New technology should be developed for more effective retention of desired properties of the protein.

The basic properties of proteins from various sources require study in order to relate their molecular properties to their physical properties, and their physical properties to their functional (or performance) properties, through theoretical or correlative techniques. The critical molecular properties needing identification include molecular weights and distribution, ionization properties, reactive side chains, and primary, secondary, tertiary, and quaternary structure and morphology. Physical aspects to be characterized include hydration properties such as solubility in various media, suspendability, swelling, gelling, and rheological properties such as viscosity or apparent viscosity, yield stress, time dependency, shear dependency, and creep and relaxation. Thermal aspects include gelation and coagulation, while surface properties include hydrophilic and lipophilic properties, and surface absorption and adsorption.

For data on performance characteristics, studies are required on reactivity between starch, fat, flavoring, and other food ingredients, and on relationships between performance properties and chemical and physical properties.

Examples of process research would be concentration and isolation of protein materials, and methodologies for restructuring protein materials, or combining proteins with other materials. The effect of processing conditions on physical properties, and the influence of changes in the latter on organoleptic (taste, smell) characteristics of the final product both need to be studied.

Potentials for Improving Protein Quality in Plants by Genetic Means

Geneticists and plant breeders have, in recent years, succeeded in developing or identifying mutations in corn, barley and sorghum which have significantly higher relative levels of lysine, as well as favorable alterations in levels of other amino acids, in their seed proteins (Mertz *et al.* 1964; Munck 1972). This achievement raises questions of the extent to which similar mutations are possible in other grains or crop plants, and also brings up the question of this kind of genetic manipulation's ultimate limits.

The single gene mutations which substantially increase lysine content in corn, barley, and sorghum involve changes in the proportions of the few protein species that normally constitute the storage protein complement of seeds, but apparently without affecting their amino acid composition.

Apparently the situation is different in the case of rice and wheat. Rice has a low proportion of prolamine to begin with, and no prolamine-depressing mutation would have much effect in altering amino acid composition. Wheat does contain a sizable proportion of its total seed protein as prolamine, but mutants which cause its repression are recessive. The hexaploid nature of the wheat species combines genomes of three closely related diploid species. The probability that a similar mutation would occur in all the others at the same time is obviously extremely small (Mattern *et al.* 1970).

The suppression of the synthesis of less desirable protein fractions, and the secondary increase in synthesis of fractions with greater biological value, may also be effective in legumes which, in the case of *Phaseolus*, show similar storage protein heterogeneity (Nelson 1969).

Biological Nitrogen Fixation

Enhancement of biological nitrogen fixation, whether symbiotic or nonsymbiotic, requires greater understanding of the processes themselves and of the organisms which effect the processes. The protein nature of the nitrogenase complex that catalyzes nitrogen fixation in organisms is fairly well understood. Research is needed to determine the crystalline enzyme components, the role of ATP in the system, the mechanism whereby iron and molybdenum are incorporated into the enzyme, and the energetics and kinetics of the overall reaction (Zelitch 1975).

The genetic coding for nitrogenase in bacteria has been successfully transferred from one species to another. Thus, the genetic potential exists for extending nitrogen-fixing capability to a variety of economically important organisms.

In nitrogen-fixing legumes, a better understanding is needed of the genetics of both the host species and the bacterial endophytes. In the breeding of improved legume cultivars, insufficient attention has been given to these host/endophyte relationships. Useful application of biological nitrogen fixation to grasses requires better understanding of the genetic and biochemical compatibility of host and endophytes.

Limitations to increased yield in legumes seem to be related not only to conditions affecting the availability of useful *Rhizobium* strains and the adequacy of present inoculation techniques, but, in the case of soybeans at least, to an insufficiency in photosynthate energy required to maintain optimum nitrogen inputs (Hardy 1976). A broad attack is needed on the many physiological and agronomic aspects of nitrogen-fixing systems of known or potential agronomic significance.

**POTENTIALS AND RESEARCH NEEDS
IN SPECIFIC PROTEIN RESOURCES**

Grain Crops for Food and Feed

The tremendous and relatively unexplored genetic diversity of cereal grains provides great promise for expanded food and protein production. An accelerated genetic search for increased protein quantity and quality in the major cereals requires medium- and long-term funding. These genetic questions also call for considerable research in fundamental scientific areas relating to genetics, and in this respect represent an appropriate area of NSF concern. For example, comprehensive biochemical research on the uptake and translocation of nitrogen in crop plants, as well as related seed protein metabolism in the cereal grains especially, can lead to the development of genetic or other means for increasing seed protein quantity and quality. There is a need for expanded basic studies on the biochemical aspects of plant physiology and metabolism.

Rapid, accurate procedures for screening a variety of crops for protein nutritional quality need to be developed, for use by plant breeders, to select superior lines (PAG 1975). Such research requires reevaluation or review of bioassay procedures and their pertinence to human and livestock animal requirements.

Increasing soybean yields by breeding requires use of germ plasm from mainland Asia. Another approach to the soybean yield problem is to determine the cause or mechanism of oxidative photorespiration in vegetative soybeans and other legumes. This mechanism limits carbon dioxide fixation and, therefore, grain yield (Aldrich 1973).

Symbiotic association of nitrogen-fixing bacteria with roots of certain grasses and corn suggest that similar associations with cereals can be identified or developed, so as to allow large savings in energy because of reduced need for nitrogen fertilizer.

Cereal Protein Technology

The normal abundance in the U.S. of low-cost grains and their underutilized by-products suggests the desirability of initiating greater scientific and technological efforts to upgrade their food use and nutritional value.

New technologies and processes need to be perfected for separating upgraded protein products from grains and their industrial by-products. These methods would include improved techniques for producing stable edible protein concentrates from wheat milling by-products, better methods for separation of wheat gluten by aqueous processes, and separation of edible protein isolates from other cereals and their by-products. Along with an improved technology for preparing protein concentrates, new food processing applications for proteins in formulated food products should also be developed (Inglett 1970, 1974).

Research is under way in these areas, but with low priority status, in the U.S. Department of Agriculture (USDA), the state institutions, and some food industry laboratories. To stimulate this important application of technology, an economic marketing feasibility survey is needed.

Oilseed Proteins

Soybeans represent the single largest source of oilseed protein for both animal and human consumption, and most technology for the food use of oilseed proteins relates to soybeans and a spectrum of products derived from them (World Soy Protein Conference 1973). A number of other oilseeds, *e.g.*, cottonseed, peanuts, sunflowers, rapeseed, and sesame, have potential for similar utilization; however, the necessary technology for production of food-grade products from these materials is not fully developed. Additional research is needed on all these oilseeds, including soybeans, to solve problems involving flavor, functionality, color, antinutritional factors, processing technology, and many other factors.

Research and development in six food processing and utilization areas will be essential if oilseed proteins are to achieve maximum acceptance in human foods. Specifically,

(1) Protein and nonprotein component interactions during processing must be identified;

(2) Fundamental physical and chemical characteristics of the proteins of oilseeds, as related to their functional properties, need to be studied;

(3) Screening methods must be developed for determining functional properties of oilseed proteins with respect to their utilization in food systems;

(4) Identification of components responsible for undesirable flavors and development of processing technology to eliminate these from food products is essential;

(5) Investigation should be undertaken of modification of the functionality of oilseed proteins by physical, chemical, or enzymatic means; and

(6) Assessment is needed of the level at which the presence of flatus-producing oligosaccharides in oilseed products assumes practical significance, and technology to minimize this problem needs to be developed.

The efforts being devoted to these problems in USDA, the state institutions, and the food industry require stimulation, and the desirability of a selective role on the part of NSF is clearly evident.

Food Legumes

Taking into account the major problems that need to be overcome in production, utilization and marketing of food legumes (PAG 1973), the following recommendations are offered:

Production.—Research is needed to:

(1) Overcome genetic-physiological barriers to higher seed and protein yield;

(2) Improve control of and resistance to disease and insect pests;

(3) Promote greater efficiency in cultural systems; and

(4) Reduce the deleterious effects of adverse environmental factors.

Utilization.—The most urgent priorities are to:

(1) Develop convenient and rapid quantitative methods to be used, by breeders, for the estimation, identification, and removal of undesirable or toxic constituents in legume seeds;

(2) Develop more rapid and valid methods for estimating the nutritive value of legume proteins, and apply this new technology to the

improvement of protein quality in commercial seed types (Milner 1975);

(3) Study combinations of legumes with cereal products and other staple foods, with the objective of improving the nutritional properties and digestibility of both; and

(4) Develop improved, stable, processed dry legume products that may be prepared for eating quickly, *i.e.*, with a minimum of energy (fuel) input.

Livestock Animal Production

U.S. animal industries are capable of producing most of the fresh meat consumed in the U.S., plus some for export, until the year 2000. However, in order to do so, great advances will be necessary in the productivity and efficiency of animal production enterprises. These advances can be achieved through increased research in (1) recycling of animal wastes; (2) further use of by-products, wastes, and other alternative feed sources; (3) improvement of animal reproduction and growth efficiency; (4) greater use of forages in animal rations; (5) increased efficiency of nonprotein nitrogen utilization by animals; and (6) development, through breeding programs, of high-efficiency animals which can effectively utilize more forage rations (Cunha 1974).

In regard to these problems, research sponsored by USDA and the state universities and experiment stations is inadequate in relation to the urgency of the problems. These agencies must expand and intensify their activities accordingly, and NSF can contribute to progress by selective funding of research in animal reproduction and growth efficiency, and by supporting innovative approaches in the recycling of animal wastes.

Animal Protein from Dairy Products

The supply and utilization of milk in the U.S. dropped from 56,700 million kg in the early 1960s (296 kg per capita in 1960) to about 53,500 million kg in the early 1970s (253 kg per capita in 1973). It is estimated that by 1980, per capita dairy food consumption in the U.S. will be down to 237 kg.

It is not likely, that is, that U.S. per capita milk consumption will increase. At present the American consumer perceives no nutritional need for increased dairy consumption, although this conception overlooks the importance of milk as a calcium source. It would take increased government-financed subsidies to produce more, but it is unlikely that the government will take steps to increase production capacity as long as other countries can produce surplus milk more cheaply than the U.S. can. Demands, therefore, will probably be met increasingly by imports.

This situation requires that in the next ten years, government-sponsored research on dairy products be strengthened by 30%, and industry research by 50%, over present levels. For primary dairy products and for dairy analogs, research and related efforts are needed to:

(1) Improve quality and convenience of the established products, on the basis of market research guidance;

(2) Develop new dairy products and concepts which are marketable and competitive with fruit juices and soft drinks;

(3) Perfect a rapid, automated cheese production and ripening process which would increase the competitiveness of cheese in both domestic and export markets;

(4) Develop new processing and preservation methods for milk, so as to increase its useful storage life, reduce bulk, and make milk a significant protein resource suitable for worldwide shipment and trade;

(5) Develop dairy analogs based on vegetable proteins;

(6) Conduct nutritional studies to clarify the role of nonlipid dietary components, as well as milkfat and dietary cholesterol, with respect to coronary heart disease;

(7) Optimize the food technological utility and functionality of dairy proteins as replacement proteins in fabricated foods, taking into account safety as well as nutritional factors; and

(8) Undertake economic and marketing studies to determine both the feasibility of increasing U.S. dairy exports, and the most favorable product mix of natural and analog dairy products.

Areas suggested for research into secondary dairy products include (1) chemical modification and upgrading of whey proteins into useful new products; (2) whey protein fractionation and characterization; (3) fundamental investigation into improving sensory and organoleptic properties of whey for applications in foods; (4) clarification of galactose metabolism in humans as a first step to wider utilization of lactose; and (5) development of whey products for the world market as protein supplements to upgrade protein-deficient diets.

All the agencies concerned with dairy research—*i.e.*, industry, USDA, and the state universities and experiment stations—have roles and responsibilities in attacking these problems more vigorously. The call for innovations in food science also suggests that NSF could usefully support selected projects in the areas indicated.

Animal Protein from Meat, Poultry, and Eggs

There is great potential for expanded development, production and utilization in the meat and egg industries (Cox *et al.* 1971). The utilization of this potential will depend on research and development inputs, and on the realism of regulatory constraints in terms of risk-cost-benefit considerations. It is also apparent that energy availability and cost, environmental issues and governmental policy decisions will have far-reaching effects on the future availability of meat and eggs.

A significant expansion of the supply of animal protein for human consumption could be brought about as a result of research into use of animal tissues which are at present diverted from the market by health or palatability considerations, or by government restrictions on otherwise safe and nutritious products.

Many packinghouse raw materials, such as visceral organs, are currently processed into livestock and pet feeds. Processes similar to those used to produce fish protein concentrates have been developed for the production of edible meat protein concentrates from the various offal tissues. However, further studies are needed in this area, as well as review of regulations to permit their use in the food supply.

Technical knowledge now available to the meat industry could result in increasing the output of processed meat products by over 50%, through appropriate extension or supplementation with plant proteins.

Consumers are concerned with the possible relationship of diet to coronary heart disease. Attention centers on cholesterol and animal fats. Because the consumption of nutritious foods may be inhibited by adverse publicity based on inadequate scientific evidence, it is imperative that well-planned, long-range studies be implemented to clarify these questions.

Rigid and frequently outdated standards of identity discourage the development of new or improved products, which could extend the quality and variety of the national protein supply. Other standards—e.g., microbiological standards—increase cost to the consumer without demonstrating improvements in product quality or safety as a result of their enforcement. Overregulation, in short, adds to product cost, channeling much industry research funding into matters concerned primarily with regulatory compliance.

Broad areas of research have been receiving attention by industry, USDA, and the state universities and experiment stations, but such activity requires far greater support from funding sources than has been the case heretofore. As regards relatively basic research, there are a number of areas where NSF support would help stimulate the solution of serious and longstanding problems.

Aquatic Proteins

Most of the U.S. fishery stocks traditionally used by Americans for food are either fully exploited or depleted. There are, however, species even within the present, limited economic control zone—not to mention the larger resources within the proposed 200-mile zone—which have remained largely un- or underexploited. It is estimated, for example, that between 100,000 and 500,000 metric tons of capelin could be landed yearly on the northwestern Atlantic coast. This fish could be offered in the form of engineered or fabricated foods which would be more familiar to the public. Squid, likewise, is most suitably utilized in textured and engineered food products.

Much more needs to be known about the biology and behavior patterns of aquatic animals, and techniques need to be developed for locating, harvesting, preserving, and processing this resource (Bardach 1976). Polyculture, which relies on the growing-together of a number of fish species of varying feeding types, presents opportunities for attaining very high yields per hectare by taking advantage of symbiotic relationships among the different species. The high yields in such systems can be maintained without competing with non-ruminant farm animals for feed. Species with maximum productivity potential in warm freshwater pond polyculture systems include carp, buffalofish, tilapia, white amur, mullet, milkfish and catfish.

Capture fisheries represent the major focus of aquatic protein resource development. Antarctic krill, for example, is the single largest natural source of animal protein on this globe, and probably 45 million metric tons of it could be harvested yearly without compromising the availability of the resource. In view of the lead taken principally by Russia, and, to a lesser extent, by Japan, in krill harvesting and utilization, the U.S. must now decide what commitments it should undertake in the management and utilization of this resource. It seems prudent to advise allocation of funds at levels sufficient at least to undertake a technical and economic feasibility study of U.S. participation in and utilization of the krill harvest.

As regards support for research and development, the financial and political decisions will have to be made by government, since industry will not initially invest the hundreds of millions of dollars required to perfect the tools needed to produce and utilize increased tonnage of raw aquatic material. New types of suitably equipped fishing vessels will have to be designed and constructed. New methods of locating, harvesting, preserving, and processing must be developed. New toxicological and nutritional information will be necessary to develop regulations for monitoring the quality and safety of the new aquatic food products. Investigations will be needed in biology, genetics, animal nutrition, and ecology of aquatic organisms. The National Science Foundation could most appropriately provide funding for studies in oceanography, ecology, physiology, basic biology, and other background areas.

Potatoes

The potato has protein levels comparable to those of cereal grains; indeed, cultivars containing 17–18% protein (dry weight basis) have been identified. The crop, clearly, does not merit the status of a "poor man's" energy food.

With the introduction of new potato varieties, this crop's yields per hectare have increased nearly 300% in the past 20 yr (Kaldy 1972). Through intensified breeding studies, exploration of hybridization should be continued, in order to expand adaptation, and improve disease and insect resistance, nutritive value, and yield; the ideal polyploid level for optimum productivity needs to be determined.

Systematic study is needed on factors affecting productivity and nutritional quality. These factors include optimum stage of maturity, effects of fertilization and irrigation, and possible benefits of multiple cropping with early and late maturing varieties, and with other crops. (This last factor is relevant especially to potato culture in warmer environments.)

Researchers need to analyze potatoes for quantity and quality of various proteins, evaluate their amino acid composition, and determine whether or not breeding can alter protein ratios, so as to improve the overall protein nutritive quality. Simple screening methods for protein quality and quantity are needed to assist breeders in analyzing large potato populations.

Much work also needs to be done on the reclaiming and upgrading of potato processing wastes, with emphasis on recovery of proteins and amino acids.

Nonphotosynthetic Single-Cell Protein

The role of single-cell protein as a supplement to the world protein supply is now well established. Research and development on SCP production have been intense for over a decade (Mateles *et al.* 1968), and as a consequence there exist a number of large SCP plants in operation for animal feed production; several more are under construction. The problems being addressed today relate to second and third generation processes for SCP production for both animal feed and human food (Tannenbaum and Wang 1975).

No single raw material or organism will provide *the* ultimate SCP process. Consequently, it is essential to evaluate, in parallel, several process alternatives, to appreciate fully the flexibility available in the use of SCP. There is no specific recommendation of substrates, but it is clearly the overall view of the experts that alcohols and cellulose hold the greatest potential for the U.S., and therefore should receive priority consideration.

It is believed that cell yield is the single most important economic factor in the fermentation process, and that great stress should be placed, in practice, on approaching the theoretical fermentation yield. Improved processes for RNA removal would involve study of RNA levels in cells, the role and activation of endogenous ribonucleases, and their control by means of genetic and cell physiology factors.

More efficient use could be made of the proteins within the cell if economical means for their recovery were available. Selective isolation of proteins would also be a method of avoiding the RNA problem. Studies are needed of SCP engineering properties, interactions with food constituents, and use in structure-forming operations. Dewatering and drying comprise one of the most significant cost factors in the overall process. Study is needed on membrane processes, improved methods of mass transfer, and the effect of drying on functional properties.

Research on nonphotosynthetic SCP clearly needs much wider support in the U.S., in order to increase production of protein and other useful products, especially from waste. At present in the U.S., such research and development activities are confined to industry and a few universities. The growing urgency in the U.S. for waste recovery and recycling—the need to reprocess livestock animal waste and manure for refeeding is only one example—suggests a priority area for NSF funding.

Photosynthetic Single-Cell Protein

Photosynthetic single-cell protein is produced in the cells of microscopic plants (algae) which grow in suspension in the waters of shallow, illuminated ponds. These ponds contain culture medium which consists of simple salts—such as carbonates, nitrates, and phosphates—dissolved in water. Depending on the genus, the cells may be separated from the growth medium by screening, filtration, coagulation either by flotation or sedimentation and centrifugation (Myers 1963; Waslien 1975).

During the next decade, steps should be taken to:

(1) Study the relationship of algal species to algal nutritional characteristics, including protein quality, and the influence of ratios of critical substrate nutrients, including consideration of water quality factors;

(2) Study productivity as a function of species, predators and pathogens, nutrient mix, physical parameters such as detention period, mixing, recirculation, and waste as a source of nutrients; and

(3) Study processing of algae to produce various products, including decolorized and bleached algae, algal pigments, and spun algal

protein; and study algal preservation by dehydration, canning, drying, freezing and freeze-drying.

Leaf Protein

Green leaves are the largest producers of protein in the world, supplying protein to other plant tissues including the crop seeds which nourish humans and animals. Indeed, the protein in the 1973–1974 U.S. alfalfa crop was almost twice as much as that in the 12.5 million metric tons of soybean meal utilized in the U.S. that same year.

Process research is well developed for a commercial system which would be added to a conventional dehydration plant to recover whole leaf protein (Pirie 1971). When economic feasibility studies are completed by USDA, a full-scale demonstration plant will be recommended for construction at the site of a commercial dehydration plant.

Process research for an on-the-farm-type leaf protein recovery system (in connection with silage or hay production from the press cake) has been undertaken by workers at the University of Wisconsin and elsewhere (Twelfth Technical Alfalfa Conference Proc. 1975). This work involves low-cost equipment development, and further research on production and quality of the silage and/or hay produced.

Research applicable to both types of systems is needed for optimization of unit operations, for higher-value utilization of the residual juice solubles fraction, and on economic analysis of the several possible variants in the systems.

Denatured white (insoluble) products are made by heat coagulation of chlorophyll-free alfalfa juice after prior removal of the green protein fraction. A great deal of further research is needed to increase yields of this white protein, to increase its purity, and to explore its utilization possibilities in various types of food.

When *Nicotiana* (tobacco) species are used as raw leaf materials, a white protein (Fraction I protein) can be obtained readily in pure crystalline form (Kawashima and Wildman 1970). This tasteless, pure-white product should have special value where high purity is needed. Research is needed on growing these plants on an intensive scale, on determination of the economic value of the crystalline protein, and on the economics of its production.

It may be concluded that leafy plants have tremendous potential for supplying an important proportion of the protein requirements of both monogastric animals and man by the years 1985–2000. What is needed is a coherent, well-financed, multidisciplinary research effort in the agronomic, genetic, chemical, engineering, nutritional, toxicological, microbiological, and food technological aspects of leaf protein production.

Chemical Synthesis of Nutrients

The possibility exists of filling some of humankind's protein needs by nontraditional means, particularly by chemical synthesis. Synthetic vitamins and amino acids are, of course, already in common use. While the cost of their chemical synthesis is relatively high compared to the cost of products made by fermentation, the use of mixtures made from a common intermediate might considerably reduce the cost.

Synthesis of industrial fatty acids from carbon monoxide was conducted in wartime Germany, and increased efficiency is possible. The synthesis of glycerol for human food has also been investigated, particularly by the National Aeronautics and Space Administration, as a potential energy source for a planetary base. A more advanced approach involves 1,3-butanediol, which is metabolized with 50% greater energy output than are sugars or glycerol.

Recent studies have shown that normal paraffins can provide some metabolic energy to chickens. There is a strong possibility that an efficient and relatively high-yielding synthesis of ATP can be developed using HCN as a starting material for adenine. All such developments, however, are clearly in the area of long-range prospects.

RESEARCH RECOMMENDATIONS

The reports of the various working groups identify a large number of important research areas and projects, from which only a limited number have been selected for the recommendations which follow here. Most research which is already of major ongoing concern to USDA and the state agricultural institutions in the U.S. has been excluded only because of the special purpose of this report, and not because of any judgment as to its relative importance. The principal criterion for selection was the research's applicability to (1) an area of importance which complements or underlies traditional agricultural research and which is at present inadequately supported by USDA, the state agricultural institutions, or any other source; or (2) an unconventional area not now receiving an emphasis commensurate with its potential. No meaningful further subdivision of priority rankings can be assigned to the basic and long-term research recommended because too many uncontrollable and unpredictable factors are involved.

Fundamental Mission-Oriented Research

Nutrition.—(a) Evaluate protein requirements of human subjects of various ages and physiological states, by using improved metabolic balance techniques.

(b) Improve existing biochemical and biological methods for the evaluation of protein quality to reflect more closely the nutritional requirements of humans and livestock, and apply these methods to new or novel protein sources as well as processed protein foods of all types. The urgent needs of plant breeders and regulatory agencies for rapid methods of protein quality evaluation should be taken into account in this research.

Toxicology.—Undertake basic research on the toxicological hazards associated with new protein sources, and with conventional proteins processed in new ways. This basic research should be followed, when appropriate, by clinical trials, and by efforts to remove, through processing or other means, toxic factors identified in the animal or clinical trials.

Biological Nitrogen Fixation.—Study comprehensively the subject of nitrogen fixation in crops. This study should consist of basic biochemical and genetic research on the nitrogen and carbon metabolism of plants and microorganisms, with the objective of reducing the need for synthetic nitrogen fertilizer.

Mission-Oriented Research in Specific Resource Areas

Grain Crops for Food and Feed.—(a) Investigate the metabolic process of oxidative photorespiration in cereals, legumes, and leguminous oilseeds, which seriously limits fixed carbon accumulation and thus reduces potential crop yields. The objective should be to identify genetic or other means to control this undesirable metabolic process.

(b) Expand research on symbiotic and nonsymbiotic nitrogen fixation in cereals, with emphasis on development of nitrogen-fixing microorganisms compatible with host cereal species.

(c) Intensify basic research to provide new sexual and asexual methods for breeding more productive crops with higher protein potentials. Such studies would involve identification of specific genes and gene groups, use of cell culture, somatic hybridization and chemical stimulation of gene compatibility for broad sexual crosses, as well as other innovative techniques.

(d) Study more extensively the uptake of nitrogen in crop plants, and the related protein metabolism of seeds, particularly in the cereal grains, in order to identify genetic and other means for increasing protein quantity and quality without seriously compromising crop productivity and other important qualities. This will require improved screening methods to assist plant breeders in the rapid identification of desirable cultivars in protein improvement breeding programs.

Cereal Protein Technology.—Develop new technologies for separating, recovering, and concentrating the protein fractions of cereal grains (wheat, corn, sorghum, barley, oats, rice, triticale, rye), and processed cereal by-products. Attention should also be given to development of new processes for reconstituting or combining cereal protein fractions with other constituents to produce attractive new protein foods such as meat analogs, meat extenders, and beverages.

Oilseed and Legume Proteins.—(a) Study modification of the functionality of oilseed and legume proteins, by using physical, chemical, and enzymatic methods, to facilitate development of protein materials with wide versatility and acceptability in formulated foods.

(b) Intensify studies of the basic cellular and subcellular structure of oilseeds and legumes, with attention to the location and form of primary constituents (proteins, oils, carbohydrates, etc.), in order to permit development of more efficient technologies for protein and oil recovery with retention of optimal functional and nutritional characteristics.

Livestock Animal Production.—(a) Expand research on the feeding value of various fractions of animal and vegetable wastes for different livestock species (beef, swine, and poultry), and develop technologies, such as fermentation with appropriate microorganisms, or silage treatment, to minimize toxicity and unpalatability, and to upgrade the nutritive value of waste materials.

(b) Study the carry-over, into animal tissues used for human food, of possible toxic or pathogenic factors related to feeding recycled wastes of various kinds to livestock.

Dairy Products, Meat, Poultry, and Eggs.—Develop sanitary and microbiologically safe technologies for reclaiming food products from dairy by-products, defective livestock animals, muscle and organ tissues, and poultry and eggs rejected for food use and used only partially (for livestock and pet feeding) under present food laws and grading standards.

Aquatic Proteins.—(a) Investigate more thoroughly the basic problems affecting the development and economic efficiency of aquaculture. Studies should include further identification of optimum species for monoculture and polyculture systems, and examination of the problems in breeding and culturing these species, including their metabolic and nutritional requirements and diseases to which they may be susceptible. Polyculture, which can take advantage of symbiotic relationships among several different species, should be further developed to increase efficiency of feed utilization and thus reduce competition with nonruminant farm animals for feed.

(b) Undertake a comprehensive feasibility study on the harvesting and utilization, for human comsumption, of Antarctic krill. The study should assess integrated harvesting technology, acceptability of possible food products, use of by-products such as chitin, and potential profitability. Since most krill resources occur primarily in international waters, apparently beyond the present or future jurisdictional limits of most countries, and because the U.S.S.R. now leads in the development and application of the relevant technology, NSF should develop a collaborative effort with the Soviet Union, if an appropriate agreement can be negotiated.

Nonphotosynthetic Single-Cell Protein (SCP).—(a) Clarify the nature, occurrence, and mode of action of substances giving adverse reactions in human feeding, either occurring naturally in, or produced during the processing of, bacteria, yeasts, or fungi. In addition, more economical methods are needed to reduce nucleic acid content of SCP.

(b) Develop efficient methods for the selective isolation of proteins free from nucleic acids and other undesirable constituents, and for producing such proteins with desirable functional and organoleptic properties.

(c) Expand application of SCP technology to waste processing and recovery for animal feeding, utilizing materials such as cellulosic and other crop and animal wastes, sewage, and industrial food wastes.

(d) Improve cell yields and increase the efficiency of heat and oxygen transfer in fermentors, taking into account cooling difficulties in tropical environments, and cell yield limitations due to inadequate oxygen supply. Improve efficiency of cell harvesting or collection, and of dewatering and drying, taking into consideration effects on protein functionality and nutritional properties.

Photosynthetic Single-Cell Protein.—(a) Gather more information on the toxicology, food/feed safety, nutritional value, and acceptability as food of algal sources.

(b) Develop economically feasible techniques to process algae as human food, including means for decolorizing or bleaching, recovery of algal pigments, isolation or concentration of protein constituents, and application of algae to food uses; also develop means for preservation by common technologies.

Leaf Proteins.—(a) Undertake comprehensive research, technological development, and testing of an integrated on-the-farm-type system for recovery and direct feeding, to nonruminants, of leaf protein products, including evaluation of the nutritional and toxicological implications of such feed use.

(b) Intensify development of technology for recovery of leaf protein concentrates and isolates for food use. Attention should be directed toward evaluation of nutritional and food safety aspects, as well as functionality and organoleptic qualities, of concentrates and isolates used as ingredients in formulated or fabricated foods. This work should include comprehensive study of the unique properties of *Nicotiana* (tobacco) leaves, which permit easy separation and recovery of relatively pure protein, and exploration of agronomic and other factors which relate to the optimal production of tobacco for such use.

Advanced Science and Technology

Innovative Technology for Protein Utilization.—Undertake research to contribute to the basic understanding of physical and chemical properties of protein molecules. Study the separation and restructuring of plant proteins and the application of chemical engineering technology to the development of edible protein.

Targets of Opportunity.—Make available funds for innovative exploratory research yet to be identified.

CONCLUSION

In closing, may we indicate that, notwithstanding current expenditures of over $1,000 million a year in the U.S. on relevant agricultural research, it is our conviction that many useful and important opportunities still exist for productive complementary support, by NSF, of research in the basic biological and biochemical sciences which are the foundations for all agricultural technology, and for innovative food science applications. In addition, it would certainly reflect the national interest to explore, through a modest long-term investment on the part of NSF, the contributions which various unconventional food sources might make to U.S. and world food supplies late in this century by supplementing agricultural food production.

BIBLIOGRAPHY

ALDRICH, R. J. 1973. National Soybean Research Needs. A Report of the National Soybean Research Coordinating Committee. University of Missouri, Columbia, Mo.

BARDACH, J. 1976. Fisheries prospects for the future: Freshwater and marine. *In* Nutrition and Agricultural Development: Significance and Potential for the Tropics, N. S. Scrimshaw and M. Behar (Editors). Plenum Publishing Corp., New York.

COX, C. B., BINKERD, E. F., and CRIGLER, T. P. 1971. Animal protein: Keystone of foods. Food Technol. *25*, 808–812.

CUNHA, T. J. 1974. The future of animal agriculture. Feedstuffs *46*, No. 21, 96, 98, 110–111.

FAO/WHO (JOINT FAO/WHO AD HOC EXPERT COMMITTEE ON PROTEIN REQUIREMENTS). 1973. Energy and Protein Requirements. WHO Technical Report Series No. 522. World Health Organization, Geneva.

HARDY, R. W. F. 1976. Photosynthate as a major factor limiting N_2 fixation by field grown legumes with emphasis on soybeans. *In* Symbiotic Nitrogen Fixation in Plants, P. Nutman (Editor). Cambridge Univ. Press, London.

INGLETT, G. E. 1970. Corn: Culture, Processing, Products. AVI Publishing Co., Westport, Conn.

INGLETT, G. E. 1974. Wheat: Production and Utilization. AVI Publishing Co., Westport, Conn.

KALDY, M. S. 1972. Protein yield of various crops as related to protein value. Econ. Bot. *26*, 142–144.

KAWASHIMA, N., and WILDMAN, S. G. 1970. Fraction I protein. Ann. Rev. Plant Physiol. *21*, 325–358.

LIENER, I. E. 1969. Toxic Constituents of Plant Protein Foodstuffs. Academic Press, New York.

MATELES, R. I., and TANNENBAUM, S. R. 1968. Single Cell Protein. M.I.T. Press, Cambridge, Mass.

MATTERN, P. J., SCHMIDT, J. W., and JOHNSON, V. A. 1970. Screening for high lysine content in wheat. Cereal Sci. Today *15*, 409–411.

MERTZ, E. T., BATES, L. S., and NELSON, O. E. 1964. Mutant gene that increases lysine content of maize endosperm. Science *145*, 279–280.

MILNER, M. 1975. Nutritional Improvement of Food Legumes by Breeding. John Wiley and Sons, New York.

MUNCK, L. 1972. Improvements of nutritional value in cereals. Hereditas *72*, 1–128.

MYERS, J. 1963. Algal cultures. *In* Encyclopedia of Chemical Technology, 2nd Edition, Vol. 1, H. F. Mark, J. J. McKetta, Jr., D. F. Othmer, and A. Standen (Editors). Interscience Publishers. New York.

NAS/NRC, COMMITTEE ON FOOD PROTECTION. 1973. Toxicants Occurring Naturally in Foods, 2nd Edition. National Academy of Sciences, National Research Council, Washington, D.C.

NELSON, O. E. 1969. Genetic modification of protein quality in plants. Advan. Agron. *21*, 171–194.

PAG (PROTEIN-CALORIE ADVISORY GROUP OF THE UNITED NATIONS SYSTEM). 1973. PAG statement (No. 22) on upgrading human nutrition through the improvement of food legumes. PAG Bull. *3*, No. 2, 1–24.

PAG (PROTEIN-CALORIE ADVISORY GROUP OF THE UNITED NATIONS SYSTEM). 1974A. PAG guideline (No. 15) on nutritional and safety aspects of novel protein sources for animal feeding. PAG Bull. *4*, No. 3, 11–17.

PAG (PROTEIN-CALORIE ADVISORY GROUP OF THE UNITED NATIONS SYSTEM). 1974B. PAG guideline (No. 6) for preclinical testing of novel sources of protein. PAG Bull. *4*, No. 3, 17–31.

PAG (PROTEIN-CALORIE ADVISORY GROUP OF THE UNITED NATIONS SYSTEM). 1975. PAG guideline (No. 16) on protein methods for cereal breeders as related to human nutritional requirements. PAG Bull. *5*, No. 2, 22–48.

PIRIE, N. W. 1971. Leaf Protein: Its Agronomy, Preparation, Quality and Use. Blackwell, Oxford.

SCRIMSHAW, N. S. WANG, D. I. C., and MILNER, M. 1975. Protein Resources and Technology: Status and Research Needs. Research Recommendations and Summary. National Science Foundation, Research Applied to National Needs. NSF-RA-T-75-037.

TANNENBAUM, S. R., and WANG, D. I. C. 1975. Single-Cell Protein II. M.I.T. Press, Cambridge, Mass.

TWELFTH TECHNICAL ALFALFA CONFERENCE PROCEEDINGS. 1975. USDA Western Regional Research Service, Berkeley, Calif.

WASLIEN, C. I. 1975. Unusual sources of protein for man. CRC Crit. Rev. Food Sci. and Nutr. *6*, 77–151.

WORLD SOY PROTEIN CONFERENCE. 1974. Complete proceedings of conference, Munich, Nov. 11–14, 1973. J. Am. Oil Chem. Soc. *51*, 47A–207A.

ZELITCH, I. 1975. Improving the efficiency of photosynthesis. Science, *188*, 628–633.

Index

603